STUDY GUIDE
AND
SOLUTIONS MANUAL

PAULA YURKANIS BRUICE
University of California
Santa Barbara

With Contributions by Jess Jones
Saint Leo University

ORGANIC CHEMISTRY

Seventh Edition

PEARSON

Boston   Columbus   Indianapolis   New York   San Francisco   Upper Saddle River
Amsterdam   Cape Town   Dubai   London   Madrid   Milan   Munich   Paris   Montréal   Toronto
Delhi   Mexico City   São Paulo   Sydney   Hong Kong   Seoul   Singapore   Taipei   Tokyo

Editor in Chief: Adam Jaworski

Executive Editor: Jeanne Zalesky

Senior Marketing Manager: Jonathan Cottrell

Assistant Editor: Erin Kneuer

Managing Editor, Chemistry and Geosciences: Gina M. Cheselka

Project Manager: Pat Brown

Compositor: PreMediaGlobal

Cover Designer: Integra Software Services Pvt. Ltd.

Cover Image Credit: Fuse/Getty Images

Credits and acknowledgments borrowed from other sources and reproduced, with permission, in this textbook appear on the appropriate page within the text.

www.pearsonhighered.com

1 2 3 4 5 6 7 8 9 10—EBM—17 16 15 14 13

ISBN-10: 0-321-82659-0 ISBN-13: 978-0-321-82659-6

CONTENTS

Spectroscopy Problems   1
Answers to Spectroscopy Problems   27

CHAPTER 1   Remembering General Chemistry: Electronic Structure and Bonding   54
                Important Terms   54
                Solutions to Problems   58
                Practice Test   72

CHAPTER 2   Acids and Bases: Central to Understanding Organic Chemistry   74
                Important Terms   74
                Solutions to Problems   75
                Practice Test   89

SPECIAL TOPIC I   pH, pK_a, and Buffers   91
                Problems on pH, pK_a, and Buffers   102
                Solutions to Problems on pH, pK_a, and Buffers   104

CHAPTER 3   An Introduction to Organic Compounds: Nomenclature,
                Physical Properties, and Representation of Structure   111
                Important Terms   111
                Solutions to Problems   115
                Practice Test   135

CHAPTER 4   Isomers: The Arrangement of Atoms in Space   137
                Important Terms   137
                Solutions to Problems   140
                Practice Test   160

CHAPTER 5   Alkenes: Structure, Nomenclature, and an Introduction to
                Reactivity • Thermodynamics and Kinetics   163
                Important Terms   163
                Solutions to Problems   166
                Practice Test   180

CHAPTER 6 The Reactions of Alkenes: The Stereochemistry of Addition Reactions 182
Important Terms 182
Solutions to Problems 185
Practice Test 216

CHAPTER 7 The Reactions of Alkynes • An Introduction to Multistep Synthesis 219
Important Terms 219
Solutions to Problems 221
Practice Test 234

CHAPTER 8 Delocalized Electrons and Their Effect on Stability, pK_a, and
the Products of a Reaction 236
Important Terms 236
Solutions to Problems 240
Practice Test 277

SPECIAL TOPIC II Molecular Orbital Theory 281

CHAPTER 9 Substitution Reactions of Alkyl Halides 288
Important Terms 288
Solutions to Problems 290
Practice Test 306

CHAPTER 10 Elimination Reactions of Alkyl Halides • Competition Between
Substitution and Elimination 308
Important Terms 308
Solutions to Problems 309
Practice Test 332

CHAPTER 11 Reactions of Alcohols, Ethers, Epoxides, Amines, and Thiols 334
Important Terms 334
Solutions to Problems 335
Practice Test 370

CHAPTER 12 Organometallic Compounds 372
Important Terms 372
Solutions to Problems 373
Practice Test 385

CHAPTER 13 Radicals • Reactions of Alkanes 387
      Important Terms 387
      Solutions to Problems 389
      Practice Test 405

CHAPTER 14 Mass Spectrometry, Infrared Spectroscopy, and Ultraviolet/Visible Spectroscopy 407
      Important Terms 407
      Solutions to Problems 409
      Practice Test 432

CHAPTER 15 NMR Spectroscopy 434
      Important Terms 434
      Solutions to Problems 437
      Practice Test 457

CHAPTER 16 Reactions of Carboxylic Acids and Carboxylic Acid Derivatives 459
      Important Terms 459
      Solutions to Problems 462
      Practice Test 492

CHAPTER 17 Reactions of Aldehydes and Ketones • More Reactions of Carboxylic Acid Derivatives • Reactions of α,β-Unsaturated Carbonyl Compounds 495
      Important Terms 495
      Solutions to Problems 497
      Practice Test 534

CHAPTER 18 Reactions at the α-Carbon of Carbonyl Compounds 536
      Important Terms 536
      Solutions to Problems 538
      Practice Test 575

CHAPTER 19 Reactions of Benzene and Substituted Benzenes 577
      Important Terms 577
      Solutions to Problems 579
      Practice Test 630

CHAPTER 20 More About Amines • Reactions of Heterocyclic Compounds 632
      Important Terms 632
      Solutions to Problems 633
      Practice Test 651

CHAPTER 21   The Organic Chemistry of Carbohydrates   653
                 Important Terms   653
                 Solutions to Problems   656
                 Practice Test   678

CHAPTER 22   The Organic Chemistry of Amino Acids, Peptides, and Proteins   680
                 Important Terms   680
                 Solutions to Problems   683
                 Practice Test   707

CHAPTER 23   Catalysis in Organic Reactions and in Enzymatic Reactions   709
                 Important Terms   709
                 Solutions to Problems   711
                 Practice Test   730

CHAPTER 24   The Organic Chemistry of the Coenzymes, Compounds Derived
                 from Vitamins   731
                 Important Terms   731
                 Solutions to Problems   733
                 Practice Test   749

CHAPTER 25   The Organic Chemistry of the Metabolic Pathways •
                 Terpene Biosynthesis   751
                 Important Terms   751
                 Solutions to Problems   753
                 Practice Test   769

CHAPTER 26   The Chemistry of the Nucleic Acids   771
                 Important Terms   771
                 Solutions to Problems   773
                 Practice Test   784

CHAPTER 27   Synthetic Polymers   786
                 Important Terms   786
                 Solutions to Problems   789

CHAPTER 28   Pericyclic Reactions   807
                 Important Terms   807
                 Solutions to Problems   809

Answers to Practice Tests   823 *to my students*

I am enormously grateful to Jordan Fantini of Denison University, who reworked all the problems to make this book as error free as possible and who gave me many important suggestions along the way. I am solely responsible for any errors that may remain. If you find any, please email me so they can be fixed in a future printing.

Try to work as many problems as possible, so you can truly enjoy the wonderful world of organic chemistry.

Paula Yurkanis Bruice
pybruice@chem.ucsb.edu

Spectroscopy Problems

1.   Determine the structure of the straight-chain pentanol that produces the mass spectrum shown here.

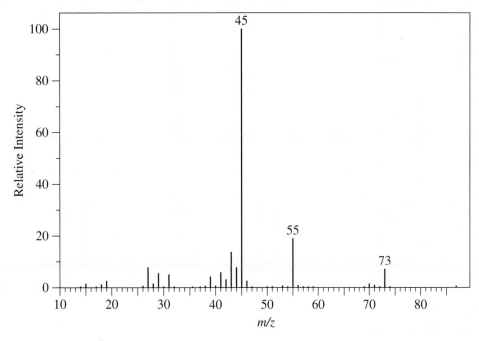

2.   The mass spectrum of an ether is shown here. Determine the molecular formula of the ether that would produce this spectrum and then draw possible structures for it.

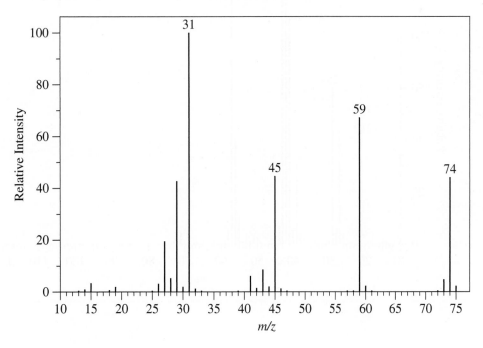

1

3.   The mass spectra of pentane and isopentane are shown here. Determine which spectrum belongs to which compound.

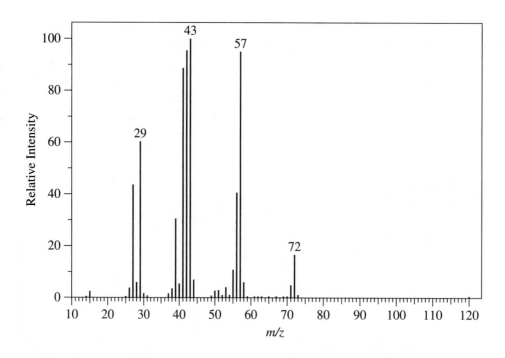

4.   Which of the following compounds gives the mass spectrum shown here?

5.   An unknown acid underwent a reaction with 1-butanol. The product of the reaction gave the mass spectrum shown here. What is the product of the reaction, and what acid was used?

6.   Identify the compound with molecular formula $C_9H_{10}O_3$ that gives the following IR and ^1H NMR spectra.

7.   Identify the compound with molecular formula $C_5H_{11}Br$ that gives the following ^1H NMR spectrum.

8.   Identify the compound with molecular formula $C_6H_{12}O$ that gives the following IR and ^1H NMR spectra.

9.   A methyl-substituted benzene is treated with Cl_2 in the presence of $AlCl_3$, and a ^1H NMR spectrum is taken of one of the monochlorinated products that is isolated from the reaction mixture. Identify this compound.

10. Identify the compound with molecular formula $C_6H_{14}O$ that gives the following IR and ^1H NMR spectra.

11. Identify the compound with molecular formula C_4H_9Br that gives the following ^1H NMR spectrum.

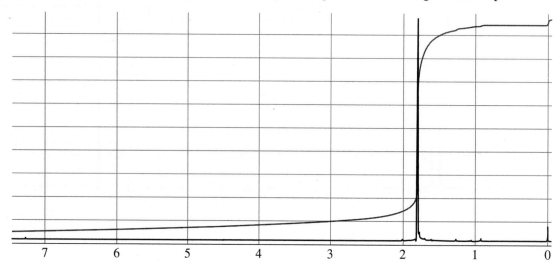

12.   Identify the compound with molecular formula C_6H_{12} that gives the following IR and ^1H NMR spectra.

13.   Identify the compound with molecular formula $C_8H_{10}O$ that gives the following ^1H NMR spectrum.

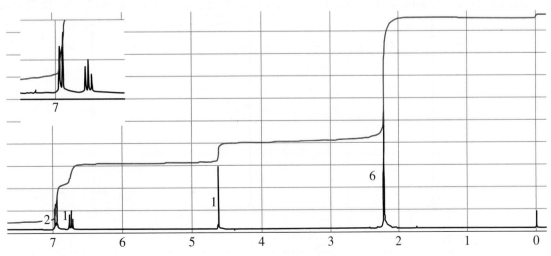

14.   Identify the compound with molecular formula C_4H_7ClO that gives the following IR and ^1H NMR spectra.

WAVENUMBERS

15.   Identify the compound with molecular formula $C_8H_8Br_2$ that gives the following ^1H NMR spectrum.

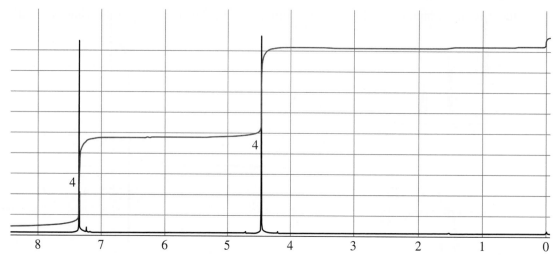

16.   Identify the compound with molecular formula C_4H_6O that gives the following IR and ^1H NMR spectra.

17.   Identify the compound with molecular formula C_7H_8BrN that gives the following ^1H NMR spectrum.

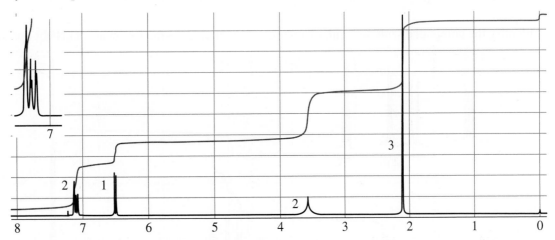

18. The ^{1}H NMR spectra of 1-chloro-3-iodopropane and 1-bromo-3-chloropropane are shown here. Which compound gives which compound spectrum?

a.

b.

19. Identify the compound with molecular formula C_4H_8O that gives the following ^{1}H NMR spectrum.

20.   Identify the compound with molecular formula $C_5H_{10}O_2$ that gives the following ^{1}H NMR spectrum.

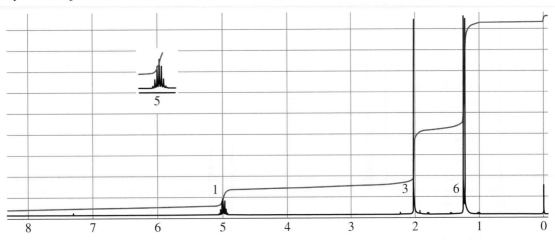

21.   Identify the compound with molecular formula $C_8H_7O_2Br$ that gives the following IR and ^{1}H NMR spectra.

WAVENUMBERS

Offset: 0.4 ppm.

22.   Identify the compound with molecular formula C_7H_6O that gives the following ^1H NMR spectrum.

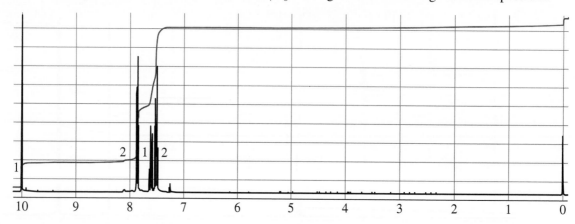

23.   The two ^1H NMR spectra shown here are given by constitutional isomers with molecular formula C_3H_7Br. Identify each isomer.

a.

b.

24. An unknown alkene with molecular formula C_8H_{16} undergoes ozonolysis (Section 6.11 in the text). Only one product is formed. The IR and ^1H NMR spectra of the product are shown here. What is the product? What alkene produced this product?

25.   Identify the compound with molecular formula $C_4H_7BrO_2$ that gives the following IR and ^1H NMR spectra.

26.   Identify the compound with molecular formula $C_3H_6Cl_2$ that gives the following ^1H NMR spectrum.

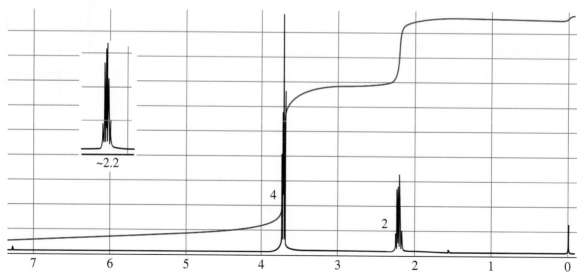

27.   A compound with the following IR spectrum was formed by a reaction with 1-propyne. If the number of carbons in the reactant and product is the same, what compound is formed and what reaction conditions produced this compound?

28.   Identify the compound with molecular formula $C_4H_8O_2$ that gives the following ^{1}H and ^{13}C NMR spectra.

29.   Identify the compound with molecular formula $C_4H_8O_2$ that gives the following ^{1}H NMR spectrum.

30.   Identify the compound with molecular formula $C_9H_{11}NO$ that gives the following ^{1}H NMR spectrum.

31.   Identify the compound with molecular formula $C_9H_{10}O_2$ that gives the following ^{1}H NMR spectrum.

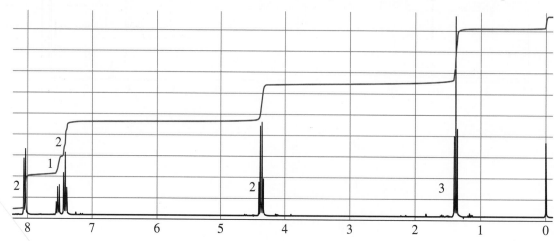

32. The ^{13}C NMR and ^1H NMR spectra of 1, 2-, 1, 3-, and 1, 4-ethylmethylbenzene are shown here. Determine which spectrum belongs to which compound.

a.

b.

c.

33.   Identify the compound with molecular formula $C_7H_{14}O$ that gives the following ^1H NMR spectrum.

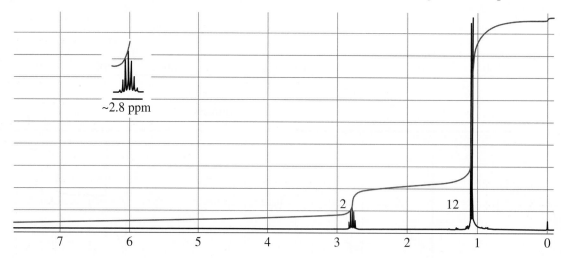

~2.8 ppm

34.   Identify the compound with molecular formula C_5H_{10} that gives the following ^1H NMR spectrum.

35.   Identify the compound with molecular formula C_3H_4O that gives the following IR and ^1H NMR spectra.

36.   Identify the compound with molecular formula C_4H_7Cl that gives the following ^1H NMR spectrum.

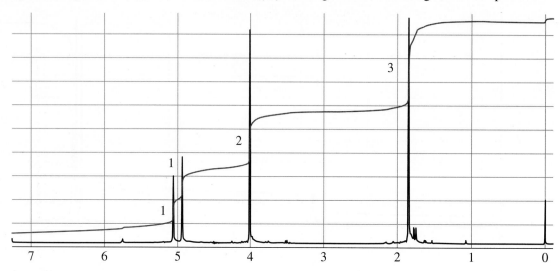

37.   Identify the alcohol that gives the following ^1H NMR spectrum.

38. An unknown alcohol gives the following ^{1}H NMR spectrum. Identify the alcohol. (HINT: Because this is a spectrum of a pure alcohol, the OH proton is split by adjacent protons.)

39. Identify the compound with molecular formula $C_5H_{12}O_2$ that gives the following ^{1}H NMR spectrum.

40.   Identify the compound with molecular formula C_3H_7NO that gives the following IR and ^1H NMR spectra.

41.   The ^1H NMR spectra shown here are given by constitutional isomers of propylamine (C_3H_9N). Identify the isomer that gives each spectrum.

a.

b.

42.   Identify the compound with molecular formula C_3H_7N that gives the following IR and ^1H NMR spectra.

43.   Identify the compound with molecular formula C_4H_9BrO that gives the following ^1H NMR spectrum.

44.   Identify the compound with molecular formula $C_7H_8O_2$ that gives the following IR and NMR spectra.

45. Identify the alcohol that gives the following ^1H NMR spectrum.

46. Identify the compound with molecular formula $C_6H_{12}O_2$ that gives the following ^1H NMR data. The number of hydrogens responsible for each signal is given in parentheses.

1.1 ppm (6H) doublet   2.2 ppm (2H) quartet

1.7 ppm (3H) triplet   3.8 ppm (1H) septet

47. Identify the compound with molecular formula $C_9H_{10}O$ that gives the following ^1H NMR data. The number of hydrogens responsible for each signal is given in parentheses.

1.4 ppm (2H) multiplet   3.8 ppm (2H) triplet

2.5 ppm (2H) triplet   6.9–7.8 ppm (4H) multiplet

48. Identify the compound with molecular formula $C_{11}H_{22}O$ that gives the following ^1H NMR data.

1.1 ppm (18H) singlet   2.2 ppm (4H) singlet

49. Identify the compound with molecular formula C_4H_8O that gives the following ^1H NMR data.

1.6 ppm (4H) multiplet   3.8 ppm (4H) triplet

50. Identify the compound with molecular formula $C_{15}H_{14}O_3$ that gives the following ^1H NMR data.

3.8 ppm (6H) singlet   7.7 ppm (4H) doublet

7.3 ppm (4H) doublet

Answers to Spectroscopy Problems

1. There are three straight-chain pentanols: 1-pentanol, 2-pentanol, and 3-pentanol. Because the most stable fragment for an alcohol is the one formed by α-cleavage, we can see which of the alcohols will form the base peak shown in the spectrum (that is, a base peak with $m/z = 45$) as a result of α-cleavage.

For 1-pentanol, only one α-cleavage is possible. It will form a cationic fragment with $m/z = 31$.

$$CH_3CH_2CH_2CH_2-CH_2-OH \xrightarrow{\alpha\text{-cleavage}} CH_3CH_2CH_2\overset{\cdot}{C}H_2 + H\overset{+}{O}=CH_2$$
$$m/z = 31$$

For 2-pentanol, two α-cleavages are possible. One will form a cationic fragment with $m/z = 73$ and a methyl radical. The second will form a cationic fragment with $m/z = 45$ and a propyl radical. Because a propyl radical is more stable than a methyl radical, the base peak is expected to have $m/z = 45$.

$$\overset{\displaystyle OH}{\underset{\displaystyle |}{CH_3-CHCH_2CH_2CH_3}} \xrightarrow{\alpha\text{-cleavage}} \overset{\cdot}{C}H_3 + H\overset{+}{O}=CHCH_2CH_2CH_3$$
$$m/z = 73$$

$$\overset{\displaystyle OH}{\underset{\displaystyle |}{CH_3CH-CH_2CH_2CH_3}} \xrightarrow{\alpha\text{-cleavage}} CH_3CH_2\overset{\cdot}{C}H_2 + H\overset{+}{O}=CHCH_3$$
$$m/z = 45$$

For 3-pentanol, only one α-cleavage is possible because of the symmetry of the molecule. α-Cleavage will form a cationic fragment with $m/z = 59$.

$$\overset{\displaystyle OH}{\underset{\displaystyle |}{CH_3CH_2-CH-CH_2CH_3}} \xrightarrow{\alpha\text{-cleavage}} CH_3\overset{\cdot}{C}H_2 + H\overset{+}{O}=CHCH_2CH_3$$
$$m/z = 59$$

The base peak of the given mass spectrum has $m/z = 45$. Thus, the mass spectrum is that of **2-pentanol**. We also see a significant fragment at $m/z = 73$, the m/z value of the other α-cleavage product.

2.   First, we must first identify the molecular ion. The molecular ion, the peak that represents the intact starting compound, has an $m/z = 74$. Now we can use the rule of 13 to determine the molecular formula.

$$\frac{74}{13} = 5 \text{ carbons with 9 left over}$$

From the rule of 13, we end up with a molecular formula of C_5H_{14}. Because the compound is an ether, we know that it has one oxygen, so we must add one O and subtract one C and four Hs from the molecular formula. The resulting molecular formula is:

$$C_4H_{10}O$$

There are three ethers that have this molecular formula: methyl propyl ether, diethyl ether, and isopropyl methyl ether.

3.   First, we need to determine what the most abundant cationic fragments would be for each compound.

The possible fragments for **pentane** are:

1    $\left[CH_3-CH_2CH_2CH_2CH_3\right]^{\cdot +}$ ⟶ $\cdot CH_3$ + $^+CH_2CH_2CH_2CH_3$

        $m/z = 57$

2    $\left[CH_3CH_2-CH_2CH_2CH_3\right]^{\cdot +}$ ⟶ $\cdot CH_2CH_3$ + $^+CH_2CH_2CH_3$

        $m/z = 43$

3    $\left[CH_3CH_2-CH_2CH_2CH_3\right]^{\cdot +}$ ⟶ $\cdot CH_2CH_2CH_3$ + $^+CH_2CH_3$

        $m/z = 29$

4    $\left[CH_3-CH_2CH_2CH_2CH_3\right]^{\cdot +}$ ⟶ $\cdot CH_2CH_2CH_2CH_3$ + $^+CH_3$

        $m/z = 15$

The most abundant fragments are going to be the result of bond cleavages that produce the most stable cations and radicals. Fragments from patterns 2 and 3 should be the most common because both the cations and the radicals formed are primary.

Fragmentation pattern 2 is expected to give the base peak (the most stable fragment). The cation formed in pattern 2 ($m/z = 43$) will be more stable than the cation formed in pattern 3 ($m/z = 29$), because the longer fragment will result in greater inductive stabilization (the longer chain will do a better job of stabilizing the carbocation).

Fragments from fragmentation patterns 1 and 4 are expected to be less common. They each form a primary species, like fragmentation patterns 2 and 3, but the second species they form is a methyl fragment (either a radical or a carbocation), which is less sable than the second species formed in fragmentation patterns 2 and 3.

Only four fragmentation patterns are shown for **isopentane**. Fragmentations that would result in only a primary fragment and a methyl fragment have been excluded because these would be less abundant than those shown here.

$$1 \quad \left[\begin{array}{c} CH_3 \\ | \\ CH_3CHCH_2CH_3 \end{array} \right]^{\cdot +} \longrightarrow \cdot CH_3 \; + \; CH_3\overset{+}{C}HCH_2CH_3$$

$$m/z = 57$$

$$2 \quad \left[\begin{array}{c} CH_3 \\ | \\ CH_3CHCH_2CH_3 \end{array} \right]^{\cdot +} \longrightarrow CH_3\overset{\cdot}{C}HCH_2CH_3 \; + \; ^+CH_3$$

$$m/z = 15$$

$$3 \quad \left[\begin{array}{c} CH_3 \\ | \\ CH_3CH-CH_2CH_3 \end{array} \right]^{\cdot +} \longrightarrow \cdot CH_2CH_3 \; + \; \begin{array}{c} CH_3 \\ | \\ CH_3\overset{+}{C}H \end{array}$$

$$m/z = 43$$

$$4 \quad \left[\begin{array}{c} CH_3 \\ | \\ CH_3CH-CH_2CH_3 \end{array} \right]^{\cdot +} \longrightarrow \begin{array}{c} CH_3 \\ | \\ CH_3\overset{\cdot}{C}H \end{array} \; + \; ^+CH_2CH_3$$

$$m/z = 29$$

The four major fragments occur at the same m/z values (57, 43, 29, and 15) as those found for pentane, but their relative intensities are different.

Fragmentation patterns 1 and 3 are expected to produce the most abundant fragments because they both form a secondary cation, and the stability of the cation is more important than the stability of the radical in determining the most abundant fragments. Therefore, we expect a base peak with $m/z = 43$ (because the secondary cation is accompanied by a primary radical) and a less intense peak with $m/z = 57$ (because the secondary cation is accompanied by a methyl radical).

Both spectra show a base peak at $m/z = 43$. The major difference in the two spectra is the intensity of the peak with $m/z = 57$. The spectrum of isopentane should show a more intense peak because it is due to a secondary cation, whereas the peak with $m/z = 57$ in the spectrum of pentane is due to a primary cation.

Thus, **pentane** gives the first mass spectrum and **isopentane** gives the second.

4.   The molecular ion for this compound has an $m/z = 73$. The nitrogen rule states that if a molecular ion has an odd value, then the structure must have an odd number of nitrogens. Therefore, we can eliminate the alkane, the ketone, and the ether.

Now we can determine the molecular formula of the compound using the rule of 13. When we subtract 14 (the mass of nitrogen) from 73, we get 59.

$$\frac{59}{13} = 4 \text{ carbons with 7 left over}$$

Therefore, the molecular formula is $C_4H_{11}N$. Both amines given as possible structures have this formula.

To determine which of the amines is the source of the spectrum, we can take clues from how ethers and alcohols cleave and apply them to amines. Oxygen-containing species undergo α-cleavage. If nitrogen behaves similarly, then we would expect a fragment to form by cleaving a $C-C$ bond alpha to the nitrogen in each compound. Given the relative stability of this fragment, we can anticipate that it will be the base peak of the mass spectrum.

α-Cleavage of *N,N*-dimethylethylamine gives a cation with $m/z = 58$. α-Cleavage of butylamine gives a cation with $m/z = 30$. The spectrum shows a base peak with $m/z = 58$, indicating that the compound that gives the spectrum is ***N,N*-dimethylethylamine**.

5.   The mass spectrum has two peaks of identical height with m/z values $= 136$ and 138, indicating the presence of bromine in the product. (Recall that bromine has two isotopes of equal abundance with weights of 79 and 81 amu.)

Now we need to think about the type of reaction that occurred.

Under acidic conditions, the starting material (1-butanol) will be protonated.

The protonated alcohol now has a leaving group that can be replaced by a nucleophile. Because we know that bromine is present in the product, we can assume that bromide ion is the incoming nucleophile.

We now know that the product of the reaction is **1-bromobutane**. The acid, which must be the source of the nucleophile, is **HBr**.

6. The two signals near 7 and 8 ppm are due to the hydrogens of a benzene ring. Because these signals integrate to 4 protons, the benzene ring must be disubstituted. Since both signals are doublets, the protons that give each signal must each be coupled to one proton ($N + 1 = 1 + 1 = 2$). Thus, the substituents must be at the 1- and 4-positions. The proton or set of protons that gives the signal at the lowest frequency is labeled *a*, the next lowest *b*, and so on.

By subtracting the six Cs and four Hs of the benzene ring from the molecular formula, we know that the two substituents contain three Cs, six Hs, and three Os ($C_9H_{10}O_3 - C_6H_4 = C_3H_6O_3$).

A triplet (1.4 ppm) that integrates to 3 protons and a quartet (4.2 ppm) that integrates to 2 protons are characteristic of an ethyl group. Because the signal for the CH_2 group of the ethyl substituent appears at a relatively high frequency, we know that it is attached to an electronegative atom (in this case, an O).

The presence of the CH_3CH_2O group consumes more of the remaining molecular formula ($C_3H_6O_3 - C_2H_5O = CHO_2$). There is one remaining NMR signal, a singlet (9.8 ppm) that integrates to 1 proton.

To help with the identification, we turn to the IR spectrum. The broad absorption near 3200 cm^{-1} indicates the O—H stretch of an alcohol; the proton of the OH group would give the broad NMR signal at 9.8 ppm. The strong absorption at 1680 cm^{-1} indicates the presence of a carbonyl C=O group.

Now that all the fragments of the compound have been identified, we can put them together. The compound is **ethyl 4-hydroxybenzoate**.

7.   The signals at 1.1 and 1.8 ppm have been magnified and are shown as insets on the spectrum (the **2** and **1** represent the ppm scale) so you can better see the splitting. The triplet (1.1 ppm) that integrates to 3 protons and the quartet (1.8 ppm) that integrates to 2 protons are characteristic of an ethyl group. (The peak to the right of the quartet is actually the beginning of the adjacent signal that integrates to 6 protons.)

The singlet (1.7 ppm) that integrates to 6 protons indicates that there are two methyl groups in the same environment. Because the signal is a singlet, the carbon to which they are attached cannot be bonded to any hydrogens. The only atom not accounted for in the molecular formula is Br.

$$
\begin{array}{c}
CH_3 \\
| \\
-C- \\
| \\
CH_3
\end{array}
$$

Therefore, the ethyl group and the bromine must be the two substituents that are attached to the carbon. Thus, the compound is **2-bromo-2-methylbutane**.

$$
\begin{array}{c}
CH_3 \\
| \\
CH_3-C-CH_2CH_3 \\
| \\
Br
\end{array}
$$

8.   A major clue comes from the IR spectrum. The strong absorption at $\sim 1710 \text{ cm}^{-1}$ indicates the presence of a carbonyl (C=O) group. Because the compound has only one oxygen, we know that it must be an aldehyde or a ketone. The absence of absorptions at 2820 and 2720 cm^{-1} tells us that the compound is not an aldehyde.

The absorptions at 2880 and 2970 cm^{-1} are due to C—H stretches of hydrogens attached to sp^3 carbons.

The ^{1}H NMR spectrum has two unsplit signals. One integrates to 9 protons, the other to 3 protons. A signal that integrates to 9 protons suggests a *tert*-butyl group, and a signal that integrates to 3 protons suggests a methyl group. The fact that they are both singlets indicates that they are on either side of the carbonyl group. Thus, the compound is **3,3-dimethyl-2-butanone**.

$$
\begin{array}{c}
\quad CH_3 \; O \\
\quad | \quad || \\
CH_3-C-C-CH_3 \\
\quad | \\
\quad CH_3
\end{array}
$$

That the methyl group shows a signal at ~ 2.1 ppm reinforces this conclusion because that is where a methyl group attached to a carbonyl group is expected to occur.

9.   From the reaction conditions provided, we know that the product is a monochlorinated toluene.

The singlet (2.3 ppm) that integrates to 3 protons is due to the methyl group.

The signals in the 7–8 ppm region that integrate to 4 protons are due to the protons of a disubstituted benzene ring. Because both signals are doublets, we know that each proton is coupled to one adjacent proton. Thus, the compound has a 1,4-substituted benzene ring.

Therefore, the compound is **4-chloromethylbenzene**.

10.   The strong and broad absorption in the IR spectrum at 3400 cm^{-1} indicates a hydrogen-bonded O—H group. The absorption bands between 2800 and 3000 cm^{-1} indicate hydrogens bonded to sp^3 carbons.

There is only one signal in the ^{1}H NMR spectrum that integrates to 1 proton, so it must be due to the hydrogen of the OH group. The singlet that integrates to 3 protons can be attributed to a methyl group that is attached to a carbon that is not attached to any hydrogens.

Since the other two signals show splitting, they represent coupled protons (that is, protons on adjacent carbons). The quartet and triplet combination indicates an ethyl group. Since the quartet and triplet integrate to 6 and 4 protons, respectively, the compound must have two ethyl groups.

The identified fragments of the molecule are:

When these fragments are subtracted from the molecular formula, only one carbon remains. Therefore, this carbon must connect the four identified fragments. The compound is **3-methyl-3-pentanol**.

11.   The ^{1}H NMR spectrum contains only one signal, so only one type of hydrogen is present in the molecule. Because the compound has 4 carbons and 9 identical hydrogens, the compound must be ***tert*-butyl bromide**.

12.   The molecular formula indicates that the compound is a hydrocarbon with one degree of unsaturation. The IR spectrum can tell us whether the degree of unsaturation is due to a cyclic system or a double bond. The absorption of moderate intensity near 1660 cm^{-1} indicates a C=C stretch. The absorption at $\sim 3100 \text{ cm}^{-1}$, due to C—H stretches of hydrogens attached to sp^2 carbons, reinforces the presence of the double bond.

The two relatively high-frequency singlets (4.7 ppm) is given by vinylic protons. Because the signal integrates to 2 protons, we know that the compound has two vinylic protons. Because the signals are not split, the vinyl protons must not be on adjacent carbons. Thus, they must be on the same carbon.

The singlet (1.8 ppm) that integrates to 3 protons must be a methyl group. Because it is a singlet, the methyl group must be bonded to a carbon that is not attached to any protons.

The doublet (1.1 ppm) that integrates to 6 protons and the septet (2.2 ppm) that integrates to 1 proton is characteristic of an isopropyl group.

isopropyl group

Now that we know that the compound has a methyl group, an isopropyl group, and two vinylic hydrogens attached to the same carbon, the compound must be **2,3-dimethyl-1-butene**.

13.   The signals in the ^{1}H NMR spectrum between 6.7 and 6.9 ppm indicate the presence of a benzene ring. Because the signals integrate to 3 protons, it must be a trisubstituted benzene ring.

The triplet (6.7 ppm) that integrates to 1 proton and the doublet (6.9 ppm) that integrates to 2 protons tell us that the three substituents are adjacent to one another. (The H_d protons are split into a doublet by the H_c proton, and the H_c proton is split into a triplet by the two H_d protons.)

Subtracting the trisubstituted benzene (C_6H_3) from the molecular formula leaves C_2H_7O unaccounted for. The singlet (2.2 ppm) that integrates to 6 protons indicates that there are two methyl groups in identical environments. Now only OH is left from the molecular formula. The singlet at 4.6 ppm is due to the proton of the OH group. The compound is **2,6-dimethylphenol**.

14. A major clue to the compound's structure comes from the IR spectrum. The strong absorption at ~1740 cm^{-1} indicates the presence of a carbonyl (C=O) group. Since the compound has only one oxygen, the compound must be an aldehyde or a ketone. The absence of absorptions at 2820 and 2720 cm^{-1} tells us that the compound is not an aldehyde.

The NMR spectrum shows a singlet (2.3 ppm) that integrates to 3 protons, indicating that it is due to a methyl group. The chemical shift of the signal (hydrogens attached to carbons adjacent to carbonyl carbons typically have shifts between 2.1 and 2.3 ppm) and the fact that the signal is a singlet suggest that the methyl group is directly attached to the carbonyl group.

The two remaining signals are split, indicating that the protons that give these signals are attached to adjacent carbons. Because the signal at 4.3 ppm is a quartet, we know that the proton that gives this signal is bonded to a carbon that is attached to a methyl group. The other signal (1.6 ppm) is a doublet, so the proton that gives this signal is bonded to a carbon that is attached to one hydrogen.

When these two fragments are subtracted from the molecular formula, only a Cl remains.

The only possible arrangement has the alkyl group directly bonded to the other side of the carbonyl group and the chlorine on the last available bond. The relatively high-frequency chemical shift of the quartet (4.3 ppm) reinforces this assignment because it must be attached to an electronegative atom. Thus, the compound is **3-chloro-2-butanone**.

15. Given the simplicity of the ^{1}H NMR spectrum, the product must be highly symmetrical.

The singlet (7.4 ppm) that integrates to 4 protons is due to benzene-ring protons. Because there are four aromatic protons, we know that the benzene ring is disubstituted. Because the signal is a singlet, we know that the four protons are chemically equivalent. Therefore, the two substituents must be the same and they must be on the 1- and 4-positions of the benzene ring.

Subtracting the disubstituted benzene ring from the molecular formula, only $C_2H_4Br_2$ remains. Thus, each substituent must contain 1 carbon, 2 hydrogens, and 1 bromine. The compound that gives the spectrum, therefore, is the one shown here.

$$BrCH_2-\!\!\left\langle\ \right\rangle\!\!-CH_2Br$$

16.   The molecular formula indicates that the compound has two degrees of unsaturation. The weak absorption at ~2120 cm^{-1} is due to a carbon-carbon triple bond, which accounts for the two degrees of unsaturation. The intense and sharp absorption at 3300 cm^{-1} is due to the C—H stretch of a hydrogen attached to an *sp* carbon. The intensity and shape of this absorption distinguishes it from an alcohol (intense and broad) and an amine (weaker and broad). Thus, we know that the compound is a terminal alkyne.

The absorptions between 2800 and 3000 cm^{-1} are due to the C—H stretch of hydrogens attached to sp^3 carbons.

All three signals in the ^{1}H NMR spectrum are singlets, indicating that none of the protons that give these signals have any neighboring protons. The singlet (2.4 ppm) that integrates to 1 proton is the proton of the terminal alkyne.

$$-C\equiv C-H$$

The two remaining signals (3.4 and 4.1 ppm) that integrate to 3 protons and 2 protons, respectively, can be attributed to a methyl group and a methylene group. When the alkyne fragment and the methyl and methylene groups are subtracted from the molecular formula, only an oxygen remains.

$$-O- \qquad \overset{\displaystyle H}{\underset{\displaystyle H}{-\overset{|}{\underset{|}{C}}-}} \qquad \overset{\displaystyle H}{\underset{\displaystyle H}{-\overset{|}{\underset{|}{C}}-H}}$$

The arrangement of these groups can be determined by the splitting and the chemical shift of the signals. Since each signal is a singlet, the methyl and methylene groups cannot be adjacent or they would split each other's signal. Since the terminal alkyne and the methyl group must be on the ends of the molecule, the only possible arrangement is shown below. Thus, the compound is **3-methoxy-1-propyne**.

$$CH_3OCH_2C\equiv CH$$

Notice that both the methyl and methylene groups show strong deshielding because of their direct attachment to the oxygen. The methylene hydrogens are also deshielded by the neighboring alkyne.

17.   The signals in the ^{1}H NMR spectrum between 6.5 and 7.2 ppm indicate the presence of a benzene ring. Since the signals integrate to 3 protons, it must be a trisubstituted benzene ring.

The singlet (2.1 ppm) that integrates to three protons must be a methyl group; 2.1 ppm is characteristic of protons bonded to a benzylic carbon.

When the trisubstituted benzene ring (C_6H_3) and the methyl group (CH_3) are subtracted from the molecular formula, NH_2Br is all that remains. Thus, the three substituents must be a methyl group, bromine, and an amino group (NH_2). The amino group gives the broad singlet (3.6 ppm) that integrates to 2 protons. Hydrogens attached to nitrogens and oxygens typically give broad signals.

The substitution pattern for the trisubstituted benzene can be determined from the splitting patterns. Because the signal (6.5 ppm) that integrates to 1 proton is a doublet, we know that the proton that gives this signal has only one neighboring proton. Looking at the magnification of the signal at 7.1 ppm, we see that it is actually two separate signals. One is a singlet, and, therefore, it is attached to a carbon that is separated by substituents from the carbons that are attached to protons. The other signal is a doublet that integrates to 1 proton; because it gives a doublet, we know that it is next to the proton that gives the doublet at 6.5 ppm.

To determine the relative positions of the substituents, the chemical shifts must be analyzed. Bromine is the most electronegative substituent and, therefore, must be adjacent to the two protons that give signals at 7.1 ppm. Thus, Z is Br. The amino group donates its lone-pair electrons into the ring, so it shields benzene-ring protons. Thus, the signal at 6.5 ppm is from a proton in close proximity to the amino group. Therefore, X must be the amino group.

The compound is that gives the spectrum is shown here.

18. The two compounds that produce the spectra have the following structures.

$$ClCH_2CH_2CH_2Br \qquad ClCH_2CH_2CH_2I$$

1-bromo-3-chloropropane   1-chloro-3-iodopropane

The number of signals (three) and the splitting patterns are identical for each compound. The only difference is variations in the chemical shift due to the different electronegativities of bromine and iodine.

Because chlorine is more electronegative than bromine or iodine, the protons bonded to the carbon that is attached to chlorine will have the most deshielded signal (that is, the signal that occurs at the highest frequency). This is the triplet that occurs at 3.7 ppm in both spectra.

The spectra differ in the signal that occurs at 3.4 ppm in the top spectrum and the one that appears at 3.6 ppm in the second spectrum. Because bromine is more electronegative than iodine, the protons bonded to the carbon that is attached to bromine will occur at a higher frequency than the protons bonded to the carbon that is attached to iodine.

Thus, **1-chloro-3-iodopropane** gives the top spectrum, and **1-bromo-3-chloropropane** gives the bottom spectrum.

19. The molecular formula shows that the compound has one degree of unsaturation, indicating a cyclic compound, an alkene, or a carbonyl group.

A cyclic system containing an oxygen (a cyclic ether) would have the most deshielded signal at ~3.5 ppm, which would be due to the hydrogens attached to the carbon adjacent to the oxygen. Therefore, a cyclic ether would not give a signal at 6.4 ppm, so it can be ruled out.

Protons attached to a carbon adjacent to a carbonyl group show a signal at ~2.1 ppm. Since there is no signal in that region, a carbonyl group can also be ruled out.

Vinylic protons would account for the signals in the 3.9–4.2 ppm range that integrate to 2 protons, so we can conclude that the compound is an alkene. Because a highly deshielding oxygen is also present, the high-frequency signal (6.4 ppm) is not unexpected.

The triplet (1.3 ppm) that integrates to 3 protons and the quartet (3.8 ppm) that integrates to 2 protons indicate the presence of an ethyl group. The fact that the quartet is deshielded suggests that the ethyl's methylene group is attached to the oxygen.

The highly deshielded doublet of doublets (6.4 ppm) that integrates to 1 proton suggests that the proton that gives this signal is attached to an sp^2 carbon that is attached to the oxygen. The fact that the signal is a doublet of doublets indicates that it is split by each of two nonidentical protons on the adjacent carbon. Thus, the compound is **ethyl vinyl ether**.

The identification is confirmed by the two doublets (~4.0 and 4.2 ppm) that each integrate to 1 proton. When those signals are magnified, we can see that each is really a doublet of doublets. The doublets of doublets are not well defined because of the small coupling constant (J value) for geminal coupling on sp^2 carbons.

20. The doublet (1.2 ppm) that integrates to 6 protons and the septet (5.0 ppm) that integrates to 1 proton are characteristic of an isopropyl group. (The two methyl groups are split by a single proton, and the single proton is split by six protons.)

The remaining signal (a singlet at a tiny bit more than 2.0 ppm) that integrates to 3 protons indicates an unsplit methyl group.

When the isopropyl and methyl groups are subtracted from the molecular formula, one carbon and two oxygens are left over. Thus, the compound has the following fragments:

There are two ways these fragments can be pieced together. Because the most deshielded signal in the spectrum (the one at 5.0 ppm) is the proton bonded to the central carbon of the isopropyl group, that carbon must be directly attached to the oxygen. Thus, the compound is **isopropyl acetate**.

methyl 2-methylpropanoate   isopropyl acetate

21.   The IR spectrum shows an absorption at ~ 1700 cm^{-1} for a C=O stretch and a very broad absorption (2300–3300 cm^{-1}) for an O—H stretch, indicating that the compound is a carboxylic acid. Intermolecular hydrogen bonding explains the broad nature of this peak and also the broader-than-expected carbonyl peak. The proton of the carboxylic acid gives a singlet at 12.4 ppm in the NMR spectrum.

The two doublets (7.5 and 7.9 ppm) that each integrate to 2 protons indicate a 1,4-disubstituted benzene ring.

Subtracting the disubstituted benzene ring and the COOH group from the molecular formula leaves CH$_2$Br, so we know that the second substituent is a bromomethyl group; it gives the singlet at ~ 4.7 ppm.

Therefore, the compound that gives the spectrum is the one shown here.

22.   The signals with chemical shifts in the range of 7–8 ppm are due to benzene-ring protons. Because the three signals integrate to a total of 5 protons, we know that the benzene ring is monosubstituted.

The singlet at 10 ppm indicates the hydrogen of an aldehyde or a carboxylic acid. Since there is only one oxygen in the molecular formula, we know the compound is an aldehyde. Thus, the compound is **benzaldehyde**—a compound with a monosubstituted benzene ring and an attached aldehyde.

23.   In the first spectrum, the doublet (~ 1.7 ppm) that integrates to 6 protons and the septet (~ 4.2 ppm) that integrates to 1 proton indicate an isopropyl group. When the isopropyl group is subtracted from the molecular formula, only a Br remains. Thus, the compound is **2-bromopropane**.

In the second spectrum, the triplet (~1.0 ppm) that integrates to 3 protons is a methyl group that is attached to a methylene group. The triplet (~3.4 ppm) that integrates to 2 protons is a methylene group that is also attached to a methylene group; the highly deshielded nature of the signal indicates that the carbon is attached to an electronegative group. Thus, the compound is **1-bromopropane**.

The structure is confirmed by the multiplet (~1.8) that integrates to 2 protons; the signal is split by both the adjacent methyl and methylene groups.

Notice that the pattern of a triplet that integrates to 3 protons, a multiplet that integrates to 2 protons, and a triplet that integrates to 2 protons is characteristic of a propyl group.

24.   A strong and sharp absorption in the IR spectrum at ~1730 cm^{-1} indicates a carbonyl (C=O) group. The two absorptions at 2710 and 2810 cm^{-1} tell us that the product of ozonolysis is an aldehyde. The aldehydic proton is also visible in the NMR as a singlet (9.0 ppm) that integrates to 1 proton.

The NMR spectrum has two additional signals. One is a doublet (1.1 ppm) that integrates to 6 protons and the other is a septet (2.4 ppm) that integrates to 1 proton. This is characteristic of an isopropyl group. Therefore, we know that the product of ozonolysis is **2-methylpropanal**.

Because only one product is formed, we know that the alkene that formed the aldehyde must be symmetrical. The identification of the aldehyde also agrees with the molecular formula of the alkene that underwent ozonolysis—that is, an eight-carbon symmetrical alkene will form a four-carbon carbonyl compound.

There are two symmetrical alkenes that will form 2-methylpentanal—*trans*-**2,5-dimethyl-3-hexene** and *cis*-**2,5-dimethyl-3-hexene**. We are not given any information that would distinguish between the two stereoisomers. Thus, the unknown alkene can be either of the two stereoisomers.

or

25.   The strong and sharp absorption in the IR spectrum at ~1720 cm^{-1} indicates the presence of a carbonyl group. The broad absorption centered at 3000 cm^{-1} tells us that the carbonyl-containing compound is a carboxylic acid. The broad singlet (12.0 ppm) in the NMR spectrum (shown as offset by 0.2 ppm from where it is placed on the spectrum) confirms the presence of a carboxylic acid group.

The only other signal in the NMR spectrum is a singlet (2.0 ppm) that integrates to six protons, indicating two methyl groups in the same environment. Because the signal is a singlet, the methyl groups must be attached to a carbon that is not attached to a proton. Because we know that the compound has only four carbons and contains a bromine, the compound must be **2-bromo-2-methylpropanoic acid**.

$$CH_3-C(CH_3)(Br)-C(=O)-OH$$

26. The quintet (~2.2 ppm) that integrates to 2 protons indicates that the protons that give this signal have four identical neighboring protons. A carbon cannot be bonded to four protons and still be able to bond to anything else, so the two protons that give the quintet must be bonded to a carbon that is attached to two methylene groups in the same environment.

The triplet (~3.8 ppm) that integrates to 4 protons must be the signal for the four protons of the two methylene groups. The two methylene groups must be on either side of a carbon that is bonded to two protons (that is, the protons that give the quintet).

$$-C(H_2)-C(H_2)-C(H_2)-$$

There are two bonds left unaccounted for, so this is where the two chlorines shown in the molecular formula go. Therefore, the compound is **1,3-dichloropropane**. The highly deshielded nature of the signal at 3.8 ppm for the protons bonded to the carbons that are attached to chlorines is further evidence that the chlorines are attached to these carbons.

$$ClCH_2CH_2CH_2Cl$$

27. The IR spectrum shows a strong and sharp absorption at ~1720 cm^{-1}, indicating a carbonyl (C=O) group. The two absorptions at 2720 and 2820 cm^{-1} are characteristic of an aldehyde; they are due to the C—H stretch of the bond between the carbonyl carbon and the aldehydic hydrogen. Because the reactant has three carbons, the aldehyde that produces the IR spectrum must also have three carbons. Therefore, the product of the reaction is **propanal**.

$$CH_3CH_2-C(=O)H$$

Thus, the reaction that occurred was the conversion of 1-propyne to propanal. This reaction can occur by hydroboration-oxidation of the alkyne (Section 7.8 in the text).

$$H_3C-C\equiv CH \xrightarrow{\text{1. R}_2\text{BH, THF}}_{\text{2. H}_2\text{O}_2, \text{HO}^-, \text{H}_2\text{O}} CH_3CH_2-C(=O)H$$

28.   The short signal at ~185 ppm in the ^{13}C NMR spectrum suggests the presence of the carbonyl group of a carboxylic acid.

The broad singlet (12.2 ppm) in the ^{1}H NMR spectrum that integrates to 1 proton confirms that the compound contains a carboxylic acid group.

The doublet (1.2 ppm) that integrates to 6 protons and the septet (2.6 ppm) that integrates to 1 proton are characteristic of an isopropyl group.

$$
\begin{array}{ccccc}
 & H & & H & \\
 & | & & | & \\
H-&C-&C-&C-&H \\
 & | & | & | & \\
 & H & H & H &
\end{array}
$$

Therefore, the compound is **2-methylpropanoic acid**.

29.   The breath of the singlet (11.8 ppm) that integrates to 1 proton indicates a hydrogen that is attached to an oxygen. The chemical shift of the signal indicates that it is due to the OH group of a carboxylic acid.

The triplet (~0.9 ppm) that integrates to 3 protons is a methyl group that is attached to a methylene group. The triplet (~2.3 ppm) that integrates to 2 protons indicates a methylene group that is also attached to a methylene group; the chemical shift of this signal indicates that the protons that give this signal are closest to the electron-withdrawing carboxylic acid group. The multiplet at 1.7 ppm that integrates to 2 protons is given by the two protons that split the other two signals into triplets.

We can conclude that the compound responsible for the spectrum is **butanoic acid**.

30.   The singlet (~9.7 ppm) that integrates to 1 proton and the molecular formula that contains one oxygen suggest that an aldehyde is present.

The signals at 7.7 and 6.7 ppm are due to benzene-ring protons. The fact that they are both doublets that integrate to 2 protons tells us that substituents are on the 1- and 4-positions of the benzene ring.

If the aldehyde group and the disubstituted ring are subtracted from the molecular formula, we find that the second substituent contains 2 carbons, 6 hydrogens, and 1 nitrogen. The remaining NMR signal (~3.0 ppm) is a singlet that integrates to 6 hydrogens. These must be due to two methyl groups in the same environment. The nitrogen must be between the two methyl groups or else they would split each other's signals. The nitrogen causes the signal for the methyl groups to appear at a higher frequency than where methyl groups normally appear.

Thus, the compound is **4-(dimethylamino)benzaldehyde**.

31. The three signals between 7.4 and 8.1 ppm that together integrate to 5 protons indicate a monosubstituted benzene ring. Subtracting the monosubstituted ring (C_6H_5) from the molecular formula leaves $C_3H_5O_2$ to be accounted for.

The two oxygens in the molecular formula tells us that the compound is an ester because a broad singlet between 10 and 12 ppm that would indicate a carboxylic acid is not present. The remainder of the molecule contains two carbons and five hydrogens.

The two remaining signals, a triplet (1.4 ppm) that integrates to 3 protons and a quartet (4.4 ppm) that integrates to 2 protons, are characteristic of an ethyl group. The three known segments can now be joined in one of two ways:

ethyl benzoate

phenyl propanoate

The choice between the two compounds can be made by looking at the chemical shift of the methylene protons. In the ethyl ester, the signal will be highly deshielded by the adjacent oxygen. In the phenyl ester, the signal will be at ~2.1 ppm because the methylene protons would be next to the carbonyl group. Because the chemical shift of the methylene protons is 4.4 ppm, we know that the compound is **ethyl benzoate**.

32. Because all three spectra are given by ethylmethylbenzenes, the low-frequency signals in both the ^{1}H NMR and ^{13}C NMR spectra can be ignored because they belong to the methyl and ethyl substituents. The key to determining which spectrum belongs to which ethylmethylbenzene can be found in the aromatic region of the ^{1}H NMR and ^{13}C NMR spectra.

4-ethylmethylbenzene   3-ethylmethylbenzene   2-ethylmethylbenzene

The aromatic region of the ^{13}C NMR spectrum of 4-ethylmethylbenzene will show four signals because it has four different ring carbons.

The aromatic region of the ^{13}C NMR spectrum of 3-ethylmethylbenzene will show six signals because it has six different ring carbons.

The aromatic region of the ^{13}C NMR spectrum of 2-ethylmethylbenzene will also show six signals because it has six different ring carbons.

We now know that spectrum **(b)** is the spectrum of 4-ethylmethylbenzene because its ^{13}C NMR spectrum has four signals and the other two compounds will show six signals.

To distinguish between 2-ethylmethylbenzene and 3-ethylmethylbenzene, we need to look at the splitting patterns in the aromatic regions of the ^{1}H NMR spectra. Analysis of the aromatic region for spectrum **(c)** is difficult because the signals are superimposed. Analysis of the aromatic region for spectrum **(a)** provides the needed information. A triplet (7.2 ppm) that integrates to 1 proton is clearly present. This means that spectrum **(a)** is 3-ethylmethylbenzene because 2-ethylmethylbenzene would not show a triplet. Therefore, spectrum **(c)** is 2-ethylmethylbenzene.

splitting pattern for 3-methylethyl benzene   splitting pattern for 2-methylethyl benzene

The final assignments are:

(a) 3-ethylmethylbenzene   (b) 4-ethylmethylbenzene   (c) 2-ethylmethylbenzene

33. The simplicity of the NMR spectrum of a compound with 7 carbons and 14 hydrogens indicates that the compound must be symmetrical. From the molecular formula, we see that it has one degree of unsaturation. The absence of signals near 5 ppm rules out an alkene. Because the compound has an oxygen, the degree of unsaturation may be due to a carbonyl group.

The doublet (1.1 ppm) that integrates to 12 protons and the septet (2.8 ppm) that integrates to 2 protons suggest the presence of two isopropyl groups.

If two isopropyl groups are subtracted from the molecular formula, we find that the remainder of the molecule is composed of one carbon and one oxygen. Thus, the compound is the one shown here.

34. The molecular formula tells us that compound has one degree of unsaturation. The multiplet (5.2 ppm) that integrates to 1 proton is due to a vinylic proton (that is, it is attached to an sp^2 carbon). Thus, the degree of unsaturation is due to a carbon-carbon double bond. Because there is only one vinylic proton, we can assume that the alkene is trisubstituted.

Three additional signals are present that each integrate to 3 protons, suggesting that all three signals are due to methyl groups. The alkene, therefore, is **2-methyl-2-butene**.

Notice that two of the three signals given by the methyl groups are singlets and one is a doublet. The methyl group that gives the doublet is bonded to the carbon that is attached to the vinylic proton. The other two methyl groups are bonded to the other sp^2 carbon.

35.   A medium-intensity absorption at ~2120 cm^{-1} indicates the presence of a carbon-carbon triple bond. The sharp absorption at 3300 cm^{-1} is due to the C—H stretch of a hydrogen bonded to an *sp* carbon. Thus, the compound is a terminal alkyne.

$$-C\equiv C-H$$

The intense and broad peak centered at 3300 cm^{-1} is evidence of an O—H group.

The NMR spectrum can be used to determine the connectivity between the groups. The signal (2.5 ppm) that integrates to 1 proton is due to the proton of the terminal alkyne.

The singlet (3.2 ppm) that integrates to 1 proton must be due to the proton of the OH group.

The singlet (4.2 ppm) that integrates to 2 protons must be due to a methylene group that connects the triply bonded carbon to the OH group. The compound is **2-propyn-1-ol**.

$$H-C\equiv C-CH_2-OH$$

This arrangement explains the absence of any splitting and the highly deshielded nature of the signal for the methylene group.

36.   The two singlets (4.9 and 5.1 ppm) that each integrate to 1 proton are vinylic protons. Therefore, we know that the compound is an alkene.

The singlet (2.8 ppm) that integrates to 3 protons is a methyl group. The deshielding results from its being attached to an sp^2 carbon.

If we subtract the two vinylic protons, the two sp^2 carbons of the alkene, and the methyl group from the molecular formula, we are left with CH_2Cl. Thus, a chloromethyl group is the fourth substituent of the alkene and gives the singlet (4.9 ppm) that integrates to 2 protons. Its deshielding is due to the proximity to the electronegative chlorine.

Now we need to determine the substitution pattern of the alkene. The absence of splitting indicates that the two vinylic protons must be attached to the same carbon. If these protons were cis or trans to one other, they would give doublets with significant *J* values. Geminal protons attached to sp^2 carbons have very small *J* values, so splitting is typically not observed (see Table 15.2 on page 676 of the text).

Thus, the compound is **3-chloro-2-methyl-1-propene**.

37.   The only signal that integrates to 1 proton is the singlet at 2.2 ppm. This must be due to the OH group of the alcohol.

The signals centered around 7.3 ppm are those given by benzene-ring protons. Because they integrate to 5 protons, the benzene ring must be monosubstituted.

The two triplets (2.8 and 3.8 ppm) that each integrate to 2 protons suggest two adjacent methylene groups. Both signals are fairly deshielded, indicating an electronegative atom nearby.

The fragments identified at this point are a monosubstituted benzene ring, an OH group, and two adjacent methylene groups.

There are no other signals in the NMR spectrum, so the compound must be the one shown here.

38.   The doublet (0.9 ppm) that integrates to 6 protons and the multiplet (1.8 ppm) that integrates to 1 proton suggest the presence of an isopropyl group.

Because we are told that the compound is an alcohol, the other signal that integrates to 1 proton (the triplet at 2.4 ppm) must be due to the OH proton. The fact that the signal a triplet indicates that the OH group is probably attached to a methylene group.

The signal for the methylene group must be the remaining signal at 3.4 ppm because it integrates to 2 protons. The relatively high-frequency chemical shift confirms that the methylene group is attached to the oxygen.

Putting together the isopropyl group and the methylene group that is attached to an OH group identifies the compound as **2-methyl-1-propanol**.

39.   The protons that are responsible for the doublet (1.2 ppm) that integrates to 6 protons must be adjacent to a carbon that is attached to only one proton. Because the spectrum does not have a signal that integrates to one proton, the compound must have two methyl groups in the same environment, and each must be adjacent to a carbon that is attached to one proton and those two single protons must also be in identical environments.

Because the compound must be symmetrical, the two oxygens in the compound must be due to two OH groups in identical environments. The hydrogens of the OH groups give a singlet (3.8 ppm) that integrates to 2 protons.

The protons that give the triplet (2.6 ppm) must be bonded to a carbon that is adjacent to a total of two protons. Because the triplet integrates to 2 protons, it must be due to a methylene group that connects the two pieces.

This structure is confirmed by the relatively high-frequency multiplet (4.2 ppm) that is given by the protons attached to the carbons that are attached to the OH groups. The signal for these protons is split by both the adjacent methyl group and the adjacent methylene group.

40. The absorption in the IR spectrum at $\sim 1650 \text{ cm}^{-1}$ could be due to either a carbonyl group or an alkene. Its strength and breadth tells us that it is probably due to a carbonyl (C=O) group. The strong and broad absorption at $\sim 3300 \text{ cm}^{-1}$ that contains two broad peaks suggests two N—H bonds; thus, an NH_2 group is present. When these two groups are subtracted from the molecular formula, all that is left is C_2H_5.

The triplet (~ 1.1 ppm) that integrates to 3 protons and the quartet (2.2 ppm) that integrates to 2 protons indicate the presence of an ethyl group; this accounts for the C_2H_5 fragment. Thus, all the fragments of the compound have been identified: C=O, NH_2, and CH_3CH_2. The compound, therefore, is **propanamide**.

The presence of an amide explains the lower-than-normal frequency of the C=O stretch in the IR spectrum. The breadth of the N—H stretches confirms that these are amide N—H stretches and not amine N—H stretches. The broad singlets (6.2 and 6.6 ppm) in the NMR spectrum are given by the protons attached to the nitrogen. The protons resonate at different frequencies because the C-N bond has partial double bond character, which causes the protons to be in different environments.

41. The singlet (2.3 ppm) in the first spectrum that integrates to 3 hydrogens must be due to an isolated methyl group.

The triplet (1.1 ppm) that integrates to 3 protons and the quartet (2.5 ppm) that integrates to 2 protons are characteristic of an ethyl group.

The singlet (4.8 ppm) that integrates to 1 proton must be due to a single hydrogen attached to nitrogen.

$$\begin{array}{ccc} \text{H} & \text{H} \;\; \text{H} & \\ | & | \;\;\;\; | & \\ -\text{C}-\text{H} & -\text{C}-\text{C}-\text{H} & -\text{N}- \\ | & | \;\;\;\; | & | \\ \text{H} & \text{H} \;\; \text{H} & \text{H} \end{array}$$

Now that the three fragments have been identified, we know that the compound is **ethylmethylamine**.

Now that the three fragments have been identified, we know that the compound is **ethylmethylamine**.

The second spectrum shows that a broad singlet (2.8 ppm) must be due to hydrogens that are attached to nitrogens. Because the signal integrates to 2 protons, we know that the compound is a primary amine.

The triplet (0.8 ppm) that integrates to 3 protons is due to a methyl group that is adjacent to a methylene group. The triplet (2.7 ppm) that integrates to 2 protons must also be adjacent to a methylene group. The multiplet (1.5 ppm) that integrates to 2 protons is the methylene group that splits both the methyl and methylene groups. The two triplets and multiplet are characteristic of a propyl group.)

Therefore, the compound is **propylamine**.

42. The relatively weak absorption in the IR spectrum at $\sim 1650 \text{ cm}^{-1}$ tells us it is probably due to a carbon-carbon double bond. This is reinforced by the presence of absorptions at $\sim 3080 \text{ cm}^{-1}$, indicating $\text{C}-\text{H}$ bond stretches of hydrogens attached to sp^2 carbons.

The shape of the two absorptions at $\sim 3300 \text{ cm}^{-1}$ suggests the presence of an NH_2 group of a primary amine. (Compare these to the shape of the $\text{N}-\text{H}$ stretches of an NH_2 group of an amide in Problem 39.)

The three signals in the NMR spectrum between 5.0 and 6.0 ppm that integrate as a group to 3 protons indicate that there are three vinylic protons. Therefore, we know that the alkene is monosubstituted.

The two remaining signals in the NMR spectrum are a doublet (3.3 ppm) and a singlet (1.3 ppm) that each integrate to 2 protons. Because splitting is not typically seen with protons attached to nitrogens, we can identify the singlet at 1.3 ppm as due to the two amine protons. The doublet must be due to a methylene group that is attached to an sp^2 carbon and split by a vinylic proton that is attached to the same carbon. The compound, therefore, is **allylamine**.

$$\begin{array}{ccc} \text{H} & & \text{H} \\ \backslash & & / \\ & \text{C}=\text{C} & \\ / & & \backslash \\ \text{H} & & \text{CH}_2\text{NH}_2 \end{array}$$

Now we can understand why the signal at 5.9 ppm is a multiplet. This vinylic proton is split by the methylene group and two unique vinylic protons. The signals for the other two vinylic protons are doublets because each is split by the single proton attached to the adjacent sp^2 carbon. Notice that the higher-frequency doublet has the larger J value. This is the signal for the proton that is trans to its coupled proton.

43. The molecular formula tells us that the compound does not have any degrees of unsaturation. Therefore, the oxygen must be the oxygen of either an ether or an alcohol. Because there are no signals that integrate to one proton, we can conclude that the compound is an ether.

The triplet (1.2 ppm) that integrates to 3 protons and the quartet (3.5 ppm) that integrates to 2 protons suggests the presence of an ethyl group. The high-frequency chemical shift of the ethyl's methylene group and the fact that it shows splitting only by the three protons of the methyl group indicate that the ethyl group is next to the oxygen.

The two remaining signals are both triplets (3.4 and 3.5 ppm) and each integrates to 2 protons. Thus, the signals are due to two adjacent methylene groups. Because both signals occur at high frequencies, both must be attached to electron-withdrawing atoms.

Because the molecular formula tells us that the compound contains a bromine, we can conclude that the compound is **2-bromoethyl ethyl ether**.

44. The IR spectrum shows a strong and broad absorption at $\sim 3300 \text{ cm}^{-1}$, indicating that the compound is an alcohol.

The signals in the NMR spectrum between 6 and 7 ppm indicate the presence of a benzene ring. Because these signals integrate to a total of 3 protons, the benzene ring must be trisubstituted.

Because the signal at 6.3 ppm is a doublet, it must be adjacent to one proton; and since the signal at 6.7 ppm is a triplet, it must be adjacent to two protons. Thus, the three benzene-ring protons must be adjacent to one other.

The singlet at 8.7 ppm is the only signal in the spectrum that can be attributed to the proton of the OH group. Because the signal integrates to 2 protons, the compound must have two OH groups in the same environment.

The singlet (2.0 ppm) that integrates to 3 protons indicates that the compound has a methyl group that is not adjacent to a carbon that is attached to any hydrogens.

Therefore, we know that the three substituents that are attached to adjacent carbons on the benzene ring are two OH groups and a methyl group. Because the OH groups are in the same environment, the compound must be the one shown here.

45.   We are told that the compound is an alcohol. Because the singlet (1.4 ppm) is the only signal that integrates to 1 proton, it must be the signal given by the OH group.

We know that a triplet that integrates to 3 protons and a quartet that integrates to 2 protons are characteristic of an ethyl group. In this case, the triplet (0.8 ppm) integrates to 6 protons and the quartet integrates to 4 protons. Therefore, the compound must have two ethyl groups in identical environments.

The only other signal in the spectrum is the singlet (1.2 ppm) that integrates to 3 protons. This signal must be due to a methyl group that is bonded to a carbon that is not attached to any hydrogens.

Because the NMR spectrum does not show any additional signals, the compound must be **3-methyl-3-pentanol**.

46.   The doublet (1.1 ppm) that integrates to 6 protons and the septet (3.8 ppm) that integrates to 1 proton suggests an isopropyl group. The triplet (1.7 ppm) that integrates to 3 protons and the quartet (2.2 ppm) that integrates to 2 protons suggests an ethyl group. When these two groups are subtracted from the molecular formula, all that remains is CO_2. The observed splitting patterns tell us that the isopropyl and ethyl groups are isolated from one another. We can conclude then that the compound is an ester. There are two possibilities:

isopropyl propanoate   or   ethyl 2-methylpropanoate

Because the highest-frequency signal (the septet) is given by the CH of the isopropyl group, we know that the CH is attached to an oxygen. Therefore, we know that the compound that gives the NMR data is **isopropyl propanoate**.

47.   The multiplet (6.9–7.8 ppm) that integrates to 4 protons indicates a disubstituted benzene ring. Because the signal is a multiplet, we know that the substituents are on either the 1- and 2-positions or the 1- and 3-positions. If the substituents were on the 1- and 4-positions, either one singlet (if the two substituents were identical) or two doublets (if the substituents were not identical) would be observed.

Three signals (1.4, 2.5, 3.8 ppm) each integrate to 2 protons. The fact that the signals are two triplets and a multiplet suggests that the compound has three adjacent methylene groups. (The methylene groups on the ends would be triplets and the one in the middle would be a multiplet.)

$—CH_2CH_2CH_2—$

From the molecular formula, we know that the compound has an oxygen. The triplet at 3.8 ppm indicates that that particular methylene group is next to the oxygen. The two fragments that we have identified account for the entire molecular formula. Thus, the compound must have the structure shown here.

48.   The molecular formula indicates that the compound has one degree of unsaturation. The NMR spectrum does not show any signals in the area expected for vinylic protons, so the compound must be either a ketone or a cyclic ether. A cyclic ether would be expected to have protons on adjacent carbons, so the signals would show splitting. Because the two signals in the spectrum are both singlets, the compound must be a ketone.

The fact that the compound has 11 carbons and 22 hydrogens but gives only two singlets in the NMR spectrum indicates that the compound must be symmetrical.

We know that a *tert*-butyl group gives a singlet that integrates to 9 protons. The symmetry of the molecule leads us to conclude that the singlet that integrates to 18 protons is due to two *tert*-butyl groups. We can then assume that the singlet that integrates to 4 protons is due to two nonadjacent methylene groups.

These fragments account for all atoms in the molecular formula. Therefore, the compound must be **2,2,4,4-tetramethyl-4-heptanone**.

2,2,6,6-tetramethyl-4-heptanone

49.   The molecular formula has one degree of unsaturation, so it must have a carbon-carbon double bond, a cyclic structure, or a carbonyl group.

The signal (3.8 ppm) that integrates to 4 hydrogens suggests the presence of two methylene groups in identical environments because, other than the carbon in methane, a single carbon cannot be attached to four hydrogens. The chemical shift suggests that each methylene group must be attached to an oxygen. Because there is only one oxygen in the compound, the two methylene groups must be attached to the same oxygen.

$$-CH_2-O-CH_2-$$

The signal (1.6 ppm) that integrates to 4 hydrogens also suggests the presence of two methylene groups in identical environments. The four methylene groups and the oxygen account for all the atoms in the molecular formula. Thus, the compound must be a cyclic ether.

The fact that the signal at the higher frequency is a triplet and the other signal is a multiplet confirms this structure.

50.   The doublet (7.3 ppm) that integrates to 4 protons and the doublet (7.7 ppm) that also integrates to 4 protons are given by benzene-ring protons. Because a benzene ring does not have 8 protons, there must be two benzene rings in the compound. The doublets indicate that the benzene rings have substituents at the 1- and 4-positions and, because each doublet integrates to 4 protons, the two substituents on each of the benzene rings must be the same.

The singlet (3.8 ppm) that integrates to 6 protons suggests the compound has two methyl groups in an identical environment. The chemical shift of the singlet indicates that each is attached to an electronegative atom. The molecular formula indicates that the electronegative atom is an oxygen.

$$2 \quad -OCH_3$$

When the two disubstituted benzene rings and the two CH_3O groups are subtracted from the molecular formula, all that remains is CO. Thus, a carbonyl group must connect the two benzene rings.

CHAPTER 1
Remembering General Chemistry: Electronic Structure and Bonding

Important Terms

antibonding molecular orbital	a molecular orbital that results when two atomic orbitals with opposite phases interact. Electrons in an antibonding orbital decrease bond strength.
atomic number	a number that tells us how many protons (or electrons) a neutral atom has.
atomic orbital	an orbital associated with an atom; the three-dimensional area around a nucleus where electrons are most likely to be found.
atomic weight	the average mass of the atoms in the naturally occurring element.
aufbau principle	a principle that states that an electron will always go into the available orbital with the lowest energy.
bond dissociation energy	the amount of energy required to break a bond in a way that allows each of the atoms to retain one of the bonding electrons, or the amount of energy released when a bond is formed.
bonding molecular orbital	a molecular orbital that results when two atomic orbitals with the same phase interact. Electrons in a bonding orbital increase bond strength.
bond length	the internuclear distance between two atoms at minimum energy (maximum stability).
carbanion	a species containing a negatively charged carbon.
carbocation	a species containing a positively charged carbon.
condensed structure	a structure that does not show some (or all) of the covalent bonds.
core electrons	electrons in filled shells.
covalent bond	a bond created as a result of sharing electrons.
degenerate orbitals	orbitals that have the same energy.
dipole	a separation of positive and negative charges.
dipole moment (μ)	a measure of the separation of charge in a bond or in a molecule.
double bond	a bond composed of a sigma bond and a pi bond.
electronegative	describes an element that readily acquires an electron.
electronegativity	the tendency of an atom to pull electrons toward itself.

electrostatic attraction	an attractive force between opposite charges.
electrostatic potential map (potential map)	a map that allows you to see how electrons are distributed in a molecule.
equilibrium constant	the ratio of products to reactants at equilibrium.
excited-state electronic configuration	the electronic configuration that results when an electron in the ground state has been moved to a higher-energy orbital.
formal charge	the number of valence electrons − (the number of nonbonding electrons + half the number of bonding electrons).
free radical (radical)	a species with an unpaired electron.
ground-state electronic configuration	a description of which orbitals the electrons of an atom occupy when they are all in their lowest available energy orbitals.
Heisenberg uncertainty principle	a principle that states that both the precise location and the momentum of an atomic particle cannot be simultaneously determined.
Hund's rule	a rule that states that when there are degenerate orbitals, an electron will occupy an empty orbital before it will pair up with another electron.
hybrid orbital	an orbital formed by hybridizing (mixing) atomic orbitals.
hydride ion	a negatively charged hydrogen (a hydrogen atom with an extra electron).
hydrogen ion (proton)	a positively charged hydrogen (a hydrogen atom without its electron).
ionic bond	a bond formed as a result of the attraction of opposite charges.
ionic compound	a compound composed of a positive ion and a negative ion.
ionization energy	the energy required to remove an electron from an atom.
isotopes	atoms with the same number of protons but a different number of neutrons.
Kekulé structure	a model that represents the bonds between atoms as lines.
Lewis structure	a model that represents the bonds between atoms as lines or dots and the lone-pair electrons as dots.
lone-pair electrons (nonbonding electrons)	valence electrons not used in bonding.

mass number	the number of protons plus the number of neutrons in an atom.
molecular orbital	an orbital associated with a molecule that results from the combination of atomic orbitals.
molecular orbital (MO) theory	a theory that describes a model in which the electrons occupy orbitals as they do in atoms but the orbitals extend over the entire molecule.
node	a region within an orbital where there is zero probability of finding an electron.
nonbonding electrons	valence electrons not used in bonding.
nonpolar covalent bond	a bond formed between two atoms that share the bonding electrons equally.
octet rule	a rule that states that an atom will give up, accept, or share electrons in order to achieve a filled outer shell (or an outer shell that contains eight electrons) and no electrons of higher energy. Because a filled second shell contains eight electrons, this is known as the octet rule.
orbital	the volume of space around the nucleus where an electron is most likely to be found.
orbital hybridization	mixing of atomic orbitals.
organic compound	a compound that contains carbon.
Pauli exclusion principle	a principle that states that no more than two electrons can occupy an orbital and that the two electrons must have opposite spin.
pi (π) bond	a bond formed as a result of side-to-side overlap of p orbitals.
polar covalent bond	a bond formed between two atoms that do not share the bonding electrons equally.
potential map (electrostatic potential map)	a map that allows you to see how electrons are distributed in a molecule.
proton (hydrogen ion)	a positively charged hydrogen ion.
quantum mechanics	the use of mathematical equations to describe the behavior of electrons in atoms or molecules.
radical (free radical)	a species with an unpaired electron.
sigma (σ) bond	a bond with a symmetrical distribution of electrons about the internuclear axis.
single bond	a single pair of electrons shared between two atoms.
tetrahedral bond angle	the bond angle (109.5°) formed by an sp^3 hybridized central atom.

tetrahedral carbon	an sp^3 hybridized carbon; a carbon that forms covalent bonds using four sp^3 hybrid orbitals.
trigonal planar carbon	an sp^2 hybridized carbon.
triple bond	composed of a sigma bond and two pi bonds.
valence electron	an electron in an outermost shell.
valence-shell electron-pair repulsion (VSEPR) model	a model for the prediction of molecular geometry based on the minimization of electron repulsion between regions of electron density (that is, between bonds and nonbonding electrons) around an atom.
wave equation	an equation that describes the behavior of each electron in an atom or a molecule.
wave functions	a series of solutions to a wave equation.

Solutions to Problems

1.   The atomic number = the number of protons.
        The mass number = the number of protons + the number of neutrons.
        All isotopes have the same atomic number; in the case of oxygen it is 8. Therefore:

> The isotope of oxygen with a mass number of 16 has 8 protons and 8 neutrons.
> The isotope of oxygen with a mass number of 17 has 8 protons and 9 neutrons.
> The isotope of oxygen with a mass number of 18 has 8 protons and 10 neutrons.

2.   (percentage of naturally occurring ^{35}Cl × atomic mass of ^{35}Cl) + (percentage of naturally occurring ^{37}Cl × atomic mass of ^{37}Cl)

$$(.7577 \times 34.969) + (.2423 \times 36.966)$$
$$(26.496 + 8.957) = 35.453$$

3.   All four atoms have two core electrons (in their filled shell); the valence electrons are in the outer shell. Notice that because the four atoms in the question are in the same row of the periodic table, they have the same number of core electrons.
        a. 3   **b.** 5   **c.** 6   **d.** 7

4.   **a.** Use the aufbau principle (electrons go into available orbitals with the lowest energy) and the Pauli exclusion principle (no more than two electrons are in each atomic orbital). The order in which the orbitals are filled is:

$$1s < 2s < 2p < 3s < 3p < 3d < 4s < 4p < 4d < 5s < 5p$$

> The question can be answered if you remember that there are one s atomic orbital and three degenerate p atomic orbitals in each shell. To write the electronic configurations for Br and I, you need to remember that there are five degenerate d atomic orbitals in the third and fourth shell.

> Cl $1s^2\, 2s^2\, 2p^6\, 3s^2\, 3p^5$
> Br $1s^2\, 2s^2\, 2p^6\, 3s^2\, 3p^6\, 3d^{10}\, 4s^2\, 4p^5$
> I $1s^2\, 2s^2\, 2p^6\, 3s^2\, 3p^6\, 3d^{10}\, 4s^2\, 4p^6\, 4d^{10}\, 5s^2\, 5p^5$

   b. They each have seven electrons in their outer shell (two are in an s orbital and five are in p orbitals). Thus, they each have seven valence electrons. Notice that because they all are in the same column of the periodic table, they all have the same number of valence electrons.

5.   The atomic numbers can be found in the periodic table on the last page of the text. Notice that elements in the same column of the periodic table have the same number of valence electrons and their valence electrons are in similar orbitals.

   a. carbon (atomic number = 6; 2 core, 4 valence): $1s^2\, 2s^2\, 2p^2$
          silicon (atomic number = 14; 10 core, 4 valence): $1s^2\, 2s^2\, 2p^6\, 3s^2\, 3p^2$

   b. oxygen (atomic number = 8; 2 core, 6 valence): $1s^2\, 2s^2\, 2p^4$
          sulfur (atomic number = 16; 10 core, 6 valence): $1s^2\, 2s^2\, 2p^6\, 3s^2\, 3p^4$

    c. nitrogen   (atomic number $=$ 7; 2 core, 5 valence): $1s^2\,2s^2\,2p^3$
        phosphorus (atomic number $=$ 15; 10 core, 5 valence): $1s^2\,2s^2\,2p^6\,3s^2\,3p^3$

    d. magnesium (atomic number $=$ 12; 10 core, 2 valence): $1s^2\,2s^2\,2p^6\,3s^2$
        calcium   (atomic number $=$ 20; 18 core, 2 valence): $1s^2\,2s^2\,2p^6\,3s^2\,3p^6\,4s^2$

6.   **a.** Potassium is in the first column of the periodic table; therefore, like lithium and sodium that are also in the first column, potassium has one valence electron.
      b. It occupies a $4s$ orbital.

7.   The polarity of a bond can be determined by the difference in the electronegativities (given in Table 1.3 on page 11 of the text) of the atoms sharing the bonding electrons. The greater the difference in electronegativity, the more polar the bond.

    a. Cl—CH$_3$ **b.** H—OH   **c.** H—F   **d.** Cl—CH$_3$

8.   **a.** KCl has the most polar bond because its two bonded atoms have the greatest differences in electronegativity. The electronegativity differences in the four listed compounds are as follows:

$$
\begin{aligned}
\text{KCl} \quad & 3.0 - 0.8 = 2.2 \\
\text{LiBr} \quad & 2.8 - 1.0 = 1.8 \\
\text{NaI} \quad & 2.5 - 0.9 = 1.6 \\
\text{Cl}_2 \quad & 3.0 - 3.0 = 0
\end{aligned}
$$

      b. Cl$_2$ has the least polar bond because the two chlorine atoms share the bonding electrons equally.

9.   Solved in the text.

10.   To answer this question, compare the electronegativities of the two atoms sharing the bonding electrons using Table 1.3 on page 11 of the text.

            $\delta-$ $\delta+$ $\delta-$ $\delta+$ $\delta+$ $\delta-$ $\delta+$ $\delta-$
    a. HO—H   **b.** F—Br   **c.** H$_3$C—NH$_2$ **d.** H$_3$C—Cl

            $\delta-$ $\delta+$ $\delta-$ $\delta+$ $\delta+$ $\delta-$ $\delta+$ $\delta-$
    e. HO—Br   **f.** H$_3$C—Li   **g.** I—Cl   **h.** H$_2$N—OH

    (Notice that if the atoms being compared are in the same row of the periodic table, the atom farther to the right is the more electronegative; if the atoms being compared are in the same column, the one closer to the top of the column is the more electronegative.)

11.   Solved in the text.

12.   The dipole moment is the magnitude of the charge times the distance between the charges. Because fluorine is more electronegative than Cl, the charge on H and F in HF is larger than the charge on H and Cl in HCl. The larger charge on F compared to the charge on Cl is more than enough to make up for the fact that H—F is a shorter bond than H—Cl.

13.   **a.** LiH and HF are polar (they have a red end and a blue end).
      b. A potential map marks the edges of the molecule's electron cloud. The electron cloud is largest around the H in LiH, because that H has more electrons around it than the Hs in the other molecules.
      c. Because the hydrogen of HF is blue, we know that this compound has the most positively charged hydrogen and, therefore, it will be most apt to attract a negatively charged species.

14.   By answering this question you will see that a formal charge is a book-keeping device. It does *not necessarily* tell you which atom has the greatest electron density or is the most electron deficient.

    a. oxygen   **b.** oxygen (it is the more red)

    c. oxygen   **d.** hydrogen (it is the deepest blue)

Notice that in hydroxide ion, the atom with the formal negative charge **is** the atom with the greater electron density. In the hydronium ion, however, the atom with the formal positive charge **is not** the most electron-deficient atom.

15.   formal charge = number of valence electrons

                    − (number of lone-pair electrons + half the number of bonding electrons)

In all four structures, every H is singly bonded and thus has a formal charge $= 1 - (0 + 1) = 0$. Similarly, all CH_3 carbon atoms have four bonds and a formal charge $= 4 - (0 + 4) = 0$. The formal charges on the remaining atoms follow:

a. formal charge on O
$6 - (2 + 3) = +1$

b. formal charge on C
$4 - (2 + 3) = -1$

c. formal charge on N
$5 - (0 + 4) = +1$

d. formal charge on
N: $5 - (0 + 4) = +1$
B: $3 - (0 + 4) = -1$

16.   The bond between two atoms can be shown by a pair of dots or by a line, so there are two ways each of the answers can be written.

17. a.

$$H\text{:}\overset{\displaystyle H}{\underset{\displaystyle H}{C}}\text{:}\overset{\displaystyle H}{\underset{\displaystyle H}{C}}\text{:}\overset{..}{\underset{..}{O}}\text{:}H \qquad and \qquad H\text{:}\overset{\displaystyle H}{\underset{\displaystyle H}{C}}\text{:}\overset{..}{\underset{..}{O}}\text{:}\overset{\displaystyle H}{\underset{\displaystyle H}{C}}\text{:}H$$

or

$$H{-}\underset{\displaystyle H}{\overset{\displaystyle H}{C}}{-}\underset{\displaystyle H}{\overset{\displaystyle H}{C}}{-}\overset{..}{\underset{..}{O}}{-}H \quad and \quad H{-}\underset{\displaystyle H}{\overset{\displaystyle H}{C}}{-}\overset{..}{\underset{..}{O}}{-}\underset{\displaystyle H}{\overset{\displaystyle H}{C}}{-}H$$

b.

$$H\text{:}\overset{\displaystyle H}{\underset{\displaystyle H}{C}}\text{:}\overset{\displaystyle H}{\underset{\displaystyle H}{C}}\text{:}\overset{\displaystyle H}{\underset{\displaystyle H}{C}}\text{:}\overset{..}{\underset{..}{O}}\text{:}H \qquad and \qquad H\text{:}\overset{\displaystyle H}{\underset{\displaystyle H}{C}}\text{:}\overset{\displaystyle H}{\underset{\displaystyle H}{C}}\text{:}\overset{..}{\underset{..}{O}}\text{:}\overset{\displaystyle H}{\underset{\displaystyle H}{C}}\text{:}H \qquad and \qquad H\text{:}\overset{\displaystyle H}{\underset{\displaystyle H}{C}}\text{:}\overset{\displaystyle H}{\underset{..}{C}}\text{:}\overset{\displaystyle H}{\underset{..}{C}}\text{:}H$$
$$H\text{:}\overset{..}{\underset{..}{O}}\text{:}H$$
$$H$$

or

$$H{-}\underset{\displaystyle H}{\overset{\displaystyle H}{C}}{-}\underset{\displaystyle H}{\overset{\displaystyle H}{C}}{-}\underset{\displaystyle H}{\overset{\displaystyle H}{C}}{-}\overset{..}{\underset{..}{O}}{-}H \quad and \quad H{-}\underset{\displaystyle H}{\overset{\displaystyle H}{C}}{-}\underset{\displaystyle H}{\overset{\displaystyle H}{C}}{-}\overset{..}{\underset{..}{O}}{-}\underset{\displaystyle H}{\overset{\displaystyle H}{C}}{-}H \quad and \quad H{-}\underset{\displaystyle H}{\overset{\displaystyle H}{C}}{-}\underset{\displaystyle\underset{..}{O}}{\overset{\displaystyle H}{C}}{-}\underset{\displaystyle H}{\overset{\displaystyle H}{C}}{-}H$$
$$H$$

18. Because the compounds are neutral, a halogen will have three lone pairs, an oxygen will have two, a nitrogen will have one, and carbon or a hydrogen will have no lone pairs.

a. $CH_3CH_2\overset{..}{N}H_2$ **c.** $CH_3CH_2\overset{..}{\underset{..}{O}}H$ **e.** $CH_3CH_2\overset{..}{\underset{..}{C}}l\text{:}$

b. $CH_3\overset{..}{N}HCH_3$ **d.** $CH_3\overset{..}{\underset{..}{O}}CH_3$ **f.** $H\overset{..}{\underset{..}{O}}\overset{..}{N}H_2$

19. a. $CH_3CH_2CH_2Cl$ **b.** $CH_3\overset{\displaystyle O}{\overset{\|}{C}}OCH_2CH_3$ **c.** $CH_3CH_2\underset{\displaystyle CH_3}{\overset{\displaystyle O}{\overset{\|}{C}N}}CH_2CH_3$ **d.** $CH_3CH_2C{\equiv}N$

20. a. the (green) chlorine atom

b. the (red) oxygen atoms

c. the (blue) nitrogen atoms

d. the (black) carbon atoms and (gray) hydrogen atoms

21. a.

$$H{-}\underset{\displaystyle H}{\overset{\displaystyle H}{C}}{-}\underset{\displaystyle H}{\overset{..}{N}}{-}\underset{\displaystyle H}{\overset{\displaystyle H}{C}}{-}\underset{\displaystyle H}{\overset{\displaystyle H}{C}}{-}\underset{\displaystyle H}{\overset{\displaystyle H}{C}}{-}H$$

c.

$$H{-}\underset{\displaystyle H{-}\overset{\displaystyle H}{\underset{\displaystyle H}{C}}{-}H}{\overset{\displaystyle H{-}\overset{\displaystyle H}{\underset{}{C}}{-}H}{C}}{-}\overset{\displaystyle H}{\underset{\displaystyle H}{C}}{-}\overset{..}{\underset{..}{B}}r\text{:}$$

b.

$$H{-}\underset{\displaystyle H}{\overset{\displaystyle H}{C}}{-\!-\!-}\underset{\displaystyle H{-}\overset{\displaystyle H}{\underset{\displaystyle H}{C}}{-}H}{\overset{\displaystyle H}{C}}{-}\overset{..}{\underset{..}{C}}l\text{:}$$

d.

$$H{-}\underset{\displaystyle H}{\overset{\displaystyle H}{C}}{-\!-\!-}\underset{\displaystyle H{-}\overset{\displaystyle H}{\underset{\displaystyle H}{C}}{-}H}{\overset{\displaystyle H{-}\overset{\displaystyle H}{\underset{}{C}}{-}H}{C}}{-}\underset{\displaystyle H}{\overset{\displaystyle H}{C}}{-}\underset{\displaystyle H}{\overset{\displaystyle H}{C}}{-}\underset{\displaystyle H}{\overset{\displaystyle H}{C}}{-\!-\!-}\overset{\displaystyle :O:}{\overset{\|}{C}}{-}H$$

22.   **a.** 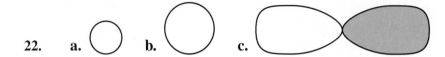 **b.** **c.**

23.   He$_2^+$ has three electrons. Using Figure 1.4 on page 25 of the text, two electrons would be in a bonding molecular orbital and one electron would be in an antibonding molecular orbital. Because there are more electrons in the bonding molecular orbital than in the antibonding molecular orbital, He$_2^+$ exists.

24.   **a.** π^* This involves out-of-phase overlap of atomic orbitals (the overlapping orbitals have different colors), leading to an antibonding molecular orbital. Because this example involves the side-to-side overlap of p orbitals, this results in a π^* antibonding molecular orbital.

   b. π This involves in-phase overlap of atomic orbitals (the overlapping orbitals have the same color), leading to a bonding molecular orbital. Because this example involves the side-to-side overlap of p orbitals, this results in a π bonding molecular orbital.

   c. σ^* This involves out-of-phase overlap of atomic orbitals (the overlapping orbitals have different colors), leading to an antibonding molecular orbital. Because this example involves the end-on overlap of atomic orbitals, this results in a σ^* antibonding molecular orbital.

   d. σ This involves in-phase overlap of atomic orbitals (the overlapping orbitals have the same color), leading to a bonding molecular orbital. Because this example involves the end-on overlap of atomic orbitals, this results in a σ bonding molecular orbital.

25.   The carbon-carbon bonds form as a result of $sp^3 — sp^3$ overlap.
   The carbon-hydrogen bonds form as a result of $sp^3 — s$ overlap.

26.   The electron density of the large lobe of an sp^3 orbital is greater than the electron density of a lobe of a p orbital. Therefore, the overlap of an s orbital with an sp^3 orbital forms a stronger bond than the overlap of an s orbital with a p orbital.

27.   Solved in the text.

28.   **a.** **One** s orbital and **three** p orbitals form **four** sp^3 orbitals.
   b. **One** s orbital and **two** p orbitals form **three** sp^2 orbitals.
   c. **One** s orbital and **one** p orbital form **two** sp orbitals.

29.   **a(1).** Solved in the text.
   b(1). Solved in the text.

   a(2). The carbon forms four bonds, and each chlorine forms one bond.

$$
\begin{array}{c}
\ddot{\text{C}}\text{l}: \\
| \\
:\ddot{\text{C}}\text{l}-\text{C}-\ddot{\text{C}}\text{l}: \\
| \\
:\ddot{\text{C}}\text{l}:
\end{array}
$$

   b(2). The carbon uses sp^3 orbitals to form the bonds with the chlorine atoms, so the bond angles are all 109.5°.

$$
\begin{array}{c}
:\ddot{\text{C}}\text{l}: \\
| \quad 109.5° \\
:\ddot{\text{C}}\text{l}-\text{C}\text{'''}\ddot{\text{C}}\text{l}: \\
:\ddot{\text{C}}\text{l}:
\end{array}
$$

a(3). The first attempt at drawing a Lewis structure shown below shows that carbon does not have a complete octet and does not form the needed number of bonds.

Using one of oxygen's lone pairs to put a double bond between the carbon and oxygen solves both problems.

b(3). The sp^3 hybridized CH_3 carbon has 109.5° bond angles and the sp^2 hybridized CH=O carbon has 120° bond angles.

a(4). In order to fill their octets and form the required number of bonds, carbon and nitrogen must form a triple bond.

$$H—C\equiv N:$$

b(4). Because the carbon is sp hybridized, the bond angle is 180°.

30.   a. 120°   **b.**   120°
c. Because the carbon is sp^3 hybridized and it has a lone pair, you can predict that the bond angle is similar to that in NH_3 (107.3°).

31. The nitrogen atom has the greatest electron density.
The hydrogens are the bluest atoms. Therefore, they have the least electron density. In other words, they have the most positive (least negative) electrostatic potential.

32. Water is the most polar—it has a deep red area and the most intense blue area.
Methane is the least polar—it is all nearly the same color (green) with no red or blue areas.

33. Solved in the text.

34. Electrons in atomic orbitals farther from the nucleus form **longer** bonds; they also form **weaker** bonds due to less electron density in the region of orbital overlap. Therefore:
a. **relative lengths** of the bonds in the halogens are: $Br_2 > Cl_2$
relative strengths of the bonds are: $Cl_2 > Br_2$
b. **relative lengths**: $CH_3—Br > CH_3—Cl > CH_3—F$
relative strengths: $CH_3—F > CH_3—Cl > CH_3—Br$

35. **a.** **longer:** **1.** C—I   **2.** C—Cl   **3.** H—Cl

      b. **stronger:** **1.** C—Cl   **2.** C—C   **3.** H—F

36. **a. CCl$_4$**

The carbon in CCl_4 is bonded to four atoms, so it uses four sp^3 orbitals.

Each carbon-chlorine bond is formed by the overlap of an sp^3 orbital of carbon with an sp^3 orbital of chlorine. Because the four sp^3 orbitals of carbon orient themselves to get as far away from each other as possible, the bond angles are all **109.5°**.

bond angles = 109.5°

b. CO$_2$

The carbon in CO_2 is bonded to two atoms, so it uses two sp orbitals. Each carbon-oxygen bond is a double bond. One of the bonds of each double bond is formed by the overlap of an sp orbital of carbon with an sp^2 orbital of oxygen. The second bond of the double bond is formed as a result of side-to-side overlap of a p orbital of carbon with a p orbital of oxygen. Because carbon's two sp orbitals orient themselves to get as far away from each other as possible, the bond angle in CO_2 is **180°**.

O=C=O

bond angle = 180°

c. HCOOH

The double-bonded carbon and the double-bonded oxygen in HCOOH each uses sp^2 orbitals; thus, the bonds around the double-bonded carbon are all 120°. The single-bonded oxygen forms bonds using sp^3 orbitals, whereas each hydrogen uses an s orbital. The predicted C—O—H bond angle is based on an sp^3 oxygen that has a reduced bond angle due to the two lone pairs. The bond angle is predicted to be similar to that in H_2O (104.5°).

d. N$_2$

The triple bond consists of one σ bond and two π bonds. Each nitrogen has two sp orbitals; one is used to form the σ bond, and the other contains the lone pair. Each nitrogen has two p orbitals that are used to form the two π bonds. A bond angle is the angle formed by three atoms. Therefore, there are no bond angles in this two-atom containing compound.

:N≡N: the σ bond is formed by sp—sp overlap
each π bond is formed by p—p overlap

e. BF$_3$

Promotion gives boron three unpaired electrons, and hybridization gives it three sp^2 orbitals.

Each sp^2 orbital of boron overlaps an sp^3 orbital of fluorine. The three sp^2 orbitals orient themselves to get as far away from each other as possible, resulting in bond angles of **120°**.

bond angles = 120°

37.   Solved in the text.

38.   We know that the σ bond is stronger than the π bond, because the σ bond in ethane has a bond dissociation energy of 90.2 kcal/mol, whereas the bond dissociation energy of the double bond ($\sigma + \pi$) in ethene is 174.5 kcal/mol, which is less than twice as strong.
Because the σ bond is stronger, we know that it has more effective orbital-orbital overlap.

39.   Because electrons in an s orbital are closer on average to the nucleus than those in a p orbital, the greater the s character in the interacting orbitals, the stronger (and shorter) will be the bond. Therefore, the carbon-carbon σ bond formed by sp^2—sp^2 overlap is stronger (and shorter), because an sp^2 orbital has 33.3% s character, whereas an sp^3 orbital has 25% s character.

40.

41.   **a.**

b.

c.   the nitrogen and chlorine atoms

42.   The bond angle depends on the central atom.

    a. sp^3 nitrogen with no lone pair: 109.5°

    b. sp^3 nitrogen with a lone pair: 107.3°

    c. sp^3 carbon with no lone pair: 109.5°

    d. sp^3 carbon with no lone pair: 109.5°

43.   The one with the greater electronegativity will decrease the electron flow toward the F, giving the compound a smaller dipole moment. Since CH_3F has a smaller dipole moment than CD_3F, we know that hydrogen is more electronegative than deuterium.

44.   The electrostatic potential map of ammonia is not symmetrical in the distribution of the charge—the nitrogen is more electron rich and therefore more red than the three hydrogens. Therefore, its shape, which indicates charge distribution, is not symmetrical.

The electrostatic potential map of the ammonium ion is symmetrical in the distribution of the charge, so its shape is symmetrical. Its symmetry results from the fact that nitrogen forms a bond with each of the four hydrogens and the four bonds point to the corners of a regular tetrahedron. The nitrogen in the ammonium ion has significantly lower electron density than the nitrogen in ammonia since the lone pair has formed a bond to hydrogen.

45.   **a, e, g,** and **h** have a dipole moment of zero because they are symmetrical molecules.

46.   **a.** $H\!:\!\ddot{O}\!:\!\ddot{C}\!:\!\ddot{O}\!:\!H$ **or**

    c. $H\!:\!\ddot{C}\!:\!H$ **or**

    b. $:\!\ddot{O}\!:\!\ddot{C}\!:\!\ddot{O}\!:^{-}$ **or**

    d. $\ddot{O}\!::\!C\!::\!\ddot{O}$ **or** $\ddot{O}\!=\!C\!=\!\ddot{O}$

47.   **a.** CH_3CH_3 **b.** CH_3F

48.   If the central atom is sp^3 hybridized, the bond angle will depend on the number of lone pairs it has: none = 109.5°; one = 107.3°; two = 104.5°

    a. sp^3, 107.3°   **e.** sp^3, 109.5°   **i.** sp^3, 107.3°

    b. sp^2, 120°   **f.** sp^2, 120°   **j.** sp^2, 120°

    c. sp^3, 107.3°   **g.** sp, 180°

    d. sp^2, 120°   **h.** sp^3, 109.5°

49.   **a.** $CH_3CH_2CH_3$ **b.** $CH_3CH\!=\!CH_2$ **c.** $CH_3C\!\equiv\!CCH_3$ **or** $CH_3CH_2C\!\equiv\!CH$

50.   The hybridization of the central atom determines the bond angle. If the hybridization is sp^3, the number of lone pairs on the central atom determines the bond angle.

    a. 109.5°   **b.** 104.5°*   **c.** 107.3°   **d.** 107.3°

*104.5° is the correct prediction based on the bond angle in water.
However, the bond angle is actually somewhat larger (108.2°) because the bond opens up to minimize the interaction between the electron cloud of the relatively bulky CH_3 group.

51.

$$1s \quad 2s \quad 2p_x \quad 2p_y \quad 2p_z \quad 3s \quad 3p_x \quad 3p_y \quad 3p_z$$

a. Mg   ↿⇂ ↿⇂ ↿⇂ ↿⇂ ↿⇂ ↿⇂ $1s^2\,2s^2\,2p^6\,3s^2$

b. Ca^{2+} ↿⇂ ↿⇂ ↿⇂ ↿⇂ ↿⇂ ↿⇂ ↿⇂ ↿⇂ ↿⇂ $1s^2\,2s^2\,2p^6\,3s^2\,3p^6$

c. Ar   ↿⇂ ↿⇂ ↿⇂ ↿⇂ ↿⇂ ↿⇂ ↿⇂ ↿⇂ ↿⇂ $1s^2\,2s^2\,2p^6\,3s^2\,3p^6$

d. Mg^{2+} ↿⇂ ↿⇂ ↿⇂ ↿⇂ ↿⇂ $1s^2\,2s^2\,2p^6$

52.

a. H:C:N:H or H—C—N—H (with H's)

b. H—Ö—N̈=Ö

c. H:N:N:H or H—N̈—N̈—H (with H's)

d. H:N:Ö:⁻ or H—N̈—Ö:⁻ (with H's)

53.

54. The greater the electronegativity between the two bonded atoms, the more polar the bond. (See Table 1.3 on page 11.)

a. C—F   >   C—O   >   C—N

b. C—Cl   >   C—Br   >   C—I

c. H—O   >   H—N   >   H—C

d. C—N   >   C—H   >   C—C

55.

a.
$$\text{H—C—C—H}$$
with H, H on first C, and :O: double bonded to second C

b.
$$\text{H—C—Ö—C—H}$$
with H, H on each C

c.
$$\text{H—C—C—O—H}$$
with H, H on first C, and :O: double bonded to second C

56.

a. $CH_3CH{=}CH_2$ (sp^2)

b. CH_3CCH_3 with O double bonded (sp^2)

c. CH_3CH_2OH (sp^3)

d. $CH_3C{\equiv}N$ (sp)

e. $CH_3CH{=}NCH_3$ (sp^2)

f. $CH_3OCH_2CH_3$ (sp^3)

57. **a.** 120°   **b.** 109.5°   **c.** 180°   **d.** 109.5°

58. **a.** H₃C—Br   **c.** HO—NH₂   **e.** H₃C—OH

   b. H₃C—Li   **d.** I—Br   **f.** (CH₃)₂N—H

59. formal charge = number of valence electrons
               − (number of lone-pair electrons + half the number of bonding electrons)

In all four compounds, H has a single bond and is neutral and each C has four bonds and is neutral. Thus, the indicated formal charge is for O or N.

a. H—C—Ö—H (with H above and below C)

b. H—C—Ö⁺—H (with H above and below each)

c. H—C—Ö:⁻ (with H above and below C)

d. H—C—N̈—H (with H above and below, H on N)

   formal charge   formal charge   formal charge   formal charge
   6 − (4 + 2) = 0   6 − (2 + 3) = +1   6 − (6 + 1) = −1   5 − (2 + 3) = 0

60. The open arrow in the structures points to the shorter of the two indicated bonds in each compound.

1. sp^3 sp^2 sp^2 ⇓
   CH₃CH=CHC≡CH
              sp sp

3. ⇓
   CH₃NHCH₂CH₂N=CHCH₃
   sp^3 sp^3 sp^3 sp^3 sp^2 sp^2 sp^3

2. O sp^2
   ⟹‖ sp^3
   CH₃CCH₂OH
   sp^3 ⟍ sp^3
      sp^2

4. H⟍ sp
         C=CHC≡C—⇓H
      H⟋ sp^2 sp^2 sp

5. H⟍ ⇗ sp sp sp^3 CH₃
         C=CHC≡C—C—H
      H⟋ sp^2 sp^2 |
                          CH₃
                          sp^3

      sp^3 CH₃ (above C)

6. Br—CH₂CH₂CH₂—⇓Cl
      sp^3 sp^3 sp^3

For **1**, **2**, and **3**:   A triple bond is shorter than a double bond, which is shorter than a single bond.

For **4** and **5**:   The greater the *s* character in the hybrid orbital, the shorter is the bond formed using that orbital, because an *s* orbital is closer than a *p* orbital to the nucleus. Therefore, the bond formed by a hydrogen and an *sp* carbon is shorter than the bond formed by a hydrogen and an sp^2 carbon, which is shorter than the bond formed by a hydrogen and an sp^3 carbon. (See Table 1.7 on page 44 of the text.)

For **6**:   Cl forms a bond using a $3sp^3$ orbital and Br forms a bond using a $4sp^3$ orbital. Therefore, the C—Cl bond is shorter.

61.   a.

b.

c.

The three carbons are all sp^3 hybridized.
All the bond angles are 109.5°.

d.

The four carbons are all sp^2 hybridized.
All the bond angles are 120°.

62.   a.

b.

c.

63.

1
highest dipole
moment

2

3

4
lowest dipole
moment

64.

65.

66.   In an alkene, six atoms are in the same plane: the two sp^2 carbons and the two atoms that are bonded to each of the two sp^2 carbons. The other atoms in the molecule will not be in the same plane with these six atoms.

If you put stars next to the six atoms that lie in a plane in each molecule, you will be able to see more clearly whether the indicated atoms lie in the same plane.

67.   **a.**   If the central atom is sp^3 hybridized and it does not have a lone pair, the molecule will have tetrahedral bond angles ($109.5°$). Therefore, only $^+NH_4$ has tetrahedral bond angles. The following species are close to being perfectly tetrahedral: H_2O, H_3O^+, NH_3, $^-CH_3$. However, they all have bond angles slightly smaller than $109.5°$.

   b. $^+CH_3$ and BF_3

68. CH_3CH_2Cl has the longer C—Cl bond because it is formed by the overlap of an sp^3 orbital of Cl with an sp^3 orbital of C, whereas the C—Cl bond in CH_2=CHCl is formed by the overlap of an sp^3 orbital of Cl with an sp^2 orbital of C. (The more the s character, the shorter and stronger the bond.)

69. $CHCl_3$ has the larger dipole moment because the three chlorines are withdrawing electrons in the same general direction, whereas in CH_2Cl_2 two chlorines are withdrawing electrons in the same general direction.

70.   The dipole moment is 2.95 D because the two Cls are withdrawing electrons in the same general direction, so their electron withdrawing effect is additive.

The dipole moment is 0 D because the two Cls are withdrawing electrons in opposite directions, so the electron withdrawing effect of one cancels the electron withdrawing effect of the other.

71. The bond angles at the triple-bonded carbons, when the bonding orbitals overlap maximally, are 180°. A 180° angle cannot fit into the ring structure. Therefore, the overlap between the *sp* orbital and the adjacent *sp*³ orbital becomes distorted from the ideal end-on overlap. This poor overlap causes the compound to be unstable. (Compare the structure shown here with Figure 3.8 on page 125 of the text.)

72. The dipole moment depends on the size of the charge and the distance between the bonded electrons. The longer C—Cl bond more than makes up for the greater charge on fluorine.

73.

a.
$$\text{H}-\overset{\displaystyle \overset{\textstyle \text{H}}{|}}{\underset{\displaystyle \underset{\textstyle \text{H}}{|}}{\text{C}}}-\overset{+}{\text{N}}\equiv\text{N:}$$

b. :N≡N:

c. :N̈=N̈⁺=N̈:

d.
$$\text{H}-\overset{\displaystyle \overset{\textstyle \text{H}}{|}}{\underset{\displaystyle \underset{\textstyle \text{H}}{|}}{\text{C}}}-\overset{\displaystyle \overset{\textstyle :\ddot{\text{O}}:}{||}}{\text{C}}-\overset{\displaystyle}{\underset{\displaystyle \underset{\textstyle \text{H}}{|}}{\ddot{\text{N}}}}-\text{H}$$

Chapter 1 Practice Test

1. Answer the following:

    a. Which bond has a greater dipole moment, a carbon-oxygen bond or a carbon-fluorine bond?

    b. If He_2^+ has three electrons in its molecular orbitals, how many electrons are in an antibonding molecular orbital?

    c. Which is the longer bond, a carbon-hydrogen bond in ethene or a carbon-hydrogen bond in ethane?

    d. Which is larger, the bond angle in water or the bond angle in ammonia?

    e. Which is stronger, a sigma bond or a pi bond?

2. What is the hybridization of the carbon atom in each of the following compounds?

$$^+CH_3 \qquad ^-CH_3 \qquad \cdot CH_3$$

3. Draw the Lewis structure for HCO_3^-.

4. Circle the compounds below that have a dipole moment $= 0$.

$$CH_2Cl_2 \qquad CH_3CH_3 \qquad CH_3Cl \qquad H_2C{=}O \qquad CCl_4$$

5. Which compound has greater bond angles, H_3O^+ or $^+NH_4$?

6. Draw the structure for each of the following:

    a. methyl carbocation

    b. a hydride ion

    c. a bromine radical

    d. an alkane with only primary carbons

7. Draw the structure of a compound that contains five carbons, two of which are sp^2 hybridized and three of which are sp^3 hybridized.

8. What is the hybridization of each of the indicated atoms?

$$CH_3CH_2C{\equiv}N \qquad CH_3C{=}NCH_3 \qquad CH_3\overset{O}{\underset{}{C}}CH_3 \qquad O{=}C{=}O$$
$$\underset{CH_3}{|}$$

9. a. What orbitals do carbon's electrons occupy before promotion?

    b. What orbitals do carbon's electrons occupy after promotion and before hybridization?

10.   Answer the following:

    a.   What is the H—C—O bond angle in CH_3OH?

    b.   What is the H—Be—H bond angle in BeH_2?

    c.   What is the H—B—H bond angle in BH_3?

    d.   What is the C—O—H bond angle in CH_3OH?

11.   For each of the following compounds, indicate the hybridization of the atom to which the arrow is pointing.

$$\overset{O}{\overset{\|}{HCOH}} \quad HC\equiv N \quad CH_3OCH_3 \quad CH_3CH=CH_2$$

12.   Indicate whether each of the following statements is true or false.

    a.   A pi bond is stronger than a sigma bond.   T   F

    b.   A triple bond is shorter than a double bond.   T   F

    c.   The oxygen-hydrogen bonds in water are formed by the overlap
of an sp^2 orbital of oxygen with an s orbital of hydrogen.   T   F

    d.   A double bond is stronger than a single bond.   T   F

    e.   A tetrahedral carbon has bond angles of 107.5°.   T   F

ANSWERS TO ALL THE PRACTICE TESTS CAN BE FOUND AT THE END OF
THE SOLUTIONS MANUAL.

CHAPTER 2
Acids and Bases: Central to Understanding Organic Chemistry

Important Terms

acid (Brønsted acid)	a species that donates a proton.
acid-base reaction	a reaction in which an acid donates a proton to a base.
acid dissociation constant	a measure of the degree to which an acid dissociates.
acidity	a measure of how easily a compound gives up a proton.
base (Brønsted base)	a species that accepts a proton.
basicity	a measure of the tendency of a compound to share its electrons with a proton.
Brønsted acid	a species that donates a proton.
Brønsted base	a species that accepts a proton.
buffer solution	solution of a weak acid and its conjugate base.
conjugate acid	the species formed when a base accepts a proton.
conjugate base	the species formed when an acid loses a proton.
delocalized electrons	electrons that are shared by three or more atoms (that is, do not belong to a single atom nor are they shared in a bond between two atoms).
equilibrium constant	the ratio of products to reactants at equilibrium.
Henderson–Hasselbalch equation	$pK_a = pH + \log[HA]/[A^-]$
inductive electron withdrawal	the pull of electrons through sigma bonds by an atom or by a group of atoms.
Lewis acid	a species that accepts an electron pair.
Lewis base	a species that donates an electron pair.
pH	the pH scale used to describe the acidity of a solution ($pH = -\log[H^+]$).
pK_a	a measure of the tendency of a compound to lose a proton ($pK_a = -\log K_a$, where K_a is the acid dissociation constant).
proton	a positively charged hydrogen ion.
proton transfer reaction	a reaction in which a proton is transferred from an acid to a base.
resonance	having delocalized electrons.
resonance contributors	structures with localized electrons that together approximate the true structure of a compound with delocalized electrons.
resonance hybrid	the actual structure of a compound with delocalized electrons.

74

Solutions to Problems

1. CO_2 and CCl_4 are not acids because neither has a proton that it can lose.

2. **a.** $HCl + NH_3 \rightleftharpoons Cl^- + {}^+NH_4$

   b. $H_2O + {}^-NH_2 \rightleftharpoons H_3O^+ + NH_3$

3. The conjugate acid is obtained by adding an H^+ to the species.

   a. **(1)** ${}^+NH_4$ **(2)** HCl **(3)** H_2O **(4)** H_3O^+

   The conjugate base is obtained by removing an H^+ from the species.

   b. **(1)** ${}^-NH_2$ **(2)** Br^- **(3)** NO_3^- **(4)** HO^-

4. **a.** The lower the pK_a, the stronger the acid, so the compound with $pK_a = 5.2$ is the stronger acid.
   b. The greater the dissociation constant, the stronger the acid, so the compound with a dissociation constant $= 3.4 \times 10^{-3}$ is the stronger acid.

5. Because we know that $K_a = K_{eq}[H_2O]$

$$K_{eq} = \frac{K_a}{[H_2O]} = \frac{4.53 \times 10^{-6}}{55.5} = 8.16 \times 10^{-8}$$

6. Its K_a value is 1.51×10^{-5}. It is a weaker acid than vitamin C whose K_a value was determined to be 6.8×10^{-5} in the Problem-Solving Strategy.

7. **a.** $HO^- + H^+ \rightleftharpoons H_2O$

   b. $HCO_3^- + H^+ \rightleftharpoons H_2CO_3 \rightleftharpoons H_2O + CO_2$

   c. $CO_3^{2-} + 2H^+ \rightleftharpoons H_2CO_3 \rightleftharpoons H_2O + CO_2$

8. If the pH is <7, the body fluid is acidic; if the pH is >7, the body fluid is basic.

   a. basic   **b.** acidic   **c.** basic

9. Remember that a proton will be picked up by an atom that has one or more lone pairs. Notice that two oxygens have lone pairs in part **e**. To see which one gets the proton, see Table 2.1 on page 60 of the text.

   a. $CH_3CH_2\overset{+}{O}H_2$ **b.** CH_3CH_2OH **c.** $CH_3\underset{}{\overset{O}{\underset{}{C}}}OH$ **d.** $CH_3CH_2\overset{+}{N}H_3$ **e.** $CH_3CH_2\underset{}{\overset{\overset{+}{O}H}{\underset{}{C}}}OH$

10. In each of the following reactions, the position of equilibrium is established by analyzing the relative strengths of the acids on either side of the reaction—the equilibrium favors reaction of the stronger acid to form the weaker acid (see Section 2.5). (Note that HCl is a stronger acid than H_3O^+; see Appendix I in the text.)

If the lone pairs are not shown:

a. CH$_3$OH as an acid: CH$_3$OH + NH$_3$ $\rightleftharpoons$ CH$_3$O$^-$ + $\overset{+}{\text{N}}$H$_4$

CH$_3$OH as a base: CH$_3$OH + HCl $\rightleftharpoons$ CH$_3\overset{+}{\text{O}}$H + Cl$^-$
$\quad\quad\quad\quad\quad\quad\quad\quad\quad\quad\quad\quad\quad\quad$ |
$\quad\quad\quad\quad\quad\quad\quad\quad\quad\quad\quad\quad\quad\quad$ H

b. NH$_3$ as an acid: NH$_3$ + CH$_3$O$^-$ $\rightleftharpoons$ $^-$NH$_2$ + CH$_3$OH

NH$_3$ as a base: NH$_3$ + HBr $\rightleftharpoons$ $\overset{+}{\text{N}}$H$_4$ + Br$^-$

If the lone pairs are shown:

a. CH$_3$OH as an acid: CH$_3\overset{\cdot\cdot}{\text{O}}$H + $\overset{\cdot\cdot}{\text{N}}H_3$ $\rightleftharpoons$ CH$_3\overset{\cdot\cdot}{\text{O}}$:$^-$ + $\overset{+}{\text{N}}$H$_4$

CH$_3$OH as a base: CH$_3\overset{\cdot\cdot}{\text{O}}$H + H$\overset{\cdot\cdot}{\text{C}}$l: $\rightleftharpoons$ CH$_3\overset{+}{\overset{\cdot\cdot}{\text{O}}}$H + :$\overset{\cdot\cdot}{\text{C}}$l:$^-$
$\quad\quad\quad\quad\quad\quad\quad\quad\quad\quad\quad\quad\quad\quad$ |
$\quad\quad\quad\quad\quad\quad\quad\quad\quad\quad\quad\quad\quad\quad$ H

b. NH$_3$ as an acid: $\overset{\cdot\cdot}{\text{N}}H_3$ + CH$_3\overset{\cdot\cdot}{\text{O}}$:$^-$ $\rightleftharpoons$ $^-$:$\overset{\cdot\cdot}{\text{N}}H_2$ + CH$_3\overset{\cdot\cdot}{\text{O}}$H

NH$_3$ as a base: $\overset{\cdot\cdot}{\text{N}}H_3$ + H$\overset{\cdot\cdot}{\text{B}}$r: $\rightleftharpoons$ $\overset{+}{\text{N}}$H$_4$ + :$\overset{\cdot\cdot}{\text{B}}$r:$^-$

11.   **a.** ~40   **b.** ~15   **c.** ~5   **d.** ~10

12.   **a.** CH$_3$COO$^-$ is the stronger base.
Because CH$_3$COOH is the weaker acid, it has the stronger conjugate base.

b. $^-$NH$_2$ is the stronger base.
Because NH$_3$ is the weaker acid, it has the stronger conjugate base.

c. H$_2$O is the stronger base.
Because H$_3$O$^+$ is the weaker acid, it has the stronger conjugate base.

13.   The conjugate acids of the given bases have the following relative strengths:

$$\text{CH}_3\overset{+}{\text{O}}\text{H}_2 > \text{CH}_3\overset{\displaystyle \text{O} \atop \displaystyle \|}{\underset{}{\text{C}}}\text{OH} > \text{CH}_3\overset{+}{\text{N}}\text{H}_3 > \text{CH}_3\text{OH} > \text{CH}_3\text{NH}_2$$

The bases, therefore, have the following relative strengths, since the weakest acid has the strongest conjugate base.

$$\text{CH}_3\overset{-}{\text{N}}\text{H} > \text{CH}_3\text{O}^- > \text{CH}_3\text{NH}_2 > \text{CH}_3\overset{\displaystyle \text{O} \atop \displaystyle \|}{\underset{}{\text{C}}}\text{O}^- > \text{CH}_3\text{OH}$$

14.   Methanol is the acid because it has a pK_a value of ~15, whereas methylamine is a much weaker acid with a pK_a value of about 40.

15.   Notice that in each case, the equilibrium favors reaction of the stronger acid to form the weaker acid.

a. CH_3OH + HO^- $\rightleftharpoons$ CH_3O^- + H_2O

$pK_a = 15.5$ $pK_a = 15.7$

CH_3OH + H_3O^+ $\rightleftharpoons$ $CH_3\overset{+}{O}H_2$ + H_2O

$pK_a = -1.7$ $pK_a = -2.5$

$CH_3\overset{O}{\overset{\|}{C}}OH$ + HO^- $\rightleftharpoons$ $CH_3\overset{O}{\overset{\|}{C}}O^-$ + H_2O

$pK_a = 4.8$ $pK_a = 15.7$

$CH_3\overset{O}{\overset{\|}{C}}OH$ + H_3O^+ $\rightleftharpoons$ $CH_3\overset{+OH}{\overset{\|}{C}}OH$ + H_2O

$pK_a = -1.7$ $pK_a = -6.1$

CH_3NH_2 + HO^- $\rightleftharpoons$ $CH_3\bar{N}H$ + H_2O

$pK_a = 40$ $pK_a = 15.7$

CH_3NH_2 + H_3O^+ $\rightleftharpoons$ $CH_3\overset{+}{N}H_3$ + H_2O

$pK_a = -1.7$ $pK_a = 10.7$

b. HCl + H_2O $\rightleftharpoons$ H_3O^+ + Cl^-

$pK_a = -7$ $pK_a = -1.7$

NH_3 + H_2O $\rightleftharpoons$ $^+NH_4$ + HO^-

$pK_a = 15.7$ $pK_a = 9.4$

16.   Because a strong acid is more likely to lose a proton than a weak acid, the equilibrium will favor loss of a proton from the strong acid and formation of the weak acid.

a. $HC\equiv CH$ + HO^- $\rightleftharpoons$ $HC\equiv C^-$ + H_2O

$pK_a = 25$ $pK_a = 15.7$

b. $HC\equiv CH$ + $^-NH_2$ $\rightleftharpoons$ $HC\equiv C^-$ + NH_3

$pK_a = 25$ $pK_a = 36$

c. $^-NH_2$ would be a better base because when it removes a proton, the equilibrium favors the products. When HO^- removes a proton, the equilibrium favors the reactants.

17.   Each of the following bases will remove a proton from acetic acid in a reaction that favors products, because each of these bases will form an acid that is a weaker acid than acetic acid.

HO^- CH_3NH_2 $HC\equiv C^-$

The other three choices will form an acid that is a stronger acid than acetic acid.

18. $pK_{eq} = pK_a$ (reactant acid) $- pK_a$ (product acid)

For **a** the reactant acid is HCl and the product acid is H_3O^+.

For **b** the reactant acid is CH_3COOH and the product acid is H_3O^+.

For **c** the reactant acid is H_2O and the product acid is $CH_3\overset{+}{N}H_3$.

For **d** the reactant acid is $CH_3\overset{+}{N}H_3$ and the product acid is H_3O^+.

a. $pK_{eq} = -7 - (-1.7) = -5.3$ **c.** $pK_{eq} = 15.7 - (10.7) = 5.0$
   $K_{eq} = 2.0 \times 10^5$ $K_{eq} = 1.0 \times 10^{-5}$

b. $pK_{eq} = 4.8 - (-1.7) = 6.5$ **d.** $pK_{eq} = 10.7 - (-1.7) = 12.4$
   $K_{eq} = 3.2 \times 10^{-7}$ $K_{eq} = 4.0 \times 10^{-13}$

19.   Recall that the weakest acid has the strongest conjugate base.

$$^-CH_3 \; > \; ^-NH_2 \; > \; HO^- \; > \; F^-$$

20.   Again, recall that the weakest acid has the strongest conjugate base.

$$CH_3\bar{C}H_2 \; > \; H_2C{=}\bar{C}H \; > \; HC{\equiv}C^-$$

21.   The protonated ketone is the stronger acid because its hydrogen is attached to an sp^2 oxygen, which is more electronegative than the sp^3 oxygen to which the hydrogen in the protonated alcohol is attached.

22.   **a.** **A** $HC{\equiv}CH \; + \; CH_3\bar{C}H_2 \; \rightleftharpoons \; HC{\equiv}C^- \; + \; CH_3CH_3$

         B $H_2C{=}CH_2 \; + \; HC{\equiv}C^- \; \rightleftharpoons \; H_2C{=}\bar{C}H \; + \; HC{\equiv}CH$

         C $CH_3CH_3 \; + \; H_2C{=}\bar{C}H \; \rightleftharpoons \; CH_3\bar{C}H_2 \; + \; H_2C{=}CH_2$

     b. Only A, because only A has a reactant that is a stronger acid than the acid that is formed in the product.

23.   The smaller the ion, the stronger it is as a base.

$$F^- \; > \; Cl^- \; > \; Br^- \; > \; I^-$$

24.   **a.** oxygen   **b.** H_2S **c.** CH_3SH

The size of an atom is more important than its electronegativity in determining stability. Therefore, even though oxygen is more electronegative than sulfur, H_2S is a stronger acid than H_2O and CH_3SH is a stronger acid than CH_3OH.

Because the sulfur atom is larger, the electrons in its conjugate base are spread out over a greater volume, which stabilizes it. The more stable the base, the stronger its conjugate acid.

25.   The stronger acid will have its proton attached to the more electronegative atom (if the atoms are about the same size) or to the larger atom (if the atoms are not the same size).

a. HBr   **b.** $CH_3CH_2CH_2\overset{+}{O}H_2$ **c.** $CH_3CH_2CH_2OH$ **d.** $CH_3CH_2CH_3SH$

26.   Remember that the stronger the acid, the weaker (or more stable) its conjugate base,

a. Because HI is the strongest acid, I^- is the most stable (weakest) base.
b. Because HF is the weakest acid, F^- is the least stable (strongest) base.

27.   Compare the acid strengths of the conjugate acids, recalling that a weaker acid has a stronger conjugate base.

   a. HO^- because H_2O is a weaker acid than H_3O^+.

   b. NH_3 because $^+NH_4$ is a weaker acid than H_3O^+.

   c. CH_3O^- because CH_3OH is a weaker acid than CH_3COOH.

   d. CH_3O^- because CH_3OH is a weaker acid than CH_3SH.

28.   **a.** $CH_3OCH_2CH_2OH$ because its conjugate base has its negative charge stabilized by electron withdrawal by the CH_3O group.

   b. $CH_3CH_2CH_2\overset{+}{O}H_2$ because oxygen is more electronegative than nitrogen.

   c. $CH_3CH_2OCH_2CH_2OH$ because the electron-withdrawing oxygen is closer to the OH group.

   d. $CH_3CH_2\overset{\displaystyle O}{\overset{\|}{C}}OH$ because the electron-withdrawing C=O is closer to the OH group.

29. $\underset{\underset{F}{|}}{CH_3CHCH_2OH} > \underset{\underset{Cl}{|}}{CH_3CHCH_2OH} > \underset{\underset{Cl}{|}}{CH_2CH_2CH_2OH} > CH_3CH_2CH_2OH$

   The first two listed compounds are the most acidic because they have the electron-withdrawing substituent that stabilizes the conjugate base closest to the O—H bond. The first is more acidic than the second because fluorine is more electronegative than chlorine.
   The second listed compound is a stronger acid than the third listed compound because the chlorine in the third compound is farther away from the O—H bond.
   The compound on the far right does not have a substituent that withdraws electrons inductively, so it is the least acidic of the four compounds.

30.   The weaker acid has a stronger conjugate base.

   a. $\underset{\underset{Br}{|}}{CH_3\overset{\displaystyle O}{\overset{\|}{C}}HCO^-}$ **b.** $\underset{\underset{Cl}{|}}{CH_3CHCH_2\overset{\displaystyle O}{\overset{\|}{C}}O^-}$ **c.** $CH_3CH_2\overset{\displaystyle O}{\overset{\|}{C}}O^-$ **d.** $CH_3\overset{\displaystyle O}{\overset{\|}{C}}CH_2CH_2O^-$

31.   Solved in the text.

32.   **a.**   Because the atom (P) to which each of the OH groups is attached is also attached to two electronegative oxygens.

   b.   The middle OH group is the weakest of the remaining acidic groups. It is an alcohol ($pK_a \sim 15$) and the atom (C) to which it is attached is not attached to any strongly electronegative atoms. (The protonated amino group has a pK_a value of ~ 10.)

33.   When a sulfonic acid loses a proton, the electrons left behind are shared by three oxygens. In contrast, when a carboxylic acid loses a proton, the electrons left behind are shared by only two oxygens. The sulfonate ion, therefore, is more stable than the carboxylate ion.

a sulfonate ion

a carboxylate ion

The more stable the base, the stronger is its conjugate acid. Thus, the sulfonic acid is a stronger acid than the carboxylic acid.

34. a.

b.

35.   Remember, the smaller the pK_a, the stronger the acid.

a. $CH_3C{\equiv}\overset{+}{N}H$

b. CH_3CH_3

c. F_3CCOH (with O double-bonded above the C)

d. an sp^2 oxygen

$pK_a = -7.3$ $pK_a = -3.6$

e. $CH_3C{\equiv}\overset{+}{N}H$ $>$ $CH_3\overset{+}{C}{=}NHCH_3$ $>$ $CH_3CH_2\overset{+}{N}H_3$

$\qquad\qquad\qquad\qquad\qquad\;\;\;|$
$\qquad\qquad\qquad\qquad\qquad CH_3$

$pK_a = -10.1$ $pK_a = 5.5$ $pK_a = 11.0$

36.   If the solution is more acidic than the compound's pK_a value, the compound will be in its acidic form (with its proton).

If the solution is more basic than the compound's pK_a value, the compound will be in its basic form (without its proton).

a. CH_3COO^- **c.** H_2O **e.** $\overset{+}{N}H_4$ **g.** NO_2^-

b. $CH_3CH_2\overset{+}{N}H_3$ **d.** Br^- **f.** $HC{\equiv}N$ **h.** NO_3^-

37.   pH 10.4

(As long as the pH is greater than the pK_a value of the compound, at least 50% of the compound will be in its basic form.)

38. **a.** **1.** neutral   **b.** **1.** charged   **c.** **1.** neutral
  2. neutral   **2.** charged   **2.** neutral
  3. charged   **3.** charged   **3.** neutral
  4. charged   **4.** charged   **4.** neutral
  5. charged   **5.** neutral   **5.** neutral
  6. charged   **6.** neutral   **6.** neutral

39. **a.** The $^+NH_3$ group withdraws electrons, which increases the acidity of the COOH group.

  b. Both the COOH group and the $^+NH_3$ group will be in their acidic forms, because the solution is more acidic than both of their pK_a values.

$$CH_3CH(^+NH_3)-C(=O)-OH$$

  c. The COOH group will be in its basic form because the solution is more basic than its pK_a value. The $^+NH_3$ group will be in its acidic form because the solution is more acidic than its pK_a value.

$$CH_3CH(^+NH_3)-C(=O)-O^-$$

  d. Both the COOH group and the $^+NH_3$ group will be in their basic forms because the solution is more basic than both of their pK_a values.

$$CH_3CH(NH_2)-C(=O)-O^-$$

  e. No, alanine can never be without a charge. To be without a charge would require a group with a pK_a value of 9.69 to lose a proton before a group with a pK_a value of 2.34. This clearly cannot happen. A weak acid cannot have a stronger tendency to lose a proton than a stronger acid has.

  f. As the pH becomes more basic than 2.34, the COOH group will become more negatively charged.

   As the pH becomes more acidic than 9.69, the $^+NH_3$ group will become more positively charged.

   Therefore, the amount of negative change will be the same as the amount of positive charge at the pH that is equidistant from the two pK_a values.

$$\frac{2.34 + 9.69}{2} = 6.02$$

40. **a.** 10.4 (two log units more basic than the pK_a)

  b. 2.7 (one log unit more acidic than the pK_a)

  c. 6.4 (two log units more acidic than the pK_a)

  d. 7.3 (when the pH is equal to the pK_a)

  e. 5.6 (one log unit more basic than the pK_a)

41. **a.** **1.** pH = 4.9   When the pH = pK_a, half the compound will be in its acidic form (with its proton) and half will be in its basic form (without its proton).

    2. pH = 10.7

  b. **1.** pH > 6.9   Because the basic form is the form in which the compound is charged, the pH needs to be more than two units more basic than the pK_a value.

    2. pH < 8.7   Because the acidic form is the form in which the compound is charged, the pH needs to be more than two units more acidic than the pK_a value.

42. **a.** Solved in the text.

    b. For the carboxylic acid to dissolve in water, it must be charged (in its basic form), so the pH will have to be greater than 6.8. For the amine to dissolve in ether, it will have to be neutral (in its basic form), so the pH will have to be greater than 12.7 to have essentially all of it in the neutral form. Therefore, the pH of the water layer must be greater than 12.7.

    c. To dissolve in ether, the carboxylic acid will have to be neutral, so the pH will have to be less than 2.8 to have essentially all the carboxylic acid in the acidic (neutral) form. To dissolve in water, the amine will have to be charged, so the pH will have to be less than 8.7 to have essentially all the amine in the acidic form. Therefore, the pH of the water layer must be less than 2.8.

43. **a.** The basic form of the buffer removes a proton from the solution.

$$CH_3COO^- + H^+ \rightleftharpoons CH_3COOH$$

    b. The acidic form of the buffer donates a proton in order to remove an hydroxide from the solution.

$$CH_3COOH + HO^- \rightleftharpoons CH_3COO^- + H_2O$$

44. Solved in the text.

45. **a.** $ZnCl_2 + CH_3\overset{..}{\underset{..}{O}}H \rightleftharpoons$

$$\overset{-}{ZnCl_2}$$
$$|+$$
$$H\overset{..}{\underset{..}{O}}CH_3$$

    b. $FeBr_3 + :\overset{..}{\underset{..}{Br}}:^- \rightleftharpoons Br-\bar{F}eBr_3$

    c. $AlCl_3 + :\overset{..}{\underset{..}{Cl}}:^- \rightleftharpoons Cl-\bar{A}lCl_3$

46. **a, b, c,** and **h** are Brønsted acids (proton-donating acids). Therefore, they react with HO^- by giving a proton to it. **d, e, f,** and **g** are Lewis acids. They react with HO^- by accepting a pair of electrons from it.

    a. $CH_3OH + HO^- \rightleftharpoons CH_3O^- + H_2O$

    b. $\overset{+}{N}H_4 + HO^- \rightleftharpoons NH_3 + H_2O$

    c. $CH_3\overset{+}{N}H_3 + HO^- \rightleftharpoons CH_3NH_2 + H_2O$

    d. $BF_3 + HO^- \rightleftharpoons HO-\bar{B}F_3$

    e. $\overset{+}{C}H_3 + HO^- \rightleftharpoons CH_3OH$

    f. $FeBr_3 + HO^- \rightleftharpoons HO-\bar{F}eBr_3$

    g. $AlCl_3 + HO^- \rightleftharpoons HO-\bar{A}lCl_3$

    h. $CH_3COOH + HO^- \rightleftharpoons CH_3COO^- + H_2O$

47. **a.** $CCl_3CH_2OH > CHCl_2CH_2OH > CH_2ClCH_2OH$

    b. The greater the number of electron-withdrawing chlorine atoms equidistant from the OH group, the stronger the acid. (Notice that the larger the K_a, the stronger the acid.)

48. The stronger base has the weaker conjugate acid.

    a. HO^- **b.** $CH_3\bar{N}H$ **c.** CH_3O^- **d.** Cl^-

    e. CH_3COO^- **f.** $CH_3CHBrCOO^-$

49. **a.** $\overset{..}{N}H_3$ + $H\overset{..}{\underset{..}{Cl}}$: $\rightleftharpoons$ $\overset{+}{N}H_4$ + :$\overset{..}{\underset{..}{Cl}}$:$^-$

b. $H_2\overset{..}{O}$: + FeBr$_3$ $\rightleftharpoons$ $H_2O\overset{+}{-}\overset{-}{F}eBr_3$

c. [carboxylic acid structure] + $H\overset{..}{\underset{..}{Cl}}$: $\rightleftharpoons$ [protonated carboxylic acid] + :$\overset{..}{\underset{..}{Cl}}$:$^-$

50. **a.** $CH_3CH_2CHCOOH$ > CH_3CHCH_2COOH > $ClCH_2CH_2CH_2COOH$ > $CH_3CH_2CH_2COOH$
           | |
           Cl   Cl

b. An electron-withdrawing substituent makes the carboxylic acid more acidic, because it stabilizes its conjugate base by decreasing the electron density around the oxygen atom. (Remember that the larger the K_a, the stronger the acid.)

c. The closer the electron-withdrawing chloro substituent is to the acidic proton, the more it can decrease the electron density around the oxygen atom, so the more it stabilizes the conjugate base and increases the acidity of its conjugate acid.

51. **a.** $CH_3\overset{..+}{O}-\overset{-}{B}F_3$ **b.** $CH_3\overset{..+}{O}-H$ + Cl^- **c.** $CH_3\overset{+}{N}-\overset{-}{A}lCl_3$
                | | |
                CH_3 CH_3 H (top), H (bottom)

52. **a.** $HOCH_2CH_2\overset{+}{N}H_3$ **b.** $^-OCH_2CH_2NH_2$

53. O is more electronegative than N, which is more electronegative than C. Therefore, the alcohol is more acidic than the amine, which is more acidic than the alkane.
S is larger than O, so CH_3CH_2SH is more acidic than CH_3CH_2OH.

$$CH_3CH_2SH > CH_3CH_2OH > CH_3CH_2NH_2 > CH_3CH_2CH_3$$

54. If the solution is more acidic than the pK_a of the compound, the compound will be in its acidic form (with the proton).
If the solution is more basic than the pK_a of the compound, the compound will be in its basic form (without the proton).

a. at pH = 3 CH_3COOH **b.** at pH = 3 $CH_3CH_2\overset{+}{N}H_3$ **c.** at pH = 3 CF_3CH_2OH

at pH = 6 CH_3COO^- at pH = 6 $CH_3CH_2\overset{+}{N}H_3$ at pH = 6 CF_3CH_2OH

at pH = 10 CH_3COO^- at pH = 10 $CH_3CH_2\overset{+}{N}H_3$ at pH = 10 CF_3CH_2OH

at pH = 14 CH_3COO^- at pH = 14 $CH_3CH_2NH_2$ at pH = 14 $CF_3CH_2O^-$

55.   In all four reactions, the products are favored at equilibrium. (Recall that the equilibrium favors formation of the weaker acid.)

a. $CH_3COOH + CH_3O^- \rightleftharpoons CH_3COO^- + CH_3OH$

b. $CH_3CH_2OH + {}^-NH_2 \rightleftharpoons CH_3CH_2O^- + NH_3$

c. $CH_3COOH + CH_3NH_2 \rightleftharpoons CH_3COO^- + CH_3\overset{+}{N}H_3$

d. $CH_3CH_2OH + HCl \rightleftharpoons CH_3CH_2\overset{+}{O}H_2 + Cl^-$

56. **a.** $HC{\equiv}CCH_2OH > CH_2{=}CHCH_2OH > CH_3CH_2CH_2OH$

  b. These three compounds differ only in the group that is attached to CH_2OH. The more electronegative the group attached to CH_2OH, the stronger the acid because inductive electron withdrawal stabilizes the conjugate base, and the more stable the base, the stronger its conjugate acid. An *sp* carbon is more electronegative than an sp^2 carbon, which is more electronegative than an sp^3 carbon.

57.   In each compound, the nitrogen atom is the most apt to be protonated because it is the stronger base.

a. $CH_3{-}\underset{\underset{OH}{|}}{CH}{-}CH_2\overset{+}{N}H_3$ **b.** $CH_3{-}\underset{\underset{+NH_3}{|}}{\overset{\overset{CH_3}{|}}{C}}{-}OH$ **c.** $CH_3{-}\underset{\underset{+NH_3}{|}}{\overset{\overset{CH_3}{|}}{C}}{-}CH_2OH$

58.   The log of $10^{-4} = -4$, the log of $10^{-5} = -5$, the log of $10^{-6} = -6$, and so on. Because the $pK_a = -\log K_a$, the pK_a of an acid with a K_a of 10^{-4} is $-(-4) = 4$. An acid with a K_a of 4.0×10^{-4} is a stronger acid than one with a K_a of 1.0×10^{-4}. Therefore, the pK_a can be estimated as being between 3 and 4.

a. **1.** between 3 and 4   **b.** **1.** $pK_a = 3.4$
   2. between -2 and -1 **2.** $pK_a = -1.3$
   3. between 10 and 11   **3.** $pK_a = 10.2$
   4. between 9 and 10   **4.** $pK_a = 9.1$
   5. between 3 and 4   **5.** $pK_a = 3.7$

c. Nitric acid (HNO_3) is the strongest acid because it has the lowest pK_a value. (The lower the pK_a value, the stronger the acid.)

59.   The direction of the dipole will be toward the more electronegative of the two atoms that are sharing the bonding electrons.

a. $CH_3{-}C{\equiv}CH$ **b.** $CH_3{-}CH{=}CH_2$

60.   The hydrogen bonded to the oxygen is the most acidic hydrogen in Tenormin.

61.   **A** and **C** because, in each case, the acid is stronger than the acid (H_2O) that will be formed as a product.

62.   The reaction with the more favorable equilibrium constant will be the one with the smallest difference between the pK_a value of the reactant acid and the pK_a value of the product acid, because $pK_{eq} = pK_a$ (reactant acid) $- pK_a$ (product acid) and the smaller the pK_{eq}, the larger the K_{eq}.

a. **1.** CH_3CH_2OH has a pK_a value $= 15.9$
   CH_3OH has a pK_a value $= 15.5$
   $^+NH_4$ has a pK_a value $= 9.4$

$$pK_{eq} = pK_a \,(\text{reactant acid}) - pK_a \,(\text{product acid})$$
$$= 15.9 - 9.4 = 6.5; \, K_{eq} = 3.2 \times 10^{-7}$$
$$= 15.5 - 9.4 = 6.1; \, K_{eq} = 7.9 \times 10^{-7}$$

Thus, the reaction of CH_3OH with NH_3 has the more favorable equilibrium constant.

2. CH_3CH_2OH has a pK_a value $= 15.9$
    $^+NH_4$ has a pK_a value $= 9.4$
    $CH_3\overset{+}{N}H_3$ has a pK_a value $= 10.7$

$$pK_{eq} = pK_a \,(\text{reactant acid}) - pK_a \,(\text{product acid})$$
$$= 15.9 - 9.4 = 6.5; \, K_{eq} = 3.2 \times 10^{-7}$$
$$= 15.5 - 10.7 = 5.2; \, K_{eq} = 6.3 \times 10^{-6}$$

Thus, the reaction of CH_3CH_2OH with CH_3NH_2 has the more favorable equilibrium constant.

b. Because the reaction of CH_3CH_2OH with CH_3NH_2 has the smallest difference between the pK_a values of the reactant and product acids, it has most favorable equilibrium constant.

63. If the reaction is producing protons, the basic form of the buffer will pick up the protons. At the pH at which the reaction is carried out (pH $= 10.5$), a protonated methylamine/methylamine buffer with a $pK_a = 10.7$ will have a larger percentage of the buffer in the needed basic form than will a protonated ethylamine/ethylamine buffer with a pK_a of 11.0.

64.   **a.** $CH_2\!=\!CHCOOH$ because an sp^2 carbon is more electronegative than an sp^3 carbon

   b.   because an oxygen can withdraw electrons inductively

   c. $HC\!\equiv\!CCOOH$ because an sp carbon is more electronegative than an sp^2 carbon

   d.   because an sp^2 nitrogen is more electronegative than an sp^3 nitrogen

65.   **a.** The first pK_a is lower than the pK_a of acetic acid because citric acid has additional oxygen-containing groups that withdraw electrons inductively and thereby stabilize the conjugate base.

   b. The third pK_a is greater than the pK_a of acetic acid because loss of the third proton puts a third negative charge on the molecule. Increasing the number of charges on a species destablizes it.

66.

$$K_a = \frac{[H^+][HO^-]}{[H_2O]}$$

Since $[H^+] = [H^-]$, both must be 1×10^{-7} M

$$K_a = \frac{(1 \times 10^{-7})(1 \times 10^{-7})}{55.5}$$

$$K_a = 1.80 \times 10^{-16}$$
$$pK_a = -\log 1.80 \times 10^{-16}$$
$$pK_a = 15.7$$

The answer can also be obtained in the following way:

$$K_a = \frac{[H^+][HO^-]}{[H_2O]}$$

$$K_a[H_2O] = [H^+][HO^-]$$

take the log of both sides

$$\log K_a + \log[H_2O] = \log[H^+] + \log[HO^-]$$

multiply both sides by -1

$$-\log K_a - \log[H_2O] = -\log[H^+] - \log[HO^-]$$

$$pK_a - \log[H_2O] = pH + pOH$$

$$pK_a - \log[H_2O] = 14$$

$$pK_a = 14 + \log[H_2O]$$

$$pK_a = 14 + \log 55.5$$

$$pK_a = 14 + 1.7$$

$$pK_a = 15.7$$

67. Charged compounds will dissolve in water, and uncharged compounds will dissolve in ether. The acidic forms of carboxylic acids and alcohols are neutral, and the basic forms are charged. The acidic forms of amines are charged, and the basic forms are neutral.

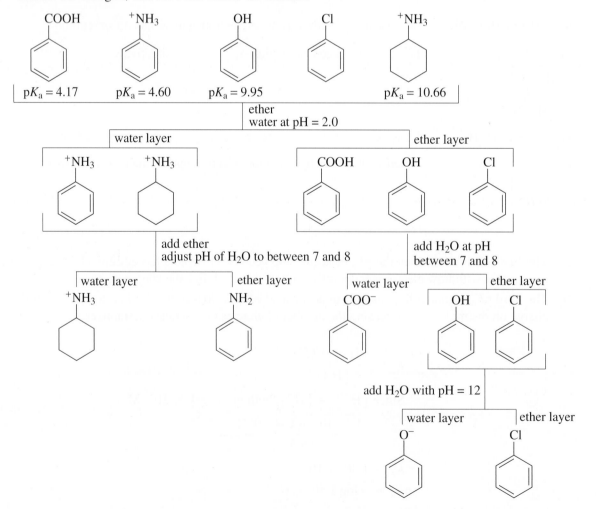

68.   For a discussion of how to do problems such as Problems 68–70, see **Special Topic I** (pH, pK_a, and Buffers).

$$pK_a = pH + \log\frac{[HA]}{[A^-]}$$

The above equation, called the Henderson–Hasselbalch equation, shows that:

1.   When the value of the pH is equal to the value of the pK_a, the concentration of buffer in the acidic form [HA] equals the concentration of buffer in the basic form $[A^-]$.

2.   When the solution is more acidic than the value of the pK_a, more buffer species are in the acidic form than in the basic form.

3.   When the solution is more basic than the value of the pK_a, more buffer species are in the basic form than in the acidic form.

Because the pH of the blood (~ 7.3) is greater than the pK_a value of the buffer (6.1), more buffer species are in the basic form than in the acidic form. Therefore, the buffer is better at neutralizing excess acid.

69.   **a.**   fraction present in the acidic form $= \dfrac{\text{amount in the acidic form}}{\text{amount in the acidic form} + \text{amount in the basic form}}$

$$= \frac{[HA]}{[HA] + [A^-]}$$

Because there are two unknowns, we must define one in terms of the other.
By using the definition of the acid dissociation constant, we can define $[A^-]$ in terms of $[HA]$, $[H^+]$, and K_a, so we have only one unknown.

$$K_a = \frac{[H^+][A^-]}{[HA]}$$

$$[A^-] = \frac{K_a[HA]}{[H^+]}$$

Substituting the definition of $[A^-]$ into the equation for the fraction present in the acidic form gives:

$$\frac{[HA]}{[HA] + [A^-]} = \frac{[HA]}{[HA] + \dfrac{K_a[HA]}{[H^+]}} = \frac{1}{1 + \dfrac{K_a}{[H^+]}} = \frac{[H^+]}{[H^+] + K_a}$$

Therefore, the percentage that is present in the acidic form is given by:

$$\frac{[H^+]}{[H^+] + K_a} \times 100$$

Because the pK_a of the acid is given as 5.3, we know that K_a is 5.0×10^{-6} (because $pK_a = -\log K_a$). Because the pH of the solution is given as 5.7, we know that $[H^+]$ is 2.0×10^{-6} (because $pH = -\log [H^+]$).

Substituting into the equation for the percentage present in the acidic form gives:

$$\frac{[H^+]}{[H^+] + K_a} = \frac{2.0 \times 10^{-6}}{2.0 \times 10^{-6} + 5.0 \times 10^{-6}} \times 100$$

$$\frac{2.0 \times 10^{-6}}{7.0 \times 10^{-6}} \times 100 = 29\%$$

b. Fraction present in the acidic form $= \dfrac{[H^+]}{[H^+] + K_a} = 0.80$

$$[H^+] = 0.80([H^+] + K_a)$$

$$[H^+] = 0.80[H^+] + 0.80 K_a$$

$$0.20[H^+] = 0.80 K_a$$

$$[H^+] = 4 K_a$$

$$[H^+] = 4 \times 5.0 \times 10^{-6}$$

$$[H^+] = 20 \times 10^{-6}$$

$$pH = 4.7$$

70. In each problem, we define $[H^+] = x$. Then, $[A^-]$ is also x. In part **a**, since we have a 1.0 M solution, $[HA] = 1.0 - x$; in part **b**, since we have a 0.1 M solution, $[HA] = 0.1 - x$.

a.

$$K_a = \frac{[H^+][A^-]}{[HA]}$$

$$1.74 \times 10^{-5} = \frac{x^2}{1.0 - x}$$

$$1.74 \times 10^{-5} = x^2$$

$$x = 4.16 \times 10^{-3}$$

$$pH = 2.38$$

b.

$$K_a = \frac{[H^+][A^-]}{[HA]}$$

$$2.00 \times 10^{-11} = \frac{x^2}{0.1 - x}$$

$$2.00 \times 10^{-12} = x^2$$

$$x = 1.41 \times 10^{-6}$$

$$pH = 5.85$$

c. This question can be answered by plugging the given concentrations into the Henderson–Hasselbalch equation.

$$pK_a = pH + \log\frac{[acid]}{[base]}$$

$$3.76 = pH + \log\frac{0.3}{0.1}$$

$$3.76 = pH + \log 3$$

$$3.76 = pH + 0.48$$

$$pH = 3.76 - 0.48 = 3.28$$

Chapter 2 Practice Test

1.   Which compound is the stronger acid?

  a. CH_3CHCH_2OH or CH_3CHCH_2OH **c.** CH_3CCH_2OH or CH_2CHCH_2OH
        | | | |
        Cl   F   Cl   Cl

  (with Cl substituents shown above the carbons in c.)

  b. HBr   or   HI   **d.** CH_4 or NH_3

2.   Which compound is the stronger base?

  a. $CH_3CH_2NH_2$ or CH_3CH_2OH **b.** F^- or I^-

3.   Draw a circle around the strongest base and draw a square around the weakest base.

$$\underset{CH_3CO^-}{\overset{O}{\overset{\|}{}}}\quad CH_3O^-\quad CH_3OH\quad CH_3\overset{..}{N}H\quad CH_3NH_2$$

4.   The following compounds are drawn in their acidic forms, and their pK_a values are given. Draw the form in which each compound would predominantly exist at pH = 8.

CH_3COOH	CH_3CH_2OH	$CH_3\overset{H}{\underset{+}{O}H}$	$CH_3CH_2\overset{+}{N}H_3$
$pK_a = 4.8$	$pK_a = 15.9$	$pK_a = -2.5$	$pK_a = 11.2$

5.   **a.** Write the acid-base reaction that occurs when methylamine is added to water.

  b. Does the above reaction favor reactants or products?

6.   **a.** What is the conjugate base of NH_3?

  b. What is its conjugate acid?

7.   What is the pK_a of a compound that at pH = 7.2 has a 10 times greater concentration in its basic form than in its acidic form?

8.   **a.** What products would be formed from the following reaction?

$$CH_3OH \;+\; \overset{+}{N}H_4 \;\rightleftharpoons$$

  b. Does the reaction favor reactants or products?

9.   A compound has a $K_a = 6.3 \times 10^{-9}$. What is its approximate pK_a (that is, between what two integers)?

10.   Label the compounds in order of decreasing acidity. (Label the most acidic compound #1.)

$$CH_3CH_2OH \qquad CH_3CH_2NH_2 \qquad CH_3CH_2SH \qquad CH_3CH_2CH_3$$

11.   You are planning to carry out a reaction at pH = 4 that releases protons. Would it be better to use a 1.0 M formic acid/sodium formate buffer or a 1.0 M acetic acid/sodium acetate buffer? (The pK_a of formic acid is 3.75; the pK_a of acetic acid is 4.76.)

12.   Indicate whether each of the following statements is true or false.

a.	HO^- is a stronger base than $^-NH_2$.	T	F
b.	A Lewis acid is a compound that accepts a share in a pair of electrons.	T	F
c.	CH_3CH_3 is more acidic than $H_2C{=}CH_2$.	T	F
d.	The weaker the acid, the more stable the conjugate base.	T	F
e.	The larger the pK_a, the weaker the acid.	T	F
f.	The weaker the base, the more stable it is.	T	F

ANSWERS TO ALL THE PRACTICE TESTS CAN BE FOUND AT THE END OF
THE SOLUTIONS MANUAL.

pH, pK_a, and Buffers

This is a continuation of the discussion on acids and bases found in Chapter 2 of the text. First, we will see how the pH of solutions of acids and bases can be calculated. We will look at three different kinds of solutions.

1. A solution made by dissolving a strong acid or a strong base in water.

2. A solution made by dissolving a weak acid or a weak base in water.

3. A solution made by dissolving a weak acid and its conjugate base in water. Such a solution is known as a **buffer solution**.

Before we start, we need to review a few terms.

An acid is a compound that loses a proton, and a base is a compound that gains a proton.

The degree to which an acid (HA) dissociates is described by its acid dissociation constant (K_a).

$$ \text{HA} \rightleftharpoons \text{H}^+ + \text{A}^- $$

$$ K_a = \frac{[\text{H}^+][\text{A}^-]}{[\text{HA}]} $$

The strength of an acid can be indicated by its acid dissociation constant (K_a) or by its pK_a value.

$$ pK_a = -\log K_a $$

The stronger the acid, the **larger** its acid dissociation constant and the **smaller** its pK_a value.

For example, an acid with an acid dissociation constant of 1×10^{-2} (pK_a = 2) is a stronger acid than one with an acid dissociation constant of 1×10^{-4} (pK_a = 4).

While pK_a is used to describe the strength of an acid, pH is used to describe the acidity of a solution. In other words, pH describes the concentration of hydrogen ions in a solution.

$$ pH = -\log [\text{H}^+] $$

The smaller the pH, the more acidic the solution:
acidic solutions have pH values <7; a neutral solution has a pH = 7; basic solutions have pH values >7.

A solution with a pH = 2 is more acidic than a solution with a pH = 4.
A solution with a pH = 12 is more basic than a solution with a pH = 8.

Determining the pH of a Solution

To determine the pH of a solution, the concentration of hydrogen ion $[\text{H}^+]$ in the solution must be determined.

Strong Acids

A strong acid is one that dissociates completely in solution. Strong acids have pK_a values <1.

Because a strong acid dissociates completely, the concentration of hydrogen ions is the same as the concentration of the acid: a 1.0 M HCl solution contains 1.0 M $[H^+]$; a 1.5 M HCl solution contains 1.5 M $[H^+]$. Therefore, to determine the pH of a strong acid, the $[H^+]$ value does not have to be calculated; it is the same as the molarity of the strong acid.

Solution	$[H^+]$	pH
1.0 M HCl	1.0 M	0
1.0×10^{-2} M HCl	1.0×10^{-2} M	2.0
6.4×10^{-4} M HCl	6.4×10^{-4} M	3.2

Strong Bases

Strong bases are compounds such as NaOH or KOH that dissociate completely in water.

Because they dissociate completely, the $[HO^-]$ is the same as the molarity of the strong base.

pOH describes the basicity of a solution. The smaller the pOH, the more basic the solution; just like the smaller the pH, the more acidic the solution.

$$pOH = -\log [HO^-]$$

$[HO^-]$ and $[H^+]$ are related by the ionization constant for water (K_w).

$$K_w = [H^+][HO^-] = 10^{-14}$$
$$pH + pOH = 14$$

Solution	$[HO^-]$	pOH	pH
1.0 M NaOH	1.0 M	0	$14.0 - 0 = 14.0$
1.0×10^{-4} M NaOH	1.0×10^{-4} M	4.0	$14.0 - 4.0 = 10.0$
7.8×10^{-2} M NaOH	7.8×10^{-2} M	1.1	$14.0 - 1.1 = 12.9$

Weak Acids

A weak acid does not dissociate completely in solution. Therefore, $[H^+]$ must be calculated before the pH can be determined.

Acetic acid (CH_3COOH) is an example of a weak acid. It has an acid dissociation constant of 1.74×10^{-5} ($pK_a = 4.76$). The pH of a 1.00 M solution of acetic acid can be calculated as follows:

$$CH_3COOH \rightleftharpoons H^+ + CH_3COO^-$$

$$K_a = \frac{[H^+][CH_3COO^-]}{[CH_3COOH]}$$

Each molecule of acetic acid that dissociates forms one proton and one acetate ion. Therefore, the concentration of protons in solution equals the concentration of acetate ions. Each has a concentration that can be represented by x. The concentration of acetic acid, therefore, is the concentration we started with minus x.

$$1.74 \times 10^{-5} = \frac{(x)(x)}{1.00 - x}$$

The denominator $(1.00 - x)$ can be simplified to 1.00 because 1.00 is much greater than x. (When we actually calculate the value of x, we see that it is 0.004. And $1.00 - 0.004 = 1.00$.)

$$1.74 \times 10^{-5} = \frac{x^2}{1.00}$$

$$x = 4.17 \times 10^{-3}$$

$$pH = -\log 4.17 \times 10^{-3}$$

$$pH = 2.38$$

Formic acid (HCOOH) has a pK_a value of 3.75. The pH of a 1.50 M solution of formic acid can be calculated as follows:

$$HCOOH \rightleftharpoons H^+ + HCOO^-$$

$$K_a = \frac{[H^+][HCOO^-]}{[HCOOH]}$$

A compound with a $pK_a = 3.75$ has an acid dissociation constant of 1.78×10^{-4}.

$$1.78 \times 10^{-4} = \frac{(x)(x)}{1.50 - x} = \frac{x^2}{1.50}$$

$$x^2 = 1.50\,(1.78 \times 10^{-4})$$

$$x^2 = 2.67 \times 10^{-4}$$

$$x = 1.63 \times 10^{-2}$$

$$pH = -\log(1.63 \times 10^{-2})$$

$$pH = 1.79$$

Weak Bases

When a weak base is dissolved in water, it accepts a proton from water, creating hydroxide ion.

Determining the concentration of hydroxide allows the pOH to be determined, and this in turn allows the pH to be determined.

The pH of a 1.20 M solution of sodium acetate can be calculated as follows:

$$CH_3COO^- + H_2O \rightleftharpoons CH_3COOH + HO^-$$

$$\frac{K_w}{K_a} = \frac{[HO^-][CH_3COOH]}{[CH_3COO^-]}$$

$$\frac{1.00 \times 10^{-14}}{1.74 \times 10^{-5}} = \frac{(x)(x)}{1.20 - x}$$

$$5.75 \times 10^{-10} = \frac{x^2}{1.20}$$

$$x^2 = 6.86 \times 10^{-10}$$

$$x = 2.62 \times 10^{-5} = [HO^-]$$

$$pOH = -\log 2.62 \times 10^{-5}$$

$$pOH = 4.58$$

$$pH = 14.00 - 4.58$$

$$pH = 9.42$$

Notice that by setting up the equation equal to K_w/K_a, we can avoid the introduction of a new term (K_b), because $K_w/K_a = K_b$.

Buffer Solutions

A buffer solution is a solution that maintains nearly constant pH in spite of the addition of small amounts of H^+ or HO^-. That is because a buffer solution contains both a weak acid and its conjugate base. The weak acid can donate a proton to any HO^- added to the solution, and the conjugate base can accept any H^+ that is added to the solution, so the addition of small amounts of HO^- or H^+ does not significantly change the pH of the solution.

A buffer can maintain nearly constant pH in a range of one pH unit on either side of the pK_a of the conjugate acid. For example, an acetic acid/sodium acetate mixture can be used as a buffer in the pH range 3.76–5.76 because acetic acid has a $pK_a = 4.76$; methylammonium ion/methylamine can be used as a buffer in the pH range 9.7–11.7 because the methylammonium ion has a $pK_a = 10.7$.

The pH of a buffer solution can be determined from the Henderson-Hasselbalch equation. This equation comes directly from the expression defining the acid dissociation constant. Its derivation is found on page 74 of the text.

Henderson-Hasselbalch equation

$$pK_a = pH + \log \frac{[HA]}{[A^-]}$$

The pH of an acetic acid/sodium acetate buffer solution (pK_a of acetic acid $= 4.76$) that is 1.00 M in acetic acid and 0.50 M in sodium acetate is calculated as follows:

$$pK_a = pH + \log \frac{[HA]}{[A^-]}$$

$$4.76 = pH + \log \frac{1.00}{0.50}$$

$$4.76 = pH + \log 2$$

$$4.76 = pH + 0.30$$

$$pH = 4.46$$

Remember from Section 2.10 that compounds exist primarily in their acidic forms in solutions that are more acidic than their pK_a values and primarily in their basic forms in solutions that are more basic than their pK_a values. Therefore, it could have been predicted that the above solution will have a pH less than the pK_a of acetic acid, because there is more conjugate acid than conjugate base in the solution.

There are three ways a buffer solution can be prepared:

1. Weak Acid and Weak Base

A buffer solution can be prepared by mixing a solution of a weak acid with a solution of its conjugate base.

The pH of a formic acid/sodium formate buffer (pK_a of formic acid $= 3.75$) solution prepared by mixing 25 mL of 0.10 M formic acid and 15 mL of 0.20 M sodium formate is calculated as follows:

The equation below shows that the number of millimoles (mmol) of each of the buffer components can be determined by multiplying the number of milliliters (mL) by the molarity (M).

$$M = \text{molarity} = \frac{\text{moles}}{\text{liters}} = \frac{\text{millimoles}}{\text{milliliters}}$$

Therefore:

$$25 \text{ mL} \times 0.10 \text{ M} = 2.5 \text{ mmol formic acid}$$

$$15 \text{ mL} \times 0.20 \text{ M} = 3.0 \text{ mmol sodium formate}$$

Notice that in the following equation we use mmol for $[HA]$ and $[A^-]$ rather than molarity (mmol/mL) because both the acid and the conjugate base are in the same solution so they have the same volume. Therefore, volumes will cancel in the equation.

$$pK_a = pH + \log \frac{[HA]}{[A^-]}$$

$$3.75 = pH + \log \frac{2.5}{3.0}$$

$$3.75 = pH + \log 0.83$$

$$3.75 = pH - 0.08$$

$$pH = 3.83$$

It could have been predicted that the above solution would have a pH greater than the pK_a of formic acid, because there is more conjugate base than conjugate acid in the solution.

2. Weak Acid and Strong Base

A buffer solution can be prepared by mixing a solution of a weak acid with a strong base such as NaOH. The NaOH reacts completely with the weak acid, thereby creating the conjugate base needed for the buffer solution. For example, if 20 mmol of a weak acid and 5 mmol of a strong base are added to a solution, the 5 mmol of strong base will react with 5 mmol of weak acid, creating 5 mmol of weak base and leaving behind 15 mmol of weak acid.

The pH of a solution prepared by mixing 10 mL of a 2.0 M solution of a weak acid with a pK_a of 5.86 with 5.0 mL of a 1.0 M solution of sodium hydroxide can be calculated as follows:

When the 20 mmol of HA and the 5.0 mmol of HO^- are mixed, the 5.0 mmol of strong base will react with 5.0 mmol of HA, with the result that 5.0 mmol of A^- will be formed and 15 mmol (20 mmol $-$ 5.0 mmol) of HA will be left unreacted.

$$10 \text{ mL} \times 2.0 \text{ M} = 20 \text{ mmol HA} \longrightarrow 15 \text{ mmol HA}$$

$$5.0 \text{ mL} \times 1.0 \text{ M} = 5.0 \text{ mmol HO}^- \longrightarrow 5.0 \text{ mmol A}^-$$

$$pK_a = pH + \log \frac{[HA]}{[A^-]}$$

$$5.86 = pH + \log \frac{15}{5}$$

$$5.86 = pH + \log 3$$

$$5.86 = pH + 0.48$$

$$pH = 5.38$$

3. Weak Base and Strong Acid

A buffer solution can be prepared by mixing a solution of a weak base with a strong acid such as HCl. The strong acid will react completely with the weak base, thereby forming the conjugate acid needed for the buffer solution. The pH of an ethylammonium ion/ethylamine buffer (pK_a of $CH_3CH_2\overset{+}{N}H_3 = 11.0$) prepared by mixing 30 mL of 0.20 M ethylamine with 40 mL of 0.10 M HCl can be calculated as follows:

$$30 \text{ mL} \times 0.20 \text{ M} = 6.0 \text{ mmol RNH}_2 \longrightarrow 2.0 \text{ mmol RNH}_2$$

$$40 \text{ mL} \times 0.10 \text{ M} = 4.0 \text{ mmol H}^+ \longrightarrow 4.0 \text{ mmol R}\overset{+}{N}H_3$$

Notice that 4.0 mmol H^+ reacts with 4.0 mmol RNH_2, forming 4.0 mmol $R\overset{+}{N}H_3$, leaving behind 2.0 mmol RNH_2.

$$pK_a = pH + \log \frac{[HA]}{[A^-]}$$

$$11.0 = pH + \log \frac{4.0}{2.0}$$

$$11.0 = pH + \log 2.0$$

$$11.0 = pH + 0.30$$

$$pH = 10.7$$

Fraction Present in the Acidic or the Basic Form

A common question asked is what fraction of a buffer will be in a particular form at a given pH; either what fraction will be in the acidic form or what fraction will be in the basic form. This is an easy question to answer if you remember the following formulas that are derived at the end of this section:

$$\text{fraction present in the acidic form } = \frac{[H^+]}{K_a + [H^+]}$$

$$\text{fraction present in the basic form } = \frac{K_a}{K_a + [H^+]}$$

What fraction of an acetic acid/sodium acetate buffer $(pK_a$ of acetic acid $= 4.76; K_a = 1.74 \times 10^{-5})$ is present in the acidic form at pH $= 5.20$; $[H^+] = 6.31 \times 10^{-6}$?

$$\frac{[H^+]}{K_a + [H^+]} = \frac{6.31 \times 10^{-6}}{(1.74 \times 10^{-5}) + (6.31 \times 10^{-6})}$$

$$= \frac{6.31 \times 10^{-6}}{(17.4 \times 10^{-6}) + (6.31 \times 10^{-6})}$$

$$= \frac{6.31 \times 10^{-6}}{23.7 \times 10^{-6}} = \frac{6.31}{23.7}$$

$$= 0.26$$

What fraction of a formic acid/sodium formate buffer $(pK_a$ of formic acid $= 3.75; K_a = 1.78 \times 10^{-4})$ is present in the basic form at pH $= 3.90$; $[H^+] = 1.26 \times 10^{-4}$?

$$\frac{K_a}{K_a + [H^+]} = \frac{1.78 \times 10^{-4}}{(1.78 \times 10^{-4}) + (1.26 \times 10^{-4})}$$

$$= \frac{1.78 \times 10^{-4}}{3.04 \times 10^{-4}} = \frac{1.78}{3.04}$$

$$= 0.586$$

$$= 0.59$$

The formulas describing the fraction present in the acidic or basic form are obtained from the definition of the acid dissociation constant.

$$K_a = \frac{[H^+][A^-]}{[HA]}$$

To derive the equation for the fraction present in the acidic form, we need to define $[A^-]$ in terms of $[HA]$, so we will have only one unknown in the equation.

$$[A^-] = \frac{K_a[HA]}{[H^+]}$$

$$\text{fraction present in the acidic form } = \frac{[HA]}{[HA] + [A^-]} = \frac{[HA]}{[HA] + \dfrac{K_a[HA]}{[H^+]}} = \frac{1}{1 + \dfrac{K_a}{[H^+]}}$$

$$= \frac{[H^+]}{K_a + [H^+]}$$

To derive the equation for the fraction present in the basic form, we need to define $[HA]$ in terms of $[A^-]$, so we can get rid of the $[HA]$ term.

$$K_a = \frac{[H^+][A^-]}{[HA]}$$

$$[HA] = \frac{[H^+][A^-]}{K_a}$$

$$\text{fraction present in the basic form } = \frac{[A^-]}{[HA] + [A^-]} = \frac{[A^-]}{[A^-] + \dfrac{[H^+][A^-]}{K_a}}$$

$$= \frac{1}{1 + \dfrac{[H^+]}{K_a}}$$

$$= \frac{K_a}{K_a + [H^+]}$$

Preparing Buffer Solutions

The type of calculations just shown can be used to determine how to make a buffer solution.

For example, how can 100 mL of a 1.00 M buffer solution with a pH = 4.24 be prepared if you have available to you 1.50 M solutions of acetic acid ($pK_a = 4.76$; $K_a = 1.74 \times 10^{-5}$), sodium acetate, HCl, and NaOH?

First, we need to determine what fraction of the buffer will be present in each form at pH = 4.24; $[H^+] = 5.75 \times 10^{-5}$. We will start by calculating the fraction of the buffer present in the acidic form.

$$\frac{[H^+]}{K_a + [H^+]} = \frac{5.75 \times 10^{-5}}{(1.74 \times 10^{-5}) + (5.75 \times 10^{-5})}$$

$$= \frac{5.75 \times 10^{-5}}{7.49 \times 10^{-5}}$$

$$= 0.77$$

If a 1.00 M buffer solution is desired, the buffer must be 0.77 M in acetic acid and 0.23 M in sodium acetate.

There are three ways to make such a buffer solution:

1. **By mixing the appropriate amounts of acetic acid and sodium acetate in water, and adding water to obtain a final volume of 100 mL.**

The amount of acetic acid needed: $[CH_3COOH] = 0.77\ M$

$$M = \frac{mmol}{mL} = \frac{x\ mmol}{100\ mL} = 0.77\ M$$

$$x = 77\ mmol$$

Therefore, we need to have 77 mmol of acetic acid in the final solution.

To obtain 77 mmol of acetic acid from a 1.50 M solution of acetic acid:

$$\frac{77\ mmol}{y\ mL} = 1.50\ M$$

$$y = 51.3\ mL$$

Notice that the formula M = mmol/mL was used twice. The first time it was used to determine the number of mmol of acetic acid that is needed in the final solution. The second time it was used to determine how that number of mmol can be obtained from an acetic acid solution of a known concentration.

The amount of sodium acetate needed: $[CH_3COO^-] = 0.23\ M$

$$\frac{x\ mmol}{100\ mL} = 0.23$$

$$x = 23\ mmol$$

To obtain 23 mmol of sodium acetate from a 1.50 M solution of sodium acetate:

$$\frac{23\ mmol}{y\ mL} = 1.50\ M$$

$$y = 15.3\ mL$$

The desired buffer solution can be prepared using: 51.3 mL 1.50 M acetic acid

15.3 mL 1.50 M sodium acetate

$$\underline{33.4\ mL\ H_2O}$$

100.0 mL

2. **By mixing the appropriate amounts of acetic acid and sodium hydroxide, and adding water to obtain a final volume of 100 mL.**

Sodium hydroxide is used to convert some of the acetic acid into sodium acetate. This means that acetic acid will be the source of both acetic acid and sodium acetate.

The concentrations needed are: $[CH_3COOH] = 1.00\ M$

$$[NaOH] = 0.23\ M$$

The amount of acetic acid needed: $[CH_3COOH] = 1.00\ M$

$$\frac{x\ mmol}{100\ mL} = 1.00\ M$$

$$x = 100\ mmol$$

To obtain 100 mmol of acetic acid from a 1.50 M solution of acetic acid:

$$\frac{100\ mmol}{y\ mL} = 1.50\ M$$

$$y = 66.7\ mL$$

The amount of sodium hydroxide needed: $[NaOH] = 0.23\ M$

$$\frac{x\ mmol}{100\ mL} = 0.23\ M$$

$$x = 23\ mmol$$

To obtain 23 mmol of sodium hydroxide from a 1.50 M solution of NaOH:

$$\frac{23\ mmol}{y\ mL} = 1.50\ M$$

$$y = 15.3\ mL$$

The desired buffer solution can be prepared using: 66.7 mL 1.50 M acetic acid

15.3 mL 1.50 M NaOH

$$\underline{18.0\ mL\ H_2O}$$

100.0 mL

3. **By mixing the appropriate amounts of sodium acetate and hydrochloric acid, and adding water to obtain a final volume of 100 mL.**

Hydrochloric acid is used to convert some of the sodium acetate into acetic acid.

This means that sodium acetate will be the source of both acetic acid and sodium acetate.

The concentrations needed are: $[CH_3COONa] = 1.00\ M$

$$[HCl] = 0.77\ M$$

The amount of sodium acetate needed: $[CH_3COONa] = 1.00$ M

$$\frac{x \text{ mmol}}{100 \text{ mL}} = 1.00 \text{ M}$$

$$x = 100 \text{ mmol}$$

To obtain 100 mmol of sodium acetate from a 1.50 M solution of sodium acetate:

$$\frac{100 \text{ mmol}}{y \text{ mL}} = 1.50 \text{ M}$$

$$y = 66.7 \text{ mL}$$

The amount of hydrochloric acid needed:

$$[HCl] = 0.77 \text{ M}$$

$$\frac{x \text{ mmol}}{100 \text{ mL}} = 0.77 \text{ M}$$

$$x = 77 \text{ mmol}$$

To obtain 77 mmol of hydrochloric acid from a 1.50 M solution of HCl:

$$\frac{77 \text{ mmol}}{y \text{ mL}} = 1.50 \text{ M}$$

$$y = 51.3 \text{ mL}$$

100 mL of a 1.00 M acetic acid/acetate buffer cannot be made from these reagents, because the volumes needed (66.7 mL + 51.3 mL) add up to more than 100 mL. To make this buffer using sodium acetate and hydrochloric acid, you would need to use a more concentrated solution (>1.50 M) of sodium acetate or a more concentrated solution (>1.50 M) of HCl.

Problems on pH, pK_a, and Buffers

1.   Calculate the pH of each of the following solutions:

   a. 1×10^{-3} M HCl

   b. 0.60 M HCl

   c. 1.40×10^{-2} M HCl

   d. 1×10^{-3} M KOH

   e. 3.70×10^{-4} M NaOH

   f. a 1.20 M solution of an acid with a pK_a = 4.23

   g. 1.60×10^{-2} M sodium acetate (pK_a of acetic acid = 4.76)

2.   Calculate the pH of each of the following buffer solutions:

   a. A buffer prepared by mixing 20 mL of 0.10 M formic acid and 15 mL of 0.50 M sodium formate (pK_a of formic acid = 3.75).

   b. A buffer prepared by mixing 10 mL of 0.50 M aniline and 15 mL of 0.10 M HCl (pK_a of the anilinium ion = 4.60).

   c. A buffer prepared by mixing 15 mL of 1.00 M acetic acid and 10 mL of 0.50 M NaOH (pK_a of acetic acid = 4.76).

3.   What fraction of a carboxylic acid with pK_a = 5.23 would be ionized at pH = 4.98?

4.   What would be the concentration of formic acid and sodium formate in a 1.00 M buffer solution with a pH = 3.12 (pK_a of formic acid = 3.75)?

5.   You have found a bottle labeled 1.00 M RCOOH. You want to determine what carboxylic acid it is, so you decide to determine its pK_a value. How would you do this?

6.   **a.** How would you prepare 100 mL of a buffer solution that is 0.30 M in acetic acid and 0.20 M in sodium acetate using a 1.00 M acetic acid solution and a 2.00 M sodium acetate solution?

   b. The pK_a of acetic acid is 4.76. Would the pH of the above solution be greater or less than 4.76?

7.   You have 100 mL of a 1.50 M acetic acid/sodium acetate buffer solution that has a pH = 4.90. How could you change the pH of the solution to 4.50?

8.   You have 100 mL of a 1.00 M solution of an acid with a pK_a = 5.62 to which you add 10 mL of 1.00 M sodium hydroxide. What fraction of the acid will be in the acidic form? How much more sodium hydroxide will you need to add in order to have 40% of the acid in its acidic form (that is, with its proton)?

9.   Describe three ways to prepare a 1.00 M acetic acid/sodium acetate buffer solution with a pH $=$ 4.00.

10. You have available to you 1.50 M solutions of acetic acid, sodium acetate, sodium hydroxide, and hydrochloric acid. How would you make 50 mL of each of the buffers described in the preceding problem?

11. How would you make a 1.0 M buffer solution with a pH $=$ 3.30?

12. You are planning to carry out a reaction that will produce protons. In order for the reaction to take place at constant pH, it will be carried in a solution buffered at pH $=$ 4.2. Would it be better to use a formic acid/formate buffer or an acetic acid/acetate buffer?

Solutions to Problems on pH, pK_a, and Buffers

1.

a. pH $= -\log (1 \times 10^{-3})$
  pH $= 3$

b. pH $= -\log 0.60$
  pH $= 0.22$

c. pH $= -\log (1.40 \times 10^{-2})$
  pH $= 1.85$

d. pOH $= -\log (1 \times 10^{-3})$
  pOH $= 3$
  pH $= 14 - 3 = 11$

e. pOH $= -\log (3.70 \times 10^{-4})$
  pOH $= 3.43$
  pH $= 10.57$

f. p$K_a = 4.23$, $K_a = 5.89 \times 10^{-5}$

$$K_a = \frac{[\text{H}^+][\text{A}^-]}{[\text{HA}]}$$

$$5.89 \times 10^{-5} = \frac{x^2}{1.20}$$

$$x^2 = 7.07 \times 10^{-5}$$

$$x = 8.41 \times 10^{-3}$$

$$\text{pH} = 2.08$$

g. $\dfrac{K_w}{K_a} = \dfrac{[\text{HO}^-][\text{HA}]}{[\text{A}^-]}$ $(K_a = 10^{-4.76} = 1.74 \times 10^{-5})$

$$\frac{1.0 \times 10^{-14}}{1.74 \times 10^{-5}} = \frac{x^2}{1.60 \times 10^{-2}}$$

$$5.75 \times 10^{-10} = \frac{x^2}{1.60 \times 10^{-2}}$$

$$x^2 = 9.20 \times 10^{-12}$$

$$x = 3.03 \times 10^{-6}$$

$$\text{pOH} = 5.52$$

$$\text{pH} = 14.00 - 5.52 = 8.48$$

2.   **a.**   formic acid: 20 mL $\times$ 0.10 M = 2.0 mmol
sodium formate: 15 mL $\times$ 0.50 M = 7.5 mmol

$$pK_a = pH + \log \frac{[HA]}{[A^-]}$$

$$3.75 = pH + \log \frac{2.0}{7.5}$$

$$3.75 = pH + \log 0.27$$

$$3.75 = pH + (-0.57)$$

$$pH = 4.32$$

b.   aniline: 10 mL $\times$ 0.50 M = 5.0 mmol $\longrightarrow$ 3.5 mmol aniline (RNH_2)
HCl: 15 mL $\times$ 0.10 M = 1.5 mmol $\longrightarrow$ 1.5 mmol anilinium hydrochloride (RNH_3^+)

$$pK_a = pH + \log \frac{[HA]}{[A^-]}$$

$$4.60 = pH + \log \frac{1.5}{3.5}$$

$$4.60 = pH + \log 0.43$$

$$4.60 = pH + (-0.37)$$

$$pH = 4.97$$

c.   acetic acid: 15 mL $\times$ 1.00 M = 15 mmol $\longrightarrow$ 10 mmol acetic acid
NaOH: 10 mL $\times$ 0.50 M = 5.0 mmol $\longrightarrow$ 5.0 mmol sodium acetate

$$pK_a = pH + \log \frac{[HA]}{[A^-]}$$

$$4.76 = pH + \log \frac{10}{5.0}$$

$$4.76 = pH + \log 2$$

$$4.76 = pH + 0.30$$

$$pH = 4.46$$

3.   The ionized form is the basic form. Therefore, we need to use the equation that allows us to calculate the fraction present in the basic form.

$$\text{fraction of buffer in the basic form} = \frac{K_a}{K_a + [H^+]} = \frac{5.89 \times 10^{-6}}{(5.89 \times 10^{-6}) + (10.47 \times 10^{-6})} = \frac{5.89 \times 10^{-6}}{16.36 \times 10^{-6}}$$

$$= 0.36$$

4.   fraction of buffer in the basic form $= \dfrac{K_a}{K_a + [H^+]} = \dfrac{1.78 \times 10^{-4}}{(1.78 \times 10^{-4}) + (7.59 \times 10^{-4})}$

$$= \dfrac{1.78 \times 10^{-4}}{9.37 \times 10^{-4}}$$

$$= 0.19$$

$$[\,\text{sodium formate}\,] = 0.19 \text{ M}$$
$$[\,\text{formic acid}\,] = 0.81 \text{ M}$$

5.   From the Henderson-Hasselbalch equation, we see that the pH of the solution is the same as the pK_a of the species in the acidic form when the concentration of the species in the acidic form is the same as the concentration of the species in the basic form.

$$pK_a = pH + \log \dfrac{[HA]}{[A^-]}$$

when $[HA] = [A^-]$,

$$pK_a = pH$$

Thus, in order to have a solution in which the pH will be the same as the pK_a, the number of mmol of acid must equal the number of mmol of conjugate base.

Preparing a solution of x mmol of RCOOH and $1/2\,x$ mmol NaOH will give a solution in which $[\,RCOOH\,] = [\,RCOO^-\,]$.

For example: 20 mL of 1.00 M RCOOH $= 20$ mmol
                        10 mL of 1.00 M NaOH $= 10$ mmol

This will give a solution that contains 10 mmol RCOOH and 10 mmol RCOO$^-$.

The pH of this solution is the pK_a of RCOOH.

6.   **a.** $\dfrac{x \text{ mmol}}{100 \text{ mL}} = 0.30 \text{ M}$ $\dfrac{x \text{ mmol}}{100 \text{ mL}} = 0.20 \text{ M}$

         $x = 30$ mmol of acetic acid $x = 20$ mmol of sodium acetate

         $\dfrac{30 \text{ mmol}}{y \text{ mL}} = 1.00 \text{ M}$ $\dfrac{20 \text{ mmol}}{y \text{ mL}} = 2.00 \text{ M}$

         $y = 30$ mL of 1.00 M acetic acid $y = 10$ mL of 2.00 M acetic acid

         The buffer solution could be prepared by mixing: 30 mL of 1.00 M acetic acid
                                                                                            10 mL of 2.00 M sodium acetate
                                                                                            $\dfrac{60 \text{ mL of water}}{100 \text{ mL}}$

    b.   Because the concentration of buffer in the acidic form (0.30 M) is greater than the concentration of buffer in the basic form (0.20 M), the pH of the solution will be less than 4.76.

7.   **Original solution**

fraction of buffer in the basic form $= \dfrac{K_a}{K_a + [H^+]} = \dfrac{1.74 \times 10^{-5}}{(1.74 \times 10^{-5}) + (1.26 \times 10^{-5})}$

$$= \dfrac{1.74 \times 10^{-5}}{3.00 \times 10^{-5}}$$

$$= 0.58$$

$$0.58 \times 1.50 \text{ M} = 0.87 \text{ M}$$

$$[A^-] = 0.87 \text{ M}$$

$$[HA] = 0.63 \text{ M}$$

Desired solution

fraction of buffer in the basic form $= \dfrac{K_a}{K_a + [H^+]} = \dfrac{1.74 \times 10^{-5}}{(1.74 \times 10^{-5}) + (3.16 \times 10^{-5})}$

$$= \dfrac{1.74 \times 10^{-5}}{4.90 \times 10^{-5}}$$

$$= 0.35$$

$$0.35 \times 1.50 \text{ M} = 0.53 \text{ M}$$

$$[A^-] = 0.53 \text{ M}$$

$$[HA] = 0.97 \text{ M}$$

The original solution contains 87 mmol of A^- ($100 \text{ mL} \times 0.87 \text{ M}$).

The desired solution with a pH $= 4.50$ must contain 53 mmol of A^-.

Therefore, 34 mmol of A^- ($87 - 53 = 34$) must be converted to HA.

This can be done by adding 34 mmol of HCl to the original solution.

If you have a 1.00 M HCl solution, you will need to add 34 mL to the original solution in order to change its pH from 4.90 to 4.50.

$$\dfrac{34 \text{ mmol}}{x \text{ mL}} = 1.00 \text{ M}$$

$$x = 34 \text{ mL}$$

Note that after adding HCl to the original solution, it will no longer be a 1.50 M buffer; it will be more dilute ($150 \text{ mmol}/134 \text{ mL} = 1.12 \text{ M}$).

The change in the concentration of the buffer solution will be less if a more concentrated solution of HCl is used to change the pH. For example, if you have a 2.00 M HCl solution:

$$\dfrac{34 \text{ mmol}}{x \text{ mL}} = 2.00 \text{ M}$$

$$x = 17 \text{ mL}$$

You will need to add 17 mL to the original solution, and the concentration of the buffer will be 1.28 M (150 mmol/117 mL = 1.28 M).

8.   acid: 100 mL $\times$ 1.00 M = 100 mmol HA $\longrightarrow$ 90 mmol HA

NaOH: 10 mL $\times$ 1.00 M = 10 mmol HO$^-$ $\longrightarrow$ 10 mmol A$^-$

Therefore, 90% is in the acidic form.

For 40% to be in the acidic form, you need:

40 mmol HA
60 mmol A$^-$

You need to have 60 mmol rather than 10 mmol in the basic form. To get the additional 50 mmol in the basic form, you would need to add 50 mL of 1.0 M NaOH.

9. $$\frac{\text{fraction of buffer}}{\text{in the basic form}} = \frac{K_a}{K_a + [\text{H}^+]} = \frac{1.74 \times 10^{-5}}{(1.74 \times 10^{-5}) + (1.00 \times 10^{-4})} = \frac{1.74 \times 10^{-5}}{(1.74 \times 10^{-5}) + (10.00 \times 10^{-5})}$$

$$= \frac{1.74 \times 10^{-5}}{11.74 \times 10^{-5}}$$

$$= 0.15$$

$$[\text{A}^-] = 0.15 \text{ M}$$

$$[\text{HA}] = 0.85 \text{ M}$$

a. $[\text{acetic acid}] = 0.85$ M   **b.** $[\text{acetic acid}] = 1.00$ M   **c.** $[\text{sodium acetate}] = 1.00$ M
$[\text{sodium acetate}] = 0.15$ M $[\text{NaOH}] = 0.15$ M $[\text{HCl}] = 0.85$ M

10. a. $$\frac{x \text{ mmol}}{50 \text{ mL}} = 0.85 \text{ M}$$

$$x = 42.5 \text{ mmol of acetic acid}$$

$$\frac{42.5 \text{ mmol}}{y \text{ mL}} = 1.50 \text{ M}$$

$$y = 28.3 \text{ mL of } 1.50 \text{ M acetic acid}$$

$$\frac{x \text{ mmol}}{50 \text{ mL}} = 0.15 \text{ M}$$

$$x = 7.5 \text{ mmol of sodium acetate}$$

$$\frac{7.5 \text{ mmol}}{y \text{ mL}} = 1.50 \text{ M}$$

$$y = 5.0 \text{ mL of } 1.50 \text{ M sodium acetate}$$

$$28.3 \text{ mL of } 1.50 \text{ M acetic acid}$$

$$5.0 \text{ mL of } 1.50 \text{ M sodium acetate}$$

$$\frac{16.7 \text{ mL of H}_2\text{O}}{50.0 \text{ mL}}$$

b. $\dfrac{x \text{ mmol}}{50 \text{ mL}} = 1.00 \text{ M}$

$x = 50 \text{ mmol of acetic acid}$

$\dfrac{50 \text{ mmol}}{y \text{ mL}} = 1.50 \text{ M}$

$y = 33.3 \text{ mL of } 1.50 \text{ M acetic acid}$

$\dfrac{x \text{ mmol}}{50 \text{ mL}} = 0.15 \text{ M}$

$x = 7.5 \text{ mmol of NaOH}$

$\dfrac{7.5 \text{ mmol}}{y \text{ mL}} = 1.50 \text{ M}$

$y = 5.0 \text{ mL of } 1.50 \text{ M NaOH}$

$\quad$ 33.3 mL of 1.50 M acetic acid

$\quad$ 5.0 mL of 1.50 M NaOH

$\quad$ $\underline{\text{11.7 mL of } H_2O}$

$\quad$ 50.0 mL

c. $\dfrac{x \text{ mmol}}{50 \text{ mL}} = 1.00 \text{ M}$

$x = 50 \text{ mmol of sodium acetate}$

$\dfrac{50 \text{ mmol}}{y \text{ mL}} = 1.50 \text{ M}$

$y = 33.3 \text{ mL of } 1.50 \text{ M sodium acetate}$

$\dfrac{x \text{ mmol}}{50 \text{ mL}} = 0.85 \text{ M}$

$x = 42.5 \text{ mmol of HCl}$

$\dfrac{42.5 \text{ mmol}}{y \text{ mL}} = 1.5 \text{ M}$

$y = 28.3 \text{ mL of } 1.5 \text{ M HCl}$

We cannot make the required buffer with these solutions, because 33.3 mL + 28.3 mL > 50 mL.

11. Because formic acid has a $pK_a = 3.75$, a formic acid/formate buffer can be a buffer at pH $= 3.30$, since this pH is within one pH unit of the pK_a value.

$$\text{fraction of buffer in the basic form} = \frac{K_a}{K_a + [H^+]} = \frac{1.78 \times 10^{-4}}{(1.78 \times 10^{-4}) + (5.01 \times 10^{-4})}$$

$$= \frac{1.78 \times 10^{-4}}{6.79 \times 10^{-4}}$$

$$= 0.26$$

The solution must have $[\text{formic acid}] = 0.74$ M and $[\text{sodium formate}] = 0.26$ M.

12. The reaction to be carried out will generate protons that will react with the basic form of the buffer in order to keep the pH constant.

Therefore, the better buffer would be the one that has the larger percentage of the buffer in the basic form.

The pK_a of formic acid is 3.74. Because the pH of the solution (4.2) is more basic than the pK_a value of the compound, formic acid will exist primarily in its basic form.

The pK_a of acetic acid is 4.76. Because the pH of the solutions is more acidic than the pK_a value of the compound, acetic acid will exist primarily in its acidic form.

Therefore, the formate buffer is preferred, because it has a greater percentage of the buffer in the basic form.

An Introduction to Organic Compounds:
Nomenclature, Physical Properties, and Representation of Structure

Important Terms

alcohol	a compound with an OH group in place of one of the hydrogens of an alkane (ROH).
alkane	a hydrocarbon that contains only single bonds.
alkyl halide	a compound with a halogen in place of one of the hydrogens of an alkane.
alkyl substituent	a substituent formed by removing a hydrogen from an alkane.
amine	a compound in which one or more of the hydrogens of NH_3 are replaced by an alkyl substituent (RNH_2, R_2NH, R_3N)
angle strain	the strain introduced into a molecule as a result of its bond angles being distorted from their ideal values.
anti conformer	the staggered conformer in which the largest substituents bonded to the two carbons are opposite each other. It is the most stable of the staggered conformers.
axial bond	a bond of the chair conformer of cyclohexane that points directly up or directly down.
boat conformer	a conformer of cyclohexane that roughly resembles a boat.
boiling point	the temperature at which the vapor pressure of a liquid equals the atmospheric pressure.
chair conformer	a conformer of cyclohexane that roughly resembles a chair. It is the most stable conformer of cyclohexane.
cis fused	two rings fused together in such a way that if the second ring were considered to be two substituents of the first ring, the two substituents would be on the same side of the first ring.
cis isomer **(for a cyclic compound)**	the isomer with two substituents on the same side of the ring.
cis-trans stereoisomers	see the definition of "cis isomer" and "trans isomer."
common name	non-systematic nomenclature.
conformation	the three-dimensional shape of a molecule at a given instant.
conformational analysis	the investigation of various conformers of a molecule and their relative stabilities.

conformers — different conformations of a molecule.

constitutional isomers (structural isomers) — molecules that have the same molecular formula but differ in the way the atoms are connected.

cycloalkane — an alkane with its carbon chain arranged in a closed ring.

1,3-diaxial interaction — the interaction between an axial substituent and one of the other two axial substituents on the same side of a cyclohexane ring.

dipole-dipole interaction — an interaction between the dipole of one molecule and the dipole of another.

eclipsed conformer — a conformer in which the bonds on adjacent carbons are parallel to each other when viewed looking down the carbon-carbon bond.

equatorial bond — a bond of the chair conformer of cyclohexane that juts out from the ring but does not point directly up or directly down.

ether — a compound in which an oxygen is bonded to two alkyl substituents (ROR).

flagpole hydrogens — the two hydrogens in the boat conformer of cyclohexane that are at the 1- and 4-positions of the ring.

functional group — the center of reactivity of a molecule.

gauche conformer — a staggered conformer in which the largest substituents bonded to the two carbons are gauche to each other; that is, their bonds have a dihedral angle of approximately 60°.

The substituents are gauche to each other.

gauche interaction — the interaction between two atoms or groups that are gauche to each other.

geometric isomers — cis-trans (or *E,Z*) isomers.

half-chair conformer — the least stable conformer of cyclohexane.

homologue — a member of a homologous series.

homologous series — a family of compounds in which each member differs from the next by one methylene group.

hydrocarbon — a compound that contains only carbon and hydrogen.

hydrogen bond — an unusually strong dipole-dipole attraction (5 kcal/mol) between a hydrogen bonded to O, N, or F and the lone pair of a different O, N, or F.

hyperconjugation	delocalization of electrons by the overlap of a σ orbital with an empty orbital.
induced dipole-induced dipole interaction	an interaction between a temporary dipole in one molecule and the dipole that the temporary dipole induces in another molecule.
IUPAC nomenclature	systematic nomenclature.
melting point	the temperature at which a solid becomes a liquid.
methylene group	a CH_2 group.
Newman projection	a way to represent the three-dimensional spatial relationships of atoms by looking down the length of a particular carbon-carbon bond.
packing	a property that determines how well individual molecules fit into a crystal lattice.
parent hydrocarbon	the longest continuous carbon chain in a molecule; if the molecule has a functional group, it is the longest continuous carbon chain that contains the functional group.
perspective formula	a way to represent the three-dimensional spatial relationships of atoms using two adjacent solid lines, one solid wedge, and one hatched wedge.
polarizability	the ease with which an electron cloud of an atom can be distorted.
primary alcohol	an alcohol in which the OH group is bonded to a primary carbon.
primary alkyl halide	an alkyl halide in which the halogen is bonded to a primary carbon.
primary amine	an amine with one alkyl group bonded to the nitrogen.
primary carbon	a carbon bonded to only one other carbon.
primary hydrogen	a hydrogen bonded to a primary carbon.
quaternary ammonium salt	a nitrogen compound with four alkyl groups bonded to the nitrogen, plus an accompanying anion.
ring-flip (chair-chair interconversion)	the conversion of a chair conformer of cyclohexane into the other chair conformer. Bonds that are axial in one chair conformer are equatorial in the other chair conformer.
sawhorse projection	a way to represent the three-dimensional spatial relationships of atoms by looking at the carbon-carbon bond from an oblique angle.
secondary alcohol	an alcohol in which the OH group is bonded to a secondary carbon.
secondary alkyl halide	an alkyl halide in which the halogen is bonded to a secondary carbon.
secondary amine	an amine with two alkyl groups bonded to the nitrogen.
secondary carbon	a carbon bonded to two other carbons.
secondary hydrogen	a hydrogen bonded to a secondary carbon.

skeletal structure

a structure that shows the carbon-carbon bonds as lines and does not show the carbon-hydrogen bonds.

skew-boat conformer

one of the conformers of a cyclohexane ring.

solubility

the extent to which a compound dissolves in a solvent.

solvation

the interaction between a solvent and another molecule (or ion).

staggered conformer

a conformer in which the bonds on one carbon bisect the bond angles on the adjacent carbon when viewed looking down the carbon-carbon bond.

steric hindrance

hindrance due to groups occupying a volume of space.

steric strain

the repulsion between the electron cloud of an atom or group of atoms and the electron cloud of another atom or group of atoms.

straight-chain alkane

an alkane in which the carbons form a continuous chain with no branches.

structural isomers (constitutional isomers)

molecules that have the same molecular formula but differ in the way the atoms are connected.

symmetrical ether

an ether with two identical alkyl substituents bonded to the oxygen.

systematic nomenclature

a system of nomenclature based on rules such as IUPAC nomenclature.

tertiary alcohol

an alcohol in which the OH group is bonded to a tertiary carbon.

tertiary alkyl halide

an alkyl halide in which the halogen is bonded to a tertiary carbon.

tertiary amine

an amine with three alkyl groups bonded to the nitrogen.

tertiary carbon

a carbon bonded to three other carbons.

tertiary hydrogen

a hydrogen bonded to a tertiary carbon.

trans-fused

two rings fused together in such a way that if the second ring were considered to be two substituents of the first ring, the two substituents would be on opposite sides of the first ring.

trans isomer (for a cyclic compound)

the isomer with two substituents on opposite sides of the ring.

twist-boat conformer

one of the conformers of a cyclohexane ring.

unsymmetrical ether

an ether with two different alkyl substituents bonded to the oxygen.

van der Waals forces

induced dipole-induced dipole interactions.

Solutions to Problems

1.   **a.** C_nH_{2n+2} If there are 17 carbons, then there are 36 hydrogens.

 b. C_nH_{2n+2} If there are 74 hydrogens, then there are 36 carbons.

2. $CH_3CH_2CH_2CH_2CH_2CH_2CH_2CH_3$ $CH_3CHCH_2CH_2CH_2CH_2CH_3$
                  octane $|$
                                                                 CH_3 isooctane

3.   **a.** propyl alcohol   **b.** butyl methyl ether   **c.** propylamine

                                    CH_3
                                      $|$
4.   **a.** $CH_3CHCH_2CH_3$ **b.** CH_3CCH_3
                  $|$ $|$
                 CH_3 CH_3
           2-methylbutane   2,2-dimethylpropane

5.   Notice that each carbon forms four bonds, and each hydrogen and bromine forms one bond.

                                                                       CH_3
                                                                         $|$
      $CH_3CH_2CH_2CH_2Br$ $CH_3CHCH_2CH_3$ CH_3CHCH_2Br CH_3CCH_3
                                        $|$ $|$ $|$
                                       Br CH_3 Br

        n-butyl bromide   *sec*-butyl bromide   isobutyl bromide   *tert*-butyl bromide
              or
        butyl bromide

6.   "Dibromomethane does not have constitutional isomers" proves that carbon is tetrahedral.

      If carbon were flat, rather than being tetrahedral, dibromomethane would have constitutional isomers because the two structures shown below would be different since the bromines would be 90° apart in one compound and 180° apart in the other compound. Only because carbon is tetrahedral are the two structures identical.

                           H   Br
                           | |
                       H—C—Br   H—C—H
                           | |
                           Br   Br

                                                                       CH_3
                                                                         $|$
7.   **a.** CH_3CHOH **c.** CH_3CH_2CHI **e.** CH_3CNH_2
                  $|$ $|$ $|$
                 CH_3 CH_3 CH_3

                                    CH_3
                                      $|$
      b. $CH_3CHCH_2CH_2F$ **d.** CH_3COH **f.** $CH_3CH_2CH_2CH_2CH_2CH_2CH_2CH_2Br$
                  $|$ $|$
                 CH_3 CH_2CH_3

8. **a.** ethyl methyl ether   **c.** *sec*-butylamine   **e.** isobutyl bromide

 b. methyl propyl ether   **d.** butyl alcohol or *n*-butyl alcohol   **f.** *sec*-butyl chloride

9. **a.** Solved in the text.

                                                                         CH_3 CH_3 CH_3

d. $CH_3CHCH_2C\!-\!CHCH_2CH_3$

                                                                       $CHCH_3$

                                                                       CH_3

            CH_3

b. $CH_3CHCHCH_2CH_2CH_3$

               CH_3

                                                  CH_3 CH_3

e. $CH_3CHCH_2CHCHCH_2CH_2CH_3$

                                                    CH_2

                                                    $CHCH_3$

                                                    CH_3

                 CH_2CH_3

c. $CH_3CH_2CH_2CCH_2CH_2CH_2CH_2CH_3$

                          CH_2CH_3

f. $CH_3CH_2CH_2CHCH_2CH_2CH_2CH_3$

                                    CH_3CCH_3

                                          CH_3

10. **a.**

#1 $CH_3CH_2CH_2CH_2CH_2CH_2CH_2CH_3$
octane

#2 $CH_3CHCH_2CH_2CH_2CH_2CH_3$
              CH_3
2-methylheptane

#3 $CH_3CH_2CHCH_2CH_2CH_2CH_3$
                 CH_3
3-methylheptane

#4 $CH_3CH_2CH_2CHCH_2CH_2CH_3$
                     CH_3
4-methylheptane

            CH_3
#5 $CH_3CCH_2CH_2CH_2CH_3$
            CH_3
2,2-dimethylhexane

                    CH_3 CH_3
#10 $CH_3CH_2CH\!-\!CHCH_2CH_3$
3,4-dimethylhexane

                 CH_3 CH_3
#11 $CH_3C\!-\!CHCH_2CH_3$
                 CH_3
2,2,3-trimethylpentane

                 CH_3 CH_3
#12 $CH_3CCH_2CHCH_3$
                 CH_3
2,2,4-trimethylpentane

                 CH_3 CH_3
#13 $CH_3CH\!-\!CCH_2CH_3$
                       CH_3
2,3,3-trimethylpentane

                 CH_3 CH_3 CH_3
#14 $CH_3CH\!-\!CH\!-\!CHCH_3$
2,3,4-trimethylpentane

#6 CH₃CH₂CCH₂CH₂CH₃ (with CH₃ above and CH₃ below central C)

3,3-dimethylhexane

#7 CH₃CH—CHCH₂CH₂CH₃ (with CH₃ above each of the first two CH)

2,3-dimethylhexane

#8 CH₃CHCH₂CHCH₂CH₃ (with CH₃ above the first and third C)

2,4-dimethylhexane

#9 CH₃CHCH₂CH₂CHCH₃ (with CH₃ above the first and fifth C)

2,5-dimethylhexane

#15 CH₃C—CCH₃ (with CH₃ CH₃ above and CH₃ CH₃ below)

2,2,3,3-tetramethylbutane

#16 CH₃CH₂CHCH₂CH₂CH₃ (with CH₂CH₃ below)

3-ethylhexane

#17 CH₃CH₂CHCHCH₃ (with CH₃ above, CH₂CH₃ below)

3-ethyl-2-methylpentane

#18 CH₃CH₂CCH₂CH₃ (with CH₃ above, CH₂CH₃ below)

3-ethyl-3-methylpentane

b. The systematic name is under each structure.

c. Only #1 (octane or *n*-octane) and #2 (isooctane) have common names.

d. #2, #7, #8, #9, #12, #13, #14, #17

e. #3, #8, #10, #11

f. #5, #11, #12, #15

11.
a. 2,2,4-trimethylhexane
b. 2,2-dimethylbutane
c. 3-methyl-4-propylheptane
d. 2,2,5-trimethylhexane
e. 3,3-diethyl-4-methyl-5-propyloctane
f. 5-ethyl-4,4-dimethyloctane
g. 3,3-diethylhexane
h. 4-(1-methylethyl)octane
i. 2,5-dimethylheptane

12.
a. CH₃CH₂CH₂CH₂CH₃

pentane

b. CH₃CCH₃ (with CH₃ above and CH₃ below)

2,2-dimethylpropane

c. CH₃CHCH₂CH₃ (with CH₃ above)

2-methylbutane

d. CH₃CHCH₂CH₃ (with CH₃ above)

2-methylbutane

13.

14. **a.**   **d.**

b.   **e.**

c.   **f.**

15. **a.**   **d.**   **g.**

b.   **e.**   **h.**

c.   **f.**   **i.**

16. **a.** CH$_3$CHCHCHCH$_2$CH$_2$CH$_3$ with CH$_3$, CH$_2$CH$_3$, and CH$_2$CH$_3$ substituents

b. CH$_3$CCH$_2$CH$_2$CHCH$_3$ with CH$_3$ and CH$_3$ substituents

17. **a.** 1-ethyl-2-methylcyclopentane
b. ethylcyclobutane
c. 4-ethyl-1,2-dimethylcyclohexane
d. 3,6-dimethyldecane
e. 2-cyclopropylhexane

f. 1-ethyl-3-isobutylcyclohexane
g. 5-(1-methylethyl)nonane
h. 1-*sec*-butyl-4-isopropylcyclohexane
i. heptane
j. 1-bromohexane

18. **a.** *sec*-butyl chloride
2-chlorobutane
secondary

b. cyclohexyl bromide
bromocyclohexane
secondary

c. isohexyl chloride
1-chloro-4-methylpentane
primary

d. isopropyl fluoride
2-fluoropropane
secondary

19. **a.**

CH₂Cl Note that the name of a CH₂Cl sustituent is "chloromethyl," because a Cl is in place of one of the Hs of a methyl substituent

chloromethylcyclohexane

b. Cl   CH₃

1-chloro-1-methylcyclohexane

c. CH₃

Cl

1-chloro-2-methylcyclohexane

CH₃

Cl

1-chloro-3-methylcyclohexane

CH₃

Cl

1-chloro-4-methylcyclohexane

20. **a.** **1.** methoxyethane
2. ethoxyethane
3. 4-methoxyoctane
4. 1-isopropoxy-3-methylbutane
5. 1-propoxybutane
6. 2-isopropoxyhexane

b. No.

c. **1.** ethyl methyl ether
2. diethyl ether
3. no common name
4. isopentyl isopropyl ether
5. butyl propyl ether
6. no common name

21. CH₃OH
common = methyl alcohol
systematic = methanol

CH₃CH₂OH
common = ethyl alcohol
systematic = ethanol

CH₃CH₂CH₂OH
common = propyl alcohol or *n*-propyl alcohol
systematic = 1-propanol

CH₃CH₂CH₂CH₂OH
common = butyl alcohol or *n*-butyl alcohol
systematic = 1-butanol

CH₃CH₂CH₂CH₂CH₂OH
common = pentyl alcohol or *n*-pentyl alcohol
systematic = 1-pentanol

CH₃CH₂CH₂CH₂CH₂CH₂OH
common = hexyl alcohol or *n*-hexyl alcohol
systematic = 1-hexanol

22. **a.** 1-pentanol
primary

b. 5-chloro-2-methyl-2-pentanol
tertiary

c. 5-methyl-3-hexanol
secondary

d. 7-methyl-3,5-octanediol (Notice that because there are two OH groups, the suffix is "diol.")
both alcohol groups are secondary

23.

CH₃
|
CH₃CCH₂CH₂CH₃
|
OH

2-methyl-2-pentanol

CH₃
|
CH₃CH₂CCH₂CH₃
|
OH

3-methyl-3-pentanol

CH₃
|
CH₃C—CHCH₃
| |
OH CH₃

2,3-dimethyl-2-butanol

24. **a.** 4-chloro-3-ethylcyclohexanol
secondary

c. 1-bromo-5,5-dimethyl-3-heptanol
secondary

b. 7,8-dimethyl-3-nonanol
secondary

d. 4-methylcyclohexanol
secondary

25. **a.** tertiary alkyl halide   **b.** tertiary alcohol   **c.** primary amine

26. **a.** hexylamine
1-hexanamine
primary

d. butylpropylamine
N-propyl-1-butanamine
secondary

b. sec-butylisobutylamine
N-isobutyl-2-butanamine
secondary

e. diethylpropylamine
N,N-diethyl-1-propanamine
tertiary

c. diethylmethylamine
N-ethyl-N-methylethanamine
tertiary

f. N-ethyl-3-methylcyclopentanamine
no common name
secondary

27. **a.** CH₃CH₂CH₂NHCH₂CHCH₃
                                    |
                                   CH₃

d. CH₃CH₂CH₂NCH₂CH₂CH₃
                      |
                     CH₃

b. CH₃CH₂NHCH₂CH₃

e. CH₃CH₂CHCH₂CH₃
                |
                N
              /   \
           CH₃   CH₃

c. CH₃CHCH₂CH₂CH₂CH₂NH₂
            |
           CH₃

f.

NCH₂CH₃
|
CH₃

28. **a.** 6-methyl-1-heptanamine
isooctylamine
primary

c. 4-methyl-N-propyl-1-pentanamine
isohexylpropylamine
secondary

b. cyclohexanamine
cyclohexylamine
primary

d. 2,5-dimethylcyclohexanamine
no common name
primary

29. **a.** The bond angle is predicted to be similar to the bond angle in water $(104.5°)$.
b. The bond angle is predicted to be similar to the bond angle in ammonia $(107.3°)$.
c. The bond angle is predicted to be similar to the bond angle in water $(104.5°)$.
d. The bond angle is predicted to be similar to the bond angle in the ammonium ion $(109.5°)$.

30. To be a liquid at room temperature, the compound must have a boiling point that is greater than room temperature.

pentane or 2-methylbutane

31. **a.** 1, 4, and 5
b. 1, 2, 4, 5, and 6

32. **a.** Each water molecule has two hydrogens that can form hydrogen bonds, whereas each alcohol molecule has only one hydrogen that can form a hydrogen bond. Therefore, there are more hydrogen bonds between water molecules than between alcohol molecules.

b. Each water molecule has two hydrogens that can form hydrogen bonds and two lone pairs that can accept hydrogen bonds, so both hydrogens of a water molecule can engage in hydrogen bonding. In contrast, ammonia has three hydrogens that can form hydrogen bonds but only one lone pair that can accept hydrogen bonds, so only one hydrogen of an ammonia molecule can engage in hydrogen bonding, so it will have a lower boiling point than water.

c. Each water molecule has two hydrogens that can form hydrogen bonds and two lone pairs that can accept hydrogen bonds, so both lone pairs of a water molecule can engage in hydrogen bonding. In contrast, HF has three lone pairs that can accept hydrogen bonds but only one hydrogen that can form a hydrogen bond, so only one lone pair of a HF molecule can engage in hydrogen bonding, so it will have a lower boiling point than water.

d. HF and ammonia can each form only one hydrogen bond, but HF has a higher boiling point because the hydrogen bond formed by HF is stronger since fluorine is more electronegative than nitrogen.

33.

34. **a.**

b. ⟋⟍⟋⟍⟋⟍⟋⟍ > ⟋⟍⟋⟍⟋⟍⟍ > ⟋⟍⟋⟍⟍ > ✕

c. ⟋⟍⟋⟍OH > ⟋⟍⟋OH > ⟋⟍⟋Cl > ⟋⟍⟋

35.   **a.** (structure with OH, OH, O) > (structure with OH, O) > (structure with OH) > ⟋⟍⟋

   b. $HOCH_2CH_2CH_2OH > CH_3CH_2CH_2OH > CH_3CH_2CH_2CH_2OH > CH_3CH_2CH_2CH_2Cl$

36.   Because cyclohexane is a nonpolar compound, it will have the lowest solubility in the most polar solvent, which, of the solvents given, is ethanol.

$CH_3CH_2CH_2CH_2CH_2OH$ $CH_3CH_2OCH_2CH_3$ CH_3CH_2OH $CH_3CH_2CH_2CH_2CH_2CH_3$
1-pentanol   diethyl ether   ethanol   hexane

37.   Start with the least stable conformer and then obtain the others by keeping the front carbon constant and rotating the back carbon clockwise.

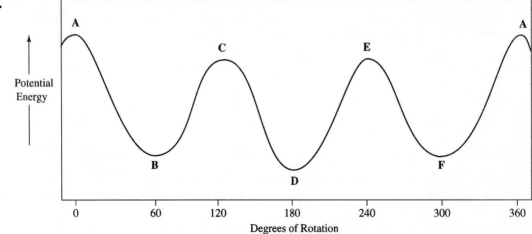

38.   a. The Newman projection shows rotation about the C-2—C-3 bond.

3-ethyl-2-pentanol

b. The Newman projection shows rotation about the C-2—C-3 bond.

2-methyl-2-pentanamine

39. To draw the most stable conformer: put the largest group on the front carbon opposite the largest group on the back carbon.

a.

b.

c.

40.   a. $180° - \dfrac{360°}{8}$

$180° - 45° = 135°$

b. $180° - \dfrac{360°}{9}$

$180° - 40° = 140°$

41. You can get the total strain energy of cycloheptane by subtracting the strainless heat of formation from the actual heat of formation:

The "strainless" heat of formation of cycloheptane is $7\,(-4.92) = -34.4\ \text{kcal/mol}$.
The actual heat of formation of cycloheptane is $-28.2\ \text{kcal/mol}$ (from Table 3.8 on page 127 of the text).

Therefore, the total strain energy of cycloheptane is $-28.2 - (-34.4) = 6.2\ \text{kcal/mol}$.

42. Hexethal would be expected to be the more effective sedative because it is less polar than barbital since hexethal has a hexyl group in place of the ethyl group of barbital. Being less polar, hexethal will be better able to penetrate the nonpolar membrane of the cell.

43.   a.

b.

44. Two 1,3-diaxial (gauche) interactions cause the chair conformer of fluorocyclohexane to be 0.25 kcal/mol less stable when the fluoro substituent is in the axial position than when it is in the equatorial position.

The gauche conformer of 1-fluoropropane has one gauche interaction (see Figure 3.15 on page 131 of the text). Therefore, the gauche conformer is $(0.25/2) = 0.13\ \text{kcal/mol}$ less stable than the anti conformer that has no gauche interactions.

45.

$$K_{eq} = \frac{[\text{equatorial conformer}]}{[\text{axial conformer}]} = \frac{5.4}{1}$$

$$\% \text{ of equatorial conformer} = \frac{[\text{equatorial conformer}]}{[\text{equatorial conformer}] + [\text{axial conformer}]} \times 100$$

$$= \frac{5.4}{5.4 + 1} \times 100 = \frac{5.4}{6.4} \times 100 = 84\%$$

46.   If both substituents point downward or both point upward, it is a cis isomer.
If one substituent points upward and the other downward, it is a trans isomer.

a. cis   **b.** cis   **c.** cis   **d.** trans   **e.** trans   **f.** trans

47.   Both *trans*-1,4-dimethylcyclohexane and *cis*-1-*tert*-butyl-3-methylcyclohexane have a conformer with two substituents in the equatorial position and a conformer with two substituents in the axial position.

cis-1-*tert*-Butyl-3-methylcyclohexane will have a higher percentage of the diequatorial-substituted conformer because the bulky *tert*-butyl substituent will have a greater preference for the equatorial position than will a less bulky methyl substituent, since the larger substituent will have greater destabilizing 1,3-diaxial interactions when it is in an axial position.

48.   **a.**   **b.**

c.   *trans*-1-Ethyl-2-methylcyclohexane is more stable because both substituents can be in equatorial positions.

49.   **a.** one equatorial and one axial in each   **d.** one equatorial and one axial in each
b. both equatorial in one and both axial in the other   **e.** one equatorial and one axial in each
c. both equatorial in one and both axial in the other   **f.** both equatorial in one and both axial in the other

50.   **a.** and **b.** Solved in the text.

c.

d.   There will be equal amounts of the two conformers at equilibrium because they have the same stability—each one has one methyl group on an equatorial bond and one methyl group on an axial bond.

51.   **a.** One chair conformer of *trans*-1,4-dimethylcyclohexane has both substituents in equatorial positions, so it does not have any 1,3-diaxial interactions. The other chair conformer has both substituents in axial positions. When a substituent is in an axial position, it experiences two 1,3-diaxial interactions, so this chair conformer has a total of four 1,3-diaxial interactions.

Because the 1,3-diaxial interaction between a methyl group and a hydrogen causes a strain energy of 0.9 kcal/mol, the chair conformer with both substituents in axial positions is $4 \times 0.9 = 3.6$ kcal/mol less stable than the chair conformer with both substituents in equatorial positions.

a 1,3-diaxial interaction

b. Each of the chair conformers of *cis*-1,4-dimethylcyclohexane has one substituent in an equatorial position and one in an axial position. Therefore, the two conformers are equally stable.

52.   Both condensed and skeletal structures are shown.

a. CH₃CH₂CHOCCH₃ with CH₃ above and CH₃ CH₃ below

d. CH₃CHCH₂CH₂Br with CH₃ below

b. CH₃CHCH₂CH₂CH₂CH₂OH with CH₃ below

e. CH₃CH₂CH₂CH₂CHCH₂CH₂CH₂CH₃ with CHCH₃ above and CH₃ above that

c. CH₃CH₂CHNH₂ with CH₃ below

f. CH₃CH₂N with CH₂CH₃ above and CH₂CH₃ below

g. $CH_3CH_2CH_2CHCH_2CH_2CH_3$
$\qquad\qquad\quad CH_3CCH_3$
$\qquad\qquad\qquad\quad CH_3$

j. $CH_3CH_2CH_2CH_2CHCH_2CH_2CH_2CH_3$
$\qquad\qquad\qquad\qquad\quad CHCH_3$
$\qquad\qquad\qquad\qquad\quad CHCH_3$
$\qquad\qquad\qquad\qquad\qquad CH_3$

h. $CH_3CHCH_2CH_2CCH_2CH_2CH_3$
$\qquad\quad CH_3 \qquad\quad Br$
with Br above the C and Br, Br below

k. $CH_3CH_2CHCHCH_2CH_2CH_2CH_3$
$\qquad\qquad CH_3$
with CH_3 above

i. CH_3
$CH_3CHCHCH_2CH_2CH_3$
$\qquad\quad OCH_2CH_3$

l. CH_3CHCH_3
$CH_3CH_2CH_2CHCH_2CH_2CH_2CH_2CH_3$

53.

has two groups that form hydrogen bonds > O is more electronegative than N, so OH hydrogen bonds are stronger than NH hydrogen bonds > primary amines form stronger hydrogen bonds than secondary amines > relatively weak hydrogen bonds > no hydrogen bonds; only dipole-dipole interactions >

no hydrogen bonds; weaker dipole-dipole interactions than oxygen-containing compounds because N is less electro-negative than O > no dipole-dipole interactions

54. **a.**

1. 2,2,6-trimethylheptane
2. 5-bromo-2-methyloctane
3. 5-methyl-3-hexanol
4. 3,3-diethylpentane
5. 5-bromo-*N*-ethyl-1-pentanamine
6. 2,3,5-trimethylhexane
7. 3-ethoxyheptane
8. 1,3-dimethoxypropane
9. *N,N*-dimethylcyclohexanamine
10. 3-ethylcyclohexanol
11. 1-bromo-4-methylcyclohexane

b.

1.

5.

9.

2.

6.

10.

3.

7.

11.

4.

8.

55.   **C** and **D** are cis isomers. (In **C** both substituents are downward pointing; in **D** both substituents are upward pointing.)

56.   **a.** **1.** 3 **2.** 4   **b.** **1.** 6 **2.** 5   **c.** **1.** 3 **3.** 4

57.   The first conformer (**A**) is the most stable because the three substituents are more spread out, so its gauche interactions will not be as large—the Cl in **A** is between a CH_3 and an H, whereas the Cl in **B** and **C** is between two CH_3 groups.

58.   **a.**   **b.**   **c.**

59. **a.** 2-butanamine

b. 2-chlorobutane

c. *N*-ethyl-2-butanamine

d. 1-ethoxypropane

e. 2-methylpentane

f. 2-propanamine

g. 2-bromo-2-methylbutane

h. 4-methyl-1-pentanol

i. bromocyclopentane

j. cyclohexanol

60. **a.** 1-bromohexane (larger, so greater surface area)
b. pentyl chloride (greater surface area than the branched compound)
c. 1-butanol (fewer carbons)
d. 1-hexanol (forms hydrogen bonds)
e. hexane (greater area of contact)
f. 1-pentanol (forms hydrogen bonds)

g. 1-bromopentane (bromine larger and more polarizable)
h. butyl alcohol (forms hydrogen bonds)
i. octane (see Table 3.1)
j. isopentyl alcohol (forms stronger hydrogen bonds)
k. hexylamine (primary amines form stronger hydrogen bonds than do secondary amines)

61. **a.**

b. The one on the right would predominate at equilibrium because it is more stable since both methyl groups are in equatorial positions.

c.

d. There would be equal amounts of each one at equilibrium, because they have the same stability since each conformer has one methyl group in an equatorial position and one methyl group in an axial position.

62. Ansaid is more soluble in water. It has a fluoro substituent that can form a hydrogen bond with water. Hydrogen bonding increases its solubility in water.

63.

$$\underset{|}{CH_3} \quad \underset{|}{CH_3} \quad \underset{|}{CH_3}$$
$$H{-}\ddot{O}{:}{-}{-}H{-}\ddot{O}{:}{-}{-}H{-}\ddot{O}{:}$$

$$\underset{|}{CH_3} \quad \underset{|}{CH_3} \quad \underset{|}{CH_3}$$
$$H{-}\ddot{O}{:}{-}{-}H{-}\ddot{O}{:}{-}{-}H{-}\ddot{O}{:}$$

$$\underset{|}{CH_3} \quad \underset{|}{CH_3} \quad \underset{|}{CH_3}$$
$$H{-}\ddot{O}{:}{-}{-}H{-}\ddot{O}{:}{-}{-}H{-}\ddot{O}{:}$$

64.   The student named only one compound correctly.

a.   2-bromo-3-pentanol

g.   correct

b.   4-ethyl-2,2-dimethylheptane

h.   2,5-dimethylheptane

c.   3-methylcyclohexanol

i.   5-bromo-2-pentanol

d.   2,2-dimethylcyclohexanol

j.   3-ethyl-2-methyloctane

e.   5-(2-methylpropyl)nonane

k.   2,3,3-trimethyloctane

f.   1-bromo-3-methylbutane

l.   5-methyl-N,N-dimethyl-3-hexanamine

65.   All three compounds are diaxial-substituted cyclohexanes. **B** has the highest energy. Only **B** has a 1,3-diaxial interaction between CH_3 and Cl, which will be greater than a 1,3-diaxial interaction between CH_3 and H or between Cl and H.

66.   The only one is 2,2,3-trimethylbutane.

67.   a.   c.   e.

b.   d.   f.

68.   First draw the structure, so you know what groups to put on the bonds in the Newman projections.

$$CH_3-CH-\overset{3}{C}H_2-\overset{4}{C}H_2-CH_2-CH_3$$
$$|$$
$$CH_3$$

a.   b.

most stable   least stable

c.   Rotation can occur about all the C—C bonds. There are six carbon-carbon bonds in the compound, so there are five other carbon-carbon bonds, in addition to the C_3—C_4 bond, about which rotation can occur.

d. Three of the carbon-carbon bonds have staggered conformers that are equally stable because each is bonded to a carbon with three identical substituents.

$$CH_3-CH-CH_2-CH_2-CH_2-CH_3$$
$$|$$
$$CH_3$$

69.

$CH_3CH_2CH_2CH_2CH_2Br$	**a.** 1-bromopentane **b.** pentyl bromide	primary alkyl halide

$CH_3CH_2CH_2CHCH_3$
$|$
Br

a. 2-bromopentane
b. no common name

secondary alkyl halide

$CH_3CH_2CHCH_2CH_3$
$|$
Br

a. 3-bromopentane
b. no common name

secondary alkyl halide

CH_3
$|$
$CH_3CHCH_2CH_2Br$

a. 1-bromo-3-methylbutane
b. isopentyl bromide

primary alkyl halide

CH_3
$|$
$CH_3CH_2CHCH_2Br$

a. 1-bromo-2-methylbutane
b. no common name

primary alkyl halide

Br
$|$
$CH_3CH_2CCH_3$
$|$
CH_3

a. 2-bromo-2-methylbutane
b. *tert*-pentyl bromide

tertiary alkyl halide

Br
$|$
$CH_3CHCHCH_3$
$|$
CH_3

a. 2-bromo-3-methylbutane
b. no common name

secondary alkyl halide

CH_3
$|$
CH_3CCH_2Br
$|$
CH_3

a. 1-bromo-2,2-dimethylpropane
b. no common name, but in older literature the common name neopentyl bromide is used.

primary alkyl halide

c. Four isomers are primary alkyl halides.
d. Three isomers are secondary alkyl halides.
e. One isomer is a tertiary alkyl halide.

70.
a. butane
b. 1-propanol
c. 5-propyldecane
d. 4-propyl-1-nonanol
e. 2-methyl-5-(1-methylethyl)octane
f. 6-chloro-4-ethyl-3-methyloctane
g. 1-methoxy-5-methyl-3-propylhexane
h. 2,3-dimethyl-6-(2-methylpropyl)decane
i. 8-methyl-4-decanamine
j. 1-methyl-2-(2-methylpropyl)cyclohexane

71. **a.**

more stable

b.

CH(CH₃)₂
CH₂CH₃
CH(CH₃)₂
more stable

CH(CH₃)₂

CH₂CH₃

c.

CH₂CH₃
CH₃
more stable

CH₃

CH₂CH₃

d.

CH₂CH₃

CH₂CH₃

CH₃CH₂ CH₂CH₃

equally stable

e.

CH₂CH₃

(CH₃)₂CH
more stable

(CH₃)₂CH   CH₂CH₃

f.

CH₂CH₃

(CH₃)₂CH

(CH₃)₂CH

more stable   CH₂CH₃

72.   Alcohols with low molecular weights are more water soluble than alcohols with high molecular weights because, as a result of having fewer carbons, they have a smaller nonpolar component that has to be dragged into water.

73.   **a.**

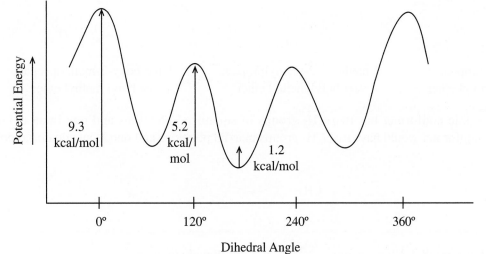

b.

Cl
H   H
H   H
Cl

c.   1.2 + 5.2 = 6.4 kcal/mol   **d.**   1.2 + 9.3 = 10.5 kcal/mol

74. The most stable isomer is the one that has a conformer with both substituents in equatorial positions. Using the following structure, you can determine easily the isomer that has both substituents in axial positions. That will be the isomer that has both groups in equatorial positions.

a. The cis isomer of a 1,3-disubstituted compound is the most stable isomer. It has a conformer with both substituents in axial positions, so its other conformer has both groups in equatorial positions.

b. The trans isomer of a 1,4-disubstituted compound is the most stable isomer. It has a conformer with both substituents in axial positions, so its other conformer has both groups in equatorial positions.

c. The trans isomer of a 1,2-disubstituted compound is the most stable isomer. It has a conformer with both substituents in axial positions, so its other conformer has both groups in equatorial positions.

75. Six ethers have the molecular formula $= C_5H_{12}O$.

$CH_3OCH_2CH_2CH_2CH_3$

$CH_3CHCH_2CH_3$
$\quad\;\; |$
$\quad\;\; OCH_3$

$CH_3CH_2OCH_2CH_2CH_3$

1-methoxybutane
butyl methyl ether

2-methoxybutane
sec-butyl methyl ether

1-ethoxypropane
ethyl propyl ether

CH_3CHCH_3
$\quad\;\; |$
$\quad\;\; OCH_2CH_3$

$\qquad\;\; CH_3$
$\qquad\;\; |$
CH_3COCH_3
$\qquad\;\; |$
$\qquad\;\; CH_3$

$CH_3CHCH_2OCH_3$
$\quad\;\; |$
$\quad\;\; CH_3$

2-ethoxypropane
ethyl isopropyl ether

2-methoxy-2-methylpropane
tert-butyl methyl ether

1-methoxy-2-methylpropane
isobutyl methyl ether

76. The most stable conformer has two CH_3 groups in equatorial positions and one in an axial position. (The other conformer would have two CH_3 groups in axial positions and one in an equatorial position.)

77. **a.** *N*-methyl-6-methyl-3-heptanamine
b. 3-ethyl-2,5-dimethylheptane
c. 1,2-dichloro-3-methylpentane

d. 2,3-dimethylpentane
e. 5-butyl-3,4-dimethylnonane
f. 5-butyl-3,3,9-trimethylundecane (undecane is an 11 carbon straight-chain hydrocarbon; Table 3.1 on page 91 of the text).

78. One chair conformer of *trans*-1,2-dimethylcyclohexane has both substituents in equatorial positions, so it does not have any 1,3-diaxial interactions. However, the figure on top of page 134 of the text shows that the two methyl substituents are gauche to one another (as they would be in gauche butane; see Figure 3.15 on page 131), giving it a strain energy of 0.87 kcal/mol.

The other chair conformer of *trans*-1,2-dimethylcyclohexane has both substituents in axial positions. When a substituent is in an axial position, it experiences two 1,3-diaxial interactions. This chair conformer, therefore, has a total of four 1,3-diaxial interactions. Each diaxial interaction is between a CH_3 and an H, so each results in a strain energy of 0.87 kcal/mol. Therefore, this chair conformer has a strain energy of 3.48 kcal/mol ($4 \times 0.87 = 3.48$).

Thus, one conformer is 2.61 kcal/mol more stable than the other.

79.

80.
  a. 5-methyl-3-hexanol
  b. 1-bromo-2-propylcyclopentane
  c. 2-methyl-3-pentanol
  d. 5-bromo-2-methyloctane

  e. 1,5-hexanediol
  f. 6-bromo-2-hexanol
  g. 4-ethyl-3-methylcyclohexanol
  h. 4-bromo-1-ethyl-2-methylcyclohexane

81.
  a. 1-Hexanol has a higher boiling point than 3-hexanol because the alkyl group in 1-hexanol can better engage in van der Waals interactions, because the OH group of 3-hexanol makes it more difficult for its six carbons to lie close to the six carbons of another molecule of 3-hexanol.

  b. The floppy ethyl groups in diethyl ether make it difficult for the water molecules to approach the oxygen in order to engage in hydrogen bonding. Therefore, it is less soluble in water than is tetrahydrofuran in which the alkyl groups are pinned back in a ring.

82. One of the chair conformers of *cis*-1,3-dimethylcyclohexane has both substituents in equatorial positions, so there are no unfavorable 1,3-diaxial interactions. The other chair conformer has three 1,3-diaxial interactions, two between a CH_3 and an H and one between two CH_3 groups.

   We know that a 1,3-diaxial interaction between a CH_3 and an H is 0.87 kcal/mol. Subtracting 1.7, for the two interactions between a CH_3 and an H, from 5.4 (the energy difference between the two conformers) results in a value of 3.7 kcal/mol for the 1,3-diaxial interaction between the two CH_3 groups.

83. Because bromine has a larger diameter than chlorine, one would expect bromine to have a greater preference for the equatorial position as would be indicated by a larger $\Delta G°$. However, Table 3.9 on page 131 of the text indicates that it has a smaller $\Delta G°$, indicating that it has less preference for the equatorial position than chlorine has. The C—Br bond is longer than the C—Cl bond, which causes bromine to be farther away than chlorine from the other axial substituents. Apparently, the longer bond more than offsets the larger diameter.

84.   **a.** 7-bromo-6-ethyl-5-decanol
      b. 5-chloro-3-ethyl-2,7-dimethylnonane
      c. 7,7-dimethyl-3-nonanol

85. Problem 78 shows that the energy difference between the two chair conformers is 2.61 kcal/mol. To calculate the equilibrium constant needed to answer the question, see Problem 22a on page 208.

$$\Delta G° = -2.61 \text{ kcal/mol}$$

$$\Delta G° = -RT \ln K_{eq}$$

$$-2.61 \text{ kcal/mol} = -1.986 \times 10^{-3} \text{ kcal/mol K} \times 298 \text{ K} \times \ln K_{eq}$$

$$-2.61 \text{ kcal/mol} = -0.5918 \text{ kcal/mol} \ln K_{eq}$$

$$\ln K_{eq} = 4.41$$

$$K_{eq} = 82.3 = \frac{\text{both equatorial}}{\text{both axial}} = \frac{82.3}{1}$$

$$\begin{array}{c}\text{percentage of molecule} \\ \text{with both groups in} \\ \text{equatorial positions}\end{array} = \frac{\text{both equatorial}}{\text{both equatorial} + \text{both axial}} \times 100 = \frac{82.3}{82.3 + 1} \times 100 = 98.8\%$$

86. The conformer on the left has two 1,3-diaxial interactions between a CH_3 and an H (2×0.87 kcal/mol) for a total strain energy of 1.7 kcal/mol.

The conformer on the right has three 1,3-diaxial interactions, two between a CH_3 and an H (1.7 kcal/mol) and one between two CH_3 groups (3.7 kcal/mol; see Problem 82) for a total strain energy of 5.4 kcal/mol. Therefore, the conformer on the left will predominate at equilibrium.

Chapter 3 Practice Test

1.   Name the following compounds

a.

b.

c.

2.   Using Newman projections, draw the following conformers of hexane considering rotation about the C_3—C_4 bond:

  a.   the most stable of all the conformers

  b.   the least stable of all the conformers

  c.   a gauche conformer

3.   What are the common and systematic names of the following compounds?

  a. $CH_3CH_2CHCH_3$ **b.** $CH_3CHCH_2CH_2CH_2OH$ **c.**
            | |
            Cl CH_3

4.   Label the three compounds in each set in order from highest boiling to lowest boiling.

  a. $CH_3CH_2CH_2CH_2CH_2Br$ $CH_3CH_2CH_2Br$ $CH_3CH_2CH_2CH_2Br$

  b. $CH_3CH_2CH_2CH_2CH_3$ $CH_3CH_2CH_2CH_2OH$ $CH_3CH_2CH_2CH_2Cl$

                 CH_3 CH_3
                  | |
  c. CH_3C——CCH_3 $CH_3CH_2CH_2CH_2CH_2CH_2CH_2CH_3$ $CH_3CHCH_2CH_2CH_2CH_2CH_3$
                  | | |
                 CH_3 CH_3 CH_3

5.   Name each of the following compounds:

  a. $CH_3CHCH_2CH_2CHCH_2CH_3$ **c.**
               | |
              CH_3 OH   Br CH_3

                                                                        Cl
                                                                        |
  b. $CH_3CH_2CHOCH_2CH_3$ **d.** $CH_3CHCHCH_2CH_2CH_2Cl$
                | |
               $CH_2CH_2CH_2CH_3$ CH_2CH_3

6. Draw the other chair conformer for the following compound:

7. Which is more stable, *cis*-1-isopropyl-3-methylcyclohexane or *trans*-1-isopropyl-3-methylcyclohexane?

8. Which of the following has:

   a. the higher boiling point: diethyl ether or butyl alcohol?
   b. the greater solubility in water: 1-butanol or 1-pentanol?
   c. the higher boiling point: hexane or isohexane?
   d. the higher boiling point: pentylamine or ethylmethylamine?
   e. the greater solubility in water: ethyl alcohol or ethyl chloride?

9. What are the common and systematic names of the following compounds?

   a. $CH_3CHCH_2CH_2Br$ **b.** $CH_3CHCH_2CH_2OH$ **c.** $CH_3CHCH_2CH_2NH_2$
       CH_3 CH_3 CH_3

10. Draw the more stable conformer of:

   a. *cis*-1-*sec*-butyl-4-isopropylcyclohexane

   b. *trans*-1-*sec*-butyl-4-isopropylcyclohexane

   c. *trans*-1-*sec*-butyl-3-isopropylcyclohexane

11. Draw the structure for each of the following:

   a. a secondary alkyl bromide that has three carbons

   b. a secondary amine that has three carbons

   c. an alkane with no secondary hydrogens

   d. a constitutional isomer of butane

   e. three compounds with molecular formula C_3H_8O

12. Name the following compounds:

   a. $CH_3CHCH_2CH_2CHCH_3$ **d.** Br, Cl
           CH_3 CH_2CH_3
                                            CH_3

           OH

   b. $CH_3CHCH_2CH_2CH_2CH_2CH_2Br$ **e.** $CH_3CH_2CH_2CHCH_2OCH_2CH_2CH_2CH_3$
                                                      CH_3

   c. $CH_3CHCH_2CHCH_2CH_2CH_3$ **f.** $CH_3CH_2CH_2NHCH_2CH_2CHCH_2CH_3$
          Cl   OH CH_3

CHAPTER 4
Isomers: The Arrangement of Atoms in Space

Important Terms

achiral	a molecule or object that contains an element (a plane or a point) of symmetry; an achiral molecule is superimposable on its mirror image.
amine inversion	results when a compound containing an sp^3 hybridized nitrogen with a lone pair rapidly inverts.
asymmetric center	an atom that is bonded to four different substituents.
chiral	a chiral molecule has a nonsuperimposable mirror image.
chiral probe	something capable of distinguishing between enantiomers.
chromatography	a separation technique in which the mixture to be separated is dissolved in a solvent and the solution is passed through a column packed with an adsorbent stationary phase.
cis isomer	the isomer with substituents on the same side of a cyclic structure, or the isomer with the hydrogens on the same side of a double bond.
cis-trans isomers	isomers that result from not being able to rotate about a carbon-carbon double bond.
configuration	the three-dimensional structure of a chiral compound. The configuration at a specific atom is designated by R or S.
configurational isomers	stereoisomers that cannot interconvert unless a covalent bond is broken. Cis-trans isomers and isomers with asymmetric centers are configurational isomers.
constitutional isomers (structural isomers)	molecules that have the same molecular formula but differ in the way the atoms are connected.
dextrorotatory	the enantiomer that rotates the plane of polarization of plane-polarized light in a clockwise direction ($+$).
diastereomers	stereoisomers that are not enantiomers.
enantiomerically pure	only one enantiomer is present in an enantiomerically pure sample.
enantiomeric excess (optical purity)	how much excess of one enantiomer is present in a mixture of a pair of enantiomers, expressed as a percentage.
enantiomers	nonsuperimposable mirror-image molecules.
erythro enantiomers	the pair of enantiomers with similar groups on the same side (in the case of stereo-isomers with adjacent asymmetric centers) when drawn in a Fischer projection.

137

Fischer projection	a method of representing the spatial arrangement of groups bonded to an asymmetric center. The asymmetric center is the point of intersection of two perpendicular lines; the horizontal lines represent bonds that project out of the plane of the paper toward the viewer, and the vertical lines represent bonds that project back from the plane of the paper away from the viewer.
isomers	nonidentical compounds with the same molecular formula.
levorotatory	the enantiomer that rotates the plane of polarization of plane-polarized light in a counterclockwise direction $(-)$.
meso compound	a compound that possesses asymmetric centers and a plane of symmetry; it is achiral, because it has a plane of symmetry.
observed rotation	the amount of rotation observed in a polarimeter.
optically active	rotates the plane of polarization of plane-polarized light.
optically inactive	does not rotate the plane of polarization of plane-polarized light.
optical purity (enantiomeric excess)	how much excess of one enantiomer is present in a mixture of a pair of enantiomers.
perspective formula	a method of representing the spatial arrangement of groups bonded to an asymmetric center. Two adjacent bonds are drawn in the plane of the paper; a solid wedge depicts a bond that projects out of the plane of the paper toward the viewer, and a hatched wedge depicts a bond that projects back from the paper away from the viewer.
plane-polarized light	light that oscillates in a single plane.
plane of symmetry	an imaginary plane that bisects a molecule into pieces that are a pair of mirror images.
polarimeter	an instrument that measures the rotation of the plane of polarization of plane-polarized light.
polarized light	light that oscillates in only one plane.
racemic mixture (racemic modification, racemate)	a mixture of equal amounts of a pair of enantiomers.
***R* configuration**	after assigning relative priorities to the four groups bonded to an asymmetric center, if the lowest-priority group is on a vertical axis in a Fischer projection (or pointing away from the viewer in a perspective formula), an arrow drawn from the highest-priority group to the next highest-priority group and then to the next highest priority group goes in a clockwise direction.
resolution of a racemic mixture	separation of a racemic mixture into the individual enantiomers.

S configuration   after assigning relative priorities to the four groups bonded to an asymmetric center, if the lowest-priority group is on a vertical axis in a Fischer projection (or pointing away from the viewer in a perspective formula), an arrow drawn from the highest-priority group to the next highest-priority group and then to the next highest priority group goes in a counterclockwise direction.

specific rotation   the amount of rotation that will be observed for a compound with a concentration given in grams per 100 mL of solution (or g/mL if it is a pure liquid) in a sample tube 1.0 dm long.

stereocenter (stereogenic center)   an atom at which the interchange of two groups produces a stereoisomer.

stereoisomers   isomers that differ in the way the atoms are arranged in space.

threo enantiomers   the pair of enantiomers with similar groups on opposite sides (in the case of stereoisomers with adjacent asymmetric centers) when drawn in a Fischer projection.

trans isomer   the isomer with substituents on the opposite sides of a cyclic structure, or the isomer with the hydrogens on the opposite sides of a double bond.

Solutions to Problems

1. **a.** $CH_3CH_2CH_2OH$ $CH_3\underset{\underset{\displaystyle CH_3}{|}}{C}HOH$ $CH_3CH_2OCH_3$

b. There are seven constitutional isomers with molecular formula $C_4H_{10}O$.

$CH_3CH_2CH_2CH_2OH$ $CH_3\underset{\underset{\displaystyle CH_3}{|}}{C}HCH_2OH$ $CH_3\overset{\overset{\displaystyle CH_3}{|}}{\underset{\underset{\displaystyle CH_3}{|}}{C}}OH$ $CH_3\underset{\underset{\displaystyle OH}{|}}{C}HCH_2CH_3$

$CH_3CH_2OCH_2CH_3$ $CH_3OCH_2CH_2CH_3$ $CH_3O\underset{\underset{\displaystyle CH_3}{|}}{C}HCH_3$

2. **a.** Br—⬡—Cl   Br—⬡···Cl   **b.** (cyclobutane with CH₃ and CH₃CH₂ cis and trans)

3. **a.** 1 and 3

b. **1.** (cis 2-hexene structure)   (trans 2-hexene structure)   **3.** (cis structure)   (trans structure)

4. **1.** (trans)   (cis)   **3.** (trans)   (cis)

2. (structure)   **4.** (structure)

5. $CH_3CH_2CH_2CH=CH_2$ $CH_3\overset{\overset{\displaystyle CH_3}{|}}{C}=CHCH_3$ $CH_3CH_2\overset{\overset{\displaystyle CH_3}{|}}{C}=CH_2$

6. Only **C** has a dipole moment of zero, because the bond dipoles cancel since they are in opposite directions.

$$\underset{\underset{\displaystyle Cl}{\diagup}}{\overset{\overset{\displaystyle H}{\diagdown}}{C}}=\underset{\underset{\displaystyle H}{\diagdown}}{\overset{\overset{\displaystyle Cl}{\diagup}}{C}}$$

7. **a, b, c, f,** and **h** are chiral.
d, e, and **g** are each superimposable on its mirror image. These, therefore, are achiral.

8. **a, c,** and **f** have asymmetric centers.

9. Solved in the text.

10. **a, c,** and **f,** because in order to be able to exist as a pair of enantiomers, the compound must have an asymmetric center (except in the case of certain compounds with unusual structures; see Problem 86).

11. **a.** It has one asymmetric center.
b. It has three stereocenters.

12. Draw the first enantiomer with the groups in any order you want. Then draw the second enantiomer by drawing the mirror image of the first enantiomer. Your answer might not look exactly like the ones shown below because the first enantiomer can be drawn with the four groups on any of the four bonds. The next one is the mirror image of the first one.

a. **1.**

2.

3.

b. **1.**

2.

3.

13. Solved in the text.

14. **a.** $-CH_2OH$ ① $-CH_3$ ③ $-H$ ④ $-CH_2CH_2OH$ ②

b. $-CH_2Br$ ② $-OH$ ① $-CH_3$ ④ $-CH_2OH$ ③

c. $-CH(CH_3)_2$ ② $-CH_2CH_2Br$ ③ $-Cl$ ① $-CH_2CH_2CH_2Br$ ④

d. $-CH=CH_2$ ② $-CH_2CH_3$ ③   ① $-CH_3$ ④

forms 2 bonds to C, so considered to be attached to 2 Cs   attached to 1 C   forms 3 bonds to C, so considered to be attached to 3 Cs   attached only to Hs

15.   **a.**   Solved in the text.

b.   *R*

c.   To determine the configuration, first add the fourth bond to the asymmetric center. Remember that it cannot be drawn between the two solid bonds. (It can be drawn on either side of the solid wedge.)

d.   *R*.

16.   **a.**   *S*   **b.**   *R*   **c.**   *S*   **d.**   *S*

17.   The easiest way to determine whether two compounds are identical or enantiomers is to determine their configurations: If both are *R* (or both are *S*), they are identical. If one is *R* and the other is *S*, they are enantiomers.

a.   identical   **b.**   enantiomers   **c.**   enantiomers   **d.**   enantiomers

18.   **a.**   **b.**

19.   **a.**   Solved in the text.   **b.**

20.   **a.**   levorotatory   **b.**   dextrorotatory

21.   specific rotation $=$ $\dfrac{\text{observed rotation (degrees)}}{\text{concentration (g/100 mL)} \times \text{length (dm)}}$

$$[\alpha] = \frac{+13.4°}{\dfrac{4\ \text{g}}{100\ \text{mL}} \times 2\ \text{dm}} = \frac{+13.4°}{8} = +1.68$$

22.   **a.** -24 **b.**   0

23.   **a.**   0 (It is a racemic mixture.)

b.   50% of the mixture is excess (+)-mandelic acid.

optical purity $= 0.50 = \dfrac{\text{observed specific rotation}}{\text{specific rotation of the pure enantiomer}}$

$0.50 = \dfrac{\text{observed specific rotation}}{+158}$

observed specific rotation $= +79$

c.   50% of the mixture is excess ($-$)-mandelic acid.
observed specific rotation $= -79$ (For the calculation, see part **b.**)

24. **a.** From the data given, you cannot determine the configuration of naproxen.

b. 97% of the commercial preparation is $(+)$-naproxen; 3% is a racemic mixture. Therefore, the commercial preparation forms 98.5% $(+)$-naproxen and 1.5% $(-)$-naproxen.

25. Solved in the text.

26. As a result of the double bond, the compound has a cis isomer and a trans isomer. Because the compound also has an asymmetric center, the cis isomer can exist as a pair of enantiomers and the trans isomer can exist as a pair of enantiomers.

cis enantiomers

trans enantiomers

27. **a.** This statement is correct. Although there are many examples where there would be fewer than 2^n stereoisomers (for example, *trans*-2-butene has two stereocenters but only two stereoisomers, there are no situations where there will be more than 2^n stereoisomers).

b. This statement is not correct. For example, the compound in Problem 26 has one asymmetric center (therefore, $2^n = 2$), but the compound has four stereoisomers.

28. **a.** enantiomers

b. identical compounds (Therefore, they are not stereoisomers.)

c. diastereomers

29. **a.** Find the sp^3 carbons that are bonded to four different substituents; these are the asymmetric centers. Cholesterol has eight asymmetric centers. They are indicated by arrows.

b. $2^8 = 256$

Only the stereoisomer shown above is found in nature.

30.   Your perspective formulas may not look exactly like the ones drawn here because you can draw the first one with the groups attached to any bonds you want. Just make certain that the second one is a mirror image of the first one.

a.   Leucine has one asymmetric center, so it will have two stereoisomers.

b.   Isoleucine has two asymmetric centers, so it has four stereoisomers. Again your perspective formulas may not look exactly like the ones drawn here. To make sure you have all four, determine the configuration of each of the asymmetric centers. You should have *R,R*, *S,S*, *R,S*, and *S,R*. Notice that the asymmetric centers in the mirror image have the opposite configurations.

31.   **B** and **D** have no symmetric centers.

A and **C** each have one asymmetric center.

E has two asymmetric centers.

32.   **a.**

H_3C—C(Cl)(H)—C(H)(OH)—$CH_2CH_2CH_3$ *R   S*

H—C(HO)...—C(CH_3)...(Cl)—$CH_3CH_2CH_2$ ... *R   S*

or

	CH_3			CH_3	
Cl—	—H	*R*	H—	—Cl	*S*
HO—	—H	*S*	H—	—OH	*R*
	$CH_2CH_2CH_3$			$CH_2CH_2CH_3$	

H_3C—C(H)...(Cl)—C(H)(OH)—$CH_2CH_2CH_3$ *S   S*

HO...—C(H)—C(CH_3)...(H)—Cl $CH_3CH_2CH_2$ *R   R*

or

	CH_3			CH_3	
H—	—Cl	*S*	Cl—	—H	*R*
HO—	—H	*S*	H—	—OH	*R*
	$CH_2CH_2CH_3$			$CH_2CH_2CH_3$	

b.

Br Cl — *S   R*

Br Cl — *R   S*

or

	CH_3			CH_3	
H—	—Br	*S*	Br—	—H	*R*
H—	—H		H—	—H	
H—	—Cl	*R*	Cl—	—H	*S*
	CH_2CH_3			CH_2CH_3	

Br Cl — *S   S*

Br Cl — *R   R*

or

	CH_3			CH_3	
H—	—Br	*S*	Br—	—H	*R*
H—	—H		H—	—H	
Cl—	—H	*S*	H—	—Cl	*R*
	CH_2CH_3			CH_2CH_3	

c.

H_3C—C(Cl)...—C(H)(Cl)—CH_2CH_3 *R   S*

H...—C(Cl)—C(CH_3)...(Cl) CH_3CH_2 *R   S*

or

	CH_3			CH_3	
Cl—	—H	*R*	H—	—Cl	*S*
Cl—	—H	*S*	H—	—Cl	*R*
	CH_2CH_3			CH_2CH_3	

H_3C—C(H)...—C(H)(Cl)—CH_2CH_3 *S   S*

H...—C(Cl)—C(CH_3)...(H)—Cl CH_3CH_2 *R   R*

or

	CH_3			CH_3	
H—	—Cl	*S*	Cl—	—H	*R*
Cl—	—H	*S*	H—	—Cl	*R*
	CH_2CH_3			CH_2CH_3	

d.

CH_2CH_2Br
CH_3CH_2—C...(H)—Br   *S*

CH_2CH_2Br
H...—C—CH_2CH_3 Br   *R*

or

	CH_2CH_2Br	
Br—	—H	
	CH_2CH_3	*S*

	CH_2CH_2Br	
H—	—Br	
	CH_2CH_3	*R*

33.

R ‖OH   HO‖ S

S CH₃   CH₃ R

cis-2-methylcyclohexanol

R ‖OH   HO‖ S

R CH₃   CH₃ S

trans-2-methylcyclohexanol

34.

Cl ....CH₃

CH₃ / Cl

CH₃ / Cl

1-chloro-1-methylcyclooctane   *cis*-1-chloro-5-methylcyclooctane   *trans*-1-chloro-5-methylcyclooctane

35.   There is more than one diastereomer for **a, b,** and **d**; **c** has only one diastereomer.

To draw a diastereomer of **a, b,** or **d**, switch any one pair of substituents bonded to one of the asymmetric centers. Because any one pair can be switched, your diastereomer may not be the one drawn here, unless you happened to switch the same pair that is switched here.

a.

CH₃

H——OH

HO——H

CH₃

c.

H₃C   H

  C=C

H   CH₃

b.

Cl   H

H····C—C—Cl

CH₃CH₂   CH₃

d.

HO   CH₃
   ⬠

36.   **A** = identical   **B** = enantiomer   **C** = diastereomer   **D** = identical

37.   **B, D,** and **F** because each has two asymmetric centers and the same four groups bonded to each of the asymmetric centers.
A has two asymmetric centers but it does not have a stereoisomer that is a meso compound, because it does not have the same four groups bonded to each of the asymmetric centers.
C and **E** do not have a stereoisomer that is a meso compound, because they do not have asymmetric centers.

38.   Solved in the text.

39.   **a.**

CH₂CH₃   CH₂CH₃   CH₂Br   CH₂Br

   C   C   **or**   H——CH₃   CH₃——H

BrCH₂ H   H CH₂Br   CH₂CH₃   CH₂CH₃
    CH₃   CH₃
    R   *S*   *R*   *S*

b.

CH$_2$CH$_3$ ... CH$_2$CH$_3$ **or**   CH$_2$CH$_2$Cl ... CH$_2$CH$_2$Cl

R   *S*   *R*   *S*

c.

CH$_3$
|
CH$_3$CHCH$_2$OH   No stereoisomers, because the compound does not have an asymmetric center.

d.

R   *S*   **or**   *R*   *S*

e.

CH$_3$
|
CH$_3$CH$_2$CCH$_2$CH$_3$ No stereoisomers, because the compound does not have an asymmetric center.
|
Cl

f.

R   S   *R   S*   **or**

S   S   *R   R*

g.

S   R   **or**

a meso compound   a meso compound

S   S   *R   R*

h.

R   S
a meso compound

or

a meso compound
(it has a plane of
symmetry)

R   R   S   S

i.

(structures with Cl, S, R substituents on chains)

S   R   R   S

or

(Fischer projections with CH₃, H, Cl, CH₂CH₂CH₃)

S   S   R   R

(Fischer projections with CH₃, H, Cl, CH₂CH₂CH₃)

j. (three cyclobutane ring structures with Cl, Cl)   The cis stereoisomer is
a meso compound.

k. (three cyclohexane ring structures with Cl, Cl)   The cis stereoisomer is
a meso compound.

l. (two cyclohexane ring structures Cl–Cl)   This compound does not have any asymmetric centers,
so it has only cis-trans isomers.

m. (four cyclobutane ring structures with Cl, Br)

n. (two cyclobutane ring structures with Br, Cl)   This compound does not have any asymmetric
centers, so it has only cis-trans isomers.

40. and **41.** How to draw perspective formulas for compounds that have two asymmetric centers is described on page 175 of the text.

a.

CH₂Cl
H —— OH
H —— Cl
CH₂CH₃
(2S,3R)-1,3-dichloro-2-pentanol

CH₂Cl
HO —— H
Cl —— H
CH₂CH₃
(2R,3S)-1,3-dichloro-2-pentanol

CH₂Cl
H —— OH
Cl —— H
CH₂CH₃
(2S,3S)-1,3-dichloro-2-pentanol

CH₂Cl
HO —— H
H —— Cl
CH₂CH₃
(2R,3R)-1,3-dichloro-2-pentanol

b.

ClCH₂   Cl
H⁗C — C⁻H
HO   CH₂CH₃
(2S,3R)-1,3-dichloro-2-pentanol

Cl   CH₂Cl
H⁻C — C⁗H
CH₃CH₂   OH
(2R,3S)-1,3-dichloro-2-pentanol

ClCH₂   H
H⁗C — C⁻Cl
HO   CH₂CH₃
(2S,3S)-1,3-dichloro-2-pentanol

H   CH₂Cl
Cl⁻C — C⁗H
CH₃CH₂   OH
(2R,3R)-1,3-dichloro-2-pentanol

42.

R   R
HO   H
C — C⁻CH₂OH
H⁗   NHCCHCl₂
                    ‖
                    O

NO₂

43. Your answer might be correct yet not look like the answers shown here. If you can get the answer shown here by interchanging **two** pairs of groups bonded to an asymmetric center on the structure you drew, then your answer is correct. If you get the answer shown here by interchanging **one** pair of groups bonded to an asymmetric center, then your answer is not correct.

a.

Cl
|
C⁗H
HOCH₂CH₂   CH₂CH₃

b.

Br,   H
H⁗C — C⁻CH₂CH₃
H₃C   Br

c.

$$H_3C \quad\quad H$$
$$H \cdots C - C - CH_2CH_3$$
$$HO \quad\quad CH_3$$

d.

$$CH_2Br$$
$$C \cdots H$$
$$Br \quad\quad CH_2CH_3$$

44.   **a.** (3*R*,4*S*)-3-chloro-4-methylhexane   **c.** (1*R*,3*S*)-3-bromocyclopentanol
  b. (2*S*,3*S*)-2-bromo-3-chloropentane   **d.** (2*R*,3*R*)-2,3-dichloropentane

45.   The carbon of the COO⁻ group is C-1. The first structure is 2*R*,3*S*. Therefore, naturally occurring threonine, with a configuration of 2*S*,3*R*, is the mirror image of the first structure. Thus, the second structure is naturally occurring threonine.

46.   **a.** Solved in the text.

  b.

$$CH_3$$
$$HO \overset{R}{\underset{}{\rule{0pt}{0pt}}} H$$
$$Br \overset{S}{\underset{}{\rule{0pt}{0pt}}} H$$
$$CH_3$$

$$HO \quad\quad H$$
$$H \cdots C - C - Br$$
$$CH_3 \;{}_R \;\; {}_S\; CH_3$$

47.   **a.** Solved in the text.

  b.

$$Cl \quad\quad H$$
$$CH_3 \cdots C - C - OH$$
$$H \;{}_S \;\; {}_S\; CH_2CH_3$$

(structure of 3-chloro-2-butanol with Cl and OH, S and S configurations)

48.   Start by naming the first stereosiomer. Finding that **A** is 2*R*,3*R* allows you to answer both questions.

  a.   **A** is D-erythrose.

  b.   D-Threose has the opposite configuration at C-2 and the same configuration at C-3. Therefore, **C** is D-threose.

49.   Solved in the text.

50.   We see that the (*R*)-alkyl halide reacts with HO⁻ to form the (*R*)-alcohol. We are told that the product (the (*R*)-alcohol) is (−). We can, therefore, conclude that the (+)-alcohol has the *S* configuration.

51.   (+)-Limonene has the *R* configuration, so it is the stereoisomer found in oranges.

52. Compound **A** has two stereoisomers, because it has an asymmetric center (at N).

Compound **B** does not have stereoisomers, because it does not have an asymmetric center.

Compound **C** has an asymmetric center at N but, because of the lone pair, the two enantiomers rapidly interconvert, so it exists as a single compound.

53. CH₃CH=CHCH₂CH₃   CH₂=CHCH₂CH₂CH₃   CH₃C=CHCH₃   CH₃CHCH=CH₂

      2 stereoisomers   no stereoisomers   no stereoisomers   no stereoisomers
      [cis and trans]

CH₃CH₂C=CH₂   CH₃/CH₃   CH₂CH₃   H₃C CH₃

no stereoisomers   no stereoisomers   no stereoisomers   3 stereoisomers

            [Cis is a meso compound.]
            [Trans is a pair of enantiomers.]

CH₃ (cyclobutane)   (cyclopentane)

no stereoisomers   no stereoisomers

54. **a.**

    Br Cl   Cl Br   Br Cl   Cl Br

b. CH₃CHCH₂ C H + H C CH₂CHCH₃ **or** H—Br + Br—H

    S   R   S   R

c. Cl Cl   Cl Cl   Cl Cl

a meso compound

d. CH₃ CH₃ CH₃ CH₃

H—Br Br—H H—Br Br—H
H—H   H—H   H—H   H—H
H—Cl Cl—H Cl—H H—Cl
CH₃ CH₃ CH₃ CH₃

e. Br—◯—Cl Br—◯···Cl

cis and trans

f. H₃C CH₃   H₃C ···CH₃   H₃C··· CH₃

g.

cis   trans

h. CH₃CH₂CCH₂CH₃ (with CH₃ above and CH₃ below the central C)   No isomers are possible for this compound, because it does not have an asymmetric center.

i.

Cl   Br   Br   Cl   Cl   Br   Br   Cl

j.

Br   Br

Cl   Cl

cis and trans only (no asymmetric centers)

55.   Only the fourth one (CHFBrCl) has an atom with four different atoms attached to a carbon, so it is the only one that has an asymmetric center.

56.   **a.**   (2R,3R)-3-chloro-2-pentanol
       b.   (S)-2-methyl-1,2,5-pentanetriol
       c.   (2S,3S)-1,2-dibromo-2-methyl-3-pentanol

57.   Mevacor has eight asymmetric centers, which are indicated by the arrows.

58.   **a.**   diastereomers (one asymmetric center has the same configuration in both compounds and the other has the opposite configuration in both compounds)
       b.   enantiomers (they are mirror images)
       c.   constitutional isomers
       d.   diastereomers (the configuration of two asymmetric centers is the same and the configuration of one asymmetric center is different)
       e.   diastereomers
       f.   identical

g. diastereomers (cis-trans isomers are diastereomers because they are stereoisomers that are not enantiomers)

h. identical (if you flip one over, you can see that they are superimposable)

59. Only the compound on the far right is optically active, because it is the only one that has one or more asymmetric centers.

60. **a.** Because there are two asymmetric centers, there are four possible stereoisomers.

b.

61. **a.** (2R,4S,6S)-4-chloro-6-methyl-2-octanol   **b.** (R)-3-bromo-2,5-dimethylhexane

62. Compounds **a**, **d**, **e**, **f**, **h**, **i**, and **j** have a stereoisomer that is achiral.

Compounds **a**, **d**, **f**, **i**, and **j** have two asymmetric centers bonded to identical substituents. Thus, there are three stereoisomers, one of which is an achiral meso compound.

Compounds **e** and **h** do not have any asymmetric centers; they have cis and trans stereosiomers, so each is achiral and, therefore, each has an achiral stereoisomer.

Compounds **b** and **g** each have two asymmetric centers bonded to different substituents. Thus, there are four stereoisomers, all of which are chiral.

Compound **c** does not have any asymmetric centers; it is, therefore, achiral but it does not have any stereoisomers.

63. **a.** One asymmetric center has the same configuration in both compounds and the other asymmetric center has the opposite configuration in both, so the compounds are diastereomers.
b. Both asymmetric centers in one compound have the opposite configuration in the other, so the compounds are enantiomers.
c. They are identical because if one is flipped over, it will superimpose on the other.
d. They are constitutional isomers because the atoms are hooked up differently; one compound is 1-chloro-2-methylcyclopentane and the other is 1-chloro-3-methylcyclopentane.

64. **a.**

(S)-citric acid

b. The reaction is catalyzed by an enzyme. Only one stereoisomer is typically formed in an enzyme-catalyzed reaction because an enzyme has a chiral binding site that allows reagents to be delivered to only one side of the functional group of the reactant.
c. The product of the reaction will be achiral because if it does not have a ^{14}C label, the two CH_2COOH groups will be identical, so it will not have an asymmetric center.

65.
  a. identical
  b. identical
  c. enantiomers
  d. constitutional isomers

  e. constitutional isomers
  f. diastereomers
  g. constitutional isomers
  h. enantiomers

66.   optical purity $= \dfrac{+1.4}{+8.7} = 0.16 = 16\%$ excess R enantiomer

$100\% - 16\% = 84\%$ is a racemic mixture
R enantiomer $= 1/2\,(84\%) + 16\% = 42\% + 16\% = 58\%$

67.   **a.** R **b.** S **c.**

$$= \qquad R$$

  d.

$$= \qquad S$$

  e.

R

  f.

S

Fisher projections show the molecule with eclipsed bonds. Therefore, to answer parts **e** and **f**, first rotate the Newman projection so it is eclipsed. Then turn the Newman projection into a Fisher projection.

68. $[\alpha] = \dfrac{\alpha}{l \times c} = \dfrac{-18°}{[2.0 \text{ dm}]\,[1.5 \text{ g}/100 \text{ mL}]} = -6.0$

69.   Butaclamol has four asymmetric centers; three of them are carbons and one is a nitrogen.

70.   The only way that R and S are related to $(+)$ and $(-)$ is that if one configuration (say, R) is $(+)$, the other one is $(-)$.

Because some compounds with the R configuration are $(+)$ and some are $(-)$, there is no way to determine whether a particular R enantiomer is $(+)$ or $(-)$ without putting the compound in a polarimeter or finding out whether someone else has previously determined how the compound rotates the plane of polarization of plane-polarized light.

71. First convert the staggered Newman projection to an eclipsed Newman projection, which can then be converted to a Fischer projection since that too is eclipsed (see page 188 of the text). Then name the Fischer projection.

a. Since one is *R* and the other is *S*, they are enantiomers

OH
H   H
H   CH₃
CH₃

H₃C OH
H   H
H   CH₃

OH
H —*R*— CH₃
CH₂CH₃

CH₃
H   CH₃
H   OH
H

H₃C CH₃
H   H
H   OH

CH₃
H —*S*— OH
CH₂CH₃

b. Since one is *R,R* and the other is *S,S*, they are enantiomers.

CH₃
H   Br
H   CH₂CH₃
Cl

Cl CH₃
H   H
Br   CH₂CH₃

CH₃
H —*R*— CH₂CH₃
Br —*R*— H
Cl

CH₂CH₃
H   Cl
H   CH₃
Br

Cl CH₂CH₃
H   Br
H   CH₃

CH₂CH₃
H —*S*— CCH₃
H —*S*— Br
Cl

72. **a.** The compound has four stereoisomers.

CH₂OH
H —— OH
H —— OH
H —— OH
CH₂OH

CH₂OH
H —— OH
HO —— H
H —— OH
CH₂OH

CH₂OH
H —— OH
H —— OH
HO —— H
CH₂OH

CH₂OH
HO —— H
HO —— H
H —— OH
CH₂OH

b. The first two stereoisomers are optically inactive because they are meso compounds. (They each have a plane of symmetry.)

73. **a.**

CH₂CH₂Br
C
Br   H
CH₂CH₂CH₃
R

b.

BrCH₂   OH
H""C—C—H
Br   CH=O
R   *R*

c.

H₃C   CH₂CH₃
Br""C—C—H
H   Br
R   *S*

74.   **a.** and **b.**

ethylcyclobutane   1,1-dimethylcyclobutane

cis-1,2-dimethylcyclobutane   *trans*-1,2-dimethylcyclobutane

cis-1,3-dimethylcyclobutane   *trans*-1,3-dimethylcyclobutane

c. **1.** ethylcyclobutane
1,1-dimethylcyclobutane
1,2-dimethylcyclobutane
1,3-dimethylcyclobutane

2. the three isomers of 1,2-dimethylcyclobutane
the two isomers of 1,3-dimethylcyclobutane

3. *cis*- and *trans*-1,2-dimethylcyclobutane
cis- and *trans*-1,3-dimethylcyclobutane

4. the two trans stereoisomers of 1,2-dimethylcyclobutane

5. all the isomers except the two trans stereoisomers of 1,2-dimethylcyclobutane

6. *cis*-1,2-dimethylcyclobutane
(Note: *cis*-1,3-dimethylcyclobutane is not a meso compound because it does not have any asymmetric centers.)

7. the two trans stereoisomers of 1,2-dimethylcyclobutane

8. *cis*-1,3-dimethylcyclobutane and *trans*-1,3-dimethylcyclobutane
cis-1,2-dimethylcyclobutane and either of the enantiomers of *trans*-1,2-dimethylcyclobutane

75. $\text{observed specific rotation} = \dfrac{\text{observed rotation}}{\text{concentration} \times \text{length}}$

$$\dfrac{-6.52°}{0.187 \text{ g/100 mL} \times 1 \text{ dm}} = -34.9$$

$$\% \text{ optical purity} = \dfrac{\text{observed specific rotation}}{\text{specific rotation of the pure enantiomer}} \times 100$$

$$= \dfrac{-34.9}{-39.0} \times 100$$

$$= 89.5\%$$

$$\% \text{ of the (+)-isomer} = \dfrac{100 - 89.5}{2} = 5.25\%$$

$$\% \text{ of the (−)-isomer} = 89.5 + 5.25 = 94.75\%$$

76. **a.** diastereomers (the configuration of all the symmetric centers is not the same in both and not opposite in both)

b. identical (by rotating one compound, you can see that it is superimposable on the other)

c. constitutional isomers

d. diastereomers (the configuration of all the stereoisomers is not the same in both and not opposite in both)

77. **a.**

b.

c.

d.

e.

78. In the transition state for amine inversion, the nitrogen atom is sp^2 hybridized, which means it has bond angles of 120°. A nitrogen atom in a three-membered ring cannot achieve a 120° bond angle, so the amine inversion that would interconvert the enantiomers cannot occur. Therefore, the enantiomers can be separated.

79. The fact that the optical purity is 72% means that there is 72% enantiomeric excess of the *S* isomer and 28% racemic mixture. Therefore, the actual amount of the *S* isomer in the sample is 72% + 1/2(28%) = 86%. The amount of the *R* isomer in the sample is 1/2(28%) = 14% (or 100% − 86% = 14%).

80. **A** = a diastereomer   **C** = a diastereomer   **E** = a diastereomer
    B = a diastereomer   **D** = a diastereomer

81.   a.

b.

This is a pair of enantiomers because they are nonsuperimposable mirror images.

c.

This is the most stable isomer because, since the chloro substituents are all trans to each other, they can all be in the more stable equatorial position. (Recall that there is less steric strain when a substituent is in the equatorial position.)

82.

83.   Yes, as long as the Fischer projection is drawn with the #1 carbon at the top of the chain of carbons.

84.   a.

b.

c.

85.   The trans compound exits as a pair of enantiomers.

As a result of ring-flip, each enantiomer has two chair conformers. In each case, the more stable conformer is the one with the larger group (the *tert*-butyl group) in the equatorial position.

more stable   more stable

86.   **a.**   The compounds do not have any asymmetric centers.

      b.   **1.**   It is not chiral.

            2.   It is chiral. Because of its unusual geometry, it is a chiral molecule, even though it does not have any asymmetric centers, because it cannot be superimposed on its mirror image. This will be easier to understand if you build models.

mirror
images are superimposable

mirror
images are not superimposable

87.   The compound is not optically active because it has a point of symmetry.

A point of symmetry is a point, and if a line is drawn to this point from an atom or group and then extended an equal distance beyond the point, the line would touch an identical atom or group.

point of symmetry

Chapter 4 Practice Test

1.   Are the following pairs of compounds identical or a pair of enantiomers?

a.   CH$_3$CH$_2$—C(—CH$_2$OH)(—CH$_3$)(H) and H—C(—CH$_2$CH$_3$)(—CH$_2$OH)(CH$_3$)

b.   (CH$_2$CH$_3$)(H$_3$C)(H)(Cl)C and (CH$_3$)(Cl)(H)(CH$_2$CH$_3$)C

2.   100 mL of a solution containing 0.80 g of a compound rotates the plane of polarized light −4.8° in a polarimeter with a 2-dm sample tube. What is the specific rotation of the compound?

3.   Which are meso compounds?

Structure 1: CH$_3$ / H—Cl / H—Cl / CH$_2$CH$_3$

Structure 2: CH$_3$ / H—Cl / Cl—H / CH$_2$CH$_3$

Structure 3: CH$_3$ / Br—Cl / Br—Cl / CH$_2$Cl

Structure 4: CH$_3$ / H—Cl / H—Cl / CH$_3$

Structure 5: CH$_3$ / Cl—H / H—Cl / CH$_3$

4.   Draw all the constitutional isomers with molecular formula C$_4$H$_9$Cl.

5.   Draw all the possible stereoisomers for each of the following compounds that has them.

a.   (cyclopentane with HO and OH)

b.   CH$_3$CHCHCH$_2$CH$_3$ with Br Br

c.   CH$_3$CH$_2$CHCH$_2$CH$_3$ with Cl

d.   CH$_3$CH$_2$CHCH$_2$CH$_2$Cl with Cl

e.   (cyclohexane with Br and OH)

f.   HO—(cyclohexane)—CH$_3$

6.   Which of the following three perspective formulas are the same as the Fischer projection shown here?

COOH / HO—H / H—OH / CH$_3$

(HO)(H)C—C(OH)(CH$_3$) with HOOC   /   (H)(HO)C—C(H)(CH$_3$) with HOOC   /   (HO)(H)C—C(OH)(H) with HOOC and CH$_3$

7. (R)-$(-)$-2-methyl-1-butanol can be oxidized to $(+)$-2-methylbutanoic acid without breaking any of the bonds to the asymmetric center. What is the configuration of $(-)$-2-methylbutanoic acid?

(R)-$(-)$-2-methyl-1-butanol $(+)$-2-methylbutanoic acid

8. $(-)$-Cholesterol has a specific rotation of -32. What would be the observed specific rotation of a solution that contains $25\%\,(+)$-cholesterol and $75\%\,(-)$-cholesterol?

9. Which of the following have the R configuration?

10. Answer the following:

a. Are the following compounds identical or a pair of enantiomers?

b. Put the remaining groups on the structure so it represents (R)-2-butanol.

c. Put the remaining groups on the Fischer projection so it represents the Newman projection shown.

d. Draw a diastereomer for each of the following:

1.

Cl   Cl

2.   CH$_3$

H——OH

H——Br

CH$_3$

11.   Indicate whether each of the following statements is true or false:

a.   Diastereomers have the same melting points.   T   F

b.   3-Chloro-2,3-dimethylpentane has two asymmetric centers.   T   F

c.   Meso compounds do not rotate the plane of polarization of
plane-polarized light.   T   F

d.   2,3-Dichloropentane has a stereoisomer that is a meso compound.   T   F

e.   All chiral compounds with the *R* configuration are dextrorotatory.   T   F

f.   A compound with three asymmetric centers can have a maximum
of nine stereoisomers.   T   F

12.   Which of the following have cis-trans isomers?

1-pentene   4-methyl-2-hexene   2-bromo-3-hexene   2-methyl-2-hexene

CHAPTER 5
Alkenes: Structure, Nomenclature, and an Introduction to Reactivity • Thermodynamics and Kinetics

Important Terms

active site	the pocket of an enzyme where all the bond-making and bond-breaking steps of an enzyme-catalyzed reaction occur.
acyclic	noncyclic.
addition reaction	a reaction in which atoms or groups are added to the reactant.
alkene	a hydrocarbon that contains a double bond.
allyl group	$CH_2\!=\!CHCH_2\!-\!$
allylic carbon	an sp^3 carbon adjacent to a vinyl carbon.
allylic hydrogen	a hydrogen bonded to an allylic carbon.
Arrhenius equation	an equation that relates the rate constant of a reaction to the energy of activation and to the temperature at which the reaction is carried out $(k = Ae^{-E_a/RT})$.
catalyst	a species that increases the rate at which a reaction occurs without being consumed in the reaction.
cis isomer	the isomer with the hydrogens on the same side of the double bond.
degree of unsaturation	the sum of the number of π bonds and rings in a hydrocarbon.
E isomer	the isomer with the high-priority groups on opposite sides of the double bond.
electrophile	an electron-deficient atom or molecule.
electrophilic addition reaction	an addition reaction in which the first species that adds to the reactant is an electrophile.
endergonic reaction	a reaction with a positive $\Delta G°$; it consumes more energy than it releases.
endothermic reaction	a reaction with a positive $\Delta H°$.
enthalpy	the heat given off (if $\Delta H° < 0$) or the heat absorbed (if $\Delta H° > 0$) during the course of a reaction.
entropy	a measure of the freedom of motion in a system.
enzyme	a protein that is a biological catalyst.
exergonic reaction	a reaction with a negative $\Delta G°$; it releases more energy than it consumes.

exothermic reaction	a reaction with a negative $\Delta H°$.
experimental energy of activation $(E_a = \Delta H^{\ddagger} + RT)$	a measure of the approximate energy barrier to a reaction. (It is approximate because it does not contain an entropy component.)
first-order rate constant	the rate constant of a first-order reaction.
first-order reaction (unimolecular reaction)	a reaction whose rate is dependent on the concentration of one reactant.
free energy of activation $(\Delta G^{\ddagger})$	the true energy barrier to a reaction.
functional group	the center of reactivity of a molecule.
geometric isomers (cis-trans stereoisomers)	cis-trans (or *E,Z*) isomers.
Gibbs free energy change $(\Delta G°)$	the difference between the free energy content of the products and the free energy content of the reactants at equilibrium under standard conditions (1M, 25 °C, 1 atm).
intermediate	a species formed during a reaction that is not the final product of the reaction.
kinetics	the field of chemistry that deals with the rates of chemical reactions.
kinetic stability	kinetic stability is indicated by $\Delta G^{\ddagger}$. If $\Delta G^{\ddagger}$ is large, the compound is kinetically stable (is not very reactive). If $\Delta G^{\ddagger}$ is small, the compound is kinetically unstable (is very reactive).
Le Châtelier's principle	a principle states that if an equilibrium is disturbed, the components of the equilibrium will adjust in a way that will offset the disturbance.
mechanism of the reaction	a description of the step-by-step process by which reactants are changed into products.
molecular recognition	the ability of one molecule to recognize another as a result of intermolecular interactions.
nucleophile	an electron-rich atom or molecule.
pheromone	a chemical substance used for the purpose of communication.
rate constant	the proportionality constant in the rate equation that quantifies the rate of a reaction.
rate-determining step or rate-limiting step	the step in a reaction that has the transition state with the highest energy.

reaction coordinate diagram	a diagram that describes the energy changes that take place during the course of a reaction.
saturated hydrocarbon	a hydrocarbon that is completely saturated with hydrogen (contains no double or triple bonds).
second-order rate constant	the rate constant of a second-order reaction.
second-order reaction	a reaction whose rate is dependent on the concentration of two reactants, or on the square of the concentration of a single reactant.
solvation	the interaction between a solvent and another molecule (or ion).
substrate	the reactant of an enzyme-catalyzed reaction.
thermodynamic stability	thermodynamic stability is indicated by $\Delta G°$. If $\Delta G°$ is negative, the products are more stable than the reactants. If $\Delta G°$ is positive, the reactants are more stable than the products.
thermodynamics	the field of chemistry that describes the properties of a system at equilibrium.
trans isomer	the isomer with the hydrogens on opposite sides of the double bond.
transition state	the energy maximum in a reaction step on a reaction coordinate diagram. In the transition state, bonds in the reactant that will break are partially broken and bonds in the product that will form are partially formed.
unsaturated hydrocarbon	a hydrocarbon that contains one or more double or triple bonds.
vinyl group	$CH_2\!=\!CH\!-$
vinylic carbon	a carbon that is doubly bonded to another carbon.
vinylic hydrogen	a hydrogen bonded to a vinylic carbon.
Z isomer	the isomer with the high-priority groups on the same side of the double bond.

Solutions to Problems

1.   **a.** C_5H_8 **b.** C_4H_6 **c.** $C_{10}H_{16}$

2.   **a.** 3   **b.** 4   **c.** 1   **d.** 3   **e.** 13

3.   **a.** degree of unsaturation = 1   **b.** degree of unsaturation = 2   **c.** degree of unsaturation = 2

$CH_3CH{=}CH_2$

$CH_3C{\equiv}CH$

$CH_2{=}C{=}CH_2$

$HC{\equiv}CCH_2CH_3$

$CH_3C{\equiv}CCH_3$

$CH_2{=}CHCH{=}CH_2$

$CH_2{=}C{=}CHCH_3$

4.   A hydrocarbon with no rings and no double bonds would have a molecular formula of $C_{40}H_{82}$. $C_{40}H_{56}$ has 26 fewer hydrogens. Therefore, β-carotene has a total of 13 rings and double bonds. Since we know that it has two rings, it has 11 double bonds.

5.   **a.**   It has two vinylic hydrogens.
      b.   It has four allylic hydrogens.

6.   **a.**

      c. $CH_3CH_2OCH{=}CH_2$

      b. $CH_3\underset{\underset{CH_3}{|}}{\overset{\overset{CH_3}{|}}{C}}{=}CCH_2CH_2CH_2Br$

      d. $CH_2{=}CHCH_2OH$

7.   **a.** 4-methyl-2-pentene
      b. 2-chloro-3,4-dimethyl-3-hexene
      c. 1-bromocyclopentene
      d. 1-bromo-4-methyl-3-hexene

      e. 1,5-dimethylcyclohexene
      f. 1-butoxy-1-propene
      g. 1-bromo-2-methyl-1,3-pentadiene
      h. 8,8-dimethyl-1-nonene

8.   **a.** 5   **b.** 4   **c.** 4   **d.** 6

9.   **a.** $-I > -Br > -OH > -CH_3$
      b. $-OH > -CH_2Cl > -CH{=}CH_2 > -CH_2CH_2OH$

10.   The high-priority groups are on same side of the double bond, so tamoxifen has the Z configuration.

11. **a.**

CH₃CH₂   CH₃   CH₃CH₂   H

C=C   C=C

H   H   H   CH₃

Z   **E**

b. CH₃CH₂   CH₂CH₃   CH₃CH₂   H

C=C   C=C

Cl   H   Cl   CH₂CH₃

E   **Z**

c.

CH₃

CH₃CH₂CH₂CH₂   CH₂Cl   CH₃CH₂CH₂CH₂   CHCH₃

C=C   C=C

CH₃CH₂   CHCH₃   CH₃CH₂   CH₂Cl

CH₃

Z   **E**

d.

CH

HOCH₂CH₂   C(CH₃)₃   HOCH₂CH₂   C

C=C   C=C

O=C   C   O=C   C(CH₃)₃

H   **Z**   CH   H   **E**

12. **a.**

c.

b.

d. HO

13.   **a.**   (*E*)-2-heptene   **b.**   (*Z*)-3,4-dimethyl-2-pentene   **c.**   (*Z*)-1-chloro-3-ethyl-4-methyl-3-hexene

14.

15.   **a.**

(*E*)-2-methyl-2,4-hexadiene   (*Z*)-2-methyl-2,4-hexadiene

b.

(2*E*,4*E*)-2,4-heptadiene   (2*E*,4*Z*)-2,4-heptadiene

(2*Z*,4*E*)-2,4-heptadiene   (2*Z*,4*Z*)-2,4-heptadiene

c.

(*E*)-1,3-pentadiene   (*Z*)-1,3-pentadiene

b has four stereoisomers because each double bond can have either the *E* or the *Z* configuration.

a and **c** have only two stereoisomers because, in each case, there are two identical substituents bonded to one of the sp^2 carbons, so only one of the double bonds can have either the *E* or the *Z* configuration.

16.   **a.** $AlCl_3$ is the electrophile and NH_3 is the nucleophile.
           b.   The $H^{\delta+}$ of HBr is the electrophile and HO^- is the nucleophile.

17.   nucleophiles: H^- CH_3O^- $CH_3C{\equiv}CH$ NH_3

           electrophiles: $CH_3\overset{+}{C}HCH_3$

18. **1.**

Drawing the arrows incorrectly leads to a bromine with an incomplete octet and a positive charge as well as an oxygen with 10 valence electrons and 2– charge.

Drawing the arrows incorrectly leads to an oxygen with an incomplete octet and a 2+ charge.

2. This one cannot be drawn because the arrow is supposed to show where the electrons move to, but there are no electrons on the H to go anywhere.

3.

The product cannot be drawn because the destination of the electrons in the breaking π bond is not clear.

4. This one cannot be drawn because the arrow is supposed to show where the electrons move to, but there are no electrons on the C to go anywhere.

19. **a.**

nucleophile   electrophile

b.

nucleophile   electrophile

c.

electrophile   nucleophile

d.

electrophile   nucleophile

20.   The labels are under the structures in Problem 19.

21.   **a.**   Because the equilibrium constants for all the monosubstituted cyclohexanes in Table 3.9 on page 131 of the text are greater than 1, all of the equilibria have negative $\Delta G°$ values.

(Recall that $\Delta G° = -RT \ln K_{eq}$.)

   b.   *tert*-butylcyclohexane

   c.   *tert*-butylcyclohexane, because it is the largest substituent

   d. $\Delta G° = -RT \ln K_{eq}$

$K_{eq} = 18$

$\Delta G° = -1.986 \times 10^{-3} \, \text{kcal/mol K} \times 298 \, \text{K} \times \ln 18$ (recall that $T = °C + 273$)

$\Delta G° = -0.59 \times \ln 18 \, \text{kcal/mol}$

$\Delta G° = -0.59 \times 2.89 \, \text{kcal/mol}$

$\Delta G° = -1.7 \, \text{kcal/mol}$

22.   **a.**   Solved in the text.

   b.
$$\Delta G° = -RT \ln K_{eq}$$
$$-2.1 = -1.986 \times 10^{-3} \times 298 \times \ln K_{eq}$$
$$\ln K_{eq} = 3.56$$
$$K_{eq} = 35$$
$$K_{eq} = \frac{[\text{isopropylcyclohexane}]_{equatorial}}{[\text{isopropylcyclohexane}]_{axial}} = \frac{35}{1}$$

$$\begin{array}{c}\text{\% of equatorial} \\ \text{isopropylcyclohexane}\end{array} = \frac{[\text{isopropylcyclohexane}]_{equatorial}}{[\text{isopropylcyclohexane}]_{equatorial} + [\text{isopropylcyclohexane}]_{axial}} \times 100$$

$$= \frac{35}{35 + 1} \times 100$$

$$= \frac{35}{36} \times 100$$

$$= 97\%$$

   c.   Isopropylcyclohexane has a greater percentage of the conformer with the substituent in the equatorial position because the isopropyl substituent is larger than the fluoro substituent. The larger the substituent, the less stable is the conformer in which the substituent is in the axial position because of the 1,3-diaxial interactions.

23. $\Delta S°$ is more significant in reactions in which the number of reactant molecules and the number of product molecules are not the same.

   a.   **1.**   A + B $\rightleftharpoons$ C

        2.   A + B $\rightleftharpoons$ C

   b.   None of the four reactions has a positive $\Delta S°$.

In order to have a positive $\Delta S°$, the products must have greater freedom of motion than the reactants. (In other words, there should be more molecules of products than molecules of reactant.)

24. **a.** **1.**

$$\Delta G° = \Delta H° - T\Delta S°$$
(recall that $T = °C + 273$)
$$\Delta G° = -12 - (273 + 30)(.01)$$
$$\Delta G° = -12 - 3 = -15 \text{ kcal/mol}$$

$$\Delta G° = -RT \ln K_{eq}$$
$$\Delta G° = -(1.986 \times 10^{-3})(303) \ln K_{eq}$$
$$-15 = -0.60 \ln K_{eq}$$
$$\ln K_{eq} = 25$$
$$K_{eq} = 7.2 \times 10^{10}$$

2.

$$\Delta G° = \Delta H° - T\Delta S°$$
$$\Delta G° = -12 - (273 + 150)(.01)$$
$$\Delta G° = -12 - 4 = -16 \text{ kcal/mol}$$

$$\Delta G° = -(1.986 \times 10^{-3})(423) \ln K_{eq}$$
$$-16 = -0.84 \ln K_{eq}$$
$$\ln K_{eq} = 19$$
$$K_{eq} = 1.8 \times 10^{8}$$

b. For this reaction: the calculations show that increasing the temperature causes $\Delta G°$ to be more negative.

c. For this reaction: the calculations show that increasing the temperature causes K_{eq} to be smaller, because $\ln K_{eq} = -\Delta G°/RT$.

25. The value for the π bond of ethene (62 kcal/mol) is given in the text on page 45.

a.

bonds broken		bonds formed	
π bond of ethene	62	C—H	101
H—Cl	103	C—Cl	85
	165 kcal/mol		186 kcal/mol $\Delta H° = 165 - 186 = -21 \text{ kcal/mol}$

b.

bonds broken		bonds formed	
π bond of ethene	62	C—H	101
H—H	104	C—H	101
	166 kcal/mol		202 kcal/mol $\Delta H° = 166 - 202 = -36 \text{ kcal/mol}$

c. Both are exothermic, because they both have a negative $\Delta H°$ value.

d. Both reactions have sufficiently negative $\Delta H°$ values to expect that they would be exergonic as well.

26. **a.** **a** and **b** because the product is more stable than the reactant.

b. **b** is the most kinetically stable product, because it has the smallest rate constant (greatest $\Delta G^{\ddagger}$) leading from the product to the transition state.

c. **c** is the least kinetically stable product, because it has the largest rate constant (smallest $\Delta G^{\ddagger}$) leading from the product to the transition state.

27. **a.** A thermodynamically **unstable** product is one that is less stable than the reactant.
A kinetically **unstable** product is one that has a large rate constant (small $\Delta G^{\ddagger}$) for the reverse reaction.

b. A kinetically **stable** product is one that has a small rate constant (large $\Delta G^{\ddagger}$) for the reverse reaction.

28.   **a.** Solved in the text.

b. Decreasing the concentration of methyl chloride (by a factor of 10) will decrease **the rate of the reaction** (by a factor of 10) to $1 \times 10^{-8}\,\text{M s}^{-1}$.

c. Changing the concentration will not affect **the rate constant (k) of the reaction**, because rate constants do not depend on the concentration of the reactants.

29.   The rate constant for a reaction can be increased by **decreasing** the stability of the reactant (increasing its energy) or by **increasing** the stability of the transition state (decreasing its energy).

30.   Taking the logarithm of both sides of the Arrhenius equation gives the following equation (where k is the rate constant), which we can use to answer the questions:

$$\ln k = \ln A - \frac{E_a}{RT}$$

a. Increasing the experimental activation energy (E_a) will decrease the rate constant of a reaction (will cause the reaction to be slower).

b. Increasing the temperature (T) will increase the rate constant of a reaction (will cause the reaction to go faster).

31.   **a.** The first stated reaction has the greater equilibrium constant:

$$K_{eq} = \frac{1 \times 10^{-3}}{1 \times 10^{-5}} = 1 \times 10^{2} \qquad K_{eq} = \frac{1 \times 10^{-2}}{1 \times 10^{-3}} = 10$$

b. Because both reactions start with the same concentration, the first stated reaction will form the most product because it has the greater equilibrium constant.

32.

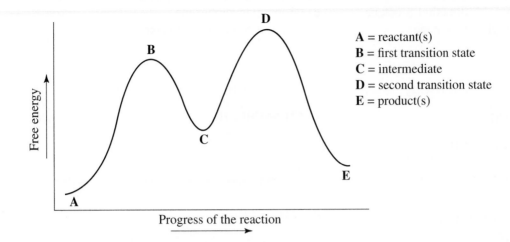

A = reactant(s)
B = first transition state
C = intermediate
D = second transition state
E = product(s)

33.   **a.**   The first step in the forward direction $(A \rightarrow B)$ has the greatest free energy of activation.
b.   The first-formed intermediate (**B**) is more apt to revert to reactants, because the free energy of activation for **B** to form **A** (the reactants) is less than the free energy of activation for **B** to form **C**.
c.   The second step $(B \rightarrow C)$ is the rate-determining step because it has the transition state with the highest energy.

Notice that the second step is rate-determining even though the first step has the greater energy of activation (steeper hill to climb). That is because it is easier for the intermediate that is formed in the first step to go back to starting material than to undergo the second step of the reaction. So the second step is the rate-limiting step.

34.

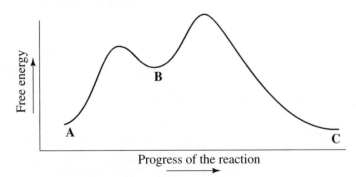

a.   one (B)
b.   two
c.   the second step (k_2), B to form A (In this particular diagram, $k_2 > k_1$: if you had made the transition state for the second step a lot higher, you could have had a diagram in which $k_1 > k_2$.)
d.   the second step in the reverse direction (k_{-1}) **f.** B to C
e.   the second step in the reverse direction (k_{-1}), B to form A   **g.** C to B

35.   A catalyst will change the energy difference between the reactants and the transition state, but it will not change the energy difference between the reactants and the products.

$$\Delta H^{\ddagger}, E_a, \Delta S^{\ddagger}, \Delta G^{\ddagger}, k$$

36.   **a.** 3,8-dibromo-4-nonene
       b. (Z)-4-ethyl-3,7-dimethyl-3-octene
       c. 1,5-dimethylcyclopentene

       d. 3-ethyl-2-methyl-2-heptene
       e. 4-methylcyclohexene
       f. 4-ethyl-5-methylcyclohexene

37.   **a.** $CH_2=CHCH_2CH_2CH_2CH_3$

       b. $CH_2=CHCHCH_2CH_3$ with CH_3 substituent

       c. $CH_2=CHCCH_3$ with two CH_3 substituents

38.   **a.**

       b.

       c.

       d. $CH_2=CHBr$

       e.

       f. $CH_2=CHCH_2NHCH_2CH=CH_2$

39.   **a.**

       b.

       c.

       d.

       e.

       f.

40. **a.** CH$_2$=CHCH$_2$CH$_2$CH$_2$CH$_3$ CH$_3$CH=CHCH$_2$CH$_2$CH$_3$ CH$_3$CH$_2$CH=CHCH$_2$CH$_3$

1-hexene   2-hexene   3-hexene

CH$_2$=CCH$_2$CH$_2$CH$_3$
|
CH$_3$
2-methyl-1-pentene

CH$_2$=CHCHCH$_2$CH$_3$
|
CH$_3$
3-methyl-1-pentene

CH$_2$=CHCH$_2$CHCH$_3$
|
CH$_3$
4-methyl-1-pentene

CH$_3$C=CHCH$_2$CH$_3$
|
CH$_3$
2-methyl-2-pentene

CH$_3$CH=CCH$_2$CH$_3$
|
CH$_3$
3-methyl-2-pentene

CH$_3$CH=CHCHCH$_3$
|
CH$_3$
4-methyl-2-pentene

CH$_3$
|
CH$_2$=CCHCH$_3$
|
CH$_3$
2,3-dimethyl-1-butene

CH$_3$
|
CH$_3$CCH=CH$_2$
|
CH$_3$
3,3-dimethyl-1-butene

CH$_3$CH$_2$C=CH$_2$
|
CH$_2$CH$_3$
2-ethyl-1-butene

CH$_3$
|
CH$_3$C=CCH$_3$
|
CH$_3$
2,3-dimethyl-2-butene

b. Of the compounds shown in part **a**, the following can have *E* and *Z* isomers:

2-hexene, 3-hexene, 3-methyl-2-pentene, 4-methyl-2-pentene

41. **a.** (*E*)-3-methyl-3-hexene
b. *trans*-8-methyl-4-nonene or (*E*)-8-methyl-4-nonene
c. *trans*-9-bromo-2-nonene or (*E*)-9-bromo-2-nonene

d. 2,4-dimethyl-1-pentene
e. 2-ethyl-1-pentene
f. *cis*-2-pentene or (*Z*)-2-pentene

42.

43.

44. $\Delta G° = -RT \ln K_{eq}$
$\ln K_{eq} = -\Delta G°/RT$
$\ln K_{eq} = -\Delta G°/0.59 \text{ kcal/mol}$

a. $\ln K_{eq} = -2.72/0.59 \text{ kcal/mol}$
$\ln K_{eq} = -4.6$
$K_{eq} = [B]/[A] = 0.01$

c. $\ln K_{eq} = 2.72/0.59 \text{ kcal/mol}$
$\ln K_{eq} = 4.6$
$K_{eq} = [B]/[A] = 100$

b. $\ln K_{eq} = -0.65/0.59 \text{ kcal/mol}$
$\ln K_{eq} = -1.10$
$K_{eq} = [B]/[A] = 0.33$

d. $\ln K_{eq} = 0.65/0.59 \text{ kcal/mol}$
$\ln K_{eq} = 1.10$
$K_{eq} = [B]/[A] = 3.0$

45. **a.** The C—Cl bond is a stronger bond because Cl uses a $3sp^3$ orbital to overlap the $2sp^3$ orbital of carbon, whereas Br uses a $4sp^3$ orbital. A $4sp^3$ has a greater volume than a $3sp^3$ and therefore has less electron density in the region of orbital-orbital overlap, so it forms a weaker bond.

b. The Br—Br bond is a stronger bond because Br uses a $4sp^3$ orbital to overlap the $4sp^3$ orbital of the other bromine, whereas I uses a $5sp^3$ orbital. A $5sp^3$ has a greater volume than a $4sp^3$ and therefore has less electron density in the region of orbital-orbital overlap, so it forms a weaker bond.

46. **a.** *Z*   **b.** *E*   **c.** *E*   **d.** *Z*   **e.** *E*   **f.** *E*

47. If the number of carbons is 30, $C_nH_{2n+2} = C_{30}H_{62}$. A compound with molecular formula $C_{30}H_{50}$ is missing 12 hydrogens. Because it has no rings, squalene has 6 π bonds $(12/2 = 6)$.

48. **a.** $-CH{=}CH_2$ > $-CH(CH_3)_2$ > $-CH_2CH_2CH_3$ > $-CH_3$

b. $-OH$ > $-NH_2$ > $-CH_2OH$ > $-CH_2NH_2$

c. $-Cl$ > $-C({=}O)CH_3$ > $-C{\equiv}N$ > $-CH{=}CH_2$

49. **a.**

(*E*)-2-methyl-2,4-hexadiene   (*Z*)-2-methyl-2,4-hexadiene

b.

(*E*)-1,5-heptadiene   (*Z*)-1,5-heptadiene

c.

1,4-pentadiene

50. **a.** $CH_3CH_2\overset{+}{N}H_3 + Br^-$ **b.**

c.

51. Only one name is correct.
  a. 2-pentene
  b. correct
  c. 3-methyl-1-hexene (the parent hydrocarbon is the longest chain that contains the functional group)
  d. 3-heptene
  e. 4-ethylcyclohexene
  f. 2-chloro-3-hexene
  g. 3-methyl-2-pentene
  h. 2-methyl-1-hexene (it does not have *E,Z* isomers)
  i. 1-methylcyclopentene

52. **a.**

c.

b.

d.

53. **a.** 2
  b. B, D, F
  c. E to G (the fastest step has the smallest energy of activation to overcome)
  d. G
  e. A
  f. C
  g. endergonic
  h. exergonic
  i. C
  j. E to G (the largest rate constant corresponds to the lowest energy of activation)
  k. G to E (the smallest rate constant corresponds to the highest energy of activation)

54.   **a.** B will have the larger $\Delta S°$ value because, unlike A, the number of reactants is not the same as the number of products.

b. $\Delta S° =$ (the freedom of motion of the products) $-$ (the freedom of motion of the reactants). Because the three products have a greater freedom of motion than the two reactants, $\Delta S°$ will be positive.

55. $\Delta G° = \Delta H° - T\Delta S°$

a. $\Delta G° = 20 - (298)(0.05)$
$\Delta G° = 20 - 14.9 = 5.1$

$\Delta G° = -RT \ln K_{eq}$
$5.1 = -1.986 \times 10^{-3} \times 298 \times \ln K_{eq}$
$5.1 = -0.59 \ln K_{eq}$
$-8.6 = \ln K_{eq}$
$K_{eq} = 1.8 \times 10^{-4}$

b. $\Delta G° = 20 - (398)(0.05)$
$\Delta G° = 20 - 20 = 0$
$K_{eq} = 1.0$

56.

57.   **a.** $\Delta G° = -RT \ln K_{eq}$
$\Delta G° = -RT \ln 10$
$\Delta G° = -1.986 \times 10^{-3} \times 298 \times 2.3$
$\Delta G° = -1.36 \, \text{kcal/mol}$

b. $\Delta G° = \Delta H° - 0$
$-1.36 = \Delta H° - 0$
$\Delta H° = -1.36 \, \text{kcal/mol}$

c. $\Delta G° = 0 - T\Delta S°$
$-1.36 = 0 - 298 \Delta S°$
$\Delta S° = 1.36/298 = 4.56 \times 10^{-3} \, \text{kcal/(mol deg)}$

58. $\Delta G° = -RT \ln K_{eq}$
$\ln K_{eq} = -\Delta G°/RT$
$\ln K_{eq} = -\Delta G°/0.59 \, \text{kcal/mol}$
$\ln K_{eq} = -5.3/0.59 \, \text{kcal/mol}$
$\ln K_{eq} = -9.0$
$K_{eq} = [B]/[A] = 0.00013$
$[B]/[A] = 0.00013/1 = 0.13/1000$

The calculation shows that for every 1000 molecules in the chair conformation there is 0.13 molecule in a twist-boat conformation.

This agrees with the statement on page 130 of the text that for every 10,000 chair conformers of cyclohexane, there is no more than one twist-boat conformer.

59.   A step-by-step description of how to solve this problem is given in the box entitled "Calculating Kinetic Parameters" on page 153 of the text.

E_a can be determined from the Arrhenius equation ($\ln k = -E_a/RT$), because a plot of $\ln k$ versus $1/T$ gives a slope $= -E_a/R$.

$\ln 2.11 \times 10^{-5} = -10.77$	$T = 304$	$1/T = 3.29 \times 10^{-3}$
$\ln 4.44 \times 10^{-5} = -10.02$	$T = 313$	$1/T = 3.19 \times 10^{-3}$
$\ln 1.16 \times 10^{-4} = -9.06$	$T = 324.5$	$1/T = 3.08 \times 10^{-3}$
$\ln 2.10 \times 10^{-4} = -8.47$	$T = 332.8$	$1/T = 3.00 \times 10^{-3}$
$\ln 4.34 \times 10^{-4} = -7.74$	$T = 342.2$	$1/T = 2.92 \times 10^{-3}$

$$\text{slope} = -8290$$
$$E_a = -(\text{slope})\, R$$
$$E_a = -(-8290) \times 1.98 \times 10^{-3}\ \text{kcal/mol}$$
$$E_a = 16.4\ \text{kcal/mol}$$

To find $\Delta G°$:

$$-\Delta G° = RT \ln kh/Tk_b$$

From the graph used to determine E_a, one can find the rate constant (k) at 30°. (It is $1.84 \times 10^{-5}\ \text{s}^{-1}$.)

$$-\Delta G° = 1.98 \times 10^{-3} \times 303 \ln (1.84 \times 10^{-5} \times 1.58 \times 10^{-31})/(303 \times 3.30 \times 10^{-19})$$
$$-\Delta G° = 1.98 \times 10^{-3} \times 303 \ln (2.90 \times 10^{-36})/(1.00 \times 10^{-16})$$
$$-\Delta G° = 1.98 \times 10^{-3} \times 303 \ln 2.90 \times 10^{-20}$$
$$-\Delta G° = 1.98 \times 10^{-3} \times 303 \times (-3.50)$$
$$\Delta G° = 2.10\ \text{kcal/mol}$$

To find $\Delta H°$:

$$\Delta H° = E_a - RT$$
$$\Delta H° = 16.4 - 1.98 \times 10^{-3} \times 303$$
$$\Delta H° = 16.4 - 0.6$$
$$\Delta H° = 15.8\ \text{kcal/mol}$$

To find $\Delta S°$:

$$\Delta S° = (\Delta H° - \Delta G°)/T$$
$$\Delta S° = (15.8 - 2.10)/303$$
$$\Delta S° = (-13.7)/303$$
$$\Delta S° = 0.045\ \text{kcal/(mol deg)} = 45\ \text{cal/(mol deg)}$$

Chapter 5 Practice Test

1.   Name each of the following compounds:

   a. $CH_3CH_2CHCH_2CH=CH_2$
             |
             CH_3

   c. $CH_3CH_2CH=CHCH_2CH_2CHCH_3$
                                      |
                                    CH_2CH_3

   b.

   d.

2.   Label the following substituents in order from highest priority to lowest priority in the *E,Z* system of nomenclature.

   $$-\overset{\overset{\textstyle O}{\|}}{C}CH_3 \qquad -CH=CH_2 \qquad -Cl \qquad -C\equiv N$$

3.   Correct the incorrect names.

   a. 3-pentene

   b. 2-vinylpentane

   c. 2-ethyl-2-butene

   d. 2-methylcyclohexene

4.   Indicate whether each of the following statements is true or false:

   a. Increasing the energy of activation increases the rate of the reaction.   T   F

   b. Decreasing the entropy of the products compared to the entropy of the reactants makes the equilibrium constant more favorable.   T   F

   c. An exergonic reaction is one with a $-\Delta G°$.   T   F

   d. An alkene is an electrophile.   T   F

   e. The higher the energy of activation, the more slowly the reaction will take place.   T   F

   f. Another name for *trans*-2-butene is (Z)-2-butene.   T   F

   g. A reaction with a negative $\Delta G°$ has an equilibrium constant greater than one.   T   F

   h. Increasing the free energy of the reactants increases the rate of the reaction.   T   F

   i. Increasing the free energy of the products increases the rate of the reaction.   T   F

   j. The magnitude of a rate constant is not dependent on the concentration of the reactants.   T   F

5.   Do the following compounds have the *E* or the *Z* configuration?

     a.

     b.

6.   Draw structures for each of the following:

  a. allyl alcohol   **b.** 3-methylcyclohexene   **c.** *cis*-3-heptene   **d.** vinyl bromide

7.   What is the total number of π bonds and rings in a hydrocarbon with a molecular formula of C_8H_8?

8.   Using curved arrows, show the movement of electrons in the following reaction mechanism:

$$CH_3CH{=}CH_2 + H{-}\ddot{\underset{\cdot\cdot}{C}l}: \;\rightleftharpoons\; CH_3\underset{+}{CH}{-}CH_3 + :\ddot{\underset{\cdot\cdot}{C}}l:^{-} \longrightarrow CH_3CH{-}CH_3$$
$$|$$
$$:\underset{\cdot\cdot}{\overset{\cdot\cdot}{C}}l:$$

9.   A favorable (negative) $\Delta G°$ is given by:
(a positive or negative $\Delta H°$), (a positive or negative $\Delta S°$), and (a high or low temperature).

10.   Which of the following has a more favorable equilibrium constant (that is, which reaction favors products more)?

  a. A reaction with a $\Delta H°$ of 4 kcal/mol or a reaction with a $\Delta H°$ of 7 kcal/mol? (Assume a constant $\Delta S°$ value.)

  b. A reaction with a positive $\Delta S°$ value that takes place at 25 °C or the same reaction that takes place at 35 °C?

  c. A reaction in which two reactants form one product or a reaction in which one reactant forms two products? (Assume a constant $\Delta H°$ value.)

11.   Draw and label the *E* and *Z* stereoisomers of:

  a. 1-bromo-2,3-dimethyl-2-pentene   **b.** 2,3,4-trimethyl-3-hexene

12.   Draw a reaction coordinate diagram for a one-step reaction with a product that is thermodynamically unstable but kinetically stable.

CHAPTER 6
The Reactions of Alkenes: The Stereochemistry of Addition Reactions

Important Terms

acid-catalyzed reaction   a reaction catalyzed by an acid.

aldehyde   a compound with a carbonyl group that is bonded to an alkyl group and to a hydrogen (or bonded to two hydrogens).

anti addition   an addition reaction in which the two added substituents add to opposite sides of the molecule.

biochemistry   the chemistry associated with living organisms.

carbocation rearrangement   the rearrangement of a carbocation to a more stable carbocation.

catalyst   a species that increases the rate at which a reaction occurs without being consumed in the reaction. Because it does not change the equilibrium constant of the reaction, it does not change the amount of product that is formed.

catalytic hydrogenation   the addition of hydrogen to a double or a triple bond with the aid of a metal catalyst.

concerted reaction   a reaction in which all the bond-making and bond-breaking processes take place in a single step.

constitutional isomers (structural isomers)   molecules that have the same molecular formula but differ in the way the atoms are connected.

dimer   a molecule formed by joining together two identical molecules.

electrophilic addition reaction   an addition reaction in which the first species that adds to the reactant is an electrophile.

enzyme   a protein that catalyzes a biological reaction.

epoxide (oxirane)   an ether in which the oxygen is incorporated into a three-membered ring.

halohydrin   an organic molecule that contains a halogen atom and an OH group.

Hammond postulate   states that the transition state will be more similar in structure to the species (reactants or products) that it is closer to energetically.

heat of hydrogenation   the heat $(\Delta H°)$ released in a hydrogenation reaction.

heterogeneous catalyst	a catalyst that is insoluble in the reaction mixture.
hydration	addition of water to a compound.
1,2-hydride shift	the movement of a hydride ion (a hydrogen with its pair of bonding electrons) from one carbon to an adjacent carbon.
hydroboration-oxidation	the addition of borane (or R_2BH) to a double or triple bond followed by reaction with hydrogen peroxide and hydroxide ion.
hydrogenation	addition of hydrogen.
hyperconjugation	delocalization of electrons by overlap of carbon-hydrogen or carbon-carbon σ bonds with an empty orbital on an adjacent carbon.
ketone	a compound with a carbonyl group that is bonded to two alkyl groups.

Markovnikov's rule	the actual rule is, "when a hydrogen halide adds to an asymmetrical alkene, the addition occurs such that the halogen attaches itself to the carbon atom of the alkene bearing the least number of hydrogen atoms." Chemists use the rule as follows: the hydrogen adds to the sp^2 carbon that is bonded to the most hydrogens.
	A more general rule is, the electrophile adds to the sp^2 carbon that is bonded to the most hydrogens.
mechanism of the reaction	a description of the step-by-step process by which reactants are changed into products.
1,2-methyl shift	the movement of a methyl group with its bonding electrons from one carbon to an adjacent carbon.
oxidation reaction	a reaction that decreases the number of C—H bonds in the reactant or increases the number of C—O, C—N, or C—X bonds (X denotes a halogen).
oxidative cleavage	an oxidation reaction that cleaves the reactant into two or more compounds.
ozonolysis	the reaction of an alkene with ozone.
primary carbocation	a carbocation with the positive charge on a primary carbon.
receptor	a protein or other compound that binds a particular molecule.
reduction reaction	a reaction that increases the number of C—H bonds in the reactant or decreases the number of C—O, C—N, or C—X bonds (X denotes a halogen).
regioselective reaction	a reaction that leads to the preferential formation of one constitutional isomer over another.

secondary carbocation   a carbocation with the positive charge on a secondary carbon.

stereocenter   an atom at which the interchange of two groups produces a stereoisomer.
(stereogenic center)

stereochemistry   the field of chemistry that deals with the structure of molecules in three dimensions.

stereoisomers   isomers that differ in the way the atoms are arranged in space.

stereoselective reaction   a reaction that leads to the preferential formation of one stereoisomer over another.

stereospecific reaction   a reaction in which the reactant can exist as stereoisomers and each stereoisomeric reactant leads to a different stereoisomeric product.

steric effect   an effect due to the space occupied by a substituent.

steric hindrance   a hindrance due to bulky groups at the site of a reaction that make it difficult for the reactants to approach one another.

syn addition   an addition reaction in which the two added substituents add to the same side of the molecule.

tertiary carbocation   a carbocation with the positive charge on a tertiary carbon.

vicinal   vicinal substituents are substituents on adjacent carbons.

Solutions to Problems

1.

2.   **a.**   The σ bond orbitals of the carbon adjacent to the positively charged carbon are available for overlap with the vacant p orbital. Because the methyl cation does not have a carbon adjacent to the positively charged carbon, there are no σ bond orbitals available for overlap with the vacant p orbital.

   b.   An ethyl cation is more stable because the carbon adjacent to the positively charged carbon has three σ bond orbitals available for overlap with the vacant p orbital, whereas a methyl cation does not have any σ bond orbitals available for overlap with the vacant p orbital.

3.   **a.**

   b.   *sec*-butyl cation

4.   **a.**

   b.   A halogen atom decreases the stability of the carbocation because, since it is an electronegative atom, it withdraws electrons away from the positively charged carbon. This increases the concentration of positive charge on the carbocation which makes it less stable.

   Because fluorine is more electronegative than chlorine and therefore withdraws electrons more strongly, the fluorine-substituted carbocation is less stable than the chlorine-substituted carbocation.

5.   The transition state will resemble the one (reactants or products) that it is closer to on the reaction coordinate diagram; that is, the one that it is closer to in energy.

   a. products   **b.** reactants   **c.** reactants   **d.** products

6.   **a.** $CH_3CH_2CHCH_3$ with Br **c.**   **e.**

   b. $CH_3CH_2CCH_3$ with CH_3 and Br **d.** $CH_3CCH_2CH_2CH_3$ with CH_3 and Br **f.** $CH_3CH_2CHCH_3$ with Br

7.   **a.** CH_3
            $CH_2\!=\!\overset{|}{C}CH_3$

c. CH_3
            $\overset{|}{C}\!=\!CH_2$ (attached to cyclohexyl)

b.   (cyclohexyl)$-CH_2CH\!=\!CH_2$

d.   (cyclohexyl)$=CHCH_3$ or   (cyclohexenyl)$-CH_2CH_3$

(cyclohexyl)$-CH\!=\!CHCH_3$

This would not be a good choice, because it would lead to a mixture of two products.

8.   **a.** CH_3
            $CH_3CH_2\overset{|}{C}\!=\!CH_2$

In both **a** and **b**, the compound that is more highly regioselective is the one where the choice is between forming a tertiary carbocation and a primary carbocation.

b. CH_2

In both **a** and **b**, the less regioselective compound is the one where the choice is between forming a tertiary carbocation and a secondary carbocation, because the difference in the stability of the two possible carbocations, and therefore the difference in the amount of product formed, is not as great as it is when the choice is between a tertiary and a primary carbocation.

9.   As long as the pH is greater than about -2.5 and less than about 15, more than 50% of 2-propanol would be in its neutral, nonprotonated form.

$$\overset{+}{R}OH_2 \xrightleftharpoons[\;]{pK_a=-2.5} ROH \xrightleftharpoons[\;]{pK_a=\sim15} RO^-$$

Recall that when the pH $=$ pK_a, half the compound is in its acidic form and half is in its basic form. Therefore at a pH less than about -2.5, more than half of the compound will be in its positively charged protonated form. At a pH greater than about ~15, more than half of the compound will exist as the negatively charged anion.

Therefore, at a pH between -2.5 and ~15, more than half of the compound will exist in the neutral nonprotonated form.

10.   **a.**   three transition states   **b.**   two intermediates

c.   The first step is the *slowest step*, so it has the *smallest rate constant*; the second step is fast because no bonds are being broken; the third step is fast because transfer of a proton from or to an O or an N is always a fast reaction.

11.   **a.** $CH_3CH_2CH_2\overset{|}{\underset{OH}{C}}HCH_3$

c. $CH_3CH_2CH_2CH_2\overset{|}{\underset{OH}{C}}HCH_3$ and $CH_3CH_2CH_2\overset{|}{\underset{OH}{C}}HCH_2CH_3$

formed in equal amounts

b.   (cyclohexyl with OH)

d.   (cyclohexyl with CH_3 and OH)

12. **a. 1.** CH₃ĊCH₃ with CH₃ and Cl substituents

$$\text{1. } \underset{\underset{Cl}{|}}{\overset{\overset{CH_3}{|}}{CH_3CCH_3}} \qquad \text{2. } \underset{\underset{Br}{|}}{\overset{\overset{CH_3}{|}}{CH_3CCH_3}} \qquad \text{3. } \underset{\underset{OH}{|}}{\overset{\overset{CH_3}{|}}{CH_3CCH_3}} \qquad \text{4. } \underset{\underset{OCH_3}{|}}{\overset{\overset{CH_3}{|}}{CH_3CCH_3}}$$

b. 1. The first step in all the reactions is addition of an electrophilic proton (H^+) to the carbon of the CH_2 group.

  2. A *tert*-butyl carbocation is formed as an intermediate in each of the reactions.

c. 1. The nucleophile that adds to the *tert*-butyl carbocation is different in each reaction.

  2. In reactions #3 and #4, there is a third step—a proton is lost from the group that acted as the nucleophile in the second step of the reaction.

3. $$\underset{\underset{\overset{+}{O}H}{|}}{\overset{\overset{CH_3}{|}}{CH_3CCH_3}} \longrightarrow \underset{\underset{OH}{|}}{\overset{\overset{CH_3}{|}}{CH_3CCH_3}} + H_3O^+$$
    $H_2\ddot{O} \quad H$

4. $$\underset{\underset{\overset{+}{O}CH_3}{|}}{\overset{\overset{CH_3}{|}}{CH_3CCH_3}} \longrightarrow \underset{\underset{OCH_3}{|}}{\overset{\overset{CH_3}{|}}{CH_3CCH_3}} + CH_3\overset{+}{O}H_2$$
    $CH_3\ddot{O}H \quad H$

13. Solved in the text.

14. a. cyclohexene $+ CH_3OH \xrightarrow{H_2SO_4}$ cyclohexane—OCH_3 **b.** $CH_2=\underset{\underset{}{\overset{\overset{CH_3}{|}}{C}}}CH_3 + CH_3OH \xrightarrow{H_2SO_4} \underset{\underset{CH_3}{|}}{\overset{\overset{CH_3}{|}}{CH_3OCCH_3}}$

c. cyclopentene $+ H_2O \xrightarrow{H_2SO_4}$ cyclopentanol (—OH)

d. $CH_3CH=CHCH_3 + H_2O \xrightarrow{H_2SO_4} \underset{\underset{OH}{|}}{CH_3CHCH_2CH_3}$

  or

  $CH_2=CHCH_2CH_3 + H_2O \xrightarrow{H_2SO_4} \underset{\underset{OH}{|}}{CH_3CHCH_2CH_3}$

15. $ROH + H_2SO_4 \rightleftharpoons \underset{\underset{H}{|}}{R\overset{+}{O}H} + HSO_4^-$

$$CH_3\underset{\underset{CH_3}{|}}{C}=CH_2 + R\overset{+}{O}H \rightleftharpoons \underset{\underset{CH_3}{|}}{\overset{\overset{+}{}}{CH_3CCH_3}} + R\ddot{O}H \rightleftharpoons \underset{\underset{CH_3}{|}}{\overset{\overset{+OR}{|}}{CH_3CCH_3}} \quad R\ddot{O}H$$
    H

$R=\underset{\underset{CH_3}{|}}{CH_3CHCH_2CH_2OH}$

  OR

$$\underset{\underset{CH_3}{|}}{\overset{\overset{OR}{|}}{CH_3CCH_3}} + R\overset{+}{\underset{\underset{H}{|}}{O}H}$$

16.   Solved in the text.

17.   **a.**   CH₃CHCH=CH₂ $\xrightarrow{\text{HBr}}$ CH₃CHCHCH₃ $\xrightarrow[\text{shift}]{\text{1,2-hydride}}$ CH₃CCH₂CH₃ $\xrightarrow{\text{Br}^-}$ CH₃CCH₂CH₃

b.   CH₃CHCH₂CH=CH₂ $\xrightarrow{\text{HBr}}$ CH₃CHCH₂CHCH₃ $\xrightarrow{\text{Br}^-}$ CH₃CHCH₂CHCH₃

c.

d.

e.   CH₂=CHCCH₃ $\xrightarrow{\text{HBr}}$ CH₃CHCCH₃ $\xrightarrow[\text{shift}]{\text{1,2-methyl}}$ CH₃CHCCH₃ $\xrightarrow{\text{Br}^-}$ CH₃CH—CCH₃

f.

1-Bromo-3-methylcyclohexane and 1-bromo-4-methylcyclohexane will be obtained in approximately equal amounts because, in each case, the intermediate is a secondary carbocation, so the two compounds will be formed at about the same rate. A carbocation rearrangement will not occur because it would just form another secondary carbocation.

18.   The reaction with 9-BBN is more highly regioselective. 9-BBN is sterically hindered, so it will be more likely than BH₃ to add to the least sterically hindered carbon.

19. **a.**

$$CH_3\underset{\displaystyle CH_3}{C}=CHCH_3 \quad \xrightarrow[\text{2. HO}^-,\ H_2O_2,\ H_2O]{\text{1. BH}_3/\text{THF}} \quad CH_3\underset{\displaystyle OH}{\overset{\displaystyle CH_3}{CH}}CHCH_3$$

b.

20.

Addition of H^+ would form a carbocation that could rearrange.

Addition of Br^+ forms a cyclic bromonium ion rather than a carbocation, so there is no rearrangement.

21. **a.** The first step in the reaction of propene with Br_2 forms a cyclic bromonium ion, whereas the first step in the reaction of propene with HBr forms a carbocation.

b. If the bromide ion were to attack the positively charged bromine, a highly unstable compound (with a negative charge on carbon and a positive charge on bromine) would be formed.

Notice that the electrostatic potential map of the cyclic bromonium ion on page 257 of the text shows that the ring carbons are the least electron dense (most blue) atoms in the intermediate and therefore the ones most susceptible to nucleophilic attack.

22. Sodium and potassium achieve an outer shell of eight electrons by losing the single electron they have in the $3s$ (in the case of Na) or $4s$ (in the case of K) orbital, thereby becoming Na^+ and K^+.

In order to form a covalent bond, they would have to regain electrons in these orbitals, thereby losing the stability associated with having an outer shell of eight electrons and no extra electrons.

23. The nucleophile that is present in greater concentration is more apt to collide with the intermediate. Therefore, if the solvent is a nucleophile, the major product will come from reaction of the solvent with the intermediate, because the concentration of the solvent is much greater than the concentration of the other nucleophile. (For example, in part **a** the concentration of CH_3OH is much greater than the concentration of Cl^-. In part **b**, the nucleophile is most likely to be I^- because there are two equivalents of NaI and one equivalent of HBr.)

24.   Because chlorine is more electronegative than iodine, iodine will be the electrophile. Therefore, it will end up attached to the sp^2 carbon that is bonded to the most hydrogens.

25.   **a.** $CH_2CHCH_2CH_3$ **b.** $CH_2CHCH_2CH_3$ **c.** $CH_2CHCH_2CH_3$ **d.** $CH_2CHCH_2CH_3$
               Br Br   Br OH   Br OCH$_2$CH$_3$ Br OCH$_3$

26.   Look at the reagent (Cl_2) and remember what that reagent does when it reacts with an alkene (do not let the rest of the molecule confuse you): when Cl_2 adds to an alkene, it forms a cyclic chloronium ion intermediate. You know that the intermediate then reacts with a nucleophile. There are two nucleophiles in the solution that can react with the intermediate, a Cl^- and the OH group at the end of the molecule. There is a greater probability that the OH group will be the nucleophile that attacks the chloronium ion because, since it is attached to the reactant, it does not have to wander through the solution to find the chloronium ion as the Cl^- has to. Loss of a proton gives the six-membered ring ether.

27.   **a.**

     c.

     b.

     d.

28.   **a.** cyclohexene   **b.** 1-butene

29.   **a.**

     d.

     b.

     e.

     c.

     f.

30. **a.**

2,3-dimethyl-2-butene

b.

cis-4-octene   *trans*-4-octene

31.   It does not tell you whether the double bond has the *E* or the *Z* configuration.

32.   Solved in the text.

33.   Solved in the text.

34. **a.** $CH_2{=}CHCH_2CH_2CH_3$ or $CH_3CH{=}CHCH_2CH_3$

b.

35. **a.** three alkenes: 1-butene, *cis*-2-butene, *trans*-2-butene
   b. four alkenes: 3-methyl-1-pentene, (*E*)-3-methyl-2-pentene, (*Z*)-3-methyl-2-pentene, 2-ethyl-1-butene
   c. five alkenes: 1-hexene, *cis*-2-hexene, *trans*-2-hexene, *cis*-3-hexene, *trans*-3-hexene

36.   Because alkene **A** has the smaller heat of hydrogenation, it is more stable.

37. **a.**

This alkene is the most stable because it has the greatest number of alkyl substituents bonded to the sp^2 carbons.

b.

This alkene is the least stable because it has the fewest number of alkyl substituents bonded to the sp^2 carbons.

c.

This alkene has the smallest heat of hydrogenation because it is the most stable of the three alkenes.

38.

| 4 alkyl substituents | 2 trans alkyl substituents | 2 cis alkyl substituents | 2 cis alkyl substituents that cause greater steric strain than those in *cis*-3-hexene |

39.   The reactant must have *E,Z* stereoisomers or *R,S* stereoisomers.

40.   **a.**   No, because only one constitutional isomer can be formed as a product since 2-butene is a symmetrical alkene.

$$CH_3CH_2CHCH_3$$
$$|$$
$$Br$$

   b.   No, because it forms a racemic mixture.
   c.   No, because *cis*-butene and *trans*-butene form the same product.
   d.   Yes, because two constitutional isomers are possible, but only one is formed.
   e.   No, because it forms a racemic mixture.
   f.   No, because *cis*-butene and *trans*-butene form the same product.

41.   Only the stereoisomers of the major product of each reaction are shown.

   a.

   b.

   c.

   This compound does not have any stereoisomers, because it does not have any asymmetric centers.

   d.

   This compound does not have any stereoisomers, because it does not have an asymmetric center.

42.   Solved in the text.

43.   Solved in the text.

44.   Solved in the text.

45.   **a.** **1.** *trans*-3-heptene   **2.** *cis*-3-heptene

b. The enantiomer of each of the structures shown will also be formed, because the peroxyacid can approach both the top and the bottom of the plane defined by the double bond.

46.   **a.**

b.

c.

d.

47.

(* indicates an asymmetric center)

Each of the reactions forms a compound with one asymmetric center from a compound with no asymmetric centers. Therefore, each of the products will be a racemic mixture.

48.   Solved in the text.

49.   **a.**

b. CH₃CH₂CCH₂CH₃ (with CH₃ above and Br below the third carbon)   This compound does not have an asymmetric center.

c. Same as **b**.

d.

e.

f.

50. Two different bromonium ions are formed because Br₂ can add to the double bond either from the top of the plane or from the bottom of the plane defined by the double bond, and the two bromonium ions are formed in equal amounts. Attacking the less hindered carbon of one bromonium ion forms one stereoisomer, whereas attacking the less hindered carbon of the other bromonium ion forms the other stereoisomer. Because Br⁻ can attack the least sterically hindered carbon with equal ease from pathway **a** as from pathway **b**, equal amounts of the threo enantiomers will be obtained. Of course, some reaction will occur at the more hindered end of the bromonium ion, but it will occur to the same extent in both pathways.

51. **a.**

and

The addition of Br and OH are anti, so in a cyclic compound these two substituents are trans to one another.

b. __mechanism of the reaction__

52. **a.** Only anti addition occurs. Because the reactant is trans, the product would be expected to be the erythro pair of enantiomers. However, in this case, each asymmetric carbon is attached to the same four groups, so the product is a meso compound. Therefore, only one stereoisomer is obtained.

b. Only one asymmetric center is created in the product, so the product is a racemic mixture.

c. Only syn addition occurs. Because the reactant is trans, the product is the threo pair of enantiomers.

d. The product of the reactions does not have any asymmetric centers, so it does not have any stereoisomers.

$$CH_3CH_2CHCH_2CH_3$$
$$|$$
$$CH_3$$

e. Only anti addition occurs. Because the reactant is cis (a cyclopentene ring cannot exist in a trans configuration), the product is the pair of enantiomers with the bromines on opposite sides of the ring.

f. Only anti addition occurs. Because the reactant is cis, the product is the pair of enantiomers with the bromines on opposite sides of the ring.

g. Only syn addition occurs. Because the reactant is cis, the product would be the pair of enantiomers with the hydrogens on the same side of the ring, but in this case the product is a meso compound, so only one stereoisomer is obtained.

h. Only syn addition occurs. Because the reactant is cis, the product is the pair of enantiomers with the hydrogens on the same side of the ring.

53. a. CH₃CHCH₂Br
         |
         Cl

     b. The R and S enantiomers will be formed in equal amounts (a racemic mixture).

54. Greater than 98% is excess of the S enantiomer; the remainder is a racemic mixture. Therefore, greater than 99% is the S enantiomer.

55. a. A racemic mixture of (R)-malate and (S)-malate would be obtained. (A product with one asymmetric center would be formed from a reactant with no asymmetric centers.)

     b. A racemic mixture of (R)-malate and (S)-malate would again be obtained. In the absence of an enzyme, the reactions are neither stereoselective (part a) nor stereospecific (part b).

56.   Solved in the text.

57.   Less of the desired product would be formed from 3-methylcyclohexene than from 1-methylcyclohexene. Hydroboration-oxidation of 3-methylcyclohexene would form both 2-methylcyclohexanol (the desired product) and 3-methylcyclohexanol, because borane could add to either the 1-position or the 2-position of the alkene, since the two transition states would have approximately the same stability.

58.   **a.** $CH_3CH{=}CH_2 \xrightarrow[CH_3OH]{H_2SO_4} CH_3CHCH_3$ with OCH_3

b. $CH_3CH_2CH{=}CHCH_3 + Br_2 \longrightarrow CH_3CH_2CHCHCH_3$ (Br, Br)

c.

d.

e.

f.

59.   **a.** **b.** $CH_3CCH_2CH_3$ (CH_3, Br) **c.** **d.** $CH_3CH_2C{-}CHCH_3$ (CH_3; Br, CH_3)

electrophile   nucleophile

60.   **a.** $CH_3\overset{+}{C}HCH_3 + :\ddot{Cl}:^- \longrightarrow CH_3CHCH_3$ with $:\ddot{Cl}:$

b.   nucleophile   electrophile

$$CH_3CH=CH_2 \ + \ H-Br \ \longrightarrow \ CH_3\overset{+}{C}H-CH_3 \ + \ \overset{-}{Br}$$

c.   electrophile
nucleophile

$$CH_3CH=CH_2 \ \longrightarrow \ CH_3CH_2-CH_2BH_2$$

$$+ \ \ H-BH_2$$

nucleophile electrophile

61. **a.** $CH_3\underset{\underset{\displaystyle CH_3}{|}}{C}=CHCH_3 \ + \ HBr \ \longrightarrow \ CH_3\underset{\underset{\displaystyle Br}{|}}{\overset{\overset{\displaystyle CH_3}{|}}{C}}-CH_2CH_3$

b. $CH_3\underset{\underset{\displaystyle CH_3}{|}}{C}=CHCH_3 \ + \ HI \ \longrightarrow \ CH_3\underset{\underset{\displaystyle I}{|}}{\overset{\overset{\displaystyle CH_3}{|}}{C}}-CH_2CH_3$

c. $CH_3\underset{\underset{\displaystyle CH_3}{|}}{C}=CHCH_3 \ + \ Cl_2 \ \longrightarrow \ CH_3\underset{\underset{\displaystyle Cl}{|}}{\overset{\overset{\displaystyle CH_3}{|}}{C}}-\underset{\underset{\displaystyle Cl}{|}}{C}HCH_3$

d. $CH_3\underset{\underset{\displaystyle CH_3}{|}}{C}=CHCH_3 \ \xrightarrow[\text{2. }(CH_3)_2S]{\text{1. }O_3,\ -78\ °C}$ [acetone structure: $CH_3-\overset{\overset{\displaystyle O}{||}}{C}-CH_3$] $+$ [acetaldehyde structure: $CH_3-\overset{\overset{\displaystyle O}{||}}{C}-H$]

e. $CH_3\underset{\underset{\displaystyle CH_3}{|}}{C}=CHCH_3 \ \xrightarrow[\text{Pd/C}]{H_2} \ CH_3\underset{\underset{\displaystyle CH_3}{|}}{C}H-CH_2CH_3$

f. $CH_3\underset{\underset{\displaystyle CH_3}{|}}{C}=CHCH_3 \ + \ \overset{\overset{\displaystyle O}{||}}{RC}OOH \ \longrightarrow$ [epoxide: $\underset{H_3C}{\overset{H_3C}{}}\!>\!\!\overset{O}{\triangle}\!\!<\!\!\underset{CH_3}{}$]

g. $CH_3\underset{\underset{\displaystyle CH_3}{|}}{C}=CHCH_3 \ + \ H_2O \ \xrightarrow{H_2SO_4} \ CH_3\underset{\underset{\displaystyle OH}{|}}{\overset{\overset{\displaystyle CH_3}{|}}{C}}-CH_2CH_3$

h. $CH_3\underset{\underset{\displaystyle CH_3}{|}}{C}=CHCH_3 \ + \ Br_2 \ \xrightarrow{CH_2Cl_2} \ CH_3\underset{\underset{\displaystyle Br}{|}}{\overset{\overset{\displaystyle CH_3}{|}}{C}}-\underset{\underset{\displaystyle Br}{|}}{C}HCH_3$

i. $CH_3\underset{\underset{\displaystyle CH_3}{|}}{C}=CHCH_3 \ + \ Br_2 \ \xrightarrow{H_2O} \ CH_3\underset{\underset{\displaystyle HO}{|}}{\overset{\overset{\displaystyle CH_3}{|}}{C}}-\underset{\underset{\displaystyle Br}{|}}{C}HCH_3$

j.

k.

62. **a.** 2,2-diethyl-3-isopropyloxirane or 3,4-epoxy-4-ethyl-2-methylhexane
  b. 3-ethyl-2,2-dimethyloxirane or 2,3-epoxy-2-methylpentane

63. First draw the structures so you can see the number of alkyl groups bonded to the sp^2 carbons:

3,4-dimethyl-2-hexene   2,3-dimethyl-2-hexene   4,5-dimethyl-2-hexene

  a. 2,3-Dimethyl-2-hexene is the most stable of the three alkenes because it has the greatest number of alkyl substituents bonded to the sp^2 carbons.
  b. 4,5-Dimethyl-2-hexene has the fewest alkyl substituents bonded to the sp^2 carbons making it the least stable of the three alkenes. It, therefore, has the greatest heat of hydrogenation.
  c. Because it is the most stable, 2,3-dimethyl-2-hexene has the smallest heat of hydrogenation.

64. R_2BH is the electrophile and H^- is the nucleophile

forms a carbocation intermediate, so a carbocation rearrrangement can occur

65. **a.** It is an oxidation reaction because two new C—Cl bonds are formed.
  b. It is neither an oxidation reaction nor a reduction reaction because both a C—O and a C—H bond are formed.

66. **a.**

oxidation

  c.

oxidation

  b.

reduction

67.

2-bromo-3-methylbutane   2-bromo-2-methylbutane

68.   a.

This compound is the most stable.
It has four alkyl substituents
bonded to the sp^2 carbons.

b. $CH_2\!\!=\!\!CHCH_2CH_2CH_2CH_3$ $CH_2\!\!=\!\!CHCH_2CHCH_3$

These compounds are the least stable.
Each has only one alkyl substituent bonded to the sp^2 carbons.

69.   a.

b. $CH_3C\!\!\equiv\!\!\overset{..}{C}\!-\!H \ + \ ^-\!\!:\!\!\overset{..}{N}\!H_2 \longrightarrow CH_3C\!\!\equiv\!\!\overset{..}{C}\!:^- \ + \ :NH_3$

c. $CH_3CH_2\!-\!Br \ + \ CH_3\overset{..}{\underset{..}{O}}\!:^- \longrightarrow CH_3CH_2\!-\!\overset{..}{\underset{..}{O}}CH_3 \ + \ Br^-$

70.

71. **a.** [cyclohexane with methyl and Cl substituents]

c. [epoxide on cyclohexane ring with methyl]

e. [cyclohexane with methyl, OH, and Cl]

g. [cyclohexane with methyl and OCH₃]

b. [cyclohexane with methyl, OCH₃, and Br]

d. [cyclohexane with methyl and OH]

f. [open chain ketone-aldehyde]

h. [cyclohexane with methyl and two Cl]

72. **a.** [cyclohexene] $\xrightarrow[\text{Pd/C}]{\text{H}_2}$

d. $CH_3CH_2CH=CHCH_2CH_3 \xrightarrow[\text{H}_2O]{\text{Br}_2}$

b. $CH_3CH_2CH_2CH=CH_2 \xrightarrow{\text{HCl}}$

e. [cyclohexane with $CH_2CH=CH_2$] $\xrightarrow[\text{H}_2O]{\text{H}_2SO_4}$

c. [methylenecyclohexane] $\xrightarrow[\text{2. H}_2O_2,\ \text{HO}^-,\ \text{H}_2O]{\text{1. R}_2BH/\text{THF}}$

f. $CH_3CH_2CH=CHCH_2CH_3 \xrightarrow[\substack{\text{NaBr} \\ \text{excess}}]{\text{Cl}_2}$

or

$CH_3CH_2CH=CHCH_2CH_3 \xrightarrow[\substack{\text{NaCl} \\ \text{excess}}]{\text{Br}_2}$

73. **a.** [methylenecyclohexane] and [1-methylcyclohexene]

b. (Note: D stands for deuterium, an isotope of hydrogen; DBr reacts in a manner very similar to HBr.) While HBr forms the same product when it reacts with the two alkenes, DBr would form different products. They are shown here.

[cyclohexane with CH₂D and Br] and [cyclohexane with D, CH₃, and Br]

74. **a.**

$$CH_3$$
$$CH_3\overset{+}{C}CH_3$$

tertiary is more stable
than secondary

d.

has the more substituted
double bond

b.

$$CH_3\overset{+}{C}HCH_3$$

the electron with drawing
chlorine destabilizes the carbocation
by increasing the amount of positive
charge on the carbon

e.

$$CH_3$$
$$CH_3C=CHCH_2CH_3$$

has the more substituted
double bond

c.

tertiary is more stable
than secondary

f.

has the more substituted
double bond

75. **a.** **1.** Both *cis-* and *trans-*2-butene give these products; in each case a product with one asymmetric center is formed, so the product is a racemic mixture.

2. Both *cis-* and *trans-*2-butene give these products; in each case a product with one asymmetric center is formed, so the product is a racemic mixture.

3. *cis-*2-Butene forms a meso compound; the product has two asymmetric centers and only syn addition occurs.

*trans-*2-Butene forms a pair of enantiomers; the product has two asymmetric centers and only syn addition occurs.

4. *cis*-2-Butene forms the threo pair of enantiomers; a product with two asymmetric centers is formed and only anti addition of Br_2 occurs.

$$
\begin{array}{cccc}
\underset{S\quad S}{\overset{H_3C\quad H}{\underset{Br\quad CH_3}{\overset{H}{}C-C\overset{}{Br}}}} & + & \underset{R\quad R}{\overset{H\quad CH_3}{\underset{CH_3\quad Br}{BrC-CH}}} & \textbf{or}
\end{array}
$$

$$
\begin{array}{cc}
\begin{array}{c} CH_3 \\ H-\!\!\!-Br\;\;S \\ Br-\!\!\!-H\;\;S \\ CH_3 \end{array} &
\begin{array}{c} CH_3 \\ Br-\!\!\!-H\;\;R \\ H-\!\!\!-Br\;\;R \\ CH_3 \end{array}
\end{array}
$$

trans-2-Butene forms a meso compound; a product with two asymmetric centers is formed and only anti addition of Br_2 occurs.

$$
\underset{S\quad R}{\overset{H_3C\quad Br}{\underset{Br\quad CH_3}{HC-CH}}} \qquad \textbf{or} \qquad
\begin{array}{c} CH_3 \\ H-\!\!\!-Br\;\;S \\ H-\!\!\!-Br\;\;R \\ CH_3 \end{array}
$$

5. *cis*-2-Butene forms the threo pair of enantiomers; a product with two asymmetric centers is formed and only anti addition of Br and OH occurs.

$$
\begin{array}{cccc}
\underset{S\quad S}{\overset{H_3C\quad H}{\underset{HO\quad CH_3}{HC-CBr}}} & &
\underset{R\quad R}{\overset{H\quad CH_3}{\underset{CH_3\quad OH}{BrC-CH}}} & \textbf{or}
\end{array}
$$

$$
\begin{array}{cc}
\begin{array}{c} CH_3 \\ H-\!\!\!-OH\;\;S \\ Br-\!\!\!-H\;\;S \\ CH_3 \end{array} &
\begin{array}{c} CH_3 \\ HO-\!\!\!-H\;\;R \\ H-\!\!\!-Br\;\;R \\ CH_3 \end{array}
\end{array}
$$

trans-2-Butene forms the erythro pair of enantiomers; a product with two asymmetric centers is formed and only anti addition of Br and OH occurs.

$$
\begin{array}{cccc}
\underset{R\quad S}{\overset{H_3C\quad H}{\underset{H\quad CH_3}{HOC-CBr}}} & &
\underset{R\quad S}{\overset{H\quad CH_3}{\underset{CH_3\quad H}{BrC-COH}}} & \textbf{or}
\end{array}
$$

$$
\begin{array}{cc}
\begin{array}{c} CH_3 \\ HO-\!\!\!-H\;\;R \\ Br-\!\!\!-H\;\;S \\ CH_3 \end{array} &
\begin{array}{c} CH_3 \\ H-\!\!\!-OH\;\;S \\ H-\!\!\!-Br\;\;R \\ CH_3 \end{array}
\end{array}
$$

6. Both *cis*- and *trans*-2-butene form this product; a product with no asymmetric centers is formed.

$$CH_3CH_2CH_2CH_3$$

7. Both *cis*- and *trans*-2-butene form these products; a product with one asymmetric center is formed.

$$
\underset{S}{\overset{CH_2CH_3}{\underset{CH_3\quad\;\; H}{C\;\;}}\atop OH} \quad + \quad
\underset{R}{\overset{CH_2CH_3}{\underset{HO\quad CH_3}{H\;\;C}}} \quad \textbf{or} \quad
\underset{S}{\begin{array}{c} CH_3 \\ H-\!\!\!-OH \\ CH_2CH_3 \end{array}} \quad + \quad
\underset{R}{\begin{array}{c} CH_3 \\ HO-\!\!\!-H \\ CH_2CH_3 \end{array}}
$$

8. Both *cis*- and *trans*-2-butene form these products; a product with one asymmetric center is formed.

$$
\underset{S}{\overset{CH_2CH_3}{\underset{CH_3\quad\;\; H}{C\;\;}}\atop OCH_3} \quad + \quad
\underset{R}{\overset{CH_2CH_3}{\underset{H\quad CH_3}{CH_3O\;\;C}}} \quad \textbf{or} \quad
\underset{S}{\begin{array}{c} CH_3 \\ H-\!\!\!-OCH_3 \\ CH_2CH_3 \end{array}} \quad + \quad
\underset{R}{\begin{array}{c} CH_3 \\ CH_3O-\!\!\!-H \\ CH_2CH_3 \end{array}}
$$

b. For the *cis*- and *trans*-2-butene to form different products, (1) the reaction must form two new asymmetric centers in the product and (2) either syn or anti addition must occur (but not both). Therefore, the cis and trans alkenes form different products when they react with a peroxyacid, with Br_2, and with Br_2 in H_2O.

76.   **a.** To determine relative rates, the rate constant of each alkene is divided by the smallest rate constant of the series (3.51×10^{-8}).

relative rates

propene $= (4.95 \times 10^{-8})/(3.51 \times 10^{-8}) = 1.41$

(*Z*)-2-butene $= (8.32 \times 10^{-8})/(3.51 \times 10^{-8}) = 2.37$

(*E*)-2-butene $= (3.51 \times 10^{-8})/(3.51 \times 10^{-8}) = 1$

2-methyl-2-butene $= (2.15 \times 10^{-4})/(3.51 \times 10^{-8}) = 6.12 \times 10^3$

2,3-dimethyl-2-butene $= (3.42 \times 10^{-4})/(3.51 \times 10^{-8}) = 9.74 \times 10^3$

b. Both compounds form the same carbocation but, since (*Z*)-2-butene is less stable than (*E*)-2-butene, (*Z*)-2-butene has a smaller free energy of activation.

c. 2-Methyl-2-butene is more stable than (*Z*)-2-butene and it also forms a more stable carbocation intermediate (tertiary) and therefore a more stable transition state than does (*Z*)-2-butene (secondary). Knowing that 2-methyl-2-butene reacts faster tells us that the energy difference between the transition states is greater than the energy difference between the alkenes. This is what we would expect from the Hammond postulate, since the transition states look more like the carbocations than like the alkenes.

d. 2,3-Dimethyl-2-butene is more stable than 2-methylbutene, and both compounds form a tertiary carbocation intermediate. On this basis, you would predict that 2,3-dimethyl-2-butene would react more slowly than 2-methylbutane. However, 2,3-dimethyl-2-butene has two sp^2 carbons that can react with a proton to form the tertiary carbocation, whereas 2-methyl-2-butene has only one sp^2 carbon that can react with a proton to form the tertiary carbocation. The fact that 2,3-dimethyl-2-butene reacts faster in spite of being more stable tells us that the more important factor is the greater number of collisions with the proper orientation that lead to a productive reaction in the case of 2,3-dimethyl-2-butene.

77.   **a.**

has a greater dipole moment than

where the C—Cl dipoles oppose each other

b.

has a greater dipole moment than

Recall that an sp^2 carbon is more electronegative than an sp^3 carbon. Thus, the C—CH_3 and C—Cl dipoles reinforce each other, leading to a higher dipole, in the compound on the left and oppose each other in the compound on the right.

78.   **a.**

b.

$$CH_3CH_2CH_2 \overset{CH_3}{\underset{OH}{\overset{|}{C}}} {}^{\prime\prime\prime}H \quad + \quad H^{\prime\prime\prime}\overset{CH_3}{\underset{HO}{\overset{|}{C}}} - CH_2CH_2CH_3$$

c.

d.

$$CH_3CH_2CH_2 \overset{CH_2CH_3}{\underset{Br}{\overset{|}{C}}} {}^{\prime\prime\prime}CH_3 \quad + \quad CH_3 {}^{\prime\prime\prime}\overset{CH_2CH_3}{\underset{Br}{\overset{|}{C}}} - CH_2CH_2CH_3$$

79. No, he should not follow his friend's advice. Adding the electrophile to the sp^2 carbon bonded to the most hydrogens forms a secondary carbocation in preference to a primary carbocation. However, in this case, the primary carbocation is more stable than the secondary carbocation. The electron-withdrawing fluorine substituents are closer to the positively charged carbon in the secondary carbon. Therefore, they will destabilize the secondary carbocation more than the primary carbocation. So the major product will be 1,1,1-trifluoro-3-iodopropane and not 1,1,1-trifluoro-2-iodopropane, the compound that would be predicted to be the major product by following the rule.

$$F_3CCH_2 - \overset{+}{C}H_2 \qquad\qquad F_3C\overset{+}{C}H - CH_3$$

more stable   less stable because of the nearby
                     electron-withdrawing fluorines

80. **a.**

$$CH_3CH_2CH=CH_2 \quad\rightleftharpoons\quad CH_3CH_2\overset{+}{C}HCH_3 \quad\underset{CH_3\ddot{O}H}{\rightleftharpoons}\quad CH_3CH_2CHCH_3$$

b. the first step (See the answer to Problem 9.)
c. H⁺   **d.** 1-butene
e. the *sec*-butyl cation   **f.** methanol

81. **a.**

b.

c.

1. BH₃/THF
2. HO⁻, H₂O₂, H₂O

(The H and OH are added to the same side.)

d.

Br₂
CH₂Cl₂

(The two bromines are added to opposite sides of the cyclohexane ring.)

82.

CH_3CH_2, H

C=C

H, CH_2CH_3

trans-3-hexene

Br₂

CH_2CH_3

H——Br

H——Br

CH_2CH_3

a meso compound

83.

CH_3CH_2, CH_2CH_3

C=C

H, H

cis-3-hexene

Br₂

CH_2CH_3

H——OH

Br——H

CH_2CH_3

(3*S*,4*S*)-4-bromo-3-hexanol

+

CH_2CH_3

HO——H

H——Br

CH_2CH_3

(3*R*,4*R*)-4-bromo-3-hexanol

84. a.

$H_2\ddot{O}:$ H

$\overset{+}{:}\ddot{O}H$

$H_3\ddot{O}^+$

:ÖH

H

H—OH

$H_2\ddot{O}:$

b.

H—Cl

$:\ddot{C}l:^-$

Cl

85.   Five- and six-membered rings are more stable than other size rings.

a.   Either a six-membered ring or a five-membered ring can be formed. However, because the more substituted three-membered ring carbon has the larger partial positive charge since it is secondary and the less substituted one is primary, the three-membered ring will break in the direction that forms the five-membered ring.

$HO-CH_2CH_2CH_2CH-CH_2$ (with $\overset{+}{Br}$) $\longrightarrow$ (5-membered oxocyclic ring, $\overset{+}{O}$–H, with CH_2Br) $\rightleftharpoons$ (tetrahydrofuran ring, O, with CH_2Br) $+ H^+$

major product

$HO-CH_2CH_2CH_2CH-CH_2$ (with $\overset{+}{Br}$) $\longrightarrow$ (6-membered oxocyclic ring, $\overset{+}{O}$–H, with Br) $\rightleftharpoons$ (tetrahydropyran ring, O, with Br) $+ H^+$

minor product

b. A six-membered ring will form in preference to a less stable seven-membered ring. (In addition, formation of the six-membered ring involves the preferred attack on the more substituted ring carbon.)

$HO-CH_2CH_2CH_2CH_2CH-CH_2$ (with $\overset{+}{Br}$) $\longrightarrow$ (6-membered oxocyclic ring, $\overset{+}{O}$–H, with CH_2Br) $\rightleftharpoons$ (tetrahydropyran ring, O, with CH_2Br) $+ H^+$

86. **a.**

$$\underset{S}{\underset{\underset{Cl}{|}}{H_3C-\overset{\overset{CH_2CH_2CH_3}{|}}{C}\cdots H}} \quad + \quad \underset{R}{\underset{\underset{Cl}{|}}{H\cdots \overset{\overset{CH_2CH_2CH_3}{|}}{C}-CH_3}} \quad + \quad CH_3CH_2\underset{\underset{Cl}{|}}{CH}CH_2CH_3$$

or

$$\underset{S}{\underset{\underset{CH_2CH_2CH_3}{|}}{H-\overset{\overset{CH_3}{|}}{\underset{|}{C}}-Cl}} \quad + \quad \underset{R}{\underset{\underset{CH_2CH_2CH_3}{|}}{Cl-\overset{\overset{CH_3}{|}}{\underset{|}{C}}-H}} \quad + \quad CH_3CH_2\underset{\underset{Cl}{|}}{CH}CH_2CH_3$$

b.

$$\underset{S}{\underset{\underset{Cl}{|}}{H_3C-\overset{\overset{CH_2CH_2CH_3}{|}}{C}\cdots H}} \quad + \quad \underset{R}{\underset{\underset{Cl}{|}}{H\cdots \overset{\overset{CH_2CH_2CH_3}{|}}{C}-CH_3}} \quad + \quad CH_3CH_2\underset{\underset{Cl}{|}}{CH}CH_2CH_3$$

or

$$\underset{S}{\underset{\underset{CH_2CH_2CH_3}{|}}{H-\overset{\overset{CH_3}{|}}{\underset{|}{C}}-Cl}} \quad + \quad \underset{R}{\underset{\underset{CH_2CH_2CH_3}{|}}{Cl-\overset{\overset{CH_3}{|}}{\underset{|}{C}}-H}} \quad + \quad CH_3CH_2\underset{\underset{Cl}{|}}{CH}CH_2CH_3$$

c. (cyclohexane ring with OH on wedge and CH_2CH_3 on bold wedge at same carbon)

d.

CH(CH₃)₂ ... CH₃CH₂CH₂ C̈ʺH CH₃ **S** + H C̈ CH₃ CH₂CH₂CH₃ **R**   **or**   CH₃—|—H CH₂CH₂CH₃ **S** + H—|—CH₃ CH₂CH₂CH₃ **R**

e.

H Cl CH₃ CH₃ **S S** cis + Cl H CH₃ CH₃ **R R** + H CH₃ CH₃ Cl **S R** trans + CH₃ H Cl CH₃ **R S**

f.

H H D **S R** D

g.   (CH₃)₂CCH₂CH₃ C̈ʺH BrCH₂ Br **R**   (CH₃)₂CCH₂CH₃ H C̈ CH₂Br Br **S**   **or**   CH₂Br H—|—Br (CH₃)₂CCH₂CH₃ **R**   CH₂Br Br—|—H (CH₃)₂CCH₂CH₃ **S**

h.   CH₃CH₂ H H C̈ʺ—C̈ʺ—CH₃ CH₃ CH₂CH₂CH₃ **S S**   +   H CH₂CH₃ CH₃—C̈—C̈ʺH CH₃CH₂CH₂ CH₃ **R R**   **or**   CH₂CH₃ H—|—CH₃ **S** CH₃—|—H **S** CH₂CH₂CH₃ +   CH₂CH₃ CH₃—|—H **R** H—|—CH₃ **R** CH₂CH₂CH₃

i.   CH₃CH₂ H CH₃ C̈—C̈ʺ—CH₃ H CH₂CH₂CH₃ **R S**   +   H CH₂CH₃ CH₃—C̈—C̈ʺCH₃ CH₃CH₂CH₂ H **R S**   **or**   CH₂CH₃ CH₃—|—H **R** CH₃—|—H **S** CH₂CH₂CH₃ +   CH₂CH₃ H—|—CH₃ **S** H—|—CH₃ **R** CH₂CH₂CH₃

j.

H H Cl **S R** CH₂CH₃   H H CH₃CH₂ Cl **S R**

87.   **a.**   Both 1-butene and 2-butene react with HCl to form 2-chlorobutane.

   b.   Both alkenes form the same carbocation, and therefore have transition states that are close in energy, but because 2-butene is more stable than 1-butene, 2-butene has the greater free energy of activation.

   c.   Both compounds form the same carbocation, and therefore have transition states that are close in energy, but since (Z)-2-butene is less stable, it will react more rapidly with HCl.

88.   **a.**
and
b.

c. [structure: 3-bromo-3-methylhexane type with Br]

d. [structure: cyclopentane with ethyl and Br]

89. **a.** [dialdehyde/diketone structure]

b. [structure with two CHO groups and a methyl]

c. $H_2C=O$ + [structure]

d. $H_2C=O$ + [structure] + [structure]

e. $H_2C=O$ + [structure] + [structure]

90.

[structure] CH_3CH_2, CH_3 ... $C=C$ ... CH_3, CH_2CH_3

$\xrightarrow[\text{Pd/C}]{H_2}$

[Fischer projections]

trans-3,4-dimethyl-3-hexene

(3S,4S)-3,4-dimethyhexane   +   (3R,4R)-3,4-dimethylhexane

91. **a.** 4 alkenes

[four cyclopentene structures: CH_2, CH_3, CH_3, CH_3]

b. 1-Methylcyclopentene is the most stable.

c. Because 1-methylcyclopentene is the most stable, it has the smallest heat of hydrogenation.

92. **a.** [structure: H_3C, H, H, C–C, Br, Br, CH_2CH_3, S S]   +   [structure: H, CH_3, Br–C–C, CH_3CH_2, H, Br, R R]   **or**   [Fischer projections: CH_3, H—Br S, Br—H S, CH_2CH_3]   +   [CH_3, Br—H R, H—Br R, CH_2CH_3]

b.

H$_3$C, C—C, H / Br, H, Br, CH$_2$CH$_3$ (R S) + H, CH$_3$ / Br—C—C, Br / CH$_3$CH$_2$, H (R S) **or** Br——H R / Br——H S / CH$_2$CH$_3$ (CH$_3$ top) + H——Br S / H——Br R / CH$_2$CH$_3$ (CH$_3$ top)

c.

CH$_2$CH$_3$ / CH$_3$—C—H / Cl (S) + CH$_2$CH$_3$ / H—C—CH$_3$ / Cl (R) **or** H——Cl / CH$_2$CH$_3$ (S) (CH$_3$ top) + Cl——H / CH$_2$CH$_3$ (R) (CH$_3$ top)

d.

cyclohexane with CH$_3$ and Br

e.

CH$_3$CH$_2$, C—C, H / Br, H, Br, CH$_2$CH$_3$ (R S) **or** CH$_2$CH$_3$ (top) / H——Br S / H——Br R / CH$_2$CH$_3$

a meso compound

f.

CH$_3$CH$_2$, C—C, H / H, Br, Br, CH$_2$CH$_3$ (S S) + H, CH$_2$CH$_3$ / Br—C—C, H / CH$_3$CH$_2$, Br (R R) **or** CH$_2$CH$_3$ (top) / H——Br S / Br——H S / CH$_2$CH$_3$ + CH$_2$CH$_3$ (top) / Br——H R / H——Br R / CH$_2$CH$_3$

g. The initially formed carbocation is secondary. It undergoes a 1,2-methyl shift to form a tertiary carbocation that gives the products shown below.

Br / (CH$_3$)$_2$CH—C—CH$_2$CH$_3$ / CH$_3$ (R) + Br / CH$_3$CH$_2$—C—CH(CH$_3$)$_2$ / CH$_3$ (S) **or** CH$_2$CH$_3$ (top) / Br——CH$_3$ R / CHCH$_3$ / CH$_3$ + CH$_2$CH$_3$ (top) / CH$_3$——Br S / CHCH$_3$ / CH$_3$

h.

CH$_2$CH$_3$ / CH$_3$—C—H / Br (S) + CH$_2$CH$_3$ / H—C—CH$_3$ / Br (R) **or** H——Br / CH$_2$CH$_3$ (S) (CH$_3$ top) + Br——H / CH$_2$CH$_3$ (R) (CH$_3$ top)

i.

H$_3$C, C—C, H / Cl, H, Cl, CH$_3$ (R S) **or** CH$_3$ (top) / H——Cl S / H——Cl R / CH$_3$

a meso compound

j.

H₃C—C(H)(Cl)⋯C(H)(CH₃)—Cl + Cl—C(CH₃)(H)⋯C(CH₃)(Cl)(H)   **or**

Fischer projections:

CH₃ | H—Cl S | Cl—H S | CH₃   +   CH₃ | Cl—H R | H—Cl R | CH₃

k.

CH₃CH₂—C(CH₃)(H)⋯C(H)(CH₃)—CH₃   **or**

CH₂CH₃ | H—CH₃ S | H—CH₃ R | CH₂CH₃

a meso compound

l.

CH₃CH₂—C(H)(CH₃)⋯C(CH₃)—CH₂CH₃ + CH₃—C(H)(CH₃CH₂)⋯C(CH₂CH₃)(H)—CH₃   **or**

CH₂CH₃ | H—CH₃ S | CH₃—H S | CH₂CH₃   +   CH₂CH₃ | CH₃—H R | H—CH₃ R | CH₂CH₃

93. The first product would not be formed, because none of the bonds attached to C-2 were broken during the reaction. Therefore, the configuration at C-2 cannot change.

94. Diazomethane is a very reactive compound because the triple-bonded nitrogen has a strong propensity to depart from the carbon in order to form a very stable molecule of nitrogen gas. As it departs, the nucleophilic alkene attacks the electrophilic carbon, and in the same step, the lone pair is the nucleophile that adds to the other sp^2 carbon of the alkene.

CH₂=CH₂   ⟶   △   +   N₂

:CH₂—N≡N

95. The C—Zn bond of the Simmons-Smith reagent has the same polarity as the C—Mg bond of a Grignard reagent. Therefore,

I—CH₂—ZnI   reacts as if it were   I—CH₂⁻ ZnI⁺

I—CH₂⁻
CH₂=CH₂   ⟶   △ + I⁻

96. **a.** The base removes a proton. Then do what is needed to get the known product of the reaction.

Cl—C(Cl)(Cl)—H + HO:⁻   ⟶   Cl—C(Cl)(Cl):⁻   ⟶   Cl—C(Cl): + Cl⁻

+ H₂O

b. Notice that this reaction has the same mechanism as that in part **a** except that instead of a base removing a proton and leaving behind its bonding electrons, heat causes CO_2 to be removed leaving behind its bonding electrons.

97.

98.   **a.**   **b.**   **c.**

99.

100. **a.**

b. The initially formed carbocation is tertiary.

c. The rearranged carbocation is secondary; it undergoes another rearrangement to a more stable tertiary carbocation.

d. The initially formed carbocation rearranges in order to release the strain in the four-membered ring. (A tertiary carbocation with a strained four-membered ring is less stable than a secondary carbocation with an unstrained five-membered ring.)

101.   2-Methylpropene will be hydrated more rapidly.

> **1.**   It is more reactive than the chloro-substituted alkene, because the electron-withdrawing chlorine makes the alkene less nucleophilic.
>
> **2.**   The carbocation intermediate that 2-methylpropene forms (and therefore the transition state leading to its formation) is more stable, because the electron-withdrawing chlorine increases the amount of positive charge on the carbon.

102.   It tells us that the first step of the mechanism is the slow step. If the first step is slow, the carbocation will react with water in a subsequent fast step, which means that the carbocation will not have time to lose a proton to reform the alkene, so all the deuterium atoms (D) will be retained in the unreacted alkene.

If the first step were not the slow step, an equilibrium would be set up between the alkene and the carbocation and, because the carbocation could lose either H^+ or D^+ when it reforms the alkene, all the deuterium atoms would not be retained in the unreacted alkene.

103.   Because fumarate is the trans isomer and it forms an erythro product, the enzyme must catalyze the anti addition of D_2O. (Recall: CIS-SYN-ERYTHRO, which allows TRANS-ANTI-ERYTHRO but does not allow TRANS-SYN-ERYTHRO, since two terms must be changed.)

104.   When (S)-3-methyl-1-pentene reacts with Cl_2, a compound with a new asymmetric center ($*$) is formed. The relative configuration of the asymmetric center in the reactant does not change because no bonds to it are broken during the course of the reaction.

The new asymmetric center can have either the R or the S configuration. Therefore, a pair of diastereoisomers is obtained.

(2R,3S)-1,2-dichloro-3-methylpentane   (2S,3S)-1,2-dichloro-3-methylpentane

105. **a.** A proton adds to the alkene, forming a secondary carbocation, which undergoes a ring-expansion rearrangement to form a more stable tertiary carbocation.

106.

107. **a.**   **b.**

Chapter 6 Practice Test

1.   Which member of each pair is more stable?

a. $CH_3\overset{+}{C}HCH_2CH_3$ or $CH_3\overset{+}{C}CH_3$ (with CH_3 on central carbon)

d.   or

b. $CH_3CH_2CH_2\overset{+}{C}H_2$ or $CH_3CH_2\overset{+}{C}HCH_3$

c. $CH_3CH_2\overset{+}{C}H_2$ or $ClCH_2CH_2\overset{+}{C}H_2$

e.   or

2.   Which would be a better compound to use as a starting material for the synthesis of 2-bromopentane?

$CH_3CH_2CH_2CH{=}CH_2$ or $CH_3CH_2CH{=}CHCH_3$

3.   The addition of H_2 in the presence of Pd/C to alkenes **A** and **B** results in the formation of the same alkane. The addition of H_2 to alkene **A** has a heat of hydrogenation $(-\Delta H°)$ of 29.7 kcal/mol, whereas the addition of H_2 to alkene **B** has a heat of hydrogenation of 27.3 kcal/mol. Which is the more stable alkene, **A** or **B**?

4.   What is the major product of each of the following reactions?

a. $CH_2{=}\overset{\underset{|}{CH_3}}{C}CH_2CH_3$ +   HBr $\longrightarrow$

b. $CH_3\overset{\underset{|}{CH_3}}{C}HCH{=}CH_2$ +   HCl $\longrightarrow$

c. $CH_3CH_2CH{=}CH_2$ + Cl_2 $\xrightarrow{H_2O}$

d. $CH_3\overset{\underset{|}{CH_3}}{\underset{\overset{|}{CH_3}}{C}}CH{=}CH_2$ +   HBr $\longrightarrow$

e. $CH_3CH{=}CH_2$ + $\longrightarrow$

f. $\xrightarrow[\text{2. HO}^-, H_2O_2, H_2O]{\text{1. R}_2BH/THF}$

5.   What product would be obtained from the hydroboration-oxidation of the following alkenes?

a.   2-methyl-2-butene   b.   1-ethylcyclopentene

6. Indicate how each of the following compounds could be synthesized using an alkene as one of the starting materials:

a.
$$\longrightarrow \quad \underset{\underset{CH_3}{|}}{\overset{\overset{CH_3}{|}}{CH_3CCH_2CH_3}}$$

b.

c.

d.

7. Indicate the carbocations that you would expect to rearrange to give a more stable carbocation.

$$\underset{\overset{+}{}}{CH_3CH_2\underset{\underset{CH_3}{|}}{CH}CHCH_3} \qquad \underset{\overset{+}{}}{CH_3CH_2CHCH_3}$$

8. Indicate whether each of the following statements is true or false:

a. The addition of Br_2 to 1-butene to form 1,2-dibromobutane is a concerted reaction.   T   F

b. The reaction of 1-butene with HCl will form 1-chlorobutane as the major product.   T   F

c. 2,3-Dimethyl-2-pentene is more stable than 3,4-dimethyl-2-pentene.   T   F

d. The reaction of HBr with 3-methylcyclohexene is more highly regioselective than is the reaction of HBr with 1-methylcyclohexene.   T   F

e. The reaction of an alkene with a carboxylic acid forms an epoxide.   T   F

f. A catalyst increases the equilibrium constant of a reaction.   T   F

g. The addition of HBr to 3-methyl-2-pentene is a stereospecific reaction.   T   F

h. The addition of HBr to 3-methyl-2-pentene is a stereoselective reaction.   T   F

i. The addition of HBr to 3-methyl-2-pentene is a regioselective reaction.   T   F

9. Draw all the products that would be obtained from each of the following reactions, indicating which isomers are formed:

a. 1-butene + HCl

b. 2-pentene + HBr

c. *trans*-3-hexene + Br_2

d. *trans*-3-heptene + Br_2

10.   Draw the stereoisomers that would be obtained from each of the following reactions:

a. $\xrightarrow[\text{CH}_2\text{Cl}_2]{\text{Br}_2}$

c. $\xrightarrow[\text{Pd/C}]{\text{H}_2}$

b. $\xrightarrow[\text{CH}_2\text{Cl}_2]{\text{Br}_2}$

d. $\xrightarrow[\text{Pd/C}]{\text{H}_2}$

The Reactions of Alkynes • An Introduction to Multistep Synthesis

<u>**Important Terms**</u>

acetylide ion	the conjugate base of a terminal alkyne.

$$RC{\equiv}C^-$$

aldehyde
a compound with a carbonyl group that is bonded to an alkyl group and to a hydrogen (or bonded to two hydrogens).

alkylation reaction
a reaction that adds an alkyl group to a reactant.

alkyne
a hydrocarbon that contains a triple bond.

carbonyl group
a carbon doubly bonded to an oxygen.

enol
an alkene with an OH group bonded to one of the sp^2 carbons.

geminal dihalide
a compound with two halogen atoms bonded to the same carbon.

internal alkyne
an alkyne with its triple bond not at the end of the carbon chain.

keto-enol tautomers
a ketone or aldehyde and its isomeric enol. The keto and enol tautomers differ only in the location of a double bond and a hydrogen.

ketone
a compound with a carbonyl group that is bonded to two alkyl groups.

π-complex
a complex formed between an electrophile and a triple bond.

radical anion   a species with a negative charge and an unpaired electron.

retrosynthetic analysis   working backward (on paper) from a target molecule to available starting materials.
or retrosynthesis

tautomerization   interconversion of tautomers.

tautomers   constitutional isomers that are in rapid equilibrium; for example, keto and enol tautomers. The keto and enol tautomers differ only in the location of a double bond and a hydrogen.

terminal alkyne   an alkyne with its triple bond at the end of the carbon chain.

vinylic cation   a species with a positive charge on a vinylic carbon.

vinylic radical   a species with an unpaired electron on a vinylic carbon.

Solutions to Problems

1.   The general molecular formula of a noncyclic hydrocarbon is C_nH_{2n+2}. Therefore, the molecular formula for a noncyclic hydrocarbon with 14 carbons is $C_{14}H_{30}$.

Because a compound has two fewer hydrogens for every ring and π bond, a compound with one ring and 4 π bonds (2 triple bonds) would have 10 fewer hydrogens than the C_nH_{2n+2} formula. Therefore, the molecular formula is **$C_{14}H_{20}$**.

2.   **a.** $ClCH_2CH_2C\equiv CCH_2CH_3$ **c.** $CH_3CHC\equiv CH$ **e.** $HC\equiv CCH_2CCH_3$
$\overset{\displaystyle |}{CH_3}$

b. (octagon with triple bond)   **d.** $CH_3CH_2CHC\equiv CCH_2CHCH_3$ **f.** $CH_3C\equiv CCH_3$
$\overset{\displaystyle |}{CH_3}\overset{\displaystyle |}{CH_3}$

3. $HC\equiv CCH_2CH_2CH_2CH_3$ $CH_3C\equiv CCH_2CH_2CH_3$ $CH_3CH_2C\equiv CCH_2CH_3$
           1-hexyne   2-hexyne   3-hexyne
        butylacetylene   methylpropylacetylene   diethylacetylene

        $CH_3CH_2CHC\equiv CH$ $CH_3CHCH_2C\equiv CH$ $CH_3CHC\equiv CCH_3$
$\overset{\displaystyle |}{CH_3}\overset{\displaystyle |}{CH_3}\overset{\displaystyle |}{CH_3}$
        3-methyl-1-pentyne   4-methyl-1-pentyne   4-methyl-2-pentyne
        sec-butylacetylene   isobutylacetylene   isopropylmethylacetylene

                                           $\overset{\displaystyle CH_3}{\overset{\displaystyle |}{CH_3CC\equiv CH}}$
                                           $\underset{\displaystyle |}{}$
                                           CH_3
                                      3,3-dimethyl-1-butyne
                                       tert-butylacetylene

4.   **a.** 5-bromo-2-pentyne   **c.** 1-methoxy-2-pentyne
     b. 6-bromo-2-chloro-4-octyne   **d.** 3-ethyl-1-hexyne

5.   **a.** 6-methyl-2-octyne   **b.** 5-ethyl-4-methyl-1-heptyne   **c.** 2-bromo-4-octyne

6.   **a.** 1-hepten-4-yne   **d.** 3-butyn-1-ol
     b. 4-methyl-1,4-hexadiene   **e.** 1,3,5-heptatriene
     c. 5-vinyl-5-octen-1-yne   **f.** 2,4-dimethyl-4-hexen-1-ol
         one of the functional groups
         cannot be included in the
         parent hydrocarbon

7.   **a.** (*E*)-2-hepten-4-ol   **b.** 1-hepten-5-yne   **c.** (*E*)-4-hepten-1-yne

8.   alkane = pentane   alkene = 1-pentene   alkyne = 1-pentyne

9.   The cis isomer has a higher boiling point because it has a small dipole moment, whereas the dipole moment of the trans isomer is zero.

10.   **a.** sp^2-sp^2 **d.** $sp-sp^3$ **g.** sp^2-sp^3
       b. sp^2-sp^3 **e.** $sp-sp$ **h.** $sp-sp^3$
       c. $sp-sp^2$ **f.** sp^2-sp^2 **i.** sp^2-sp

11.   Solved in the text.

12.   **a.** $CH_2=CCH_3$ **c.** $CH_3C=CCH_3$ **e.** $CH_3CH_2CCH_3$
              | | |
             Br   Br   Br
                                        Br

b. Br   **d.** Br Br   **f.** Br   Br
       | | | | |
  CH_3CCH_3 $HC-CCH_3$ $CH_3CCH_2CH_2CH_3$ + $CH_3CH_2CCH_2CH_3$
       | | | | |
      Br   Br Br   Br   Br

13.   **a.**

b.   Only anti addition occurs, because the intermediate is a cyclic bromonium ion. Therefore the product will have the *E* configuration.

14.   Because the alkyne is not symmetrical, two ketones will be obtained.

$$CH_3CH_2\overset{\overset{\displaystyle O}{\|}}{C}CH_2CH_2CH_2CH_3 \quad \text{and} \quad CH_3CH_2CH_2\overset{\overset{\displaystyle O}{\|}}{C}CH_2CH_2CH_3$$

15.   **a.** $CH_3C\equiv CH$ **b.** $CH_3CH_2C\equiv CCH_2CH_3$ **c.** $HC\equiv C-$

The best answer for **b** is 3-hexyne, because it would form only the desired ketone. 2-Hexyne would form two different ketones, so only half of the product would be the desired ketone.

16.   **a.** $CH_2=CCH_3$ Because the ketone has identical substituents bonded to the
                     | carbonyl carbon, it has only one enol tautomer.
                    OH

b. $\overset{\displaystyle \text{OH}}{\underset{\displaystyle |}{\text{CH}_3\text{CH}}}{=}\text{CCH}_2\text{CH}_2\text{CH}_3$ and $\text{CH}_3\text{CH}_2\overset{\displaystyle \text{OH}}{\underset{\displaystyle |}{\text{C}}}{=}\text{CHCH}_2\text{CH}_3$

E and Z isomers are possible for each of these enols.

| E | Z | E | Z |

c. $\text{CH}_2{=}\overset{\displaystyle \text{OH}}{\underset{\displaystyle |}{\text{C}}}{-}\bigcirc$ and $\text{CH}_3\overset{\displaystyle \text{OH}}{\underset{\displaystyle |}{\text{C}}}{=}\bigcirc$ Because each enol has identical groups bonded to one of its sp^2 carbons, E and Z isomers are not possible for either enol.

17.   **a.** **(1)** $\text{CH}_3\text{CH}_2\overset{\displaystyle \text{O}}{\overset{\displaystyle ||}{\text{C}}}\text{CH}_3$ **b.** **(1)** $\text{CH}_3\text{CH}_2\overset{\displaystyle \text{O}}{\overset{\displaystyle ||}{\text{C}}}\text{CH}_3$ **c.** **(1)** $\text{CH}_3\text{CH}_2\text{CH}_2\overset{\displaystyle \text{O}}{\overset{\displaystyle ||}{\text{C}}}\text{CH}_3$ and $\text{CH}_3\text{CH}_2\overset{\displaystyle \text{O}}{\overset{\displaystyle ||}{\text{C}}}\text{CH}_2\text{CH}_3$

(2) $\text{CH}_3\text{CH}_2\text{CH}_2\overset{\displaystyle \text{O}}{\overset{\displaystyle ||}{\text{C}}}\text{H}$ **(2)** $\text{CH}_3\text{CH}_2\overset{\displaystyle \text{O}}{\overset{\displaystyle ||}{\text{C}}}\text{CH}_3$ **(2)** $\text{CH}_3\text{CH}_2\text{CH}_2\overset{\displaystyle \text{O}}{\overset{\displaystyle ||}{\text{C}}}\text{CH}_3$ and $\text{CH}_3\text{CH}_2\overset{\displaystyle \text{O}}{\overset{\displaystyle ||}{\text{C}}}\text{CH}_2\text{CH}_3$

18.   Ethyne (acetylene)

An alkyne can form an aldehyde only if the OH group adds to a terminal sp carbon. In the acid-catalyzed addition of water to a terminal alkyne, the proton adds to the terminal sp carbon. Therefore, the only way the OH group can add to a terminal sp carbon is if there are two terminal sp carbons in the alkyne. In other words, the alkyne must be ethyne.

19.   **a.** $\text{CH}_3\text{CH}_2\text{CH}_2\text{C}{\equiv}\text{CH}$ **or** $\text{CH}_3\text{CH}_2\text{C}{\equiv}\text{CCH}_3$ $\xrightarrow[\text{Pd/C}]{\text{H}_2}$ $\text{CH}_3\text{CH}_2\text{CH}_2\text{CH}_2\text{CH}_3$
    1-pentyne   2-pentyne

b. $\text{CH}_3\text{C}{\equiv}\text{CCH}_3$ $\xrightarrow[\substack{\text{Lindlar}\\\text{catalyst}}]{\text{H}_2}$
    2-butyne

c. $\text{CH}_3\text{CH}_2\text{C}{\equiv}\text{CCH}_3$ $\xrightarrow[\substack{\text{NH}_3\,(\text{liq})\\-78\,°\text{C}}]{\text{Na}}$
    2-pentyne

d. $\text{CH}_3\text{CH}_2\text{CH}_2\text{CH}_2\text{C}{\equiv}\text{CH}$ $\xrightarrow[\substack{\text{Lindlar}\\\text{catalyst}}]{\text{H}_2}$ $\text{CH}_3\text{CH}_2\text{CH}_2\text{CH}_2\text{CH}{=}\text{CH}_2$
    1-hexyne

Na/NH$_3$(liq) cannot be used to reduce terminal alkynes, because Na will remove the hydrogen that is attached to the sp carbon of the terminal alkyne.

$$2\,\text{RC}{\equiv}\text{CH} + 2\,\text{Na} \longrightarrow 2\,\text{RC}{\equiv}\text{C}^- + 2\,\text{Na}^+ + \text{H}_2$$

20. **a.**

d.

b.

e. +

c.

f. +

21.   The reaction of sodium amide with an alkane will not favor products because the acid that would be formed is a stronger acid than the alkane (the reactant acid). (Recall that the equilibrium favors reaction of the strong acid (or strong base) and formation of the weak acid (or weak base); Section 2.5.)

$$CH_3CH_3 \ + \ {}^-NH_2 \ \rightleftharpoons \ CH_3\overset{-}{C}H_2 \ + \ NH_3$$

$pK_a > 60$ $\qquad\qquad\qquad\qquad\qquad pK_a = 36$

weaker acid   weaker base $\qquad$ stronger base   stronger acid

22.   The base used to remove a proton must be stronger than the base that is formed as a result of proton removal. Therefore, the base used to remove a proton from a terminal alkyne must be a stronger base than the conjugate base of the terminal alkyne. A terminal alkyne has a $pK_a \sim 25$. In other words, any base whose conjugate acid has a pK_a greater than 25 can be used.

23. **a.** $CH_3CH_2CH_2\overset{-}{C}H_2 \ > \ CH_3CH_2CH{=}\overset{-}{C}H \ > \ CH_3CH_2C{\equiv}C^-$

   b. ${}^-NH_2 \ > \ CH_3C{\equiv}C^- \ > \ CH_3CH_2O^- \ > \ F^-$

24. **a.** Solved in the text.

   b. $H_2C{=}\overset{+}{C}H$

   A triply bonded (sp) carbon is more electronegative than a doubly bonded (sp^2) carbon. Therefore, a triply bonded carbon with a positive charge would be less stable than a doubly bonded carbon with a positive charge. Thus, the vinyl cation is more stable.

25.   Solved in the text.

26. **a.** $HC{\equiv}CH \ \xrightarrow[\text{2. } CH_3CH_2CH_2Br]{\text{1. } NaNH_2} \ CH_3CH_2CH_2C{\equiv}CH$

   b. $HC{\equiv}CH \ \xrightarrow[\text{2. } CH_3Br]{\text{1. } NaNH_2} \ CH_3C{\equiv}CH \ \xrightarrow{\text{H}_2/\text{Lindlar catalyst}} \ CH_3CH{=}CH_2$

   c. $HC{\equiv}CH \ \xrightarrow[\text{2. } CH_3Br]{\text{1. } NaNH_2} \ CH_3C{\equiv}CH \ \xrightarrow[\text{2. } CH_3Br]{\text{1. } NaNH_2} \ CH_3C{\equiv}CCH_3 \ \xrightarrow[\substack{\text{Lindlar}\\\text{catalyst}}]{\text{H}_2} \ $

   d. product of **a** $\ \xrightarrow[\text{2. } HO^-,\ H_2O_2,\ H_2O]{\text{1. } R_2BH/THF} \ CH_3CH_2CH_2CH_2\overset{\overset{\displaystyle O}{\displaystyle \|}}{C}H$

e.   product of **b** $\xrightarrow{\text{HBr}}$ $\underset{\underset{\text{Br}}{|}}{\text{CH}_3\text{CHCH}_3}$

f. $\text{HC}{\equiv}\text{CH}$ $\xrightarrow[\text{2. CH}_3\text{Br}]{\text{1. NaNH}_2}$ $\text{CH}_3\text{C}{\equiv}\text{CH}$ $\xrightarrow{\text{excess HCl}}$ $\underset{\underset{\text{Cl}}{|}}{\overset{\overset{\text{Cl}}{|}}{\text{CH}_3\text{CCH}_3}}$

27.   a. $\underset{\underset{\text{Cl}}{|}}{\overset{\overset{\text{Cl}}{|}}{\text{CH}_3\text{CH}_2\text{CCH}_3}}$ c. $\underset{\underset{\text{Cl}}{|}}{\overset{\overset{\text{Cl}}{|}}{\text{CH}_3\text{CH}_2\text{CH}_2\text{CCH}_2\text{CH}_2\text{CH}_3}}$ + $\underset{\underset{\text{Cl}}{|}}{\overset{\overset{\text{Cl}}{|}}{\text{CH}_3\text{CH}_2\text{CCH}_2\text{CH}_2\text{CH}_2\text{CH}_3}}$

equal amounts

b. $\underset{\underset{\text{Cl}}{|}}{\overset{\overset{\text{Cl}}{|}}{\text{CH}_3\text{CH}_2\text{CH}_2\text{CCH}_2\text{CH}_3}}$

28.   a. $\text{CH}_3\text{C}{\equiv}\text{CCH}_2\text{CH}_2\text{CH}_3$ g. $\text{BrC}{\equiv}\text{CCH}_2\text{CH}_2\text{CH}_3$

b. $\underset{\underset{\text{CH}_2\text{CH}_3}{|}}{\text{CH}_3\text{CH}_2\text{C}{\equiv}\text{CCHCH}_2\text{CH}_2\text{CH}_3}$ h.

c. $\text{CH}_3\text{C}{\equiv}\text{CH}$ i. $\text{CH}_3\text{CH}_2\text{C}{\equiv}\text{CCH}_2\text{CH}_3$

d. $\text{CH}_2{=}\text{CHC}{\equiv}\text{CH}$ j. $\underset{\underset{\text{CH}_3}{|}}{\overset{\overset{\text{CH}_3}{|}}{\text{CH}_3\text{CC}}}{\equiv}\underset{\underset{\text{CH}_3}{|}}{\overset{\overset{\text{CH}_3}{|}}{\text{CCCH}_3}}$

e. $\text{CH}_3\text{OC}{\equiv}\text{CH}$ k.

f. $\underset{\underset{\text{CH}_3}{|}}{\overset{\overset{\text{CH}_3}{|}}{\text{CH}_3\text{CC}}}{\equiv}\overset{\overset{\text{CH}_3}{|}}{\text{CCHCH}_2\text{CH}_3}$ l. $\underset{\underset{\text{CH}_3}{|}}{\text{CH}_3\text{C}{\equiv}\text{CCH}_2\text{CHCHCH}_3}\overset{\overset{\text{CH}_3}{|}}{}$

29.   The student named only one correctly.
a. 4-methyl-2-hexyne   b. 7-bromo-3-heptyne   c. correct   d. 2-pentyne

30.

CH₃C≡C:⁻   CH₃—Br   ⟶   CH₃C≡CCH₃   +   :Br:⁻

nucleophile   electrophile

(Methyl bromide is an electrophile because the carbon has a partial positive charge.)

31.   **a.** 5-bromo-2-hexyne   **c.** 5,5-dimethyl-2-hexyne   **e.** 1,5-cyclooctadiene
b. 5-methyl-2-octyne   **d.** 6-chloro-2-methyl-3-heptyne   **f.** 1,6-dimethyl-1,3-cyclohexadiene

32.

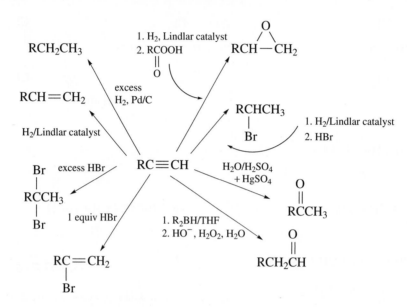

33.   First draw the straight-chain compounds with seven carbons; then draw the straight-chain compounds with six carbons and one methyl group; then draw the straight-chain compounds with five carbons and two methyl groups (or with one ethyl group). Naming them will tell you if you have drawn one compound more than once because if two compounds have the same name, they are the same compound.

HC≡CCH₂CH₂CH₂CH₂CH₃   CH₃C≡CCH₂CH₂CH₂CH₃   CH₃CH₂C≡CCH₂CH₂CH₃
1-heptyne   2-heptyne   3-heptyne
pentylacetylene   butylmethylacetylene   ethylpropylacetylene

CH₃CH₂CH₂CHC≡CH   CH₃CH₂CHCH₂C≡CH   CH₃CHCH₂CH₂C≡CH
         | | |
        CH₃   CH₃   CH₃

3-methyl-1-hexyne   4-methyl-1-hexyne   5-methyl-1-hexyne
                                                                                  isopentylacetylene

CH₃CH₂CHC≡CCH₃   CH₃CHCH₂C≡CCH₃   CH₃CHC≡CCH₂CH₃
         | | |
        CH₃   CH₃   CH₃

4-methyl-2-hexyne   5-methyl-2-hexyne   2-methyl-3-hexyne
sec-butylmethylacetylene   isobutylmethylacetylene   ethylisopropylacetylene

         CH₃   CH₃   CH₃
          | | |
CH₃CC≡CCH₃   CH₃CCH₂C≡CH   CH₃CH₂CC≡CH
          | | |
         CH₃   CH₃   CH₃

4,4-dimethyl-2-pentyne   4,4-dimethyl-1-pentyne   3,3-dimethyl-1-pentyne
tert-butylmethylacetylene   *tert*-pentylacetylene

$$\begin{array}{cc}
\underset{\substack{|\\ CH_2CH_3}}{} & \underset{\substack{|\\ CH_3}}{} \\
CH_3CH_2CHC\equiv CH & CH_3CHCHC\equiv CH \\
& \underset{\substack{|\\ CH_3}}{} \\
\text{3-ethyl-1-pentyne} & \text{3,4-dimethyl-1-pentyne}
\end{array}$$

34. Addition of Br_2 to the triple bond in the presence of H_2O forms an enol that immediately tautomerizes to a ketone.

$$CH_3CH_2CH_2C\equiv CH \xrightarrow[H_2O]{Br_2} CH_3CH_2CH_2C\!=\!CHBr \longrightarrow \underset{\substack{O\\ \|}}{CH_3CH_2CH_2CCH_2Br}$$
$$\qquad\qquad\qquad\qquad\qquad\qquad\qquad \underset{\substack{|\\ OH}}{}$$

35. **a.** $CH_3CH_2CH_2C\equiv CH \xrightarrow[\text{2. } HO^-, H_2O_2, H_2O]{\text{1. } R_2BH/THF} \underset{\substack{O\\ \|}}{CH_3CH_2CH_2CH_2CH}$

b. $CH_3CH_2CH\!=\!CH_2 \xrightarrow[\text{2. } HO^-, H_2O_2, H_2O]{\text{1. } BH_3} CH_3CH_2CH_2CH_2OH$

c. $CH_3CH_2CH_2C\equiv CCH_2CH_2CH_3 + H_2O \xrightarrow{H_2SO_4} \underset{\substack{O\\ \|}}{CH_3CH_2CH_2CCH_2CH_2CH_2CH_3}$

This symmetrical alkyne will give the greatest yield of the desired ketone.
Because the reactant is not a terminal alkyne, the reaction can take place without the mercuric ion catalyst.

36. **a.** H_2/Lindlar catalyst   **b.** $Na, NH_3(liq), -78\,°C$ **c.** excess H_2, Pd/C

37. **a.** $\underset{\substack{|\\ Br}}{CH_2\!=\!CCH_3}$ **d.** $\underset{\substack{|\\ Br}}{\overset{Br\ Br}{\overset{|\ \ |}{BrCHCCH_3}}}$ **g.** $CH_3CH_2CH_3$ **j.** $CH_3C\equiv CCH_2CH_2CH_3$

b. $\underset{\substack{|\\ Br}}{\overset{Br}{\overset{|}{CH_3CCH_3}}}$ **e.** $\underset{\substack{O\\ \|}}{CH_3CCH_3}$ **h.** $CH_3CH\!=\!CH_2$

c. $\underset{H}{\overset{Br}{}}\!\!C\!=\!C\!\!\underset{Br}{\overset{CH_3}{}}$ **f.** $\underset{\substack{O\\ \|}}{CH_3CH_2CH}$ **i.** $CH_3C\equiv C^-$

38. **a.** 1-octen-6-yne   **d.** 5-chloro-1,3-cyclohexadiene
b. *cis*-3-hexen-1-ol or (*E*)-3-hexene-1-ol   **e.** 1-methyl-1,3,5-cycloheptatriene
c. 1,5-octadiyne

39. The molecular formula of the hydrocarbon is $C_{32}H_{56}$.
With one triple bond, two double bonds, and one ring, the degree of unsaturation is 5.
Therefore, the compound is missing 10 hydrogens from $C_nH_{2n+2} = C_{32}H_{66}$.

40.   **a.** $CH_3CH{=}CCH_3$
                         |
                         Br

        e. $CH_3\overset{\overset{\displaystyle O}{\|}}{C}CH_2CH_3$

        i. no reaction

b. $CH_3CH_2\overset{\overset{\displaystyle Br}{|}}{\underset{\underset{\displaystyle Br}{|}}{C}}CH_3$

        f. $CH_3\overset{\overset{\displaystyle O}{\|}}{C}CH_2CH_3$

        j. no reaction

c. $\underset{H_3C}{\overset{Br}{\diagdown}}C{=}C\underset{Br}{\overset{CH_3}{\diagup}}$

        g. $CH_3CH_2CH_2CH_3$

d. $CH_3\overset{\overset{\displaystyle Br}{|}}{\underset{\underset{\displaystyle Br}{|}}{C}}{-}\overset{\overset{\displaystyle Br}{|}}{\underset{\underset{\displaystyle Br}{|}}{C}}CH_3$

        h. $\underset{H}{\overset{H_3C}{\diagdown}}C{=}C\underset{H}{\overset{CH_3}{\diagup}}$

41.   **a.**   (structure)

        d.   HO— (structure)

b.   (structure)

        e.   (structure, OH) + (structure, OH)

c.   (structure)

        f.   (structure, O) + (structure, O)

42.   **a.**   **1.** $CH_3\overset{\underset{\underset{\displaystyle CH_3}{|}}{}}{C}HC{\equiv}CH$ $\xrightarrow[\text{Lindlar catalyst}]{H_2}$ $CH_3\overset{\underset{\underset{\displaystyle CH_3}{|}}{}}{C}HCH{=}CH_2$ $\xrightarrow[\text{H}_2\text{SO}_4]{\text{H}_2\text{O}}$ $CH_3\overset{\overset{\displaystyle H}{+}}{C}\overset{}{C}HCH_3$
  ... $\underset{\underset{\displaystyle CH_3}{|}}{}$ ...

$\longrightarrow$ $CH_3\overset{+}{C}CH_2CH_3$
  $\underset{\underset{\displaystyle CH_3}{|}}{}$

$\downarrow H_2O$

$CH_3\overset{\overset{\displaystyle OH}{|}}{C}CH_2CH_3$
  $\underset{\underset{\displaystyle CH_3}{|}}{}$

2. $CH_3\overset{\underset{\underset{\displaystyle CH_3}{|}}{}}{C}HC{\equiv}CH$ $\xrightarrow[\text{Lindlar catalyst}]{H_2}$ $CH_3\overset{\underset{\underset{\displaystyle CH_3}{|}}{}}{C}HCH{=}CH_2$ $\xrightarrow[\text{2. H}_2\text{O}_2,\ \text{HO}^-,\ \text{H}_2\text{O}]{\text{1. BH}_3/\text{THF}}$ $CH_3\overset{\underset{\underset{\displaystyle CH_3}{|}}{}}{C}HCH_2CH_2OH$

b. 3-Methyl-2-butanol will be a minor product obtained from both **1** and **2**.

$$
\begin{array}{c}
\text{OH} \\
| \\
\text{CH}_3\text{CHCHCH}_3 \\
| \\
\text{CH}_3
\end{array}
$$

3-methyl-2-butanol

3-Methyl-2-butanol will be obtained from **1**, because occasionally water will attack the secondary carbocation before it has a chance to rearrange to the tertiary carbocation.

3-Methyl-2-butanol will be obtained from **2**, because in the second step of the synthesis boron can also add to the other sp^2 carbon; it will be a minor product because the transition state for its formation is less stable than the transition state leading to the major product. Because a carbocation is not formed as an intermediate, a carbocation rearrangement cannot occur.

(The proton cannot add to the other sp^2 carbon in the second step of part **1** because that would require the formation of a primary carbocation in the second step of the synthesis. Primary carbocations are so unstable that they can never be formed.)

43. Three of the names are correct.

a. 3-heptyne   **c.** correct   **e.** correct
b. 5-methyl-3-heptyne **d.** 6,7-dimethyl-3-octyne **f.** correct

44. Only **c** and **e** are keto-enol tautomers. Notice that an enol tautomer has an OH group bonded to an sp^2 carbon. The structures in **d** are not enol tautomers, because they do not have the oxygen on the same carbon.

45. **a.** $HC\equiv CH \xrightarrow[\text{HgSO}_4]{\text{H}_2\text{O, H}_2\text{SO}_4} CH_2=CHOH \rightleftharpoons \underset{\text{enol}}{CH_3CH} \overset{O}{\overset{||}{}}$

b.

$HC\equiv CH \xrightarrow[\text{2. CH}_3\text{CH}_2\text{Br}]{\text{1. NaNH}_2} CH_3CH_2C\equiv CH \xrightarrow[\substack{\text{Lindlar} \\ \text{catalyst}}]{\text{H}_2} CH_3CH_2CH=CH_2 \xrightarrow[\text{CH}_2\text{Cl}_2]{\text{Br}_2} CH_3CH_2CHCH_2Br \\ \qquad\qquad\qquad\qquad\qquad\qquad\qquad\qquad\qquad\qquad\qquad\qquad\qquad\qquad\qquad\qquad\qquad\quad | \\ \qquad\qquad\qquad\qquad\qquad\qquad\qquad\qquad\qquad\qquad\qquad\qquad\qquad\qquad\qquad\qquad\qquad\quad Br$

c. $HC\equiv CH \xrightarrow[\text{2. CH}_3\text{Br}]{\text{1. NaNH}_2} HC\equiv CCH_3 \xrightarrow[\text{HgSO}_4]{\text{H}_2\text{O, H}_2\text{SO}_4} CH_3CCH_3 \overset{O}{\overset{||}{}}$

d.

$HC\equiv CH \xrightarrow[\text{2. CH}_3\text{Br}]{\text{1. NaNH}_2} HC\equiv CCH_3 \xrightarrow[\text{2. CH}_3\text{CH}_2\text{Br}]{\text{1. NaNH}_2} CH_3C\equiv CCH_2CH_3 \xrightarrow[\text{NH}_3 \text{ (liq), } -78\,°\text{C}]{\text{Na}}$

e. $HC\equiv CH \xrightarrow[\text{2. CH}_3\text{Br}]{\text{1. NaNH}_2} HC\equiv CCH_3 \xrightarrow[\text{2. CH}_3\text{CH}_2\text{Br}]{\text{1. NaNH}_2} CH_3C\equiv CCH_2CH_3 \xrightarrow[\substack{\text{Lindlar} \\ \text{catalyst}}]{\text{H}_2}$

f. $HC\equiv CH \xrightarrow[\text{2. CH}_3\text{Br}]{\text{1. NaNH}_2} HC\equiv CCH_3 \xrightarrow[\text{2. CH}_3\text{CH}_2\text{Br}]{\text{1. NaNH}_2} CH_3C\equiv CCH_2CH_3 \xrightarrow[\text{Pd/C}]{\text{excess H}_2}$

46. The first equilibrium lies to the right because HOOH ($pK_a = 11.6$) is a stronger acid than H_2O ($pK_a = 15.7$).

The second equilibrium lies to the left because the alkyne ($pK_a = 25$) is a weaker acid than HOOH ($pK_a = 11.6$).

47. **a.** Syn addition of H_2 forms *cis*-2-butene; when Br_2 adds to *cis*-2-butene, the threo pair of enantiomers is formed.

b. Reaction with sodium and liquid ammonia forms *trans*-2-butene; when Br_2 adds to *trans*-2-butene, a meso compound is formed.

c. Anti addition of Cl_2 forms *trans*-2,3-dichloro-2-butene; when Br_2 adds to *trans*-2,3-dichloro-2-butene, a meso compound is formed.

48. **a.** $CH_3CH_2\overset{\overset{\textstyle O}{\|}}{C}CH_3$ **b.** $CH_3CH_2CH_2\overset{\overset{\textstyle O}{\|}}{C}CH_3$ **c.**   **d.**

49. **a.** $HC\equiv CH \xrightarrow[\text{2. } CH_3CH_2CH_2CH_2Br]{\text{1. } NaNH_2} CH_3CH_2CH_2CH_2C\equiv CH \xrightarrow[HgSO_4]{\overset{\textstyle H_2O}{H_2SO_4}} CH_3CH_2CH_2CH_2\overset{\overset{\textstyle O}{\|}}{C}CH_3$

b. $HC\equiv CH \xrightarrow[\text{2. } CH_3CH_2Br]{\text{1. } NaNH_2} CH_3CH_2C\equiv CH \xrightarrow[\substack{\text{Lindlar} \\ \text{catalyst}}]{H_2} CH_3CH_2CH=CH_2 \xrightarrow{HBr} CH_3CH_2\underset{\underset{\textstyle Br}{|}}{C}HCH_3$

c.

$HC\equiv CH \xrightarrow[\text{2. } CH_3CH_2CH_2Br]{\text{1. } NaNH_2} CH_3CH_2CH_2C\equiv CH \xrightarrow[\substack{\text{Lindlar} \\ \text{catalyst}}]{H_2} CH_3CH_2CH_2CH=CH_2 \xrightarrow[H_2SO_4]{H_2O} CH_3CH_2CH_2\underset{\underset{\textstyle OH}{|}}{C}HCH_3$

d. $-C\equiv CH \xrightarrow[\text{2. } HO^-, H_2O_2, H_2O]{\text{1. } R_2BH/THF}$ $-CH_2\overset{\overset{\textstyle O}{\|}}{C}H$

e. $-C\equiv CH \xrightarrow[HgSO_4]{H_2O, H_2SO_4}$ $-\overset{\overset{\textstyle O}{\|}}{C}-CH_3$

f.

50. The chemist can make 3-octyne by using 1-hexyne instead of 1-butyne. He would then need to use ethyl bromide (instead of butyl bromide) for the alkylation step:

$$CH_3CH_2CH_2CH_2C\equiv CH \quad \xrightarrow[\text{2. CH}_3\text{CH}_2\text{Br}]{\text{1. NaNH}_2} \quad CH_3CH_2CH_2CH_2C\equiv CCH_2CH_3$$

Or he could make the 1-butyne he needed by alkylating ethyne:

$$HC\equiv CH \quad \xrightarrow[\text{2. CH}_3\text{CH}_2\text{Br}]{\text{1. NaNH}_2} \quad CH_3CH_2C\equiv CH$$

51. a. Only one product is obtained from the hydroboration-oxidation of 2-butyne because the alkyne is symmetrical. Two different products can be obtained from hydroboration-oxidation of 2-pentyne because the alkyne is not symmetrical.

b. Only one product will be obtained from hydroboration-oxidation of any symmetrical alkyne, such as 3-hexyne or 4-octyne.

$$CH_3CH_2C\equiv CCH_2CH_3 \qquad\qquad CH_3CH_2CH_2C\equiv CCH_2CH_2CH_3$$

3-hexyne   4-octyne

52. a. The first step forms a trans alkene. Syn addition to a trans alkene forms the threo pair of enantiomers.

b. The first step forms a cis alkene. Syn addition to a cis alkene forms the erythro pair of enantiomers but, since each asymmetric carbon is bonded to the same four groups, the product here is a meso compound.

53. (3*E*,6*E*)-3,7,11-trimethyl-1,3,6,10-dodecatetraene

The configuration of the double bond at the 1-position and at the 10-position is not specified because isomers are not possible at those positions, because there are two hydrogens bonded to C-1 and two methyl groups bonded to C-11.

54. **a.** $HC\equiv CH \xrightarrow{NaNH_2} HC\equiv C^- \xrightarrow{CH_3Br} HC\equiv CCH_3 \xrightarrow{NaNH_2} {}^-C\equiv CCH_3$

$\xrightarrow{CH_3CH_2CH_2CH_2CH_2Br}$

$CH_3CH_2CH_2CH_2CH_2C\equiv CCH_3$

$\xleftarrow[\text{Lindlar catalyst}]{H_2}$

b. $HC\equiv CH \xrightarrow{NaNH_2} HC\equiv C^- \xrightarrow{CH_3CH_2Br} HC\equiv CCH_2CH_3 \xrightarrow{NaNH_2} {}^-C\equiv CCH_2CH_3$

$\xrightarrow{CH_3CH_2CH_2Br}$

$\xleftarrow[\text{NH}_3 \text{ (liq), } -78 \text{ °C}]{Na} CH_3CH_2CH_2C\equiv CCH_2CH_3$

55. If HO^- and HOOH are added at the same time as BH_3 (or R_2BH), HO^- is a better nucleophile than the alkyne (no bonds have to be broken when HO^- reacts with the electrophile), so HO^-, instead of the alkyne, will react with BH_3 (or R_2BH).

$$HO^- + BH_3 \rightleftharpoons HO^+ {-} \bar{B}H_3$$

56. **a.** $HC\equiv CH \xrightarrow{NaNH_2} HC\equiv C^- \xrightarrow{CH_3CH_2Br} HC\equiv CCH_2CH_3 \xrightarrow{NaNH_2} {}^-C\equiv CCH_2CH_3$

$\searrow CH_3CH_2Br$

$\xleftarrow[-78 \text{ °C}]{Na/NH_3(liq)} CH_3CH_2C\equiv CCH_2CH_3$

(3*S*,4*R*)-4-bromo-
3-hexanol

(3*R*,4*S*)-4-bromo-
3-hexanol

b.

$$HC\equiv CH \xrightarrow{\text{NaNH}_2} HC\equiv C^- \xrightarrow{\text{CH}_3\text{CH}_2\text{Br}} HC\equiv CCH_2CH_3 \xrightarrow{\text{NaNH}_2} {}^-C\equiv CCH_2CH_3$$

(3R,4R)-4-bromo-
3-hexanol   (3S,4S)-4-bromo-
3-hexanol

Because anti addition occurs in the last step, the threo enantiomers are formed.

If $Na/NH_3(liq)/-78\,°C$ is used instead of H_2/Lindlar catalyst in the fifth step, the trans alkene will be formed. Reaction of Br_2 and H_2O with the trans alkene will form the erythro enantiomers.

57.

a.

b.

c.

d.

e.

f.

g.

h.

Chapter 7 Practice Test

1. What reagents could be used to convert the given starting material into the desired product?

    a.

    b.

2. Draw the enol tautomer(s) of the following ketones:

    a.   b.

3. Draw the structure for each of the following:

    a. *sec*-butylisobutylacetylene

    b. 2-methyl-1,3-cyclohexadiene

4. Indicate whether each of the following statements is true or false:

a.	A terminal alkyne is more stable than an internal alkyne.	T	F
b.	Propyne is more reactive than propene toward reaction with HBr.	T	F
c.	1-Butyne is more acidic than 1-butene.	T	F
d.	An sp^2 carbon is more electronegative than an sp^3 carbon.	T	F
e.	The reactions of internal alkynes are more regioselective than the reactions of terminal alkynes.	T	F
f.	Alkenes are more reactive than alkynes.	T	F

5. What is each compound's systematic name?

    a. $CH_3CHC \equiv CCH_2CH_2Br$
        $\quad\quad\; | $
        $\quad\quad\; CH_3$

    b. $CH_3CHC \equiv CCH_2CH_2OH$
        $\quad\quad\; | $
        $\quad\quad\; CH_3$

6.   What alkyne would be the best reagent to use for the synthesis of each of the following ketones?

$$\overset{\overset{\displaystyle O}{\displaystyle \|}}{}$$

a. $CH_3CH_2CH_2CCH_3$ b. $CH_3CH_2CCH_2CH_2CH_3$

7.   Rank the following compounds in order from most acidic to least acidic:

NH_3 $CH_3C \equiv CH$ CH_3CH_3 H_2O $CH_3CH = CH_2$

8.   Give an example of a ketone that has two enol tautomers.

9.   Show how the target molecules could be prepared from the given starting materials.

a. $CH_3CH_2C \equiv CH$ $\longrightarrow$ $CH_3CH_2CH_2CH_2CH_2CH_3$

b. $CH_3CH_2C \equiv CH$ $\longrightarrow$ $CH_3CH_2CHCH_2CH_2CH_3$
                                                                                 |
                                                                                Br

c. $CH_3CH_2C \equiv CH$ $\longrightarrow$ $CH_3CH_2CCH_2CH_2CH_3$ (with carbonyl O)

Delocalized Electrons and Their Effect on Stability, pK_a, and the Products of a Reaction

<u>**Important Terms**</u>

1,2-addition **(direct addition)**	addition to the 1- and 2-positions of a conjugated system.
1,4-addition **(conjugate addition)**	addition to the 1- and 4-positions of a conjugated system.
aliphatic compound	a saturated or unsaturated organic compound that does not contain an aromatic ring.
allene	a compound with two adjacent double bonds.
allylic carbon	a carbon adjacent to an sp^2 carbon of a carbon-carbon double bond.
allylic cation	a compound with a positive charge on an allylic carbon.
antiaromatic compound	a cyclic and planar compound with an uninterrupted cloud of electrons containing an even number of pairs of π electrons.
antibonding molecular orbital	the molecular orbital formed when out-of-phase orbitals overlap.
antisymmetric molecular orbital	a molecular orbital that does not have a plane of symmetry but would have if one half of the MO were turned upside down.
aromatic compound	a cyclic and planar compound with an uninterrupted cloud of electrons containing an odd number of pairs of π electrons.
benzylic carbon	a carbon, joined to other atoms by single bonds, that is bonded to a benzene ring.
benzylic cation	a compound with a positive charge on a benzylic carbon.
bonding molecular orbital	the molecular orbital formed when in-phase orbitals overlap.
bridged bicyclic compound	a bicyclic compound in which the rings share two nonadjacent carbons.
common intermediate	an intermediate that reaction pathways have in common.
concerted reaction	a reaction in which all the bond-making and breaking processes occur in a single step.
conjugate addition	addition to the 1- and 4-positions of a conjugated system.
conjugated diene	a compound with two conjugated double bonds.

conjugated double bonds	double bonds separated by one single bond.
contributing resonance structure	a structure with localized electrons that approximates the true structure of a compound with delocalized electrons.
cumulated double bonds	double bonds that are adjacent to one another.
cycloaddition reaction	a reaction in which two π electron-containing reactants combine to form a single cyclic product.
[4 + 2] cycloaddition reaction	a cycloaddition reaction in which six π electrons participate in the transition state with four π electrons coming from one reactant and two π electrons coming from the other reactant.
delocalization energy (resonance energy)	the extra stability associated with a compound as a result of having delocalized electrons.
delocalized electrons	electrons that are not localized on a single atom or between two atoms.
Diels-Alder reaction	a $[4+2]$ cycloaddition reaction.
diene	a hydrocarbon with two double bonds.
dienophile	an alkene that reacts with a diene in a Diels-Alder reaction.
direct addition (1,2-addition)	addition to the 1- and 2-positions of a conjugated system.
donation of electrons by resonance	donation of electrons through π bonds.
fused rings	rings that share two adjacent carbons.
electron delocalization	the sharing of electrons by more than two atoms.
endo	a substituent is endo if it and the bridge are on opposite sides of a bicyclic compound.
equilibrium control	thermodynamic control.
exo	a substituent is exo if it and the bridge are on the same side of a bicyclic compound.
highest occupied molecular orbital (HOMO)	the highest energy molecular orbital that contains electrons.
heteroatom	an atom other than carbon or hydrogen.
heterocyclic compound (heterocycle)	a cyclic compound in which one or more of the atoms of the ring are heteroatoms.

Hückel's rule or the 4*n* + 2 rule	a rule that gives the number of π electrons a compound must have in its π cloud in order to be aromatic.
isolated double bonds	double bonds separated from one another by more than one single bond.
kinetic control	when a reaction is under kinetic control, the relative amounts of the products depend on the rates at which they are formed.
kinetic product	the product that is formed the fastest.
linear combination of molecular orbitals (LCAO)	the combination of atomic orbitals to produce molecular orbitals.
localized electrons	electrons that are restricted to a particular locality.
lowest unoccupied molecular orbital (LUMO)	the lowest energy molecular orbital that does not contain electrons.
pericyclic reaction	a reaction that takes place in one step as a result of a cyclic reorganization of electrons.
phenyl group	C_6H_5—
polyene	a compound that has several double bonds.
polymer	a large molecule made by linking many small molecules together.
polymerization	the process of linking up many small molecules to form a polymer.
proximity effect	an effect caused by one species being close to another.
resonance	electron delocalization.
resonance contributor (resonance structure)	a structure with localized electrons that approximates the true structure of a compound with delocalized electrons.
resonance electron donation	donation of electrons through π bonds.
resonance electron withdrawal	withdrawal of electrons through π bonds.
resonance energy (delocalization energy) (resonance stabilization energy)	the extra stability a compound possesses as a result of having delocalized electrons.

resonance hybrid	the actual structure of a compound with delocalized electrons; it is represented by two or more resonance contributors with localized electrons.
***s*-cis-conformation**	the conformation in which two double bonds of a conjugated diene are on the same side of a connecting single bond.
separated charges	a positive and a negative charge that can both be neutralized by the movement of electrons.
***s*-trans-conformation**	the conformation in which two double bonds of a conjugated diene are on opposite sides of a connecting single bond.
symmetric molecular orbital	an orbital with a plane of symmetry so that one half is the mirror image of the other half.
thermodynamic control	when a reaction is under thermodynamic control, the relative amounts of the products depend on their stabilities.
thermodynamic product	the most stable product.
withdrawal of electrons by resonance	withdrawal of electrons through π bonds.

Solutions to Problems

1.   **a.**   **1.** If stereoisomers are not included, three different monosubstituted compounds are possible.

$$BrC\equiv CC\equiv CCH_2CH_3 \qquad HC\equiv CC\equiv CCHCH_3 \qquad HC\equiv CC\equiv CCH_2CH_2Br$$
$$\mid$$
$$Br$$

If stereoisomers are included, four different monosubstituted compounds are possible, because the second listed compound has an asymmetric center so it can have both the *R* and *S* configuration.

2. If stereoisomers are not included, two different monosubstituted compounds are possible.

$$BrCH=CHC\equiv CCH=CH_2 \qquad CH_2=CC\equiv CCH=CH_2$$
$$\mid$$
$$Br$$

If stereoisomers are included, three different monosubstituted compounds are possible, because the first compound has a double bond that can have cis-trans isomers.

b.   **1.** If stereoisomers are not included, five different disubstituted compounds are possible.

$$HC\equiv CC\equiv CCH_2CHBr_2 \qquad HC\equiv CC\equiv CCBr_2CH_3 \qquad HC\equiv CC\equiv CCHCH_2Br$$
$$\mid$$
$$Br$$

$$BrC\equiv CC\equiv CCH_2CH_2Br \qquad BrC\equiv CC\equiv CCHCH_3$$
$$\mid$$
$$Br$$

2. If stereoisomers are not included, five different disubstituted compounds are possible.

$$CH_2=CC\equiv CC=CH_2 \qquad CH_2=CHC\equiv CC=CHBr \qquad CH_2=CC\equiv CCH=CHBr$$
$$\mid \quad \mid \qquad\qquad\qquad\qquad \mid \qquad\qquad\qquad\qquad \mid$$
$$Br \quad Br \qquad\qquad\qquad\qquad Br \qquad\qquad\qquad\qquad Br$$

$$CH_2=CHC\equiv CCH=CBr_2 \qquad BrCH=CHC\equiv CCH=CHBr$$

c.   **1.** If stereoisomers are included, seven different disubstituted compounds are possible, because two of the compounds have asymmetric centers so each can have either the *R* or *S* configuration.

2. If stereoisomers are included, nine different disubstituted compounds are possible, because the second and third compounds can have cis-trans isomers, and the fifth compound can have cis-cis, trans-trans, and cis-trans isomers. (Note that cis-trans is the same as trans-cis.)

2.   Ladenburg benzene is a better proposal. It would form one monosubstituted compound and three disubstituted compounds, in accordance with what early chemists knew about the structure of benzene.

Dewar benzene is not in accordance with what early chemists knew about the structure of benzene, because it would form two monosubstituted compounds and six disubstituted compounds, and it would undergo electrophilic addition reactions.

3.   **a. 2, 4,** and **5** have delocalized electrons.

   1 does not have delocalized electrons, because π electrons cannot be moved toward an sp^3 carbon.

   3, 6, and **7** do not have delocalized electrons, because lone-pair electrons cannot be moved toward an sp^3 carbon.

2. $CH_3CH{=}CH{-}CH{=}CH{-}\overset{+}{C}H_2$ $CH_3\overset{+}{C}H{-}CH{=}CH{-}CH{=}CH_2$

   $CH_3CH{=}CH{-}\overset{+}{C}H{-}CH{=}CH_2$

4.

5.

4.   The resonance contributor that makes the greatest contribution to the hybrid is labeled **A**. **B** contributes less to the hybrid than **A**, and **C** contributes less to the hybrid than **B**.

   a.   Solved in the text.

   b.   **A**   **B**

   A is more stable than **B** because **B** has separated charges and has a positive charge on an oxygen.

   c.   **B**   **A**

   A is more stable than **B** because the negative charge in **A** is on an oxygen, whereas the negative in **B** is on carbon, which is less electronegative than oxygen.

d.

A   C   B

A is more stable than **B**, because **A** does not have separated charges. **B** is more stable than **C**, because the electronegative oxygen atom is closer to the positive charge in **C**. **C** is so unstable that it can be neglected.

e.

B   A

A is more stable than **B**, because the positive charge in **A** is on a less electronegative atom. (N is less electronegative than O.)

f. $CH_3\overset{+}{C}H-CH=CHCH_3 \longleftrightarrow CH_3CH=CH-\overset{+}{C}HCH_3$

A   B

A and B are equally stable

5.   **a.** $\overset{\delta+}{CH_3}C=CH=\overset{\delta+}{CH}CH_3$ with CH_3 below

c.   (structure with $\delta-$ and $\delta-$ O)

e.   (structure with $\overset{\delta+}{OH}$, C, $\delta+$, CH_3, $NHCH_3$)

b.   (structure with O $\delta-$, C, $\delta+$, CH_3, OCH_3)

d.   (ring structure with $\delta+$ and $\delta-$ O)

f. $\overset{\delta+}{CH_3}CH=CH=\overset{\delta+}{CH}CH_3$

6.   **a.**   All the carbon-oxygen bonds in the carbonate ion should be the same length, because each carbon-oxygen bond is represented in one resonance contributor by a double bond and in two resonance contributors by a single bond.

b.   Because the two negative charges are shared equally by three oxygens, each oxygen will have two-thirds of a negative charge.

7.   **a.**

$$CH_3CH_2\overset{\overset{CH_2}{||}}{C}-\overset{+}{C}H_2 \longleftrightarrow CH_3CH_2\overset{\overset{+}{C}H_2}{|}{C}=CH_2$$

Less stable, because the positive charge is shared by two primary allylic carbons.

$$CH_3CH_2CH=CH-\overset{+}{C}H_2 \longleftrightarrow CH_3CH_2\overset{+}{C}H-CH=CH_2$$

More stable, because the positive charge is shared by a primary allylic and a secondary allylic carbon.

b.

Less stable, because the positive charge is on a primary allylic carbon.

More stable, because the positive charge is on a secondary allylic carbon.

c.

Less stable, because the negative charge is not delocalized.

More stable, because the negative charge is delocalized.

d.

Less stable, because the positive charge is shared by an O and an N.

More stable, because the positive charge is shared by two nitrogens. Nitrogen is less electronegative than oxygen, so nitrogen is more comfortable with a positive charge.

8. The second species has the greatest delocalization energy; it has three resonance contributors and none of them have separated charges. (See the answer to Problem 6.)

The first species is the next most stable has two resonance contributors and none of them has separated charges.

The third species has the smallest delocalization energy; it has two resonance contributors, one of which has separated charges.

9. **a.** In the case of 9 pairs of π electrons, there are 18 electrons. Therefore, $4n + 2 = 18$ where $n = 4$.

b. Because it has an odd number of pairs of π electrons, it will be aromatic if it is cyclic and planar, and if every atom in the ring has a p orbital.

10. **a.**

This is the only one that is aromatic; it is cyclic, planar, every ring atom has a p orbital, and it has one pair of π electrons.

The first compound is not aromatic, because one of the atoms is sp^3 hybridized and therefore does not have a p orbital.

The third compound is not aromatic, because it has two pairs of π electrons.

b. This is the only one that is aromatic; it is cyclic, planar, every ring atom has a *p* orbital, and it has three pairs of π electrons.

The first compound is not aromatic, because one of the atoms is sp^3 hybridized and therefore does not have a *p* orbital.

The third compound is not aromatic, because it has four pairs of π electrons.

11.   Solved in the text.

12.   Only **e** is aromatic.

e is aromatic, because it is cyclic, planar, every atom in the ring has a *p* orbital, and it has seven pairs of π electrons.

c is not aromatic, because it has two pairs of π electrons.

a, **b**, and **d** are not aromatic, because each compound has two pairs of π electrons and every atom in the ring does not have a *p* orbital.

f is not aromatic, because it is not cyclic.

13. **a.** In fulvene, the electrons in the exocyclic double bond (a double bond attached to a ring) move toward the five-membered ring, because the resonance contributor that results has a relatively large predicted stability since it is aromatic.

fulvene

b. In calicene, the electrons in the double bond between the two rings move toward the five-membered ring, because that results in a resonance contributor with two aromatic rings.

calicene

14. **a.** $CH_3CH_2\overset{..}{N}H_2$ sp^3 **b.** sp^2 **c.** $CH_3CH_2C\equiv N:$ sp

15.

16.   The lone-pair electrons in **A** and **D** are sp^2 electrons; therefore, they are not part of the π cloud.

The lone-pair electrons on **B** and **C** are π electrons and are part of the π cloud. Thus, **B** has three pairs of π electrons and **C** has five pairs of π electrons, making them aromatic compounds. If these π electrons are protonated, they will no longer be part of the π cloud (they cannot be delocalized) and the compounds will not be aromatic.

17.   **a.**   The nitrogen atom (the atom at the bottom of the epm) in pyrrole has a partial positive charge because it donates electrons by resonance into the ring.

b.   The nitrogen atom (the atom at the bottom of the epm) in pyridine cannot donate electrons by resonance; it withdraws electrons from the ring inductively because it is the most electronegative atom in the molecule. Thus, this nitrogen is electron rich.

c.   The relatively electronegative nitrogen atom in pyridine withdraws electrons from the ring.

18.   **a.**   Cyclopentadiene has a lower pK_a value. That is, it is a stronger acid. When cyclopentadiene loses a proton, a relatively stable aromatic compound is formed. When cycloheptatriene loses a proton, an unstable antiaromatic compound is formed (Section 8.11). Recall that the more stable the base, the stronger its conjugate acid.

b.   Cyclopropane has a lower pK_a value because a very unstable antiaromatic compound is formed when cyclopropene loses a proton.

19.   3-Bromocyclopropene is more soluble in water because it is more apt to ionize since an aromatic compound is formed when its carbon-bromine bond breaks.

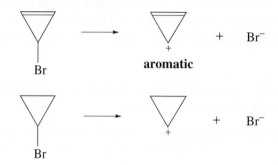

20.   **c** is antiaromatic, because it is cyclic, planar, every atom in the ring has a *p* orbital, and it has two pairs of π electrons.

When you did Problem 12, you found that **e** is aromatic.

a, **b**, **d**, and **f** are neither aromatic nor antiaromatic, **a**, **b**, and **d** have ring atoms that do not have *p* orbitals, and **f** is not cyclic.

21.   Cyclobutadiene has one bonding molecular orbital, two nonbonding molecular orbitals, and one antibonding molecular orbital. It has four π electrons; two π electrons are in the bonding π molecular orbital, and each of the two nonbonding molecular orbitals contains one π electron.

22.   It is not possible for an atom that bears a single electron to be part of the π cloud.

Remember that to be aromatic, all the bonding MOs that make up the π cloud must be filled (each containing two electrons) and there must be no partially filled orbitals. Since a radical has an unpaired electron, there will be an orbital with a single electron (a partially filled MO), so it cannot be aromatic. (A radical can, however, be attached to an aromatic compound. For example, the benzyl radical has a methylene radical attached to an aromatic benzene ring.)

23.   The Frost circles show that species with completely filled bonding MOs, and with electrons in no other orbitals, are aromatic (for example, the cycloheptatrienyl cation and the cyclopropenyl cation).

cycloheptatrienyl cation   cyclopropenyl cation

The species with unpaired electrons in degenerate MOs is antiaromatic (the cycloheptatrienyl anion).

cycloheptatrienyl anion

24.   The smaller the heat of hydrogenation (the positive value of $\Delta H°$), the more stable the compound. Therefore, the relative stabilities of the dienes are

conjugated diene > isolated diene > cumulated diene

25.

2,5-dimethyl-2,4-hexadiene   2,4-hexadiene   1,3-pentadiene   1,4-pentadiene

26.   **a.**   The compound with delocalized electrons is more stable than the compound in which all the electrons are localized.

electrons are delocalized   electrons are localized

   b.   Because nitrogen is less electronegative than oxygen, it is better able to share the positive charge.

more stable

   c.   In order for electron delocalization to occur, the atoms that share the π electrons must be in the same plane so their p orbitals can overlap. The two bulky *tert*-butyl groups do not allow enough room for the group with the positively charged carbon to be in the same plane as the benzene ring. Therefore, the carbocation cannot be stabilized by electron delocalization because the p orbital of the positively charged carbon cannot overlap the p orbitals of the benzene ring since they are not in the same plane.

more stable

27.   The ψ_3 molecular orbital of 1,3-butadiene has three nodes (two vertical and one horizontal).

   The ψ_4 molecular orbital of 1,3-butadiene has four nodes (three vertical and one horizontal).

28.   **a.** ψ_1 and ψ_2 are bonding molecular orbitals, and ψ_3 and ψ_4 are antibonding molecular orbitals.
   b. ψ_1 and ψ_3 are symmetric molecular orbitals, and ψ_2 and ψ_4 are antisymmetric molecular orbitals.
   c. ψ_2 is the HOMO and ψ_3 is the LUMO in the ground state.
   d. ψ_3 is the HOMO and ψ_4 is the LUMO in the excited state.
   e.   If the HOMO is symmetric, the LUMO is antisymmetric, and vice versa.

29.   The MO of benzene is more stable because it has six bonding interactions, whereas the MO of 1,3,5-hexatriene has five bonding interactions.

30. In each case, the compound shown is the stronger acid because the negative charge that results when it loses a proton can be delocalized. Electron delocalization stabilizes the base and the more stable the base, the more acidic its conjugate acid. Electron delocalization is not possible for the other compound in each pair.

a. $CH_3-\overset{\overset{\displaystyle O}{\|}}{C}-OH \longrightarrow H^+ + CH_3-\overset{\overset{\displaystyle O}{\|}}{C}-O^- \longleftrightarrow CH_3-\overset{\overset{\displaystyle O^-}{|}}{C}=O$

b. $CH_3CH=CHOH \longrightarrow H^+ + CH_3CH=CHO^- \longleftrightarrow CH_3\bar{C}HCH=O$

c. $CH_3CH=CH\overset{+}{N}H_3 \longrightarrow H^+ + CH_3CH=CHNH_2 \longleftrightarrow CH_3\bar{C}HCH=\overset{+}{N}H_2$

31. **a.** Ethylamine is a stronger base because when the lone pair on the nitrogen in aniline is protonated, it can no longer be delocalized into the benzene ring. Therefore, aniline is less apt to share its electrons with a proton.

b. Ethoxide ion is a stronger base because a negatively charged oxygen is a stronger base than a neutral nitrogen.

c. Ethoxide ion is a stronger base because when the phenolate ion is protonated, the pair of electrons that is protonated can no longer be delocalized into the benzene ring. Therefore, the phenolate ion is less apt to share its electrons with a proton.

32. The carboxylic acid is the most acidic because its conjugate base has greater delocalization energy than does the conjugate base of phenol. The alcohol is the least acidic because, unlike the negative charge on the conjugate base of phenol, the negative charge on the conjugate base of the alcohol cannot be delocalized.

33. Solved in the text.

34. Electron withdrawal by the methoxy group will make it a stronger acid (because it will make the conjugate base weaker).

Electron donation by the methoxy group will make it a weaker acid (because it will make the conjugate base stronger).

The pK_a values show that the methoxy substituted acid is a weaker acid (it has a larger pK_a value). Therefore, we know that resonance electron donation is a more important effect than inductive electron withdrawal.

35. Recall that, if a more stable carbocation can be formed as a result of carbocation rearrngement, rearrangement will occur.

36.   **a.**   Solved in the text.

   b.   The contributing resonance structures show that there are two sites that can be protonated, the lone pair on nitrogen or the lone pair on carbon.

37.   **a.**   The more reactive double bond is the one that forms a tertiary carbocation.

Instead of Br⁻ being the nucleophile that adds to the tertiary carbon, the π bond can be the nucleophile. In that case, a stable six-membered ring will be formed. (See the Problem-Solving Strategy on page 260 of the text.) This is expected to be a minor product because, unlike the above reaction of the carbocation with Br⁻, bond breaking is required to form the product.

   b.   The double bond is more reactive than the triple bond. The reaction forms a new asymmetric center so a pair of enantiomers is formed.

It is unlikely that the triple bond will act as a nucleophile, because it would have to form an unstable vinylic cation intermediate.

   c.   The more reactive double bond is the one that forms a tertiary carbocation.

38.   The indicated double bond is the most reactive in an electrophilic addition reaction with HBr because addition of an electrophile to this double bond forms the most stable carbocation (a tertiary allylic cation).

39.   **a.** $CH_3CH{=}CH{-}CH{=}CHCH_3$

$$CH_3\overset{+}{\underset{\underset{Cl}{|}}{CH}}{-}CH{-}CH{=}CHCH_3 \quad\longleftrightarrow\quad CH_3\underset{\underset{Cl}{|}}{CH}{-}CH{=}CH{-}\overset{+}{CH}CH_3$$

$$\underset{\text{1,2-addition product}}{CH_3\overset{\overset{Cl}{|}}{CH}\overset{\overset{Cl}{|}}{CH}CH{=}CHCH_3} \qquad + \qquad \underset{\text{1,4-addition product}}{CH_3\overset{\overset{Cl}{|}}{CH}CH{=}CH\overset{\overset{Cl}{|}}{CH}CH_3}$$

b. $\underset{\underset{CH_3}{|}}{CH_3CH_2}\overset{\overset{Br}{|}}{C}{-}\underset{\underset{CH_3}{|}}{C}{=}CHCH_3 \qquad + \qquad CH_3CH_2\underset{\underset{CH_3}{|}}{C}{=}\underset{\underset{CH_3}{|}}{\overset{\overset{Br}{|}}{C}}CHCH_3$

1,2-addition product   1,4-addition product

c.

+

1,2-addition product   1,4-addition product

d.

40.   first reaction:

$$CH_2-\overset{*}{C}H-CH=CH_2$$
$$\ \ |\ \ \ \ \ \ |$$
$$\ \ Cl\ \ \ \ Cl$$

This compound has an asymmetric center, so both the *R* and *S* stereoisomers will be obtained. (Note that *E* and *Z* stereoisomers are not possbile for this double bond.)

$$CH_2-CH=CH-CH_2$$
$$\ \ |\ \ \ \ \ \ \ \ \ \ \ \ \ \ |$$
$$\ \ Cl\ \ \ \ \ \ \ \ \ \ \ Cl$$

This compound has a double bond, so both the *E* and *Z* stereoisomers will be obtained.

second reaction:

$$CH_3\overset{*}{C}H-CH=CH_2$$
$$\ \ \ \ \ |$$
$$\ \ \ \ \ Br$$

This compound has an asymmetric center, so both the *R* and *S* stereoisomers will be obtained. (Note that *E* and *Z* stereoisomers are not possbile for this double bond.)

$$CH_3CH=CH-CH_2$$
$$\ \ \ \ \ \ \ \ \ \ \ \ \ \ \ \ \ \ |$$
$$\ \ \ \ \ \ \ \ \ \ \ \ \ \ \ \ \ \ Br$$

This compound has a double bond, so both the *E* and *Z* stereoisomers will be obtained.

41. $CH_2=CH-CH=CH-CH=CH_2 \xrightarrow{\ HBr\ }$

$$CH_3-CH-CH=CH-CH=CH_2$$
$$\ \ \ \ \ \ \ \ \ \ |$$
$$\ \ \ \ \ \ \ \ \ \ Br$$

$$CH_3-CH=CH-CH-CH=CH_2$$
$$\ |$$
$$\ Br$$

$$CH_3-CH=CH-CH=CH-CH_2$$
$$\ |$$
$$\ Br$$

42.   **a.**   The chlorine adds so that the positive charge in the resonance contributor is on a secondary allylic carbon. (If the chlorine had added to the other double bond, the positive charge would be on a primary allylic carbon.)

b. The proton adds so that the positive charge in the carbocation is shared by a tertiary allylic and a secondary allylic carbon. (If the proton had added to the other double bond, the positive charge would be shared by two secondary allylic carbons.)

43.   **a.** Addition at C-1 forms the more stable carbocation because the positive charge is shared by two secondary allylic carbons. If the deuterium had added to C-4, the positive charge would have been shared by a secondary allylic and a primary allylic carbon.

   b. DCl was used to cause the 1,2- and 1,4-products to be different. If HCl had been used, the 1,2- and 1,4-products would have been the same.

44.   She should follow her friend's advice. If she uses 2-methyl-1,3-cyclohexadiene, the product that is formed faster will be 3-chloro-3-methylcyclohexene, both if the proximity effect controls which product is formed faster *and* if the more stable transition state controls which product is formed faster, because this product is formed through a transition state in which the positive charge is primarily on a tertiary allylic carbon. Thus the experiment will not be able to establish which of the two effects controls which product is formed faster.

3-chloro-3-methylcyclohexene

If she follows her friend's advice and uses 1-methyl-1,3-cyclohexadiene, the product that is formed faster will be 3-chloro-1-methylcyclohexene only if the proximity effect controls which product is formed faster. The product will be 3-chloro-3-methylcyclohexene if the more stable transition state controls which product is formed faster, because only this product is formed through a transition state in which the positive charge is primarily on a tertiary allylic carbon.

3-chloro-1-methylcyclohexene   3-chloro-3-methylcyclohexene

45.   **a.** The rate-determining step is formation of the carbocation.

   b. The product-determining step is reaction of the carbocation with the nucleophile.

46.   **a.**   Solved in the text.

b.

HCl →

Cl
CH₃

kinetic product

+

CH₃

Cl

thermodynamic product

c.

CH=CHCH₃

HCl →

CH=CHCH₃
Cl

kinetic product

+

CH—CHCH₃
Cl

thermodynamic product

Notice that the 1,2-product is always the kinetic product.
The thermodynamic product is the product with the most highly substituted double bond.

47.

Cl

kinetic product
thermodynamic product

+

Cl

The first compound is the kinetic product because it is the 1,2-product.
The first compound is the thermodynamic product because it is more stable because the double bond is conjugated with the benzene ring.

48.   **a.**

O
‖
C
CH₃

C
CH₃
‖
O

c.

H₃C

O

O

H₃C

O

b.

C≡N

d.

H₃C CH₃ O

H₃C CH₃ O

49.   First draw the resonance contributors to determine where the charges are on the reactants. The major product is obtained by joining the negatively charged carbon of the diene with the positively charged carbon of the dienophile.

Because the reaction creates an asymmetric center, the product will be a racemic mixture.

50.   The resonance contributors show that if the electron-donating substituent is at the end of the conjugated system, it and the electron-withdrawing substituent of the dienophile will be adjacent to one another in the major product of the Diels-Alder reaction.

If the electron-donating substituent is not at the end of the conjugated system, it and the electron-withdrawing substituent of the dienophile will be opposite to one another in the major product of the Diels-Alder reaction.

51.   a.

b.

52.   **A** and **C** will not react, because they are both locked in an *s*-trans conformation.

D and **E** will react, because they are both locked in an *s*-cis conformation.

B and **F** will react, because they can rotate into an *s*-cis conformation.

53.   **a.**

   b.

54.   Solved in the text.

55.   **a.** It is not optically active, because it is a meso compound.
(It has two asymmetric centers and a plane of symmetry.)

b. It is not optically active, because it is a racemic mixture.
(Identical amounts of the enantiomers will be obtained.)

56.   **a.**   +   **c.**   +   **e.**   +

   b.   +   **d.**   +   **f.**   +

57.   **a, b, c, e, f, l, m, n, o**

58.   **a** and **b.**

   1. $\overset{-}{\ddot{C}}H_2-\overset{+}{N}\equiv N: \quad \longleftrightarrow \quad CH_2=\overset{+}{N}=\ddot{\ddot{N}}:^-$

   More stable, because the
   negative charge is on nitrogen
   rather than on carbon.

2.

More stable because the
negative charge is on oxygen
rather than on nitrogen
(oxygen is more electronegative).

3.

Both are equally stable.

Additional resonance structures could be drawn for each of these three species,
but they are relatively unstable because each has an atom with an incomplete octet.

59.   a.   **b.**

60.   a.

b.

c.

d.

61. Draw the resonance contributors. Then have the Cl pointing up in all the products, with the OH pointing up in one product and down in the other. Notice that the products in row 4 are the mirror images of the products in row 1, and that the products in row 3 are the mirror images of the products in row 2. (The mirror images would also have been obtained if the Cl pointed down in all the products with the OH pointing up in one structure and down in the other.)

62. a. different compounds   d. resonance contributors
b. different compounds   e. different compounds
c. resonance contributors

Notice that in the structures that are different compounds, both atoms and electrons have changed their locations. In contrast, in the structures that are resonance contributors, only the electrons have moved.

63. a. There are six linear dienes with molecular formula C_6H_{10}.

b. Two are conjugated dienes. $CH_2=CHCH=CHCH_2CH_3$
$CH_3CH=CHCH=CHCH_3$

c. Two are isolated dienes. $CH_2=CHCH_2CH=CHCH_3$
$CH_2=CHCH_2CH_2CH=CH_2$

There are also two cumulated dienes. $CH_2=C=CHCH_2CH_2CH_3$
$CH_3CH=C=CHCH_2CH_3$

64. a. 1.

2.

The two resonance contributors have the same stability and therefore contribute equally to the resonance hybrid.

3.

major   minor   minor   minor   major

4. $CH_3-N\overset{+}{\underset{\ddot{O}^-}{\diagdown}}^O$ ⟷ $CH_3-\overset{+}{N}\overset{O^-}{\diagup}_O$

The two resonance contributors have the same stability and therefore contribute equally to the hybrid.

5. $CH_3\bar{C}H-\overset{+}{N}\overset{O}{\underset{\ddot{O}^-}{\diagdown}}$ ⟷ $CH_3\overset{-}{\ddot{C}}H-\overset{+}{N}\overset{O^-}{\underset{O}{\diagup}}$ ⟷ $CH_3CH=\overset{+}{N}\overset{O^-}{\underset{O^-}{\diagup}}$

   minor   minor   major

6. $CH_3CH=CH-\overset{+}{CH_2}$ ⟷ $CH_3\overset{+}{CH}-CH=CH_2$

   minor   major

7.

The five contributors are equally stable and therefore contribute equally to the resonance hybrid.

8. $CH_3CH_2\overset{O}{\underset{\ddot{O}CH_2CH_3}{\diagdown C \diagup}}$ ⟷ $CH_3CH_2\overset{O^-}{\underset{\overset{+}{O}CH_2CH_3}{\diagdown C \diagup}}$

      major   minor

9. $CH_3CH=CHCH=\overset{+}{CH}CH_2$ $CH_3\overset{+}{CH}CH=CHCH=CH_2$

   minor   major
   (the positive   (the positive
   charge is on   charge is on
   a primary $CH_3CH=CH\overset{+}{CH}CH=CH_2$ a secondary
   allylic carbon)   major (the positive charge is on   allylic carbon)
                     a secondary allylic carbon)

10.

      minor   major

The major contributor has a negative charge on oxygen, which is more stable than a contributor with a negative charge on carbon.

11. $CH_3\overset{..}{\underset{-}{C}}H-C\equiv N$ ⟷ $CH_3CH=C=\overset{..}{N}\overset{-}{:}$

      minor   major

12.

  major   minor   minor   minor   major

13.

  major   minor

14.

  minor   minor   major

The major contributor has a negative charge on oxygen, which is more stable than a contributor with a negative charge on carbon.

15. Notice that the electrons on the center carbon can be delocalized onto both of the carbonyl oxygens.

           minor   major

                                     major

b. **2, 4,** and **7**

65. **a.**   **b.**   **c.**   **d.** ⬜

  It is aromatic.   It is aromatic.   It is aromatic.   It is not antiaromatic.

66. Both compounds form the same product when they are hydrogenated, so the difference in the heats of hydrogenation will depend only on the difference in the stabilities of the reactants. Because 1,2-pentadiene has cumulated double bonds and 1,4-pentadiene has isolated double bonds, 1,2-pentadiene is less stable and, therefore, will have a greater heat of hydrogenation (a more negative $\Delta H°$).

$$CH_2=C=CHCH_2CH_3 \xrightarrow[\text{Pd/C}]{H_2} CH_3CH_2CH_2CH_2CH_3$$

1,2-pentadiene

$$CH_2=CHCH_2CH=CH_2 \xrightarrow[\text{Pd/C}]{H_2} CH_3CH_2CH_2CH_2CH_3$$

1,4-pentadiene

67.   **a.** $CH_3\overset{+}{C}HCH=CH_2$
This makes the greater contribution because the positive charge is on a secondary allylic carbon.

b.   This makes the greater contribution because the negative charge is on an oxygen.

c. This makes the greater contribution because the positive charge is on a tertiary allylic carbon.

d. This makes the greater contribution because a secondary benzylic cation is more stable than a secondary alkyl cation.

68.   aromatic

antiaromatic

neither aromatic nor antiaromatic

The first five compounds has an sp^3 carbon, so they do not have an uninterrupted π cloud, and the last one is not planar, so the p orbitals cannot overlap with the p orbitlas on the adjacent carbons.

69.   **a.** The resonance contributors show that the carbonyl oxygen has the greater electron density.

carbonyl oxygen

b. The compound on the right has the greater electron density on its nitrogen, because the compound on the left has a resonance contributor with a positive charge on the nitrogen as a result of electron delocalization.

c. The compound with the cyclohexane ring has the greater electron density on its oxygen, because the lone pair on the nitrogen can be delocalized only onto the oxygen.

There is less delocalization onto oxygen by the lone pair in the compound with the benzene ring (path **a**) because the lone pair can also be delocalized away from the oxygen into the benzene ring (path **b**).

70.   Remember that an sp^2 nitrogen is more electronegative than an sp^3 nitrogen, and the more electronegative the atom to which a hydrogen is attached, the stronger the acid. The stronger the acid, the weaker its conjugate base.

The lone pair is localized and the nitrogen is sp^2 hybridized, which is not as strong a base as an sp^3 hybridized nitrogen.

These are weak bases because the lone-pair electrons are delocalized.

The strongest base because the lone pair is localized and the nitrogen is sp^3 hybridized.

71.   The methyl group on benzene can lose a proton easier than the methyl group on cyclohexane because the electrons left behind on the carbon in the former can be delocalized into the benzene ring. In contrast, the electrons left behind in the other compound cannot be delocalized.

72.   The triphenylmethyl carbocation is stable because the positive charge is shared by 10 carbons (the central carbon and three carbons of each of the three benzene rings) as a result of electron delocalization.

73.   a.

b.

74.   a.   The structure shown below is the stronger acid because it has the weaker conjugate base. When this compound loses a proton, the electrons left behind can be delocalized onto six different carbons. The electrons left behind on the conjugate base of the other compound can be delocalized onto only three different carbons; they cannot be delocalized into the second benzene ring.

b.   The first compound is the most stable (weakest) base, because it has a resonance contributor that is aromatic. Therefore, the other base (the one on the right) is the stronger base.

aromatic

75.

CH₃CH₂O, CH—C≡N

#1 most stable because the negative
charge is on oxygen

CH₃CH₂O, CH=C=N⁻

#2 because the negative
charge is on nitrogen

CH₃CH₂O, CH—C≡N

#3 because the negative
charge is on carbon

CH₃CH₂Ȯ⁺, CH—C≡N

#4 the least stable because the
negative charge is on carbon
and it has separated charges

76. The more the electrons that are left behind when the proton is removed can be delocalized, the greater the stability of the base. The more stable the base, the more acidic its conjugate acid.

The negative charge on the base on the left can be delocalized onto two other carbons; the negative charge on the base in the middle can be delocalized onto one other carbon; the negative charge on the base on the right cannot be delocalized.

  >   >

Therefore, the compounds have the relative acidities shown below:

  >   > [structure]

77. **a.** CH₃—C(=O)—O⁻   The negative charge is shared by two oxygens.

b. CH₃CH₂C̈H—C(=O)—CH₃   The negative charge is shared by a carbon and an oxygen.

c. CH₃—C(=O)—C̈H—C(=O)—CH₃   The negative charge is shared by a carbon and two oxygens.

d. [succinimide N⁻ structure]   The negative charge is shared by a nitrogen and two oxygens.

78. The stronger base is the less stable base of each pair in Problem 77.

a. CH₃—C(=O)—CH₂O⁻   Less stable because the negative charge cannot be delocalized.

b.   Less stable because the negative charge cannot be delocalized.

c.   Less stable because the negative charge can be delocalized onto only one oxygen.

d.

Less stable because the negative charge can be delocalized onto only one oxygen.

79.   The resonance contributor shown here indicates which nitrogen is most apt to be protonated (the one with the greatest negative charge) and which nitrogen is least apt to be protonated (the one with the greatest positive charge).

80.   The following compound is the strongest acid, because it is the only one that forms a conjugate base that is aromatic. Recall that the more stable (weaker) the base, the stronger is its conjugate acid.

81.   The resonance contributors of pyrrole are more stable because the positive charge is on nitrogen. In furan, the positive charge is on oxygen which, being more electronegative, is less stable with a positive charge.

82.   **A** is the most acidic because the electrons left behind when the proton is removed can be delocalized onto two oxygens. **B** is the next most acidic because the electrons left behind when the proton is removed can be delocalized onto one oxygen. **C** is the least acidic because the electrons left behind when the proton is removed cannot be delocalized.

83.   **a.** It has eight molecular orbitals.
b. ψ_1, ψ_2, ψ_3, and ψ_4 are bonding molecular orbitals; ψ_5, ψ_6, ψ_7, and ψ_8 are antibonding molecular orbitals.
c. ψ_1, ψ_3, ψ_5, and ψ_7 are symmetric molecular orbitals; ψ_2, ψ_4, ψ_6, and ψ_8 are antisymmetric molecular orbitals.
d. ψ_4 is the HOMO and ψ_5 is the LUMO in the ground state.
e. ψ_5 is the HOMO and ψ_6 is the LUMO in the excited state.
f. If the HOMO is symmetric, the LUMO is antisymmetric, and vice versa.
g. It has seven nodes between the nuclei. It also has one node that passes through the nuclei.

84.

85.   The reaction of 1,3-cyclohexadiene with Br_2 forms 3,4-dibromocyclohexene as the 1,2-addition product and 3,6-dibromocyclohexene as the 1,4-addition product. (Recall that in naming the compounds, the double bond is at the 1,2-position.) The reaction of 1,3-cyclohexadiene with HBr forms only 3-bromocyclohexene, so it is both the 1,2-addition product and the 1,4-addition product.

86.   **a.**   Only part **a** involves the reaction of two unsymmetrically substituted reactants. Therefore, only for part **a** do we need to look at the charge distribution in the reactants to determine the major product of the reaction.

We now need to join the negative end of the diene and the positive end of the dienophile.

b.

c.

d.

87.   Numbering the carbons in the conjugated system will help you determine the 1,2- and 1,4-addition products.

a.   1.

2. (structure with CH₃) + HBr ⟶ (product with Br, CH₃)

or

(structure with CH₂) + HBr ⟶ (product with Br, CH₃)

3. (structure with CH₃, CH₃) + H₂O $\xrightarrow{H_2SO_4}$ (product with CH₃, HO, CH₃)

or

(structure with CH₃, CH₃) + H₂O $\xrightarrow{H_2SO_4}$ (product with CH₃, HO, CH₃)

b. 1. (structure with HO, Br)

2. (structure with CH₃, Br)

3. (structure with HO, CH₃, CH₃)

88.   a. $CH_2\text{=}CHCH\text{=}CHCH\text{=}CH_2$ $\xrightarrow{HBr}$ $CH_3\overset{|}{\underset{Br}{C}}HCH\text{=}CHCH\text{=}CH_2$ +
          1,3,5-hexatriene   **A**

$CH_3CH\text{=}CH\overset{|}{\underset{Br}{C}}HCH\text{=}CH_2$ + $CH_3CH\text{=}CHCH\text{=}CHCH_2Br$
B   **C**

b. **A** will predominate if the reaction is under kinetic control because it is the 1,2-product and, therefore, will be the product formed most rapidly as a result of the proximity effect. In addition, **A** will be the 1,2-product regardless of which end of the conjugated system reacts with the electrophile.

c. **C** will predominate if the reaction is under thermodynamic control because it is the most stable diene. (It is the most substituted conjugated diene.)

89.   The diene is the nucleophile, and the dienophile is the electrophile in a Diels-Alder reaction.

a. An electron-donating substituent in the diene would increase the rate of the reaction, because electron donation would increase its nucleophilicity.

b. An electron-donating substituent in the dienophile would decrease the rate of the reaction, because electron donation would decrease its electrophilicity.

c. An electron-withdrawing substituent in the diene would decrease the rate of the reaction, because electron withdrawal would decrease its nucleophilicity.

90.   **a.**   Addition of an electrophile to C-1 forms a carbocation with two resonance contributors, a *tertiary allylic cation* and a *secondary allylic cation*. Addition of an electrophile to C-4 forms a carbocation with two resonance contributors, a *tertiary allylic cation* and a *primary allylic cation*. Therefore, addition to C-1 results in formation of the more stable carbocation intermediate, and the more stable intermediate leads to the major products.

kinetic product   thermodynamic product

b.   Addition of an electrophile to C-1 forms a carbocation with two resonance contributors; both are *tertiary allylic* cations. Addition of an electrophile to C-4 forms a carbocation with two resonance contributors, a *secondary allylic* cation and a *primary allylic* cation. Therefore, addition to C-1 results in formation of the more stable carbocation. Only one product is formed, because the carbocation is symmetrical.

This is the only product because
the carbocation is symmetrical.

91.   **a** and **d.**

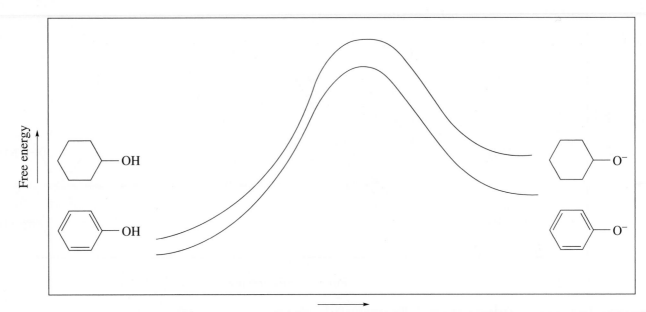

Progress of the reaction

d. The 2nd, 3rd, and 4th resonance contributors in **c** are more stable than those resonance contributors in **b**, because in **b** a positive charge is on the most electronegative atom (the oxygen) and there is charge separation. Therefore, the phenolate ion has greater electron delocalization than phenol.
Thus, as shown in the energy diagram, the difference in energy between the phenolate ion and the cyclohexoxide ion is greater than the difference in energy between phenol and cyclohexanol.

e. Because of greater electron delocalization in the phenolate ion compared to that in phenol, phenol has a larger K_a than cyclohexanol.

f. Because it has a larger K_a (a lower pK_a), phenol is a stronger acid.

92.

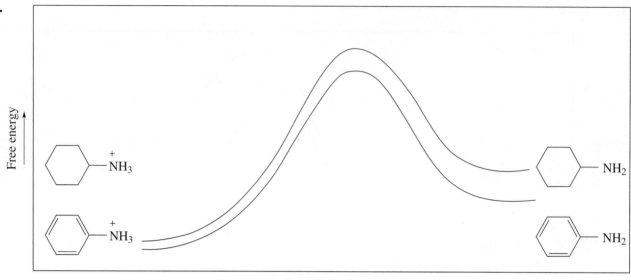

Progress of the reaction

b.

c.

d. Aniline has greater electron delocalization than the anilinium ion. Thus in the energy diagram, the difference in energy between aniline and cyclohexylamine is greater than the difference in energy between the anilinium ion and the cyclohexylammonium ion.

e. Because of greater electron delocalization in aniline compared to that in the anilinium ion, the anilinium ion has a larger K_a than the cyclohexylammonium ion.

f. Because it has a larger K_a (a lower pK_a), the anilinium ion is a stronger acid than the cyclohexylammonium ion. Therefore, cyclohexylamine is a stronger base than aniline. (The stronger the acid, the weaker its conjugate base.)

93. **a.**

Because the reaction creates an asymmetric center in the product, the product will be a racemic mixture.

b.

Even though both reactants are unsymmetrically substituted, they will be aligned primarily as shown, because of the relatively stable tertiary benzylic cation and delocalization of the π electrons of the dienophile onto the oxygen.

Because the reaction creates an asymmetric center, the product will be a racemic mixture.

c. Looking at where the partial positive charges are in the two resonance hybrids of the diene shows that the resonance hybrid with the positive charge shared by a primary and a tertiary carbon is more stable. In the dienophile, a primary carbanion is more stable than a primary carbocation would be.

Therefore, the major product is the 1,4-disubstituted product.

Because the reaction creates an asymmetric center, the product will be a racemic mixture.

94. The first pair is the preferred set of reagents because it has the more nucleophilic diene and the more electrophilic dienophile.

95. A Diels-Alder reaction is a reaction between a nucleophilic diene and an electrophilic dienophile.

a. The compound shown below is more reactive in both **1** and **2**, because electron delocalization increases the electrophilicity of the dienophile.

b. The compound shown below is more reactive, because electron delocalization increases the nucleophilicity of the diene.

$$\overset{\frown}{CH_2}=CH-\overset{\frown}{CH}=CH-\overset{\frown}{\overset{\cdot\cdot}{O}}CH_3 \longleftrightarrow \overset{-}{CH_2}=CH-CH=CH-\overset{+}{\overset{\cdot\cdot}{O}}CH_3$$

96.

exo   endo

97. **a.**

c.

b.

d.

98. **a.**

b.

99.   The electrophile can add to either end of the conjugated system. Adding the electrophile to one end forms **A** and **B**; adding the electrophile to the other end forms **C** and **B**.

a.

b.   **A** has two asymmetric centers but only two stereoisomers are obtained, because only anti addition of Br$_2$ can occur.

B has four stereoisomers because it has an asymmetric center and a double bond that can be in either the *E* or *Z* configuration.

C has two stereoisomers because it has one asymmetric center.

100.   Nine of the compounds are shown below. Since each has one asymmetric center, each can have either the *R* or *S* configuration. Therefore, 18 different products can be obtained.

101.   2-Methyl-1,3-pentadiene (with conjugated double bonds) is more stable than 2-methyl-1,4-pentadiene (with isolated double bonds). The rate-limiting step of the reaction is formation of the carbocation intermediate. 2-Methyl-1,3-pentadiene forms a more stable carbocation intermediate than does 2-methyl-1,4-pentadiene.

Since the more stable reactant forms the more stable carbocation intermediate, the relative free energies of activation for the rate-limiting steps of the two reactions depend on whether the difference in the stabilities of the reactants is greater or less than the difference in the stabilities of the transition states leading to formation of the carbocation intermediates (which depend on the difference in stabilities of the carbocation intermediates). There is a significant difference in the stabilities of the carbocation intermediates because one is stabilized by electron delocalization and one is not. The transitions states look more like the carbocation intermediates than line the alkenes. Therefore, the difference in the stabilities of the reactants is less than the difference in the stabilities of the transition states, so the rate of reaction of HBr with 2-methyl-1,3-pentadiene is the faster reaction. (If the difference in the stabilities of the reactants had been greater than the difference in the stabilities of the transition states, the rate of reaction of HBr with 2-methyl-1,4-pentadiene would have been the faster reaction.)

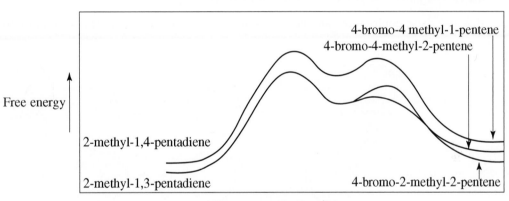

102. His recrystallization was not successful. Because maleic anhydride is a dienophile, it reacts with cyclopentadiene in a Diels-Alder reaction.

endo   exo

103. We saw in Problem 102 that maleic anhydride reacts with cyclopentadiene. The function of maleic anhydride in this reaction is to remove cyclopentadiene, since removal of a product drives the equilibrium toward products. (See Le Châtelier's principle on page 208 of the text.)

104. The bridgehead carbon cannot have the 120° bond angle required for the sp^2 carbon of a double bond, because if it did, the compound would be too strained to exist.

105.   a. Unless the reaction is being carried out under kinetic control, the amount of product obtained is not dependent on the rate at which the product is formed, so the relative amounts of products obtained will not tell you which product was formed faster.

   b. In a thermodynamically controlled reaction, the product distribution depends on the relative stabilities of the products because the products come to equilibrium. Therefore, if the distribution of products does not reflect the relative stabilities of the products, the reaction must have been kinetically controlled.

106.

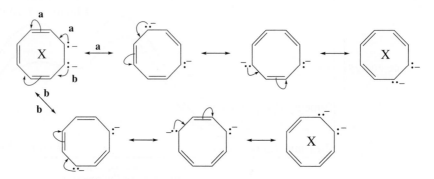

107.   a. The three resonance contributors marked with an X are the least stable, because in these contributors the two negative charges are on adjacent carbons.

   b. Because these contributors are the least stable, they make the smallest contribution to the hybrid.

108. We have classified cyclobutadiene as antiaromatic. However, the recent observation that cyclobutadiene is rectangular and the observation that there are two different 1,2-dideuterio-1,3-cyclobutadienes both indicate that the π electrons are localized rather than delocalized. Localization of the π electrons prevents the compound from being antiaromatic. Apparently, the extreme instability associated with being antiaromatic causes cyclobutadiene to be rectangular.

Chapter 8 Practice Test

1. For each of the following pairs of compounds, indicate the one that is the more stable:

a. or   **d.** $CH_2=CH\overset{+}{C}H_2$ or $CH_2=CHCH_2\overset{+}{C}H_2$

b. $CH_3\overset{-}{C}HCH_3$ or $CH_3\overset{-}{C}HC\equiv CH$

e.   or

c. $CH_3\overset{-}{C}HCH_2\overset{O}{\overset{\|}{C}}CH_3$ or $CH_3\overset{-}{C}H\overset{O}{\overset{\|}{C}}CH_3$

2. Draw resonance contributors for each of the following:

a. $CH_3CH=CH-\ddot{\overset{..}{O}}CH_3$

b. $CH_3CH=CH-CH=CH-\overset{+}{C}H_2$

c. $^{-}\ddot{C}H_2-CH=CH-\overset{O}{\overset{\|}{C}}H$

3. Which compounds do not have delocalized electrons?

$CH_3CH_2NHCH=CHCH_3$ $CH_3\overset{\overset{\displaystyle CH_3}{|}}{\underset{+}{C}}CH_2CH=CH_2$ $CH_2=CHCH_2CH=CH_2$

$CH_2=CH\overset{O}{\overset{\|}{C}}CH_3$ $CH_3CH_2NHCH_2CH=CHCH_3$

4. What are the products of each of the following reactions?

a.   + Br_2 $\longrightarrow$

b.   + $\underset{\overset{\|}{CH_2}}{\overset{\overset{O}{\|}}{CHCCH_3}}$ $\longrightarrow$

5. Which of the following pairs are resonance contributors?

a. CH_3CH_2OH and CH_3OCH_3 **c.** $CH_3\overset{O}{\overset{\|}{C}}OH$ and $CH_3\overset{O^-}{\overset{|}{C}}=\overset{+}{O}H$

b. $CH_3\overset{O}{\overset{\|}{C}}OH$ and $CH_3\overset{O}{\overset{\|}{C}}O^-$ **d.** $CH_3CH_2\overset{O}{\overset{\|}{C}}H$ and $CH_3CH=\overset{OH}{\overset{|}{C}}H$

6.   Which of the following dienes can be used in a Diels-Alder reaction?

7.   Which is a stronger base?

a.   or

b.   or

8.   Draw resonance contributors for each of the following:

a.

b.

c.

9.   Which resonance contributor makes a greater contribution to the resonance hybrid?

a.   or   **b.**   or

10.   Indicate whether each of the following statements is true or false:

a.   A compound with four conjugated double bonds has four molecular orbitals.   T   F

b. ψ_1 and ψ_2 are symmetric molecular orbitals.   T   F

c.   If ψ_3 is the HOMO in the ground state, ψ_4 will be the HOMO in the excited state.   T   F

d.   If ψ_3 is the LUMO, ψ_4 will be the HOMO.   T   F

e.   If the ground-state HOMO is symmetric, the ground-state LUMO will be antisymmetric.   T   F

    f. A conjugated diene is more stable than an isomeric-isolated diene.   T   F

    g. A single bond formed by an sp^2—sp^2 overlap is longer than a single bond formed by an sp^2—sp^3 overlap.   T   F

    h. The thermodynamically controlled product is the major product obtained when the reaction is carried out under mild conditions.   T   F

    i. 1,3-Hexadiene is more stable than 1,4-hexadiene.   T   F

11. Draw the four products that would be obtained from the following reaction. Ignore stereoisomers.

$$CH_2=\overset{\overset{\textstyle CH_3}{|}}{C}-\underset{\underset{\textstyle CH_3}{|}}{C}=CHCH_3 \ + \ HBr \ \longrightarrow$$

12. What reactants are necessary for the synthesis of the following compound via a Diels-Alder reaction?

13. Rank the following carbocations in order from most stable to least stable:

$$CH_3CH=\overset{+}{C}HCH_2 \quad CH_3CH=\overset{+}{C}HCHCH_3 \quad CH_3CH=\overset{+}{C}HCH_2CH_2 \quad CH_3CH=\overset{+}{C}H\underset{\underset{\textstyle CH_3}{|}}{C}CH_3$$

14. Two 1,2-products and two 1,4-products can be obtained from the following reaction:

$$+ \ \ HCl \ \longrightarrow$$

    a. Which are the predominant 1,2- and 1,4-products?

    b. Which of the products is the product of thermodynamic control?

15. What reagents could be used to convert the given starting material into the given product?

$$\longrightarrow \quad HO-\bigcirc-Br$$

16.   Draw the product of the following reaction, showing its configuration:

$$CH_2=CHCH=CH_2 \quad + \quad \underset{H}{\overset{HOOC}{\underset{\big|}{}}}C=C\underset{H}{\overset{COOH}{\underset{\big|}{}}} \quad \xrightarrow{\Delta}$$

17.   For each of the following reactions, give the major 1,2- and 1,4-products. Label the product of kinetic control and the product of thermodynamic control.

a. $CH_2=CH-\underset{\underset{\big|}{CH_3}}{C}=CH_2 \quad + \quad HCl \quad \longrightarrow$

b.   +   HBr   ⟶

18.   Which are aromatic compounds?

19.   Which are antiaromatic compounds?

20.   Which compound has the greater delocalization energy?

21.   Which of the following is a stronger acid?

SPECIAL TOPIC II

Molecular Orbital Theory

Chemists use models to describe such things as the bonding in molecules, the stability of molecules, and the reactions between molecules. The model used is generally the one that provides the best description of the molecule under consideration. One very powerful model is molecular orbital theory.

You were introduced to molecular orbital (MO) theory in Section 1.6 where you saw that electrons are assigned to a volume of space called an orbital. According to MO theory, covalent bonds are formed when atomic orbitals combine to form molecular orbitals. Let's review some important principles:

1. Orbitals are conserved. In other words, the number of molecular orbitals formed must equal the number of atomic orbitals combined. For example, when two atomic orbitals overlap, two molecular orbitals are formed—one lower in energy (a bonding MO) and one higher in energy (an antibonding MO, indicated by a *) than the overlapping orbitals.

2. Electrons always occupy the available *atomic orbitals* with the lowest energy, and no more than two electrons can occupy an *atomic orbital*. Similarly, electrons always occupy the available *molecular orbitals* with the lowest energy, and no more than two electrons can occupy a *molecular orbital*.

3. The relative energies of the molecular orbitals are $\sigma < \pi < \pi^* < \sigma^*$.

4. The strongest covalent bonds are formed by electrons that occupy the molecular orbitals with the lowest energy. For example, the energy of a σ MO is lower than that of a π MO, and we have seen that a σ bond is stronger than a π bond (Section 1.15).

There are two ways atomic orbitals can overlap to form molecular orbitals; these are shown in Figures 1 and 2. In Figure 1, each of the overlapping atomic orbitals contributes one electron to the bond. In Figure 2, a filled atomic orbital (it has two electrons) overlaps an empty atomic orbital. In each case, we see that the electrons are stabilized (are lower in energy) as a result of orbital overlap.

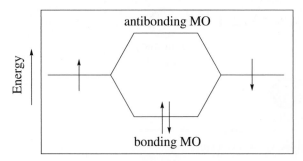

Figure 1. Each of the overlapping atomic orbitals contributes one electron to the bond.

Figure 2. A bond is formed as a result of the overlap of a filled atomic orbital with an empty atomic orbital.

Using Molecular Orbital Theory to Describe Covalent Bond Formation

I. Each of the Overlapping Atomic Orbitals Contributes One Electron to the Bond

Take a look at Figure 1.4 on page 25 of the text. There you can see that an H—H bond is formed by the overlap of a 1s atomic orbital of a hydrogen atom with a 1s atomic orbital of another hydrogen atom; each of the atomic orbitals contributes one electron to the molecular orbital.

281

Figure 3 shows that the C—C bond in ethane is formed by the overlap of an sp^3 atomic orbital of carbon with an sp^3 atomic orbital of another carbon; again, each of the overlapping atomic orbitals contributes one electron to the bond. Figure 4 shows that the C—H bond of ethane is formed by the overlap of an sp^3 atomic orbital of carbon with an s atomic orbital of hydrogen. Because an s atomic orbital of hydrogen is more stable than an sp^3 atomic orbital of carbon (see page 29 of the text), the MO formed by sp^3—s overlap is more stable than the MO formed by sp^3—sp^3 overlap. As a result, the C—H bond is stronger (and shorter) than the C—C bond.

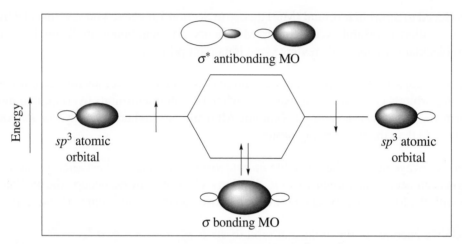

Figure 3. The C—C bond of ethane is formed by sp^3—sp^3 overlap.

Figure 4. A C—H bond of ethane is formed by sp^3—s overlap.

Figure 5 shows that the two sp^3 atomic orbitals that overlap to form the C—O bond of an alcohol or of an ether do not have the same energy. An electron is more stable in the atomic orbital of the more electronegative atom. Thus, the C—O bond is a little stronger and shorter than the C—C bond.

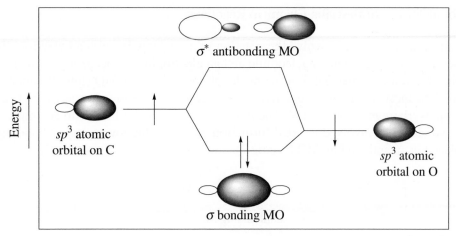

Figure 5. The C—O bond of an ether or of an alcohol is formed by sp^3—sp^3 overlap.

Figure 6 shows that the C—C σ bond of ethene is formed by the overlap of an sp^2 atomic orbital of carbon with an sp^2 atomic orbital of another carbon. The π bond of ethene is formed by the side-to-side overlap of two p orbitals (see Figure 1.5 on page 26 of the text). A π molecular orbital is less stable than a σ molecular orbital. The π bond, therefore, is weaker than the σ bond. Examine Figure 1.6 on page 27 of the text to see the molecular orbitals that are formed when the overlapping p orbitals do not each belong to the same kind of atom.

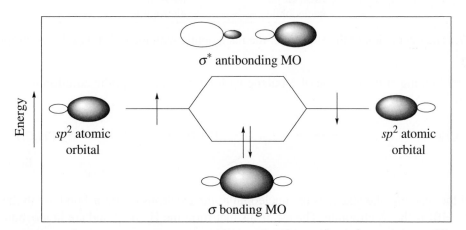

Figure 6. The C—C σ bond of ethene is formed by sp^2—sp^2 overlap.

II. A Filled Atomic Orbital Overlaps an Empty Atomic Orbital to Form a Bond

The overlap of a filled atomic orbital with an empty atomic orbital is the second way two atomic orbitals can overlap to form a bond (Figure 2).

For example, the bond formed between a Lewis base (such as NH_3) and a Lewis acid (such as $FeBr_3$) results from the base sharing a pair of electrons with the acid. Bond formation results from the overlap of a filled sp^3 orbital of nitrogen with an empty orbital of iron. This type of reaction is discussed in Section 2.12 of the text.

Using Molecular Orbital Theory to Describe Chemical Reactions

We have seen that most organic reactions involve the reaction of a nucleophile with an electrophile. Molecular orbital theory describes a reaction between a nucleophile and an electrophile as occurring as the result of the interaction of the HOMO (highest occupied MO) of the nucleophile with the LUMO (lowest unoccupied MO) of the electrophile, because the most stabilizing interaction is between orbitals closest in energy. Notice that in these examples, a filled orbital overlaps an empty orbital. The interaction of a filled orbital with an empty orbital is stabilizing, because the two electrons involved in bond formation end up in the lower-energy bonding MO and no electrons have to be placed in an antibonding MO (Figure 7).

Figure 7. The interaction of the HOMO of the nucleophile with the LUMO of the electrophile.

We will start by looking at the reaction of 2-butene with HBr, an electrophilic addition reaction we first examined in Section 5.6.

In the first step of the reaction, the alkene is the nucleophile; the electrons of the π bond are in the π bonding MO; this is the HOMO. HBr is the electrophile. The electrons that form the H—Br bond are in a σ bonding MO. Therefore, the LUMO of HBr is the σ^* antibonding MO (Figure 8).

Figure 8. Interaction of the HOMO of the alkene with the LUMO of HBr.

Figure 9. Interaction of the HOMO of Br⁻ with the LUMO of the carbocation.

In the second step of the reaction, Br^- is the nucleophile; its HOMO is a filled nonbonding sp^3 orbital. The carbocation is the electrophile; the LUMO is carbon's empty p orbital (Figure 9).

In Section 9.1 of the text, we use MO theory to explain why the alkyl halide in an S_N2 reaction undergoes backside attack, resulting in inversion of configuration. The filled nonbonding orbital of the nucleophile is the HOMO. The LUMO of the electrophilic alkyl halide is the σ^* antibonding MO, which has its largest lobe at the back of the carbon in the C—Br bond, so this is where overlap is best. (See Figure 9.1 on page 406 of the text.)

In Section 16.5 of the text, we use MO theory to explain how a nucleophile reacts with a carbonyl group to form a tetrahedral species. Once again the reaction results from the interaction of the HOMO of the nucleophile and the LUMO of the electrophile. In this case, a filled nonbonding orbital of the nucleophile is the HOMO and the LUMO is the π^* antibonding orbital of the carbonyl group. The π^* antibonding orbital is largest at the carbon atom, so this is where the nucleophile overlaps. (See Figure 16.2 on page 733 of the text.)

Molecular Orbital Theory and Delocalized Electrons

Molecular orbital theory is useful to describe compounds with delocalized electrons. For example, in Section 6.2 we saw that tertiary carbocations are more stable than secondary carbocations, which are more stable than primary carbocations, because the electrons in a filled σ bond are delocalized by overlapping the empty p orbital of the positively charged carbon. (See Figure 6.1 on page 239 of the text.) This kind of electron delocalization is known as hyperconjugation; it is one more example of the stabilization that results when a filled orbital overlaps an empty orbital.

Molecular orbital theory avoids having to use contributing resonance structures, because the electrons in the most stable MO are delocalized over the entire molecule. For example, in Section 8.14, we saw that 1,3-butadiene is more stable than 1,4-pentadiene, because the π electrons in 1,3-butadiene are delocalized over four sp^2 carbons, whereas the intervening methylene group in 1,4-pentadiene prevents the p orbitals of C-2 and C-4 from overlapping. Thus the π electrons of 1,4-pentadiene are localized, causing its molecular orbitals to have the same energy as those of ethene—another compound with localized electrons. (See page 359 of the text.)

intervening
methylene group

1,3-butadiene   1,4-pentadiene

Let's now look at the molecular orbitals of the allyl cation, the allyl radical, and the allyl anion.

$$CH_2\!\!=\!\!CH\!-\!\overset{+}{C}H_2 \qquad CH_2\!\!=\!\!CH\!-\!\overset{\cdot}{C}H_2 \qquad CH_2\!\!=\!\!CH\!-\!\overset{\cdot\cdot}{\underset{}{C}}H_2$$

the allyl cation   the allyl radical   the allyl anion

The three *p* atomic orbitals of the three allyl carbons combine to produce three π molecular orbitals (Figure 10).

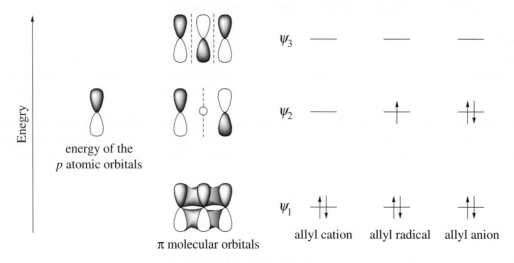

Figure 10. The distribution of the electrons in the molecular orbitals of the allyl cation, the allyl radical, and the allyl anion.

The bonding MO (ψ_1) encompasses the three carbons. In an acyclic system, the number of bonding MOs always equals the number of antibonding MOs. Therefore, when there is an odd number of MOs, one of them must be a nonbonding molecular orbital; ψ_2 is the nonbonding MO. We have seen (Section 8.14) that as the energy of the MO increases, the number of nodes increases. Consequently ψ_2 must have a node—in addition to the one that ψ_1 has—that bisects the *p* orbitals. The only symmetrical position for a node to pass through in ψ_2 is through the middle carbon. You also know that it needs to pass through the middle carbon because that is the only way ψ_2 can be fully antisymmetric, which it must be since ψ_1 and ψ_3 are symmetric. (Recall that MOs alternate between being symmetric and antisymmetric; Section 8.14.)

You can see why ψ_2 is called a nonbonding molecular orbital—there is no overlap between the *p* orbital on the middle carbon and the *p* orbital on either of the end carbons. Notice that a nonbonding MO has the same energy as the isolated *p* atomic orbitals. The third MO (ψ_3) is an antibonding MO.

The two π electrons of the allyl cation are in the bonding MO, which means they are spread out over all three carbons. Consequently, the two carbon-carbon bonds are identical, with each having some double-bond character. The resonance contributors show that the positive charge is shared equally by the end carbon atoms, which is another way of showing that the stability of the allyl cation is due to electron delocalization.

$$CH_2\!\!=\!\!CH\!\!-\!\!\overset{+}{C}H_2 \quad\longleftrightarrow\quad \overset{+}{C}H_2\!\!-\!\!CH\!\!=\!\!CH_2$$

The contributing resonance structures show that when a nucleophile such as Br^- reacts with an allyl cation, the Br^- can bond to either of the end carbons—because they share the positive charge—but cannot bond to the middle carbon. Likewise, MO theory shows that only the end carbons have an empty orbital with which the filled orbital of Br^- can overlap. The central carbon has a node so there can be no interaction with this carbon.

The allyl radical has two electrons in the bonding MO, so these electrons are spread over all three carbon atoms. The third electron is in the nonbonding MO. The MO diagram shows that the third electron is shared equally by the end carbons with none of the electron density residing on the middle carbon. This agrees with what the resonance contributors show.

$$CH_2\!\!=\!\!CH\!\!-\!\!\overset{\bullet}{C}H_2 \quad\longleftrightarrow\quad \overset{\bullet}{C}H_2\!\!-\!\!CH\!\!=\!\!CH_2$$

Finally, the allyl anion has two electrons in the nonbonding MO. These two electrons are shared equally by the end carbons. This, too, agrees with what the resonance contributors show.

$$CH_2\!\!=\!\!CH\!-\!\ddot{\overset{..}{C}}H_2 \quad \longleftrightarrow \quad \ddot{\overset{..}{C}}H_2\!-\!CH\!\!=\!\!CH_2$$

We have seen that both molecular orbital theory and contributing resonance structures can be used to explain electron delocalization. The choice is a matter of preference. Hyperconjugation is probably best shown by molecular orbital theory, because the contributing resonance structures that describe hyperconjugation would require breaking σ bonds. Contributing resonance structures are sometimes preferred when one needs to see on which atoms charges reside.

CHAPTER 9
Substitution Reactions of Alkyl Halides

Important Terms

aprotic solvent	a solvent that does not have a hydrogen bonded to an oxygen or to a nitrogen; some aprotic solvents are polar; others are nonpolar.
back-side attack	nucleophilic attack on the side of the carbon opposite to the side bonded to the leaving group.
base	a substance that accepts a proton.
basicity	the tendency of a compound to share its electrons with a proton.
bifunctional molecule	a molecule with two functional groups.
bimolecular reaction	a reaction in which two molecules are involved in the transition state of the rate-determining step.
complete racemization	formation of a pair of enantiomers in equal amounts.
dielectric constant	a measure of how well a solvent can insulate opposite charges from one another.
elimination reaction	a reaction that removes atoms or groups from the reactant to form a π bond.
first-order reaction	a reaction whose rate is proportional to the concentration of one reactant.
intermolecular reaction	a reaction that takes place between two molecules.
intimate ion pair	an ion pair that results when the covalent bond that joined the cation and anion has broken, but the cation and anion are still next to each other.
intramolecular reaction	a reaction that takes place within a molecule.
inversion of configuration	turning the carbon inside out like an umbrella so that the resulting product has a configuration opposite to that of the reactant.
ion-dipole interaction	the interaction between an ion and the dipole of a molecule.
kinetics	the field of chemistry that deals with the rates of chemical reactions.
leaving group	the group that is displaced in a substitution reaction.
nucleophile	an electron-rich atom or molecule.
nucleophilicity	a measure of how readily an atom or molecule with a lone pair attacks another atom.

nucleophilic substitution reaction	a reaction in which a nucleophile substitutes for an atom or group.
partial racemization	formation of a pair of enantiomers in unequal amounts.
protic solvent	a solvent that has a hydrogen bonded to an oxygen or to a nitrogen.
rate constant	the constant of proportionality in the rate law for a reaction; it describes how difficult it is to overcome the energy barrier of a reaction.
rate law	the equation that shows the relationship between the rate of a reaction and the concentration of the reactants.
second-order reaction	a reaction whose rate is dependent on the concentration of two reactants, or on the square of the concentration of two reactants.
S_N1 reaction	a unimolecular nucleophilic substitution reaction.
S_N2 reaction	a bimolecular nucleophilic substitution reaction.
solvent-separated ion pair	an ion pair that results when the cation and anion are separated by one or more solvent molecules.
solvolysis	reaction with a solvent.
steric effects	effects due to the fact that groups occupy a certain volume of space.
steric hindrance	caused by bulky groups at the site of a reaction that make it difficult for the reactants to approach each other.
substitution reaction	a reaction that exchanges one substituent of a reactant for another.
unimolecular reaction	a reaction in which only one molecule is involved in the transition state of the rate-determining step.

Solutions to Problems

1. DEE is formed when HCl is eliminated from DDT. (See the box on page 403.)

DDE

2. Methoxychlor has methoxy groups in place of the chlorines on the benzene rings of DDT. These methoxy groups can form hydrogen bonds with water, making methoxychlor more soluble in water and, therefore, less soluble in fatty tissues.

3.
$$\text{rate} = k \left[\text{alkyl halide} \right] \left[\text{nucleophile} \right]$$

original: $\text{rate} = k \left[1.0 \right] \left[1.0 \right]$

a. $\text{rate} = k \left[1.0 \right] \left[3.0 \right] = 3.0$ The rate is tripled.

b. $\text{rate} = k \left[0.50 \right] \left[1.0 \right] = 0.50$ The rate is cut in half.

c. $\text{rate} = k \left[0.5 \right] \left[2.0 \right] = 1.0$ The rate is not changed.

4. Increasing the height of the energy barrier decreases the magnitude of the rate constant; this causes the reaction to be slower.

5. The closer the methyl group is to the site of nucleophilic attack, the greater the steric hindrance to nucleophilic attack and the slower the rate of the reaction.

6. **a.** Solved in the text.

b.

(R)-2-bromobutane → (S)-2-methoxybutane

c.

(S)-3-chlorohexane → (R)-3-hexanol

d.

3-iodopentane → 3-pentanol

7.   **a.**   Solved in the text.

b.   I····⟨cyclohexane⟩—CH₂CH₃ $\xrightarrow{CH_3O^-}$ CH₃O—⟨cyclohexane⟩—CH₂CH₃

c. $\xrightarrow{CH_3CH_2O^-}$ ⟨cyclobutane with CH₃ and CH₃CH₂O groups⟩

8.   **a.**   ⟨structure⟩ Br   the primary alkyl halide is less sterically hindered than the secondary alkyl halide (the CH₃ group is farther away from the back side of the carbon attached to the Br)

b.   ⟨structure with Br⟩   Br is a weaker base; therefore, it is a better leaving group

c.   ⟨structure with Br⟩   with one methyl and one ethyl group, this alkyl halide is less sterically hindered than the other alkyl halide that has two ethyl groups

d.   ⟨phenyl-CH₂CH₂-Br structure⟩   the primary alkyl halide is less sterically hindered than the secondary alkyl halide

9.   A protic solvent has a hydrogen bonded to an oxygen or to a nitrogen, whereas an aprotic solvent does not have a hydrogen bonded to an oxygen or to a nitrogen.

a. aprotic   **b.** aprotic   **c.** protic   **d.** aprotic

10.   **a.** RO^- because ROH is a weaker acid than RSH since the hydrogen is attached to a smaller atom.

b. RS^- because it is less well solvated by water and sulfur is more polarizable than oxygen.

c. RO^- because, although they differ in size, they are in an aprotic solvent. Remember that the stronger base is always the better nucleophile in an aprotic solvent.

11.   Remember that the stronger base is always the better nucleophile unless they differ in size AND they are in a protic solvent.

a.   They differ in size and because they are in a protic solvent, the larger one (Br^-) is the better nucleophile.

b.   They differ in size and because they are in an aprotic solvent, the stronger base (Cl^-) is the better nucleophile.

c.   Because the oxygen is negatively charged, CH_3O^- is the better nucleophile.

d.   Because the oxygen is negatively charged, CH_3O^- is the better nucleophile.

e.   Because H_2O is a stronger acid than NH_3, $^-NH_2$ is the stronger base and the better nucleophile.

f.   Because H_2O is a stronger acid than NH_3, $^-NH_2$ is the stronger base and the better nucleophile.

g.   They differ in size and because they are in a protic solvent, the larger one (I^-) is the better nucleophile.

h.   They differ in size and because they are in an aprotic solvent, the stronger base (Br^-) is the better nucleophile.

12.   Solved in the text.

13.   **a.** $CH_3CH_2Br + HO^-$ HO^- is a better nucleophile than H_2O.

b. $CH_3CHCH_2Br + HO^-$ This alkyl halide has less steric hindrance
      | toward nucleophilic attack.
    CH_3

c. $CH_3CH_2Cl + CH_3S^-$ CH_3S^- is a better nucleophile than CH_3O^- in a protic
    solvent (a solvent that can form hydrogen bonds).

d. $CH_3CH_2Br + I^-$ Br^- is a weaker base than Cl^-,
    so Br^- is a better leaving group.

14.   These are all S_N2 reactions.

a. $CH_3CH_2OCH_2CH_2CH_3$ **c.** $CH_3CH_2\overset{+}{N}(CH_3)_3\ \ Br^-$

b. $CH_3CH_2C{\equiv}CCH_3$ **d.** $CH_3CH_2SCH_2CH_3$

15.   Solved in the text.

16.   **a.**   Reaction of an alkyl halide with ammonia gives a low yield of primary amine, because as soon as the primary amine is formed, it can react with another molecule of alkyl halide to form a secondary amine; the secondary amine can react with the alkyl halide to form a tertiary amine, which can then react with an alkyl halide to form a quaternary ammonium salt. (See Problem 15 on page 415.)

b.   The alkyl azide is not treated with hydrogen until after all the alkyl halide has reacted with azide ion. Therefore, when the primary amine is formed, there is no alkyl halide for it to react with to form a secondary amine.

17.   **a.**

one product because the leaving
group was not attached to an
asymmetric center

b.

R and S because the leaving
group was attached to an
asymmetric center

18.

19.   *trans*-4-Bromo-2-hexene (the compound on the right) is more reactive in an S_N1 solvolysis reaction, because the carbocation that is formed is stabilized by electron delocalization. (It is a secondary allylic cation.) The other alkyl halide is a secondary alkyl halide and does not undergo an S_N1 solvolysis reaction.

36. If the ato...
If the ato...
because t...

    a. HO⁻

    b. NH₃

37. The wea...

    a. H₂O

    b. H₂O

38.   **a.** HO⁻

    b. CH₃

In part **c**
is used.

39.   **a.** The
    b. The
    c. The
    d. The
    e. The

40.   **a.** The
    b. The

41.   **a.**

    b.

(A
for

    c. Ha
Th

H₂

20.   **a.** the benzyl halide is more reactive
(an aryl halide cannot form a carbocation)

    b. the allylic halide is more reactive
(a vinyl halide cannot form a carbocation)

21.   **a.** the benzyl halide is more reactive
(an aryl halide cannot undergo back-side attack)

    b. the allylic halide is more reactive
(a vinyl halide cannot undergo back-side attack)

22.

3-chloro-3-methyl-1-butene

23.   **a.**

    b.

24.   **a.** NH₃ is a good nucleophile, so this is an S$_N$2 reaction.
The product has the inverted configuration compared to that of the reactant.

    b. (CH₃)₂CH⸺ OCH₃ + (CH₃)₂CH⸺ CH₃

Once the tertiary carbocation forms, methanol can attack the sp^2 carbon from the top or the
bottom of the planar carbocation.

c. Hydride i
carbon att

BrC

d. Hydride i
carbon att

e. After the
forms an

CH₃C

42.   **a.** $CH_3CH_2S^- > CH_3CH_2O^- > CH_3\overset{\overset{\displaystyle O}{\|}}{C}O^-$ **c.** $NH_3 > H_2O$

b. (cyclohexyl)$-O^-$ > (phenyl)$-O^-$ **d.** $I^- > Br^- > Cl^-$

43.   The pK_a would increase (it would be a weaker acid) because of a decreased tendency to form a charged species in a less polar solvent. (See Problem 29.)

44.   **a.**

b. The products are obtained as a result of the nucleophiles reacting with the carbocation. 2-Bromo-2-methylpropane and 2-chloro-2-methylpropane form the same carbocation, so both alkyl halides form the same products.

45.   **a.** The S$_N$2 reaction takes place with inversion of configuration.

(R)-2-bromopentane   (S)-2-methoxypentane

or

34.   Instead of att
side of the ca

H₃

b.

(R)-3-bromo-3-methylheptane   (R)-3-methoxy-3-methylheptane   (S)-3-methoxy-3-methylheptane

or

35.   **a.** CH₃CH₂

b. CH₃CH

(Notice that
ated because
depending o

c. phenyl–CH$_2$Cl $\xrightarrow{\text{CH}_3\text{CH}_2\text{OH}}$ phenyl–CH$_2$OCH$_2$CH$_3$

d. CH$_2$=CHCH$_2$Cl $\xrightarrow{\text{CH}_3\text{OH}}$ CH$_2$=CHCH$_2$OCH$_3$

e. CH$_3$CH=CHCH$_2$Br $\xrightarrow{\text{CH}_3\text{O}^-}$ CH$_3$CH=CHCH$_2$OCH$_3$

f. CH$_3$CH=CHCH$_2$Br $\xrightarrow{\text{CH}_3\text{OH}}$ CH$_3$CH=CHCH$_2$OCH$_3$ + CH$_3$CHCH=CH$_2$
  |
  OCH$_3$

46.

a.

$\xrightarrow[\substack{\text{both the } R \text{ and } S \\ \text{stereoisomers} \\ \text{are obtained}}]{\text{CH}_3\text{CH}_2\text{OH}}$

b.

c.

This compound has an asymmetric center, so both the *R* and *S* stereoisomers will be obtained.

This compound has an asymmetric center, so both the *R* and *S* stereoisomers will be obtained.

47. Methoxide ion will be a better nucleophile in DMSO, because DMSO cannot stabilize the anion by hydrogen bonding.

48.

a.

$\xrightarrow{\text{CH}_3\text{S}^-}$ + Cl$^-$

The nucleophile is less sterically hindered.

b. $\xrightarrow[\text{H}_2\text{O}]{\text{HO}^-}$ $+$ Cl^-

The electron-withdrawing oxygen increases the electrophilicity of the carbon that the nucleophile attacks.

c. $\xrightarrow{\text{H}_2\text{O}}$ $+$ H_3O^+ $+$ Cl^-

Steric strain is decreased when the alkyl halide dissociates to form the carbocation because the hybridization of the carbon changes from sp^3 to sp^2, allowing the bond angle between the bulky groups to increase from 109.5° to 120°. Therefore, the carbocation is more stable and therefore is formed more rapidly.

d. $(\text{CH}_3)_3\text{CBr} \xrightarrow{\text{H}_2\text{O}} (\text{CH}_3)_3\text{COH} + \text{HBr}$

Because the reactants are neutral, the reaction will be faster in the more polar solvent.

49.

50. **a.** CH_3O^- **c.**

b.   **d.**

51.

52. The slow step of the solvolysis reaction is formation of the carbocation. Because the products of the slow step are charged, they will be better stabilized by formic acid than by acetic acid, because formic acid has a greater dielectric constant. Because the products are more stabilized, the transition state leading to them will also be more stabilized, so the reaction will be faster in the solvent with the greater dielectric constant. Therefore, the second reaction is faster.

53. **a.** We can predict that this is an S_N1 reaction because acetate ion is a relatively poor nucleophile and the alkyl halide is tertiary.

b. We can predict that this is an S_N1 reaction because acetate ion is a relatively poor nucleophile.

54. **a.** $CH_3CH_2CH_2Br + HC\equiv C^- \longrightarrow CH_3CH_2CH_2C\equiv CH \xrightarrow[\text{Pd/C}]{H_2} CH_3CH_2CH_2CH_2CH_3$

b. $CH_3CH_2CH_2Br + HC\equiv C^- \longrightarrow CH_3CH_2CH_2C\equiv CH \xrightarrow[\text{H}_2\text{O}]{\text{H}_2\text{SO}_4,\ \text{HgSO}_4} CH_3CH_2CH_2\overset{\overset{\displaystyle O}{\|}}{C}CH_3$

c. $CH_3CH_2CH_2Br + HC\equiv C^- \longrightarrow CH_3CH_2CH_2C\equiv CH \xrightarrow[\text{2. HO}^-,\ \text{H}_2\text{O}_2,\ \text{H}_2\text{O}]{\text{1. R}_2\text{BH/THF}}$

$CH_3CH_2CH_2CH_2\overset{\overset{\displaystyle O}{\|}}{C}H$

d. $CH_3CH_2CH_2Br + HC\equiv C^- \longrightarrow CH_3CH_2CH_2C\equiv CH \xrightarrow[\substack{\text{Lindlar}\\\text{catalyst}}]{H_2} CH_3CH_2CH_2CH=CH_2$

$\downarrow \overset{\overset{\displaystyle O}{\|}}{\text{RCOOH}}$

$CH_3CH_2CH_2\overset{\displaystyle\overset{O}{\diagup\diagdown}}{CH-CH_2}$

55. The equilbrium contstant is given by the relative stabilities of the products and reactants. Therefore, any factor that stabilizes the products will increase the equilibrium constant.

$$K_{eq} = \frac{[\text{products}]}{[\text{reactants}]}$$

Ethanol will stabilize the charged products more than will diethyl ether because ethanol has a greater dielectric constant. Therefore, the equilibrium will lie farther to the right (toward products) in ethanol.

56. **a.** The reaction with quinuclidine had the larger rate constant because quinuclidine is less sterically hindered as a result of the substituents on the nitrogen being pulled back into a ring structure.

b. The reaction with quinuclidine had the larger rate constant for the same reason given in part **a.**

c. Isopropyl iodide has the larger difference in rate constants and therefore the larger $k_{\text{quinuclidine}}/k_{\text{triethylamine}}$ ratio because, since it is more sterically hindered than methyl iodide, it is more affected by differences in the amount of steric hindrance in the nucleophile.

57. Because methanol is a poor nucleophile, the reaction will take place predominately via an S_N1 pathway. The bromine that will depart is the one that forms a secondary benzylic cation and not the one that would have to form a secondary carbocation. The nucleophile can approach from the top or the bottom of the planar carbocation.

58.   The number of atoms in the ring is given by *n*. Three- and four-membered rings are strained, so they are harder to make than five- and six-membered rings. The three-membered ring is formed faster than the four-membered ring because the compound leading to the three-membered ring has one less carbon-carbon single bond that can rotate to give a conformer in which the reacting groups are positioned too far from one another for reaction.

Even though the compound that forms the five-membered ring has one more bond that can rotate to give a conformer in which the reacting groups are positioned too far from one another for reaction (compared to the compound that forms the four-membered ring), the five-membered ring is formed faster because it is relatively strain free. So lack of strain more than makes up for the lower probability of having the reacting groups in the proper position for reaction.

The rate of the ring-forming reaction gets slower as the size of the ring being formed gets larger, because the reactant has more bonds that can rotate to give conformers in which the reacting groups are positioned too far from one another for reaction.

59.   Because all the reactions are S_N2 reactions, the configuration of the asymmetric center attached to the Br in the reactant will be inverted in the product.

60.   **a.**   **1.** $CH_3CH_2C{\equiv}CCH_2CH_3$ **2.** $CH_3CH_2CH_2OCH_3$

   b. $CH_3CH_2C{\equiv}C^-$ + $CH_3CH_2CH_2CH_2Br$ **or** $CH_3CH_2CH_2CH_2C{\equiv}C^-$ + CH_3CH_2Br

   c. $CH_3CH_2O^-$ + $\underset{\overset{|}{CH_3}}{CH_3CHCH_2CH_2Br}$ **or** $\underset{\overset{|}{CH_3}}{CH_3CHCH_2CH_2O^-}$ + CH_3CH_2Br

61.

62.   Tetrahydrofuran can solvate a charge better than diethyl ether can, because the floppy ethyl substituents of diethyl ether provide steric hindrance, making it difficult for the nonbonding electrons of the oxygen to approach the positive charge that is to be solvated.

63. CH_3CH_2Cl + K^+I^- $\xrightarrow{\text{acetone}}$ CH_3CH_2I + K^+Cl^-

   K^+Cl^- precipitates out in acetone, which drives the reaction to the right.

64.   **a.**

   b.   Two products are obtained because methanol can add to the top or the bottom of the planar double bond.

   c.   One bromine is eliminated with the help of one of oxygen's lone pairs, forming a carbocation where the positive charge is shared by a carbon and an oxygen. The oxygen cannot help the other bromine dissociate.

65.   A Diels-Alder reaction between hexachlorocyclopentadiene and 1,4-dichlorocyclopentene forms chlordane.

66. I^- is a good nucleophile in a polar solvent such as methanol, so it reacts rapidly with methyl bromide, causing the concentration of I^- to decrease rapidly.

I^- is a good leaving group so methyl iodide undergoes an S_N2 reaction with methanol.
Methanol is a poor nucleophile, so the S_N2 reaction is slow. Therefore, iodide ion returns slowly to its original concentration.

67.   It will not undergo an S_N2 reaction, because of steric hindrance to back-side attack.
It will not undergo an S_N1 reaction, because the carbocation that would be formed is unstable, since the ring structure prevents it from achieving the 120° bond angles required for an sp^2 carbon.

68.   The equation needed to calculate K_{eq} from the change in free energy is given in Section 5.7.

$$\ln K_{eq} = \frac{-\Delta G°}{RT}$$

$$\ln K_{eq} = \frac{-(-21.7 \text{ kcal mol}^{-1})}{0.001986 \text{ kcal mol}^{-1}\text{ K}^{-1} \times 303 \text{ K}} = \frac{21.7}{0.60}$$

$$\ln K_{eq} = 36.1$$

$$K_{eq} = 4.8 \times 10^{15}$$

As expected, this highly exergonic reaction has a very large equilibrium constant.

Chapter 9 Practice Test

1.   Which of the following is more reactive in an S_N1 reaction?

    a.   (benzene ring)—CHCH$_2$CH$_3$ **or**   (benzene ring)—CH$_2$CHCH$_3$
                         | |
                        Br   Br

                      Br   Br
                      | |
    b.   CH$_3$CH=CCH$_3$ **or**   CH$_3$CH=CHCHCH$_3$

2.   Which of the following is more reactive in an S_N2 reaction?

                  CH$_3$ CH$_2$CH$_3$
                  | |
    a.   CH$_3$CH$_2$CHBr   **or**   CH$_3$CH$_2$CHBr

    b.   (benzene ring)—Br   **or**   (benzene ring)—CH$_2$Br

3.   Draw the product(s) of the following reactions, showing the stereoisomers that are formed:

    a.   (cyclohexane with Br and CH$_3$) $\xrightarrow{\text{CH}_3\text{O}^-}$

    b.   (benzene ring)—CHCH$_2$CH$_3$ $\xrightarrow{\text{CH}_3\text{OH}}$
                        |
                        Br

                  CH$_3$
                  |
    c.   H——Br $\xrightarrow{\text{HO}^-}$
                  |
                CH$_2$CH$_3$

                  CH$_3$
                  |
    d.   C $\xrightarrow[\text{high concentration}]{\text{CH}_3\text{NH}_2}$
        CH$_3$CH$_2$—⟋ ‖ ⟍ H
                     Cl

    e.   CH$_3$CH=CHCH$_2$Br $\xrightarrow{\text{CH}_3\text{OH}}$

4. Indicate whether each of the following statements is true or false:

a. Increasing the concentration of the nucleophile favors an S_N1 reaction over an S_N2 reaction.   T   F

b. Ethyl iodide is more reactive than ethyl chloride in an S_N2 reaction.   T   F

c. In an S_N1 reaction, the product with the retained configuration is obtained in greater yield.   T   F

d. The rate of a substitution reaction in which none of the reactants is charged will increase if the polarity of the solvent is increased.   T   F

e. An S_N2 reaction is a two-step reaction.   T   F

f. The pK_a of a carboxylic acid is greater in water than it is in a mixture of dioxane and water.   T   F

g. 4-Bromo-1-butanol will form a cyclic ether faster than will 3-bromo-1-propanol.   T   F

5. Answer the following:

a. Which is a stronger base, CH_3O^- or CH_3S^-?

b. Which is a better nucleophile in an aqueous solution, CH_3O^- or CH_3S^-?

6. For each of the following pairs of S_N2 reactions, indicate the one that occurs with the greater rate constant (that is, occurs faster):

a. $CH_3CH_2CH_2Cl$ + HO^- **or** CH_3CHCH_3 + HO^-
                                                              |
                                                              Cl

b. $CH_3CH_2CH_2Cl$ + HO^- **or** $CH_3CH_2CH_2I$ + HO^-

c. $CH_3CH_2CH_2Br$ + HO^- **or** $CH_3CH_2CH_2Br$ + H_2O

d. CH_3CHCH_3 $\xrightarrow[H_2O/CH_3OH]{CH_3O^-}$ **or** CH_3CHCH_3 $\xrightarrow[CH_3OH]{CH_3O^-}$
        | |
        Br   Br

e. $BrCH_2CH_2CH_2CH_2NHCH_3$ **or** $BrCH_2CH_2CH_2NHCH_3$

7. Circle the aprotic solvents.

a. dimethyl sulfoxide   **b.** diethyl ether   **c.** ethanol   **d.** hexane

8. How would increasing the polarity of the solvent affect the following?

a. the rate of the S_N2 reaction of methylamine with 2-bromobutane

b. the rate of the S_N1 reaction of methylamine with 2-bromo-2-methylbutane

c. the rate of the S_N2 reaction of methoxide ion with 2-bromobutane

d. the pK_a of acetic acid

e. the pK_a of phenol

Important Terms

anti elimination	an elimination reaction in which the substituents being eliminated are removed from opposite sides of the molecule.
anti-periplanar	substituents are attached to parallel bonds on opposite sides of a molecule.
dehydrohalogenation	elimination of a proton and a halide ion.
deuterium kinetic isotope effect	ratio of the rate constant obtained for the reaction of a compound containing hydrogen and the rate constant obtained for the reaction of an analogous compound in which one or more of the hydrogens have been replaced by deuterium.
elimination reaction	a reaction that removes atoms or groups from the reactant to form a π bond.
β-elimination reaction or 1,2-elimination reaction	an elimination reaction where the groups being eliminated are bonded to adjacent carbons.
E1 reaction	a unimolecular elimination reaction.
E2 reaction	a bimolecular elimination reaction.
regioselectivity	the preferential formation of a constitutional isomer.
sawhorse projection	a way to represent the three-dimensional spatial relationships of atoms by looking at the carbon-carbon bond from an oblique angle.
syn elimination	an elimination reaction in which substituents being eliminated are removed from the same side of the molecule.
syn-periplanar	substituents are attached to parallel bonds on the same side of a molecule.
target molecule	the desired end product of a synthesis.
Williamson ether synthesis	formation of an ether from the reaction of an alkoxide ion with an alkyl halide.
Zaitsev's rule	the rule that states that the more stable alkene product of an elimination reaction is obtained by removing a proton from the β-carbon that is bonded to the fewest hydrogens.

Solutions to Problems

1.   **a.** $CH_3\overset{\displaystyle CH_3}{\underset{\displaystyle |}{C}}=CHCH_3$ **b.** $CH_3\overset{\displaystyle CH_3}{\underset{\displaystyle |}{C}}=CHCH_3$ **c.** $CH_3CH_2\overset{\displaystyle CH_3}{\underset{\displaystyle \underset{\displaystyle CH_3}{|}}{\overset{|}{C}}}CH=CH_2$

2.   **a.**

The alkene (2-butene) that is formed is more stable than the alkene (1-butene) that is formed from the other alkyl halide.

b.

Br^- is a better leaving group (weaker base) than Cl^-.

c.

The alkene (2-methyl-2-pentene) that is formed is more stable than the alkene (4-methyl-2-pentene) that is formed from the other alkyl halide.

d.

The other alkyl halide cannot undergo an E2 reaction, because it does not have any β-hydrogens.

3.   **a.** $CH_3CH=CHCH_3$

Removal of a hydrogen from the more substituted β-carbon forms the more stable "alkene-like" transition state.

b. $CH_3\overset{\displaystyle CH_3}{\underset{\displaystyle |}{C}}=CHCH_2CH_3$

Removal of a hydrogen from the more substituted β-carbon forms the more stable "alkene-like" transition state.

c. $CH_3CH=CHCH=CH_2$

The hydrogen is removed from the β-carbon that will lead to a conjugated alkene.

d. $CH_2=CHCH_2CH_3$

Removal of a hydrogen from the less substituted β-carbon forms the more stable "carbanion-like" transition state.

e.

The hydrogen is removed from the β-carbon that will lead to a conjugated alkene.

f. $CH_3\overset{\displaystyle CH_3}{\underset{\displaystyle |}{C}}HCH=CHCH_3$

Removal of a hydrogen from the less substituted β-carbon forms the more stable "carbanion-like" transition state.

4. **a.**   It forms the more stable alkene (the alkene with the most substituents bonded to the sp^2 carbons), so it has the more stable transition state.

b.   It forms the more stable alkene (the double bonds are conjugated, so it has the more stable transition state.

c.   It has four hydrogens that can be removed to form an alkene with two substituents on the sp^2 carbons, so it has a greater probability of having an effective collision with the nucleophile than the other alkyl halide that has only two such hydrogens.

d.   It forms the more stable alkene (the new double bond is conjugated with the phenyl substituent), so it has the more stable transition state.

5.

$$CH_3CH-\overset{\overset{\displaystyle CH_3}{|}}{\underset{\underset{\displaystyle Br}{|}}{C}}CH_2CH_3$$
$$\underset{\displaystyle CH_3}{}$$

3-bromo-2,3-dimethylpentane

Four alkyl substituents are bonded to the sp^2 carbons.	Three alkyl substituents are bonded to the sp^2 carbons; the largest groups are on opposite sides of the double bond.	Three alkyl substituents are bonded to the sp^2 carbons; the largest groups are on the same side of the double bond.	Two alkyl substituents are bonded to the sp^2 carbons.

6. The major product would be the one predicted by Zaitsev's rule, because the fluoride ion would dissociate in the first step, forming a carbocation. Loss of a proton from the carbocation would follow Zaitsev's rule, as it does in other E1 reactions.

7. **a.** B because it forms the more stable carbocation.
b. B because it forms the more stable alkene.
c. B because it forms the more stable carbocation.
d. A because it is less sterically hindered.

8.

9. Conjugated double bonds are more stable and, therefore, are easier to form than isolated double bonds.

a.   **b.** —CH=CHCH$_3$ **c.** CH$_2$=CHCH=CCH$_3$ (with CH$_3$ substituent)

10. Parts **a** and **c** form allylic cations and will form the same products that are formed by E2 reactions. Because the alkyl halide in part **b** is a secondary alkyl halide. it does not undergo an E1 reaction.

11. Benzyl bromide does not have a hydrogen on a β-carbon that can be removed to form an elimination product. Allyl bromide has a hydrogen on a β-carbon that, if removed, would lead to a cumulated diene. Because cumulated dienes are high energy species, allyl bromide does not form an elimination product.

benzyl bromide   allyl bromide

12.

a. E2   CH$_3$CH=CHCH$_3$

b. E2   CH$_3$CH$_2$CH=CH$_2$

c. E1   CH$_3$C=CH$_2$ (with CH$_3$ substituent)

d. E2   CH$_3$C=CH$_2$ (with CH$_3$ substituent)

e. E1   CH$_3$C=CHCH=CH$_2$ (with CH$_3$ substituent)

f. E2   CH$_3$C=CHCH=CH$_2$ (with CH$_3$ substituent)

13.   a. Solved in the text.

b. $\dfrac{E2}{E2 + E1} = \dfrac{7.1 \times 10^{-5} \times 2.5 \times 10^{-3}}{7.1 \times 10^{-5} \times 2.5 \times 10^{-3} + 1.50 \times 10^{-5}} = \dfrac{1.78 \times 10^{-7}}{1.78 \times 10^{-7} + 150 \times 10^{-7}} = \dfrac{1.78}{152} = 1.2\%$

14.   a.   1. $\overset{\overset{\displaystyle CH_3}{|}}{CH_3CH_2CH = CCH_3}$ No stereoisomers are possible.

2.   The major product is the conjugated diene with the larger group bonded to one sp^2 carbon on the opposite side of the double bond from the larger group bonded to the other sp^2 carbon.

3.   The major product is the conjugated alkene with the larger group bonded to one sp^2 carbon on the opposite side of the double bond from the larger group bonded to the other sp^2 carbon.

b. In none of the reactions is the major product dependent on whether you start with the R or S enantiomer of the reactant.

15.   a. Solved in the text.   **c.**   the larger substituent attached to one sp^2 carbon and the larger substituent attached to the other sp^2 carbon are on opposite sides of the double bond

b. $\overset{\overset{\displaystyle CH_3}{|}}{CH_3CH_2CH = CCH_3}$ **d.**

16. E2 elimination reactions from six-membered rings occur only when the substituents to be eliminated are both in axial positions.

In the cis isomer, when Br is in an axial position, it has an axial hydrogen on each of the adjacent carbons. The one bonded to the same carbon as the ethyl group will be more apt to be the one eliminated with Br because the product formed is more stable and therefore more easily formed than the product formed when the other H is eliminated with Br. (Recall that when there is a choice, a hydrogen is removed from the β-carbon bonded to the fewest hydrogens.)

In the trans isomer, when Br is in an axial position, it has an axial hydrogen on only one adjacent carbon, and it is not the carbon that is bonded to the ethyl group. Therefore, a different product is formed. (Notice in this case there is no choice; there is only one hydrogen bonded by an axial bond to a β-carbon.)

cis-1-bromo-2-ethylcyclohexane   *trans*-1-bromo-2-ethylcyclohexane

1-ethylcyclohexene   3-ethylcyclohexene

17.   In order for a six-membered ring to undergo an E2 reaction, the substituents that are to be eliminated must both be in axial positions.

When bromine and an adjacent hydrogen are both in axial positions, the large *tert*-butyl substituent is in an equatorial position in the cis isomer and in an axial position in the trans isomer. The rate constant for the reaction is $k' K_{eq}$.

Because a large substituent is more stable in an equatorial position than in an axial position, elimination of the cis isomer occurs through its more stable chair conformer (K_{eq} is large; see page 463 of the text), whereas elimination of the trans isomer has to occur through its less stable chair conformer (K_{eq} is small). The cis isomer, therefore, reacts more rapidly in an E2 reaction.

cis-1-bromo-4-*tert*-butylcyclohexane   *trans*-1-bromo-4-*tert*-butylcyclohexane

18.   **a.**   There are two stereoisomers of *trans*-1-methyl-2-methylcyclohexane.

elimination occurs through
the less stable conformer

b.   There are two stereoisomers of *cis*-l-methyl-2-methylcyclohexane.

elimination occurs through
the more stable conformer

c.

only product

d.

19.   The rate-limiting step in an E1 reaction is carbocation formation. Because the proton is removed in a subsequent fast step, the difference in the rate of removal of an H^+ versus a D^+ would not be reflected in the rate constant. Therefore, the deuterium kinetic isotope effect would be close to 1.

20. Br^- is the weakest base, so it is the best leaving group, and F^- is the strongest base and worst leaving group. It is easier to break a C—H bond than a C—D bond, so the more hydrogens on the β-carbons that are replaced by a deuterium, the slower the rate of the E2 reaction will be.

$$
\begin{array}{ccccc}
CH_3 & CH_3 & CH_3 & CD_3 & CD_3 \\
| & | & | & | & | \\
CH_3\overset{}{C}-Br > & CH_3\overset{}{C}-Cl > & CH_3\overset{}{C}-F > & CH_3\overset{}{C}-F > & CD_3\overset{}{C}-F \\
| & | & | & | & | \\
CH_3 & CH_3 & CH_3 & CD_3 & CD_3
\end{array}
$$

21.   Because CH_3S^- is a better nucleophile in the protic polar solvent and weaker base than CH_3O^-, the ratio of substitution (where Y^- reacts as a nucleophile) to elimination (where Y^- reacts as a base) will increase when the nucleophile is changed from CH_3O^- to CH_3S^-.

22.   In order to undergo an elimination reaction under E2 conditions, the substituents that are to be eliminated (H and Br) must both be in axial positions. Drawing the compound in the chair conformation shows that when Br is in an axial position, neither of the adjacent β-carbons has a hydrogen in an axial position, so an elimination reaction cannot take place.

23.   The substitution product does not require a bond to be broken in the carbocation, but the elimination product does.

24.   **a.** $CH_3CH_2CH_2Br$ This compound has less steric hindrance.

b. I^- is a better leaving group (weaker base) than Br^-.

c. $CH_3CH_2CH_2\overset{\overset{\displaystyle CH_3}{|}}{\underset{\underset{\displaystyle Br}{|}}{C}}CH_3$ The tertiary alkyl halide because a secondary alkyl halide does not undergo S_N1 reactions.

d.   When this compound undergoes an E2 reaction, it forms a conjugated double bond. The other compound cannot undergo an E2 reaction, because neither two adjacent double bonds nor a triple bond can be formed in a six-membered ring.

25.   **a.**   **1.** Because primary alkyl halides do not undergo $S_N1/E1$ reactions, they will undergo slow S_N2 reactions under $S_N1/E1$ conditions. Since the base is weak (CH_3OH), the product will be primarily the substitution product.
        2. Because primary alkyl halides do not undergo $S_N1/E1$ reactions, they will undergo slow S_N2 reactions under $S_N1/E1$ conditions. Since the base is weak (CH_3OH), the product will be primarily the substitution product.
        3. Because primary alkyl halides do not undergo $S_N1/E1$ reactions, they will undergo slow S_N2 reactions under $S_N1/E1$ conditions. Since the base is weak (CH_3OH), the product will be primarily the substitution product.
        4. substitution and elimination

b.   **1.** primarily substitution
        2. substitution and elimination (Because of the isopropyl group that slows down back-side attack and therefore formation of the substitution product, this primary alkyl halide will give more elimination product than will a primary alkyl halide such as 1-bromobutane.)
        3. substitution and elimination
        4. elimination

26.

$$CH_3CCH_2Br$$

with CH_3 above and CH_3 below the central carbon.

1-bromo-2,2-dimethylpropane

a. The bulky *tert*-butyl substituent blocks the back side of the carbon bonded to the bromine to nucleophilic attack, making an S_N2 reaction difficult. An S_N1 reaction cannot occur because it would require the formation of an unstable primary carbocation.

b. It cannot undergo an E2 reaction, because the β-carbon is not bonded to a hydrogen.
It cannot undergo an E1 reaction, because that would require the formation of a primary carbocation.

27. Because a strong base is used in the Williamson ether synthesis, the reaction is an S_N2 reaction, so a competing E2 reaction can also occur.

a. $CH_3CH_2CH_2Br + CH_3CH_2CH_2CH_2O^- \longrightarrow CH_3CH_2CH_2CH_2OCH_2CH_2CH_3 + CH_3CH=CH_2$
butyl propyl ether   propene

b. $CH_3CH_2CH_2CH_2Br + CH_3CH_2CH_2O^- \longrightarrow CH_3CH_2CH_2CH_2OCH_2CH_2CH_3 + CH_3CH_2CH=CH_2$
butyl propyl ether   1-butene

28. In order to maximize the amount of ether formed in the S_N2 reaction, make sure the less hindered group is provided by the alkyl halide. In order to convert the alcohol ($pK_a \sim 15$) to an alkoxide ion in a reaction that favors products (parts **a**, **b**, and **c**), a strong base (H^-) is needed.

a. $CH_3CH_2CHOH \xrightarrow{NaH} CH_3CH_2CHO^- \xrightarrow{CH_3CH_2CH_2Br} CH_3CH_2CHOCH_2CH_2CH_3$ (each with CH_3 substituent)

b. $CH_3CH_2CH_2CHCH_2OH \xrightarrow{Na} CH_3CH_2CH_2CHCH_2O^- \xrightarrow{CH_3CH_2Br} CH_3CH_2CH_2CHCH_2OCH_2CH_3$ (each with CH_3 substituent)

c. cyclohexane-OH $\xrightarrow{NaH}$ cyclohexane-O$^-$ $\xrightarrow{CH_3Br}$ cyclohexane-OCH$_3$

In part **d**, HO^- can be used to convert phenol ($pK_a \sim 15$) to phenoxide ion in a reaction that favors products, because phenol is a considerable stronger acid than an alcohol.

d. phenol-OH $\xrightarrow{NaOH}$ phenol-O$^-$ $\xrightarrow{C_6H_5CH_2Br}$ C$_6$H$_5$CH$_2$O-phenyl

Remember that aryl halides cannot undergo S_N2 reactions.

29. $CH_3C-Br \xrightarrow[H_2O]{CH_3CH_2OH} CH_3C-OH + CH_3C-OCH_2CH_3 + CH_3C=CH_2$ (each tert-butyl with CH_3 groups)

30.

A tertiary alkyl halide cannot undergo a substitution reaction under $S_N2/E2$ conditions.

31.   Because a cumulated diene is less stable than an alkyne, the transition state for its formation is less stable than that for the formation of the alkyne, so the cumulated diene is harder to make.

32.   In the first elimination reaction, a hydrogen will be removed from the β-carbon bonded to the fewest hydrogens as expected. In the second elimination reaction, a hydrogen will be removed from the β-carbon that results in the formation of a conjugated double bond.

major product = (E)-stereoisomer
minor product = (Z)-stereoisomer

33.

The synthesis shown on page 475 of the text will give a higher yield of the target molecule because the alkoxide ion attacks the back side of a primary alkyl halide.

In contrast, the alkoxide ion in the synthesis shown above attacks the back side of a secondary alkyl halide, which provides greater steric hindrance to back-side attack. Therefore, there will be greater competition from the E2 reaction that forms an elimination product, reducing the yield of the desired substitution product.

34.   In parts **a** and **b**, a bulky base is used to encourage elimination over substitution.

a.

b.

c.

d.

35.   a.

c.

e.

b.

d.

f.

36.   f.   this is the only one that undergoes an E1 reaction

37.   a.   1. An E1 reaction is not affected by the strength of the base, but a weak base favors an E1 reaction by disfavoring an E2 reaction.

  2. An E1 reaction is not affected by the concentration of the base, but a low concentration of a base favors an E1 reaction by disfavoring an E2 reaction.

  3. If the reactant is charged, an E1 reaction will be favored by the least polar solvent that will dissolve the reactant (generally an aprotic polar solvent). If the reactant is not charged, an E1 reaction will be favored by a protic polar solvent.

b. **1.** A strong base favors an E2 reaction.

 2. A high concentration of a base favors an E2 reaction.

 3. If one of the reactants is charged, an E2 reaction will be favored by the least polar solvent that will dissolve the reactant (generally an aprotic polar solvent). If neither of the reactants is charged, an E2 reaction will be favored by a protic polar solvent.

38. **a.** $CH_3CH_2CH_2\bar{C}H_2$

 e. $CH_3\overset{+}{C}HCH{=}CH_2$
 stabilized by electron delocalization (resonance)

 b. $CH_3\overset{+}{C}HCH_2CH_3$

 f. $CH_3CH\bar{C}HCH_3$ with CH_3

 c. $CH_3\bar{C}HCH{=}CH_2$
 stabilized by electron delocalization (resonance)

 g. $CH_3\overset{+}{C}CH_2CH_3$ with CH_3

 d. $CH_3CH_2C{=}CH_2$ with CH_3

39. He obtained only the elimination product because under the S_N2/E2 conditions he employed, the tertiary alkyl halide forms an elimination product and no substitution product.

Rather than a tertiary alkyl halide and a primary alkoxide ion, he should have used a primary alkyl halide and a tertiary alkoxide ion. They will react in an S_N2 reaction to form the desired ether.

Although the Williamson ether synthesis (an S_N2 reaction) is the preferred way to synthesize an ether because it gives higher yields, the ether also could be synthesized using an S_N1 reaction.

40.   **a.** $(CH_3)_3CBr$ $\xrightarrow[\text{H}_2\text{O}]{\text{HO}^-}$ $\underset{H_3C}{\overset{H_3C}{>}}C{=}CH_2$ + Br^-

because Br^- is a better leaving group than Cl^-

b.

This compound is the only one that can undergo an E2 elimination reaction because the other compound does not have an axial hydrogen bonded to a β-carbon.

This compound will not be able to undergo an E2 reaction because it does not have an adjacent H and Br that are both attached to axial bonds.

41.   The very minor products that would be obtained from "anti-Zaitsev" elimination (that is, the less substituted alkenes) are not shown.

a. $\underset{H}{\overset{CH_3CH_2CH_2}{>}}C{=}C\underset{CH_3}{\overset{H}{<}}$ + $\underset{H}{\overset{CH_3CH_2CH_2}{>}}C{=}C\underset{H}{\overset{CH_3}{<}}$

       major   minor

b.   The reactant has two β-carbons that are attached to two hydrogens, so two constitutional isomers can be formed. Each constitutional isomer has E and Z stereoisomers.

$\underset{H}{\overset{CH_3CH_2CH_2}{>}}C{=}C\underset{CH_3}{\overset{H}{<}}$ + $\underset{H}{\overset{CH_3CH_2CH_2}{>}}C{=}C\underset{H}{\overset{CH_3}{<}}$ + $\underset{H_3C}{\overset{CH_3CH_2}{>}}C{=}C\underset{CH_2CH_3}{\overset{H}{<}}$ + $\underset{H_3C}{\overset{CH_3CH_2}{>}}C{=}C\underset{H}{\overset{CH_2CH_3}{<}}$

     major   minor   major   minor

c.   *trans*-1-Chloro-2-methylcyclohexane has two stereoisomers.

A hydrogen cannot be removed from the β-carbon bonded to the fewest hydrogens because that hydrogen is not attached to an axial bond.

d. *trans*-1-Chloro-3-methylcyclohexane has two stereoisomers.

e.

major   minor

f.

major   minor

42.   a. 3-Bromocyclohexene forms 1,3-cyclohexadiene; 3-bromocyclohexane forms cyclohexene. 3-Bromo-cyclohexene reacts faster in an E2 reaction, because a conjugated double bond is more stable than an isolated double bond, so the transition state leading to formation of the conjugated double bond is more stable and, therefore, the conjugated double bond is easier to form.

b. 3-Bromocyclohexene, because it forms a relatively stable secondary allylic cation. The other compound is a secondary alkyl halide, so it does not undergo an E1 reaction.

43. Alkyl chlorides and alkyl iodides could also be used. It is best to avoid alkyl fluorides because they have the poorest leaving groups.

a. $CH_3CHCH_2CH_3 \xrightarrow{CH_3O^-} CH_3CHCH_2CH_3$
              | |
              Br OCH_3

b. $CH_3CH_2CH_2CH_2Br \xrightarrow{CH_3O^-} CH_3CH_2CH_2CH_2OCH_3$

c. $CH_3CH_2CH_2CH_2Br \xrightarrow{CH_3NH_2} CH_3CH_2CH_2CH_2\overset{+}{N}H_2CH_3 \xrightarrow{HO^-} CH_3CH_2CH_2CH_2NHCH_3$

44. **a.** ethoxide ion, because elimination is favored by the bulkier base and *tert*-butoxide ion is bulkier than ethoxide ion

b. ⁻SCN because elimination is favored by the stronger base and ⁻OCN is a stronger base than ⁻SCN

c. Br⁻ because elimination is favored by the stronger base and Cl⁻ is a stronger base than Br⁻

d. CH₃S⁻ because elimination is favored by the stronger base and CH₃O⁻ is a stronger base than CH₃S⁻

45. The first compound has two axial hydrogens attached to β-carbons; the second compound has one axial hydrogen attached to a β-carbon, but it cannot form the more substituted (more stable) alkene that can be formed by the first compound. The last compound cannot undergo an E2 reaction because it does not have an axial hydrogen attached to a β-carbon.

46. These are all E2 reactions, because a high concentration of a good nucleophile is employed.

a. The stereoisomer formed in greater yield is the one in which the larger group attached to one sp^2 carbon and the larger group attached to the other sp^2 carbon are on opposite sides of the double bond.

b. No stereoisomers are possible for this compound because one of the sp^2 carbons is bonded to two hydrogens.

c. No stereoisomers are possible for this compound because one of the sp^2 carbons is bonded to two methyl groups.

d. Because it is an E2 reaction and there is only one hydrogen attached to the β-carbon, the stereoisomer formed in greater yield depends on the configuration of the reactant. The reactant can have four different configurations: *S,S*; *S,R*; *R,R*; and *R,S*. To determine the product of the reaction:

1. Draw the skeleton of a perspective formula, putting the groups to be eliminated on the solid lines. Notice that on each carbon, the solid wedge is below the hatched wedge.

2. Add the remaining groups to the structure in a way that gives the asymmetric centers the desired configurations. For example, the structure below is (3*S*,4*S*)-3-bromo-3,4-dimethylhexane.

(3*S*,4*S*)-3-bromo-3,4-dimethylhexane

3. Because the groups to be eliminated are both attached to solid lines, they are anti to one another.

4. Once the groups are eliminated, you can draw the structure of the alkene product. (The groups attached to the solid wedges are on the same side of the double bond, and the groups attached by hatched wedges are on the other side of the double bond.)

(*E*)-3,4-dimethyl-3-hexene

The configuration of the elimination product obtained from each of the other three stereoisomers can be determined in the same way.

(3*S*,4*R*)-3-bromo-3,4-dimethylhexane   (Z)-3,4-dimethyl-3-hexene

(3R,4R)-3-bromo-3,4-dimethylhexane   (E)-3,4-dimethyl-3-hexene

(3R,4S)-3-bromo-3,4-dimethylhexane   (Z)-3,4-dimethyl-3-hexene

47.

this ether cannot be made by a Williamson ether synthesis

+ CH₃CBr with CH₃ groups

these reagents cannot be used because the tertiary
alkyl halide undergoes only an elimination reaction

+ CH₃CO⁻ with CH₃ groups

these reagents cannot be used because the aryl halide
cannot undergo a substitution reaction

48.

$$CH_3C-CHCH_3 \xrightarrow{base} CH_2=C-CHCH_3 + CH_3C=CCH_3$$

with CH₃ CH₃ and Br substituents

2,3-dimethyl-1-butene   2,3-dimethyl-2-butene

a. CH₃CH₂CO⁻ with CH₂CH₃ groups

Because it is the most sterically hindered base, **B** gives the highest
percentage of the less stable 1-alkene because it is easier for it to
remove the most accessible hydrogen.

b. CH₃CH₂O⁻

Because it is the least sterically hindered base, **C** gives the highest
percentage of the more stable 2-alkene.

49.   **a.**   These are $S_N2/E2$ reactions, because a good nucleophile is being used and the alkyl halide is secondary. The minor elimination products are not shown.

b.   Only the substitution products are optically active; the elimination product does not have an asymmetric center.

c.

Yes, all the products are optically active.

d.   The cis enantiomers form the substitution products more rapidly, because there is less steric hindrance from the adjacent substituent to back-side attack by the nucleophile.

e.   The cis enantiomers form the elimination products more rapidly, because the alkenes formed from the cis enantiomers are more substituted and, therefore, more stable. The more stable the alkene, the lower the energy of the transition state leading to its formation and the more rapidly it is formed.

50.

51.   In an E2 reaction, both groups to be eliminated must be in axial positions.

When the bromine is in the axial position in the cis isomer, the *tert*-butyl substituent is in the more stable equatorial position.

When the bromine is in the axial position in the trans isomer, the *tert*-butyl substituent is in the less stable axial position.

Thus, elimination takes place via the most stable conformer in the cis isomer and via the less stable chair conformer in the trans isomer, so the cis isomer undergoes elimination more rapidly.

cis-1-bromo-4-*tert*-butylcyclohexane   *trans*-1-bromo-4-*tert*-butylcyclohexane

52.   **a.**

b.

  major   minor   major   minor

c.

  major   minor   major   minor

d.

  major   minor   major   minor

e.

f.

53.   In reactions **a–d**, the β-carbon from which the hydrogen is to be removed is bonded to only one hydrogen. Therefore, the configuration of the reactant determines the configuration of the E2 elimination product.

To determine the configuration of the product, see the instructions in Problem 46.

a.

   (2S,3S)-2-chloro-3-methylpentane   (E)-3-methyl-2-pentene

b.

Cl   CH$_2$CH$_3$
| ╱CH$_3$
CH$_3$⁗C—C
| |
H   H

(2S,3R)-2-chloro-3-methylpentane

⟶

CH$_3$ CH$_2$CH$_3$
    ╲C=C╱
    ╱   ╲
H   CH$_3$

(Z)-3-methyl-2-pentene

c.

Cl   CH$_3$
| ╱CH$_2$CH$_3$
H⁗C—C
| |
CH$_3$ H

(2R,3S)-2-chloro-3-methylpentane

⟶

H   CH$_3$
 ╲C=C╱
 ╱   ╲
CH$_3$ CH$_2$CH$_3$

(Z)-3-methyl-2-pentene

d.

Cl   CH$_2$CH$_3$
| ╱CH$_3$
H⁗C—C
| |
CH$_3$ H

(2R,3R)-2-chloro-3-methylpentane

⟶

H   CH$_2$CH$_3$
 ╲C=C╱
 ╱   ╲
CH$_3$ CH$_3$

(E)-3-methyl-2-pentene

e. This reactant has two hydrogens bonded to the β-carbon, so both Z and E stereoisomers are formed.

CH$_3$CH$_2$ CH$_3$
      ╲C=C╱
      ╱   ╲
(CH$_3$)$_3$C   H

major product

+

CH$_3$CH$_2$ H
      ╲C=C╱
      ╱   ╲
(CH$_3$)$_3$C   CH$_3$

minor product

54.   Under E1 reaction conditions, both compounds will form the same product both will form the same tertiary carbocation.

a.

b.

Under E2 reaction conditions, the two compounds will from different products because in **a** the H and Br attached to the ring could be eliminated only where they are anti to one another. Because the H and Br are on the same side of the ring in **b**, a hydrogen would have to be removed from the other β-carbon (that is, from the methyl group).

a.

b.

55. In order to undergo an E2 reaction, a chlorine and a hydrogen bonded to an adjacent carbon must be trans to one another so they can both be in the required axial positions. Every Cl in the following compound has a Cl trans to it, so no Cl has a hydrogen trans to it. Thus, it is the least reactive of the isomers; it cannot undergo an E2 reaction.

56. The silver ion increases the ease of departure of the halogen atom by sharing one of bromine's lone pair, which weakens the carbon-bromine bond.

57. For a description of how to do this problem, see Problem 46.

a.

(3S,4S)-3-bromo-4-methylhexane   (E)-3-methyl-2-pentene

b.

(3R,4R)-3-bromo-4-methylhexane

(E)-3-methyl-2-pentene

+

c.

(3S,4R)-3-bromo-4-methylhexane

(Z)-3-methyl-2-pentene

+

d.

(3R,4S)-3-bromo-4-methylhexane

(Z)-3-methyl-2-pentene

+

58. **a.** $CH_3CH_2CD{=}CH_2$ and $CH_3CH_2CH{=}CH_2$

b. The deuterium-containing compound results from elimination of HBr, whereas the non-deuterium-containing compound results from elimination of DBr. The deuterium-containing compound will be obtained in greater yield, because a C—H bond is easier to break than a C—D bond.

59.

60. **a.**

b.

one equivalent reacts
in the Diels-Alder
reaction as a diene and
the other as a dienophile

61.   **a.** Both compounds form the same elimination products because they both have hydrogens bonded to the same β-carbons that are anti to the bromine.

cis-4-bromocyclohexanol

trans-4-bromocyclohexanol

b. Only the trans isomer can undergo an intramolecular substitution reaction because the S_N2 reaction requires back-side attack.

trans-4-bromocyclohexanol

The cis isomer can undergo only an intermolecular substitution reaction.

cis-4-bromocyclohexanol

c. The elimination reaction forms a pair of enantiomers because the reaction creates an asymmetric center in the product. Both substitution reactions form a single stereoisomer, because the reaction does not create an asymmetric center in the product.

62.

$$\xrightarrow[\text{S}_\text{N}2/\text{E2}]{\text{CH}_3\text{O}^-}$$

There is only a substitution product.
The reactant does not undergo elimination because when Cl is in an axial position, neither of the β-carbons is bonded to an axial hydrogen.

63. **a.**

(1S,2S)-1-bromo-1,2-diphenylpropane

(Z)-3-methyl-2-pentene

b.

(1S,2R)-1-bromo-1,2-diphenylpropane

(E)-3-methyl-2-pentene

Chapter 10 Practice Test

1.   Which of the following is more reactive in an E2 reaction?

   a.   [benzene ring]—CH$_2$CHCH$_3$ or   [benzene ring]—CH$_2$CH$_2$CH$_2$Br
                          |
                          Br

   b.   CH$_3$CH$_2$CHCH$_3$ or   CH$_2$=CHCH$_2$CHCH$_3$
             | |
             Br   Br

2.   Which of the following would give the greater amount of substitution product under conditions that favor S$_N$2/E2 reactions?

                      CH$_3$ CH$_3$
                       | |
                   CH$_3$CBr   or   CH$_3$CHBr
                       |
                      CH$_3$

3.   What products are obtained when (*R*)-2-bromobutane reacts with CH$_3$O$^-$ / CH$_3$OH under conditions that favor S$_N$2/E2 reactions? Include the configuration of the products.

4.   What alkoxide ion and what alkyl bromide should be used to synthesize the following ethers?

                 CH$_3$
                  |
   a.   CH$_3$CH$_2$COCH$_2$CH$_2$CH$_3$
                  |
                 CH$_3$

   b.   [cyclohexane ring]—O—[benzene ring]

   c.   [cyclohexane ring]—O—CH$_3$

5.   For each of the following pairs of E2 reactions, indicate the one that occurs with the greater rate constant:

   a.   CH$_3$CH$_2$CH$_2$Cl + HO$^-$ or   CH$_3$CHCH$_3$ + HO$^-$
                                                       |
                                                       Cl

   b.   CH$_3$CH$_2$CH$_2$Cl + HO$^-$ or   CH$_3$CH$_2$CH$_2$I   + HO$^-$

   c.   CH$_3$CH$_2$CH$_2$Br + HO$^-$ or   CH$_3$CH$_2$CH$_2$Br + H$_2$O

d. CH_3CHCH_3 $\xrightarrow[\text{H}_2\text{O/CH}_3\text{OH}]{\text{CH}_3\text{O}^-}$ or CH_3CHCH_3 $\xrightarrow[\text{CH}_3\text{OH}]{\text{CH}_3\text{O}^-}$
 | |
 Br   Br

e. CH_3CHCH_3 + HO^- or $CH_3\overset{\overset{\displaystyle CH_3}{|}}{C}CH_3$ + HO^-
 | |
 Br   Br

6.   What is the major product obtained from the E2 reaction of each of the following compounds with hydroxide ion?

a.

c.

b.

d.

7.   Which would be more reactive in an E2 reaction, *cis*-1-bromo-2-methylcyclohexane or *trans*-1-bromo-2-methylcyclohexane?

CHAPTER 11
Reactions of Alcohols, Ethers, Epoxides, Amines, and Thiols

Important Terms

alcohol	an organic compound with an OH functional group (ROH).
alkaloid	a natural product with a heterocyclic ring (where the heteroatom is a nitrogen) found in the leaves, bark, roots, or seeds of plants.
alkyl tosylate	an ester of *para*-toluenesulfonic acid.
antibiotic	a compound that interferes with the growth of a bacterial microorganism.
crown ether	a cyclic molecule that possesses several ether linkages around a central cavity.
dehydration	loss of water.
epoxide	an ether in which the oxygen is incorporated into a three-membered ring.
ether	a compound containing an oxygen bonded to two carbons (ROR).
inclusion compound	the complex formed when a crown ether specifically binds a metal ion or an organic molecule.
Hofmann elimination reaction	elimination of a proton and a tertiary amine from a quaternary ammonium hydroxide.
lead compound	a prototype in a search for other physiologically active compounds.
mercapto group	an SH group.
molecular modification	changing the structure of a lead compound.
molecular recognition	the recognition of one molecule by another as a result of specific interactions.
quaternary ammonium ion	a cation containing a nitrogen bonded to four alkyl groups (R_4N^+).
sulfide (thioether)	the sulfur analog of an ether (RSR).
sulfonate ester	the ester of a sulfonic acid (RSO_2OR).
sulfonium salt	$R_3S^+X^-$
thioether (sulfide)	the sulfur analog of an ether (RSR).
thiol (mercaptan)	the sulfur analog of an alcohol (RSH).

Solutions to Problems

1.   They no longer have a lone pair of electrons.

2.   The relative reactivity would be:

tertiary > primary > secondary

If secondary alcohols underwent S_N2 reactions with hydrogen halides, they would be less reactive than primary alcohols, because they are more sterically hindered than primary alcohols at the carbon attached to the OH group.

3.   **a.** Solved in the text.

      b. The conjugate acid of the leaving group of $CH_3\overset{+}{O}H_2$ is H_3O^+; its pK_a is -1.7.

         The conjugate acid of the leaving group of CH_3OH is H_2O; its pK_a is 15.5.

         Because H_3O^+ is a much stronger acid than H_2O, H_2O^+ is a much weaker base than HO^- and therefore is the better leaving group. Therefore, $CH_3\overset{+}{O}H_2$ is more reactive than CH_3OH.

4.   **a.** Solved in the text.

      b.

      c.

      d. $CH_3CH_2CH_2CH_2OH$ $\xrightarrow[\text{2. }^-C\equiv N]{\text{1. HBr, }\Delta}$ $CH_3CH_2CH_2CH_2C\equiv N$

5.   All four alcohols undergo an S_N1 reaction, because they are either secondary or tertiary alcohols. The arrows are shown for the first protonation step in part **a**, but are not shown in parts **b**, **c**, and **d**.

      a.

      b.

The carbocations that are initially formed in **c** and **d** rearrange in order to form more stable carbocations.

6.   **a.**   Solved in the text.

b.

7.   **a.**

b.

c.

d.

8.   Solved in the text.

9.   **a.**

b.

c.

d.

10.   a. $CH_3CH_2CH_2OH$ $\xrightarrow[\text{2. } CH_3CH_2S^-]{\text{1. TsCl/pyridine}}$ $CH_3CH_2CH_2SCH_2CH_3$

b. $CH_3CH_2CH_2OH$ $\xrightarrow[\substack{\text{2. } CH_3CHCH_2O^- \\ | \\ CH_3}]{\text{1. TsCl/pyridine}}$ $CH_3CH_2CH_2OCH_2CHCH_3$ with CH_3 below

11.   D because it forms a very stable tertiary allylic cation intermediate.

12.   a.

b.

c.

d.

e. $CH_2=CHCH_2CH_2OH \xrightarrow[\Delta]{H_2SO_4} CH_2=CHCH_2CH_2\overset{+}{\underset{H}{O}}H \rightleftharpoons CH_2=CHCH=CH_2$

f. In **d** and **f**, the reactant is a primary alcohol. Therefore, elimination of water takes place via an E2 reaction. Because the dehydration reaction is being carried out in an acidic solution, the alkene that is formed initially is protonated to form a carbocation. The proton that is then lost from the carbocation is the one that results in formation of the most stable alkene.

13. **a.** In order to synthesize an unsymmetrical ether (ROR') by this method, two different alcohols $(ROH$ and $R'OH)$ would have to be heated with sulfuric acid. Therefore, three different ethers would be obtained as products. Consequently, the desired ether would account for considerably less than half of the total amount of ether that is synthesized.

$$ROH + R'OH \xrightarrow[\Delta]{H_2SO_4} ROR + ROR' + R'OR'$$
$$\text{target}$$
$$\text{molecule}$$

b. It could be synthesized by a Williamson ether synthesis. (See Section 10.10 on page 471 of the text.)

$$CH_3CH_2CH_2OH \xrightarrow{\text{NaH}} CH_3CH_2CH_2O^- \xrightarrow{CH_3CH_2Br} CH_3CH_2CH_2OCH_2CH_3 \; + \; Br^-$$

14.   a.

b. Because a primary carbocation cannot be formed, the dehydration is an E2 reaction. The alkene that results is protonated, and the proton that is removed is the one that results in formation of the most stable alkene.

15.   a.

b.

16. Both alcohols form the same carbocation, so they both form the same alkene.

17. **a.**

major   +   minor

b.

major   minor

18. **a.** $CH_3CH_2C\!=\!CCHCH_3$ **b.**

19. **a.** $CH_3CH_2CHCH_2OH$ **b.** **c.** $CH_3CH_2CHCH_2CH_3$ **d.**

20. **a.** $CH_3CH_2CCH_2CH_3$ **c.** A tertiary alcohol is not oxidized to a carbonyl compound. **e.**

b. $CH_3CH_2CH_2CH_2CH$ **d.** $CH_2CCH_2CCH_2CH_3$ **f.** $HCCH_2CH_2CH$

21.   Solved in the text.

22.   We saw that HCl does not cleave ethers because Cl⁻ is not a good enough nucleophile.
F⁻ is an even poorer nucleophile, so HF cannot cleave ethers. Therefore, ethers can be cleaved only with
HBr or HI.

23.   **a.**   Solved in the text.

   b.   Cleavage occurs by an S$_N$2 pathway because a primary carbocation is too unstable to be formed.

$$HOCH_2CH_2CH_2CH_2CH_2I$$

   c.   Cleavage occurs by an S$_N$1 pathway because the benzyl cation that is formed is relatively stable;
I⁻ reacts with the benzyl cation.

   d.   Cleavage occurs by an S$_N$1 pathway because the benzyl cation that is formed is relatively stable;
I⁻ reacts with the benzyl cation.

   e.   Cleavage occurs by an S$_N$2 pathway because a primary carbocation or a vinylic cation is too unstable
to be formed.

enol   ketone

f. Cleavage occurs by an S_N1 pathway, because the tertiary carbocation that is formed is relatively stable; I^- will react with the tertiary carbocation.

24.   a. HOCH$_2$CCH$_3$ (with OCH$_3$ up, CH$_3$ down)   **b.** CH$_3$OCH$_2$CCH$_3$ (with OH up, CH$_3$ down)   **c.** CH$_3$CH—CCH$_3$ (with OH and OCH$_3$ up, CH$_3$ down)   **d.** CH$_3$CH—CCH$_3$ (with OCH$_3$ OH up, CH$_3$ down)

25. The reactivity of tetrahydrofuran is more similar to that of a noncyclic ether, because the five-membered ring does not have the strain that makes the epoxide reactive.

26.   a.

b.

27.   a.

b.

28.   a. CH$_3$C—CHCH$_2$CH$_3$ (with CH$_3$ up, OH OH down)   **b.**

29. **a.** Diol formation is a syn addition reaction. Syn addition to the trans isomer forms the threo pair of enantiomers. (See Section 6.16, particularly page 277.)

$$
\begin{array}{ccc}
\underset{\substack{HO \\ \quad S}}{\overset{H_3C}{\underset{H}{\diagdown}}}C - \underset{\substack{CH_3 \\ \quad S}}{\overset{H}{\underset{}{}}}C{\cdots}OH
&
+
&
\underset{\substack{H_3C \\ \quad R}}{\overset{H}{\underset{HO}{\cdots}}}C - \underset{\substack{OH \\ \quad R}}{\overset{CH_3}{\underset{H}{}}}C
\end{array}
\quad \text{or} \quad
$$

	CH$_3$				CH$_3$	
H	—	OH S		HO	—	H R
HO	—	H S		H	—	OH R
	CH$_3$				CH$_3$	

b. Syn addition to the cis isomer forms the erythro pair of enantiomers. In this case, the product is a meso compound because each asymmetric center is bonded to the same four substituents, so only one stereoisomer is formed.

$$
\underset{\substack{HO \\ \quad S}}{\overset{H_3C}{\underset{H}{\diagdown}}}C - \underset{\substack{CH_3 \\ \quad R}}{\overset{OH}{\underset{H}{}}}C
\quad \text{or} \quad
$$

	CH$_3$	
H	—	OH S
H	—	OH R
	CH$_3$	

c. Syn addition to the cis isomer forms the erythro pair of enantiomers.

$$
\underset{\substack{HO \\ \quad S}}{\overset{H_3C}{\underset{H}{\diagdown}}}C - \underset{\substack{CH_2CH_3 \\ \quad R}}{\overset{OH}{\underset{H}{}}}C
\quad + \quad
\underset{\substack{CH_3CH_2 \\ \quad S}}{\overset{HO}{\underset{H}{\cdots}}}C - \underset{\substack{OH \\ \quad R}}{\overset{CH_3}{\underset{H}{}}}C
\quad \text{or} \quad
$$

	CH$_3$				CH$_3$	
H	—	OH S		HO	—	H R
H	—	OH R		HO	—	H S
	CH$_2$CH$_3$				CH$_2$CH$_3$	

d. Syn addition to the trans isomer forms the threo pair of enantiomers.

$$
\underset{\substack{HO \\ \quad S}}{\overset{H_3C}{\underset{H}{\diagdown}}}C - \underset{\substack{CH_2CH_3 \\ \quad S}}{\overset{H}{\underset{}{}}}C{\cdots}OH
\quad + \quad
\underset{\substack{CH_3CH_2 \\ \quad R}}{\overset{H}{\underset{HO}{\cdots}}}C - \underset{\substack{OH \\ \quad R}}{\overset{CH_3}{\underset{H}{}}}C
\quad \text{or} \quad
$$

	CH$_3$				CH$_3$	
H	—	OH S		HO	—	H R
HO	—	H S		H	—	OH R
	CH$_2$CH$_3$				CH$_2$CH$_3$	

30. **a.** Reaction with a peroxyacid forms a product with two new asymmetric centers. Because it is a syn addition reaction, the trans alkene will form the two trans products.

Hydroxide ion can attack either of the two asymmetric centers in the epoxide.
The asymmetric center that is attacked (undergoes an S$_N$2 reaction) will have its configuration inverted. The configuration of the other asymmetric center will not change.

Therefore, when hydroxide ion attacks the asymmetric center on the left of epoxide **A**, the S,R stereoisomer will be formed: when it attacks the asymmetric center on the right of epoxide **A**, the R,S stereoisomer will be formed. The S,R and R,S stereoisomers are identical; it is a meso compound.

Attack of hydroxide ion on epoxide **B** will form the same meso compound

a meso compound

b. Reaction of the cis alkene with a peroxyacid will form only one peroxide (a meso compound). The meso compound has the *R,S* configuration, so when it reacts with hydroxide ion, the *S,S* and *R,R* products will be formed.

c. Reaction of the cis alkene with a peroxyacid will form the two cis isomers. Hydroxide ion will preferentially attack the least sterically hindered carbon of each epoxide. Thus, the *S,S* and *R,R* products will be formed.

d. Reaction of the trans alkene with a peroxyacid will form the two trans isomers. Hydroxide ion will preferentially attack the least sterically hindered carbon of each epoxide. Thus, the *S,R* and *R,S* products will be formed.

31. The carbocation leading to 1-naphthol can be stabilized by electron delocalization without destroying the aromaticity of the intact benzene ring. The carbocation leading to 2-naphthol can be stabilized by electron delocalization only by destroying the aromaticity of the intact benzene ring. Therefore, the carbocation leading to 1-naphthol is more stable.

carbocation that leads to 1-naphthol

carbocation that leads to 2-naphthol

32.   a.   with an NIH shift

b.   without an NIH shift

33.   The products obtained from the two reactions would not differ. The epoxide opens in the direction that gives the most stable carbocation. The carbocation undergoes an NIH shift, and, as a result of the NIH shift, both reactants form the same ketone intermediate. Because they form the same intermediate, they form the same products. The deuterium-containing product is the major product, because in the last step of the reaction it is easier to break a carbon-hydrogen bond than a carbon-deuterium bond.

34.   **a.**   Solved in the text.

    b.   The compound without the double bond in the second ring is more apt to be carcinogenic. It opens to form a less stable carbocation than the other compound, because it can be stabilized by electron delocalization only if the aromaticity of the benzene ring is destroyed. Because the carbocation is less stable, it is formed more slowly, giving the carcinogenic pathway a better chance to compete with ring-opening.

35.   Each arene oxide will open in the direction that forms the most stable carbocation. Thus, the methoxy-substituted arene oxide opens so the positive charge can be stabilized by electron donation from the methoxy group.

The nitro-substituted arene oxide opens in the direction that forms the most stable carbocation intermediate, the one where the positive charge is farther away from the electron-withdrawing NO_2 group.

36.   **a.**   Note that a bond shared by two rings cannot be epoxidized.

b.   The epoxide ring in phenanthrene oxides II and III can open in two different directions to give two different carbocations and, therefore, two different phenols.

c. The two different carbocations formed by phenanthrene oxides II and III differ in stability. One carbocation is more stable than the other because it can be stabilized by electron delocalization without disrupting the aromaticity of the adjacent ring. The more stable carbocation leads to the major product.

major product   minor product

major product   minor product

d. Phenanthrene oxide I is the most carcinogenic, because it is the only one that opens to form a carbocation that cannot be stabilized without disrupting the aromaticity of the other ring(s).

37. The drugs are metabolized as a result of reacting with water. Nucleophilic water adds to the electrophilic carbon of the carbonyl group. The greater the contribution from the resonance contributor on the right (the one with separated charges), the less electrophilic is the carbonyl carbon, so the less reactive is the carbonyl group.

Xylocaine's resonance contributor with separated charges is more stable than Novocaine's because it has a positive charge on nitrogen, which is less electronegative than oxygen. Therefore, Xylocaine is less reactive and consequently has a longer half-life.

38.   The strong base will immediately remove a proton from the protonated amine, and then there would be no base to carry out the elimination reaction.

39.   The major products result from removing a proton from the β-carbon that is bonded to the most hydrogens.

$$CH_3CHCH_2N \overset{CH_3}{\underset{\underset{CH_3}{|}}{|}} \quad + \quad CH_2{=}CHCH_3$$

40.   The minor products result from removing a proton from the β-carbon that is bonded to the fewest hydrogens.

$$CH_3C{=}CH_2 \underset{CH_3}{|} \quad + \quad NCH_2CH_2CH_3 \overset{CH_3}{\underset{CH_3}{|}}$$

41.   The only difference is the leaving group.

42.   In each case, a proton is removed from the β-carbon that is bonded to the most hydrogens.

a. $CH_2{=}CHCHCH_2CH_2NCH_3$ with CH_3 groups

c. $CH_3NCH_2CHCH_2CH{=}CH_2$ with CH_3 groups

b. $+$ $N(CH_3)_3$

43. **a.** Solved in the text.

b. $CH_3CH_2CH_2CHCH_3$ $\xrightarrow{(CH_3)_3N}$ $CH_3CH_2CH_2CHCH_3$ $\xrightarrow[H_2O]{Ag_2O}$ $CH_3CH_2CH_2CHCH_3$

with Br substituent; then $CH_3\overset{+}{-}N-CH_3$ Br^- with CH_3; then $CH_3\overset{+}{-}N-CH_3$ HO^- with CH_3

$\downarrow \Delta$

$(CH_3)_3N$ + $CH_3CH_2CH_2CH=CH_2$

c.

pyrrolidine (N–H ring) $\xrightarrow[K_2CO_3]{CH_3I \; excess}$ ring $\overset{+}{N}\; I^-$ with CH_3 CH_3 $\xrightarrow[H_2O]{Ag_2O}$ ring $\overset{+}{N}$ HO^- with CH_3 CH_3 $\xrightarrow{\Delta}$ ring with double bond, N with CH_3 CH_3

$\downarrow CH_3I$

$(CH_3)_3N$ + diene $\xleftarrow{\Delta}$ $\overset{+}{N}$ HO^- with $CH_3 \;|\; CH_3$, CH_3 $\xleftarrow[H_2O]{Ag_2O}$ $\overset{+}{N}$ I^- with $CH_3 \;|\; CH_3$, CH_3

44. **a.** $\diagup\!\diagdown\!SH$ $\xrightarrow{HO^-}$ $\diagup\!\diagdown\!S^-$ $\diagup\!\diagdown\!Br$ $\xrightarrow{}$ $\diagup\!\diagdown\!S\diagdown\!\diagup$ + Br^-

+ H_2O

b. The synthesis must be done with *tert*-butylthiol and a methyl halide. It cannot be done with methanethiol and *tert*-butyl bromide because a tertiary alkyl halide cannot undergo an S_N2 reaction, so it would form an elimination product rather than a substitution product.

$\underset{SH}{\times}$ $\xrightarrow{HO^-}$ $\underset{S^-}{\times}$ $\xrightarrow{CH_3Br}$ $\underset{S\diagup}{\times}$ + Br^-

+ H_2O

c. The highest yield of the target molecule is obtained by having the less substituted of the two R groups of the thioether be the alkyl halide and the more substituted be the thiol.

$$CH_2=CHCHSH \;(CH_3) \xrightarrow{HO^-} CH_2=CHCHS^- \;(CH_3) \xrightarrow{CH_3CH_2Br} CH_2=CHCHSCH_2CH_3 \;(CH_3) + Br^-$$

+ H_2O

d. The synthesis must be done with these reagents because the *sp²* carbon of the benzene ring cannot undergo back-side attack.

45. The first compound is too insoluble.

The second compound is used clinically; it has fewer carbons than the first compound, so it is more soluble in water.

The third compound is less reactive than the second compound, because the lone pair on the nitrogen can be delocalized into the benzene ring, so the lone pair is less apt to displace a chloride ion and form the three-membered ring that is needed for the compound to be an alkylating agent.

46. Melphalan is a good alkylating reagent because the sulfonate group is an excellent leaving group because the electrons left behind when the group leaves can be delocalized onto two oxygens. In addition, melphalan has two sulfonate groups, so it can react with two nucleophiles.

47. Cyclophosphamide, carmustine, and chloroambucil are less reactive than mechlorethamine because, unlike the lone-pair electrons on the nitrogen of mechlorethamine, the lone-pair electrons on the nitrogen of the other three compounds can be delocalized, and, therefore, these compounds are less apt to form the three-membered ring that is needed for the compound to be an alkylating agent.

48.

a. $CH_3CH_2CH_2O\overset{\overset{\displaystyle O}{\|}}{C}CH_3$

d. $CH_3CH=CHCH_3$

g. $CH_3\underset{\underset{\displaystyle CH_3}{|}}{C}HCH_2CH_2Cl$

b. $CH_3CH_2CH_2CH_2Br$

e. $CH_3CH_2\underset{\underset{\displaystyle CH_3O}{|}}{C}H-\overset{\overset{\displaystyle CH_3}{|}}{\underset{\underset{\displaystyle OH}{|}}{C}}CH_3$

c. $CH_3\underset{\underset{\displaystyle CH_3}{|}}{C}HCH_2CH_2O-$

f. $CH_3CH_2\underset{\underset{\displaystyle OH}{|}}{C}H-\overset{\overset{\displaystyle CH_3}{|}}{\underset{\underset{\displaystyle OCH_3}{|}}{C}}CH_3$

49.

a. CH_2CH_2OH

The other alcohol cannot undergo dehydration to form an alkene, because its *β*-carbon is not bonded to a hydrogen.

b.

The rate-limiting step in a dehydration is carbocation formation. A secondary allylic cation is more stable than a secondary carbocation and therefore is formed more rapidly.

c.

A tertiary carbocation is more stable than a secondary carbocation.

d.

A secondary benzylic cation is more stable than a secondary carbocation.

e.

A secondary benzylic cation is more stable than a primary carbocation.

50.

51. Only one S$_N$2 reaction takes place in **a** and **b**, so the product has the inverted configuration compared to the configuration of the reactant.

a.

(R)-1-deuterio-1-propanol   (S)-1-deuterio-1-propanol

or

(R)-1-deuterio-1-propanol   (S)-1-deuterio-1-propanol

b.

(R)-1-deuterio-1-propanol   (S)-1-deuterio-1-methoxypropane

or

(R)-1-deuterio-1-propanol   (S)-1-deuterio-1-methoxypropane

c.   Since the desired product has the same configuration as the starting material, it can be synthesized using reactions that do not break any of the bonds to the asymmetric center.

(R)-1-deuterio-1-propanol   (R)-1-deuterio-1-methoxypropane

It could also be synthesized using two consecutive S_N2 reactions. Since each one involves inversion of configuration, the final product will have the same configuration as the starting material. For example, treating the starting material with PBr_3 forms (S)-1-bromo-1-deuteriopropane. Treating (S)-1-bromo-1-deuteriopropane with methoxide ion forms (R)-1-deuterio-1-methoxypropane.

(R)-1-deuterio-1-propanol   (S)-1-bromo-1-deuteriopropane   (R)-1-deuterio-1-methoxypropane

52.

53.   2-hexene and 3-hexene

$$CH_3CH_2CH_2CH_2CH_2CH_2OH \xrightarrow[\Delta]{H_2SO_4} CH_3CH_2CH_2CH_2CH=CH_2$$

$$H^+ + CH_3CH_2CH_2CH=CHCH_3 \rightleftharpoons CH_3CH_2CH_2CH_2\overset{+}{C}HCH_3$$

2-hexene

$$CH_3CH_2CH_2\overset{+}{C}HCH_2CH_3 \longrightarrow CH_3CH_2CH=CHCH_2CH_3 + CH_3CH_2CH_2CH=CHCH_3 + H^+$$

3-hexene
(cis and trans)

2-hexene
(cis and trans)

54.

a. $CH_3\overset{\overset{\displaystyle CH_3}{|}}{\underset{\underset{\displaystyle CH_3}{|}}{C}}Br + CH_3CH_2OH$

b. $CH_3\overset{\overset{\displaystyle}{}}{\underset{\underset{\displaystyle CH_3}{|}}{C}}HCH_2OH + CH_3I$

c.

d.

e.

f.

g. CH_3 $C\equiv N$

h.

i. $CH_3\overset{}{\underset{\underset{\displaystyle CH_3}{|}}{C}}HCH_2\overset{\overset{\displaystyle CH_3}{|}}{\underset{\underset{\displaystyle CH_3}{|}}{N}} + CH_2=CH_2$

j.

55.   2,3-Dimethyl-2-butanol will dehydrate faster because it is a tertiary alcohol, whereas 3,3-dimethyl-2-butanol is a secondary alcohol.

$$CH_3\overset{}{\underset{\underset{\displaystyle CH_3}{|}}{C}}H-\overset{\overset{\displaystyle CH_3}{|}}{\underset{\underset{\displaystyle OH}{|}}{C}}CH_3 \qquad CH_3\overset{}{\underset{\underset{\displaystyle OH}{|}}{C}}H-\overset{\overset{\displaystyle CH_3}{|}}{\underset{\underset{\displaystyle CH_3}{|}}{C}}CH_3$$

2,3-dimethyl-2-butanol
a tertiary alcohol

3,3-dimethyl-2-butanol
a secondary alcohol

56.

2-ethyloxirane

a. CH₃CH₂CHCH₂OH
                |
               OH

0.1 M HCl is a dilute solution
of HCl in water.

c. CH₃CH₂CHCH₂OH
                |
               OH

b. CH₃CH₂CHCH₂OH
                |
               OCH₃

d. CH₃CH₂CHCH₂OCH₃
                |
               OH

57. **a.** If an NIH shift occurs, both carbocations will form the same intermediate ketone. Because it is about four times easier to break a C—H bond (k_3) compared with a C—D bond (k_4), about 80% of the deuterium will be retained.

b. If an NIH shift does not occur, 50% of the deuterium will be retained, because the epoxide can open equally easily in either direction (k_1 is equal to k_2) and subsequent loss of H^+ or D^+ is fast.

58.

unknown
alcohol →[H₂SO₄, Δ] major product + minor product

NaOCl | CH₃COOH | 0 °C →

(cyclohexanecarbaldehyde)

minor product →[1. R₂BH/THF; 2. H₂O₂, HO⁻, H₂O]→ (cyclohexylmethanol)

original unknown
alcohol

59.

a. $CH_3CH_2CH=CH_2$ →[1. R₂BH/THF; 2. H₂O₂, HO⁻, H₂O]→ $CH_3CH_2CH_2CH_2OH$

$\Updownarrow$ H₂SO₄ Δ

$2\ CH_3\overset{O}{\underset{}{C}}H$ ←[1. O₃, −78 °C; 2. (CH₃)₂S]← $CH_3CH=CHCH_3$ + H_2O

b. $CH_3CH_2CH=CH_2$ →[1. R₂BH/THF; 2. H₂O₂, HO⁻, H₂O]→ $CH_3CH_2CH_2CH_2OH$ →[NaOCl, CH₃COOH, 0 °C]→ $CH_3CH_2CH_2\overset{O}{\underset{}{C}}H$

c.

(cyclohexene) →[HBr]→ (bromocyclohexane) →[HO⁻]→ (cyclohexanol)

substitution product + (cyclohexene) elimination product

substitution product →[NaOCl, CH₃COOH, 0 °C]→ (cyclohexanone)

elimination product →[1. O₃, −78 °C; 2. (CH₃)₂S]→ (pentanedial: $H\overset{O}{\underset{}{C}}CH_2CH_2CH_2\overset{O}{\underset{}{C}}H$)

60.   When (S)-2-butanol loses water, the asymmetric center in the reactant becomes a planar sp^2 carbon. There-fore, the chirality is lost. When water attacks the carbocation, it can attack from either side of the planar carbocation, forming (S)-2-butanol and (R)-2-butanol with equal ease.

61.

62.

63.   **Diethyl ether** is the ether that would be obtained in greatest yield, because it is a symmetrical ether. Since it is symmetrical, only one alcohol is used in its synthesis. Therefore, it is the only ether that would be formed. In contrast, the synthesis of an unsymmetrical ether requires two different alcohols. Therefore, the unsymmetrical ether is one of three different ethers that would be formed.

64. **a.**

$$\text{cyclohexanol} \xrightarrow[\Delta]{H_2SO_4} \text{cyclohexene} \xrightarrow{H_2 / Pd/C} \text{cyclohexane}$$

b. $CH_3CH_2CH_2CH_2Br \xrightarrow{HO^-} CH_3CH_2CH_2CH_2OH \xrightarrow[\Delta]{H_2CrO_4} CH_3CH_2CH_2\overset{\displaystyle O}{\overset{\|}{C}}OH$

65.   *N*-Methylpiperidine forms 1,4-pentadiene.

$$CH_2{=}CHCH_2CH_2CH_2\overset{\displaystyle CH_3}{\underset{}{N}}CH_3$$

1. CH₃I
2. Ag₂O, H₂O
3. Δ

$$CH_2{=}CHCH_2CH{=}CH_2$$

2-Methylpiperidine forms 1,5-hexadiene.

β-carbon bonded
to the greatest number
of hydrogens

$$CH_2{=}CHCH_2CH_2CH_2CH_2\overset{\displaystyle CH_3}{\underset{}{N}}CH_3$$

1. CH₃I
2. Ag₂O, H₂O
3. Δ

$$CH_2{=}CHCH_2CH_2CH{=}CH_2$$

3-Methylpiperidine forms 2-methyl-1,4-pentadiene.

$$CH_2{=}CHCH_2\overset{\displaystyle CH_3}{\underset{}{C}}HCH_2\overset{\displaystyle CH_3}{\underset{}{N}}CH_3$$

1. CH₃I
2. Ag₂O, H₂O
3. Δ

$$CH_2{=}CHCH_2\overset{\displaystyle CH_3}{\underset{}{C}}{=}CH_2$$

4-Methylpiperidine forms 3-methyl-1,4-pentadiene.

$$CH_2{=}CHCH\overset{\displaystyle CH_3}{\underset{}{}}CH_2CH_2\overset{\displaystyle CH_3}{\underset{}{N}}CH_3$$

1. CH₃I
2. Ag₂O, H₂O
3. Δ

$$CH_2{=}CHCH\overset{\displaystyle CH_3}{\underset{}{}}CH{=}CH_2$$

66. 3-Methyl-2-butanol is a secondary alcohol and, therefore, will undergo an S_N1 reaction. The carbocation intermediate that is formed rearranges to a more stable tertiary carbocation.

2-Methyl-1-propanol is a primary alcohol and, therefore, will undergo an S_N2 reaction. Because carbocations are not formed in S_N2 reactions, a carbocation rearrangement cannot occur.

67. **a.**

b.

68.

69.

70. Cyclopropane does not react with HO⁻, because cyclopropane does not contain a leaving group; a carb-anion is far too basic to serve as a leaving group. Ethylene oxide reacts with HO⁻, because ethylene oxide contains an RO⁻ leaving group.

71. The alcohol must lose a proton before it attacks the sulfonium ion, otherwise there would be two positive charges on the intermediate.

72. a.

b.

(reaction mechanism showing tetrahydropyran ring opening with H—Br, forming HO—(chain)—Br, followed by reaction with H—Br and :Br:⁻ to give H₂O + Br—(chain)—Br)

73. The product of each reaction is an alkene.

$$CH_3CH_2CHCH_3 \underset{\Delta}{\overset{H_2SO_4}{\rightleftharpoons}} CH_3CH_2CHCH_3 \rightleftharpoons CH_3CH_2CHCH_3 \rightleftharpoons CH_3CH=CHCH_3 + H_3O^+$$

with OH below first, $+OH$ over H below second, and $+$ with $+ H_2O$ below third.

$$CH_3CH_2CHCH_3 \overset{HO^-}{\longrightarrow} CH_3CH=CHCH_3 + H_2O + Br^-$$

with Br below the first structure.

Recall that alkenes undergo electrophilic addition reactions, and the first step in an electrophilic addition reaction is addition of an electrophile to the alkene. The acid-catalyzed dehydration reaction is reversible, because the electrophile (H^+) needed to react with the alkene in the first step of the reverse reaction is available.

The base-promoted elimination reaction of a hydrogen halide is not reversible, because an electrophile is not available to react with the alkene in the first step of the reverse reaction.

74.

$$H_2C \overset{O}{-\!\!\!-} CH_2 + HO^- \longrightarrow CH_2CH_2OH$$

(mechanism continuing: :O:⁻ attached to CH₂CH₂OH reacting with H₂C—CH₂ epoxide, giving HOCH₂CH₂OCH₂CH₂ with :O:⁻, then reacting with another H₂C—CH₂ epoxide)

$$HOCH_2CH_2OCH_2CH_2OCH_2CH_2OH \overset{H_2O}{\longleftarrow} HOCH_2CH_2OCH_2CH_2OCH_2CH_2O^-$$

$$+\ HO^-$$

75. **a.**

b.

c. The six-membered ring is formed by attack on the more sterically hindered carbon of the epoxide. Attack on the less sterically hindered carbon is preferred.

76.

77. Ethyl alcohol is not obtained as a product, because it reacts with the excess HI and forms ethyl iodide.

$$CH_3CH_2OCH_2CH_3 \xrightarrow[\Delta]{HI} CH_3CH_2I + CH_3CH_2OH \xrightarrow[\Delta]{HI} CH_3CH_2I + H_2O$$

78. :B is any base in the solution (HSO_4^-, H_2O, ROH).

79. **a.** The primary amine is 2-methyl-3-pentanamine.

$$\text{CH}_3\text{CH}_2\text{CHCHCH}_3 \xrightarrow[\text{K}_2\text{CO}_3]{\substack{\text{CH}_3\text{I} \\ \text{excess}}} \text{CH}_3\text{CH}_2\text{CHCHCH}_3 \xrightarrow[\text{H}_2\text{O}]{\text{Ag}_2\text{O}} \text{CH}_3\text{CH}_2\text{CHCHCH}_3$$

CH₃ (top substituent); NH₂ below; 2-methyl-3-pentanamine

two β-hydrogens / one β-hydrogen

Product of Ag₂O step: CH₃NCH₃ HO⁻ with CH₃; Δ

$$\text{CH}_3\text{CH}=\text{CHCHCH}_3 \quad + \quad \text{CH}_3\text{NCH}_3$$

4-methyl-2-pentene   (CH₃)

b. The primary amine is 3-methyl-2-butanime.

$$\text{CH}_3\text{CHCHCH}_3 \xrightarrow[\text{K}_2\text{CO}_3]{\substack{\text{CH}_3\text{I} \\ \text{excess}}} \text{CH}_3\text{CHCHCH}_3 \xrightarrow[\text{H}_2\text{O}]{\text{Ag}_2\text{O}} \text{CH}_3\text{CHCHCH}_3 \xrightarrow{\Delta} \text{CH}_2=\text{CHCHCH}_3$$

NH₂; CH₃NCH₃ CH₃ I⁻; CH₃NCH₃ CH₃ HO⁻

+ CH₃NCH₃ (CH₃)

c. The secondary amine is 3-methylpyrrolidine.

3-methylpyrrolidine $\xrightarrow[\text{K}_2\text{CO}_3]{\substack{\text{CH}_3\text{I} \\ \text{excess}}}$ [pyrrolidinium] I⁻ $\xrightarrow[\text{H}_2\text{O}]{\text{Ag}_2\text{O}}$ [pyrrolidinium] HO⁻ $\xrightarrow{\Delta}$ $\text{CH}_2=\text{CHCHCH}_2\text{NCH}_3$ (CH₃ CH₃)

CH₃I ↓

$$\text{CH}_2=\text{CHC}=\text{CH}_2 \xleftarrow{\Delta} \text{CH}_2=\text{CHCHCH}_2\text{NCH}_3 \xleftarrow[\text{H}_2\text{O}]{\text{Ag}_2\text{O}} \text{CH}_2=\text{CHCHCH}_2\text{NCH}_3$$

2-methyl-1,3-butadiene   CH₃ HO⁻   CH₃ I⁻

A and **B** structures:

A: CH₃—C—C—N(CH₃)(CH₃) with H₃C, CH₃ on first C and CH₃O, CH₃ on second C

B: CH₂=C—C—N(CH₃)(CH₃) with CH₃ and CH₃ CH₃

A   **B**

80. **A** is the substitution product that forms when methoxide ion attacks a carbon of the three-membered ring and eliminates the amino group, thereby opening the ring. It does not discolor Br₂ because it does not have a double bond to which Br₂ can add.

B is the product of a Hofmann elimination reaction: methoxide ion removes a proton from a methyl group bonded to a ring carbon and eliminates the amino group. The red color disappears when Br_2 is added to **B**, because Br_2 adds to the double bond.

When the aziridinium ion reacts with methanol, only **A**, the substitution product, is formed.

81.

a.

b.

c.

82.   **B** is the fastest reaction; **A** is the slowest reaction.

In order to form the epoxide, the alkoxide ion must attack the back-side of the carbon that is bonded to Br. This means that the OH and Br substituents must both be in axial positions. To be 1,2-diaxial, they must be trans to each other.

A does not form an epoxide, because the OH and Br substituents are cis to each other.
B and **C** can form epoxides because the OH and Br substituents are trans to each other.

The rate of formation of the epoxide is given by $k'K_{eq}$, where k' is the rate constant for the substitution reaction, and K_{eq} is the equilibrium constant for the [equatorial]/[axial] conformers.

When the OH and Br substituents are in the required diaxial position, the large *tert*-butyl substituent is in the equatorial position in **B** and in the axial position in **C**.

B   **C**

Because the more stable conformer is the one with the large *tert*-butyl group in the equatorial posi-tion, the OH and Br substituents are in the required diaxial position in the more stable conformer of

B (K_{eq} is large), whereas the OH and Br substituents are in the required diaxial position in the less stable conformer of **C** (K_{eq} is small). Therefore, **B** reacts faster than **C**.

B   C

83. **a.** The reaction of 2-chlorobutane with HO⁻ is an intermolecular reaction, so the two compounds have to find one another in the solution.

$$CH_3CHCH_2CH_3 \;+\; HO^- \longrightarrow CH_3CHCH_2CH_3 \;+\; Cl^-$$

with Cl and OH substituents respectively.

The following reaction takes place in two steps. The first is an intramolecular S$_N$2 reaction; the reaction is much faster than the above reaction because the two reactants are in the same molecule and can find one another relatively easily.

The second reaction is also an S$_N$2 reaction and is fast because the strain of the three-membered ring and the positive charge on the nitrogen make the amine a very good leaving group.

b. The HO group is bonded to a different carbon, because HO⁻ attacks the least sterically hindered carbon of the three-membered ring.

84.

85. **a.**

$$CH_3\overset{\underset{|}{CH_3}}{\underset{\underset{OH}{|}}{C}}-CH_2OH \underset{}{\overset{H_2SO_4}{\rightleftharpoons}} CH_3\overset{\underset{|}{CH_3}}{\underset{\underset{\overset{|}{\overset{+}{O}H}}{|}}{C}}-CH_2OH \longrightarrow CH_3\overset{\underset{|}{CH_3}}{\underset{\overset{+}{}}{C}}-CH_2OH \; + \; H_2O$$

$$CH_3\overset{\underset{|}{CH_3}}{CH}-\overset{\overset{O}{\|}}{CH} \; \underset{\text{interconversion}}{\overset{\text{keto-enol}}{\rightleftharpoons}} \; CH_3\overset{\underset{|}{CH_3}}{C}=CHOH \; + \; H_3O^+$$

b. Dehydration of the primary alcohol group cannot occur, because it cannot lose water via an E1 pathway, since a primary carbocation cannot be formed. It cannot lose water via an E2 pathway because the β-carbon is not bonded to a hydrogen. However, dehydration of the tertiary alcohol group can occur. The product is an enol that tautomerizes to an aldehyde. (See page 311 of the text).

86.

87. Cyclopentene oxide undergoes back-side attack by the nucleophile, so the two substituents in the products are trans to each other. Therefore, both the *R,R*-isomer and the *S,S*-isomer are obtained.

88. Notice that the initially generated carbocation can undergo either a 1,2-methyl shift or a 1,2-alkyl shift.

a.

b.

89. **a.** A nitrogen is a stronger base than an oxygen, so unlike an epoxide that can be opened without the oxygen being protonated, the three-membered nitrogen-containing ring has to be protonated to improve the leaving propensity of the group.

b. A nucleophile such as an NH$_2$ group on a chain of DNA can react with the three-membered ring. If a nucleophile on another chain of DNA reacts with another of the three-membered rings in this compound, the two DNA chains will be crosslinked.

Chapter 11 Practice Test

1.   Which of the following reagents is the best one to use in order to convert methyl alcohol into methyl bromide?

$$Br^- \quad HBr \quad Br_2 \quad NaBr \quad Br^+$$

2.   Which of the following reagents is the best one to use in order to convert methyl alcohol into methyl chloride?

$$Cl_2/CH_2Cl_2 \quad Cl_2/\Delta \quad Cl^- \quad SOCl_2 \quad NaCl$$

3.   **a.** What would be the major product obtained from the reaction of the epoxide shown below in methanol containing 0.1 M HCl?

b. What would be the major product obtained from the reaction of the epoxide in methanol containing 0.1 M NaOCH₃?

4.   Draw the major product(s) of each of the following reactions:

a.

b.

c.

5. Draw the major elimination product that is obtained when each of the following alcohols is heated in the presence of H_2SO_4:

a.
$$
\begin{array}{c}
\quad\quad CH_3 \\
\quad\quad | \\
CH_3CH_2C-CHCH_3 \\
\quad\quad | \quad\; | \\
\quad\quad OH \;\; CH_3
\end{array}
$$

c. $CH_3CH_2CH_2CH_2CH_2OH$

b.
$$
\begin{array}{c}
\quad\quad\quad\quad CH_3 \\
\quad\quad\quad\quad | \\
CH_3CH_2CH_2CH-CCH_3 \\
\quad\quad\quad\quad | \quad\; | \\
\quad\quad\quad\quad OH \;\; CH_3
\end{array}
$$

d.

6. Indicate whether each of the following statements is true or false:

a.	Tertiary alcohols are easier to dehydrate than secondary alcohols.	T	F
b.	Alcohols are more acidic than thiols.	T	F
c.	Alcohols have higher boiling points than thiols.	T	F
d.	The acid-catalyzed dehydration of a primary alcohol is an S_N1 reaction.	T	F
e.	The Hofmann elimination reaction is an E2 reaction.	T	F

7. What products would be obtained from heating the following ethers with one equivalent of HI?

a.
$$
\begin{array}{c}
\quad\quad\; CH_3 \\
\quad\quad\; | \\
CH_3CH_2COCH_3 \\
\quad\quad\; | \\
\quad\quad\; CH_3
\end{array}
$$

b.

8. Draw the major product of each of the following reactions:

a.
$$CH_3CH_2CH_2NH_2 \;\; + \;\;
\begin{array}{c}
\quad\quad CH_2CH_3 \\
\quad\quad | \\
CH_3CH_2\overset{+}{S}CH_2CH_3
\end{array}
\; \longrightarrow$$

b. $CH_3CH_2CH_2OH \;\; + \;\; SOCl_2 \;\; \xrightarrow{\text{pyridine}}$

c. $CH_3CH_2CH_2CH_2CH_2CH_2OH \;\; \xrightarrow[\Delta]{H_2SO_4}$

d. $CH_3CH_2CH_2CH_2OH \;\; \xrightarrow[\substack{CH_3COOH \\ 0\,°C}]{NaOCl}$

e. $CH_3CH_2CH_2CH_2OH \;\; \xrightarrow{H_2CrO_4}$

CHAPTER 12
Organometallic Compounds

Important Terms

alkene metathesis (olefin metathesis)	a reaction that breaks the double bond of an alkene and then rejoins the fragments to form a new double bond between two carbons that were not previously bonded.
alkyne metathesis	a reaction that breaks the triple bond of an alkyne and then rejoins the fragments to form a new triple bond between two carbons that were not previously bonded.
coupling reaction	a reaction that joins two groups with a carbon-carbon bond.
dialkyl lithium cuprate	$(R)_2CuLi$; prepared by treating an organolithium reagent with cuprous iodide.
Gilman reagent (organocuprate)	a dialkyl lithium cuprate used to replace a halogen in an alkyl, aryl, or vinylic halide with an alkyl group.
Grignard reagent	the compound that results when magnesium is inserted between the carbon and the halogen of an alkyl halide ($RMgBr$, $RMgCl$).
Heck reaction	a reaction that couples an aryl halide or vinylic halide with an alkene.
olefin metathesis (alkene metathesis)	a reaction that breaks the double bond of an alkene and then rejoins the fragments to form a new double bond between two carbons that were not previously bonded.
organoboron compound	an alkyl-organoboron compound, an alkenyl-organoboron compound, or an aryl-organoboron compound: $R'B(OR)_2$.
organocuprate (Gilman reagent)	a dialkyl lithium cuprate used to replace a halogen in an alkyl, aryl, or vinylic halide with an alkyl group.
organolithium compound	RLi; prepared by adding lithium to an alkyl halide.
organomagnesium compound	$RMgBr$ or $RMgCl$; prepared by adding an alkyl halide to magnesium shavings.
organometallic compound	a compound with a carbon-metal bond.
organopalladium compound	a compound with a palladium-carbon bond.
oxidative addition	insertion of a metal between two atoms.
reductive elimination	elimination of a metal from between two atoms, resulting in a covalent bond being formed between two carbons.
Suzuki reaction	a reaction that couples an aryl halide or vinylic halide with an organoboron reagent.
transmetallation	the exchange of two groups (often alkyl and halide) between two metals.

Solutions to Problems

1. All the reactions occur because, in each case, the reactant acid is a stronger acid than the product acid (methane, $pK_a \sim 60$).

2. The greater the polarity of the carbon-metal bond, the more reactive is the organometallic compound. Polarity depends on the difference in electronegativity between the atoms forming the bond; the greater the difference in electronegativity, the more polar the bond.

$$C\text{---}Li \qquad\qquad C\text{---}Na$$
$$2.5{-}1.0 = 1.5 \qquad\qquad 2.5{-}0.9 = 1.6$$

The electronegativity difference between carbon and sodium is greater than that between carbon and lithium. Therefore, organosodium compounds (with a more polar carbon-metal bond) are more reactive than organolithium compounds.

3. Transmetallation will occur if the new metal's electronegativity is closer to that of carbon. Because gallium (1.8) is more electronegative than magnesium, (1.2), and therefore closer to that of carbon, transmetallation will occur. Notice that after transmetallation the number of alkyl groups attached to gallium is the same as the number of chorines that were attached to it in the reactant.

$$3\ CH_3MgCl\ +\ GaCl_3\ \longrightarrow\ (CH_3)_3Ga\ +\ 3\ MgCl_2$$

4. Solved in the text.

5.   **a.**   **b.**   **c.**   **d.**

6. Because the tertiary alkyl halide will undergo an elimination reaction with the strongly basic organocuprate.

7. The Br is replaced by the alkyl group of the organocuprate.

$$2\ CH_3(CH_2)_3CH_2Br\ \xrightarrow[\text{hexane}]{2\ Li}\ 2\ CH_3(CH_2)_3CH_2Li\ \xrightarrow[\text{THF}]{CuI}\ (CH_3CH_2CH_2CH_2CH_2)_2CuLi$$

8.   **a.** $CH_3CH{=}CH_2CH_2\underset{\underset{\displaystyle CH_3}{|}}{\overset{\overset{\displaystyle OH}{|}}{C}}CH_3$ **b.** $CH_3CH_2\underset{\underset{\displaystyle CH_3}{|}}{\overset{\overset{\displaystyle OH}{|}}{CH}}CHCH_3$

9.   **a.** $CH_3CH_2CH_2CH_2CH_2OH$ **b.** —CH_2CH_2OH **c.** —$CH_2CH_2CH_2OH$

10.   **a.**

b.

c.

d.

11. Solved in the text.

12. **a.** Solved in the text. **b.** Solved in the text. **c.** **d.**

13. Add the number of carbons in the chain in the starting material to the number of carbons in the organo-boron compound and then make sure that it is the same as the number of carbons in the chain in the product.

a. **b.** **c.**

14. An alkenylboron compound is prepared from an alkyne.

1-pentyne

15. **a.** **b.**

c. **d.**

16. Solved in the text.

17.   Suzuki reactions

a.

$+$ $\xrightarrow[\text{HO}^-]{\text{PdL}_2}$

$\uparrow$ HB(OR)$_2$

CH$_3$CH$_2$C≡CH

b.

$+$ $\xrightarrow[\text{HO}^-]{\text{PdL}_2}$

$\uparrow$ HB(OR)$_2$

CH$_3$CH$_2$CH=CH$_2$

Heck reactions

a.

$+$ $\xrightarrow[\text{(CH}_3\text{CH}_2)_3\text{N}]{\text{PdL}_2}$

b.

$+$ $\xrightarrow[\text{(CH}_3\text{CH}_2)_3\text{N}]{\text{PdL}_2}$ $\xrightarrow[\text{H}_2]{\text{Pd/C}}$

Notice that only a Suzuki reaction can be used to replace a halogen with an alkyl group. A Heck reaction replaces a halogen wih an alkenyl group. However, the alkenyl group can subsequently be reduced to an alkyl group.

18.   CH$_3$C(=O)—⬡—Br + CH$_3$O—⬡—CH=CH$_2$

or

CH$_3$O—⬡—Br + CH$_3$C(=O)—⬡—CH=CH$_2$

19.   a. CH$_2$=CH$_2$ + CH$_3$CH$_2$CH=CHCH$_2$CH$_3$

E and *Z*

c. CH$_2$=CH$_2$ + (H$_3$C)(H$_3$C)C=C(CH$_3$)(CH$_3$)

b. CH$_2$=CH$_2$ +

d. CH$_2$=CH$_2$ +

20.   To determine the size of the ring that is formed in ring-closing metathesis, subtract 2 from the number of carbons in the reactant.

a. $CH_2{=}CH_2$ +   **c.** $CH_2{=}CH_2$ +

b. $CH_2{=}CH_2$ +

OCH_3

21.   **a.**   Solved in the text.   **c.**

b.   Solved in the text.   **d.**

O

22.

CH_3 $-C{\equiv}C-$ $-CH_3$

23.   **a.**   **d.**

b. $CH_2CH_2CH_2OH$ **e.**

c.   **f.**

24.   **C** is the only one that can be used to form a Grignard reagent.
The Grignard reagent formed from **A** will be destroyed immediately by reacting with the proton of the alcohol group.
The Grignard reagent formed from **B** will be destroyed immediately by reacting with the proton of the carboxylic acid group.
The Grignard reagent formed from **D** will be destroyed immediately by reacting with the proton of the amino group.

25.   **A** = Li   **D** = ethylene oxide   **G** = ethylene oxide
B = CuI   **E** = H^+ **H** = H^+
C = $(CH_3)_2CuLi$ **F** = NaH

26.   a.

or

b. CH_3CHCH_2OH $\xrightarrow[\Delta]{HBr}$ CH_3CHCH_2Br $\xrightarrow[2.\ CuI]{1.\ Li}$ $\left(CH_3CHCH_2\right)CuLi$
         | | |
         CH_3 CH_3 CH_3 $_2$

$CH_3CHCH_2CH_2CH_2OH$
        |
        CH_3

c. $CH_3CH_2C\equiv CH$ $\xrightarrow{NaNH_2}$ $CH_3CH_2C\equiv C^-$ $\xrightarrow[2.\ HCl]{1.\ \triangle}$ $CH_3CH_2C\equiv CCH_2CH_2OH$

d.

e.

f.

27.   a.   **b.**   **c.**   **d.**

e.

28.   **a.**   **d.**

b. (cis-alkene structure)   **e.** (diene structure)

c. (bicyclic alkene structure)   **f.** (substituted stilbene with CH₃ structure)

CH₃

29.   **a.** (vinyl bromide structure, Br)   **b.** (bromoalkene structure, Br)   **c.** (bromoalkene structure, Br)

30.   Remember that the double bond that comes from the alkenyl-organoboron compound always has the *E* (trans) configuration.

a. (reaction scheme with cyclohexyl alkyne + H–B catechol borane → vinyl boronate, HO⁻/PdL₂, vinyl bromide → diene product)

b. (reaction scheme with cyclohexyl alkyne + H–B catechol borane → vinyl boronate, HO⁻/PdL₂, cis-vinyl bromide → diene product)

c.

31. **a.**

b.

c.

d.

32. **a.** $CH_3CH_2CH_2CH_2Br$ $\xrightarrow[\text{2. CuI}]{\text{1. Li}}$ $(CH_3CH_2CH_2CH_2)_2CuLi$ $\xrightarrow[\text{2. HCl}]{\text{1.}}$ $CH_3CH_2CH_2CH_2CH_2CH_2OH$

$\downarrow$ PCC | CH_2Cl_2 **or**   NaOCl CH_3COOH 0 °C $\downarrow$

$$CH_3CH_2CH_2CH_2CH_2\overset{O}{\overset{\|}{C}}H$$

b. $CH_3CH=CHCH_3$ $\xrightarrow{\overset{O}{\overset{\|}{R}COOH}}$ $CH_3CH\overset{O}{\overbrace{}}CHCH_3$ $\xrightarrow[\text{2. HCl}]{\text{1.}(CH_3CH_2)_2CuLi}$ $CH_3\overset{OH}{\underset{\underset{CH_3}{|}}{CH}}CHCH_2CH_3$

$\downarrow$ $\overset{H_2SO_4}{\underset{\Delta}{}}$

$$CH_3CH=\underset{\underset{CH_3}{|}}{C}CH_2CH_3$$

c. $CH_3CH=CHCH_3$ $\xrightarrow{RCOOH}$ $CH_3CH\overset{O}{-}CHCH_3$ $\xrightarrow[\text{2. HCl}]{\text{1.}(CH_3)_2CuLi}$ $CH_3\overset{OH}{\underset{CH_3}{C}}HCHCH_3$

$\downarrow$ NaOCl / CH_3COOH / 0 °C

$CH_3\overset{O}{\overset{\|}{C}}CHCH_3$ with CH_3

33. Once some bromocyclohexane has been converted to a Grignard reagent, the nucleophilic Grignard reagent can react with the bromocyclohexane that has not been converted to a Grignard reagent.

$$\text{cyclohexyl}-Br \xrightarrow[Et_2O]{Mg} \text{cyclohexyl}-MgBr + Br-\text{cyclohexyl} \longrightarrow \text{bicyclohexyl}$$

$$+ \; MgBr_2$$

34. The student did not get any of the expected product, because the Grignard reagent removed a proton from the alcohol group. Addition of HCl/H_2O protonated the alkoxide ion and opened the epoxide ring.

$$CH_3MgBr + \text{(epoxide with OH)} \longrightarrow CH_4 + \text{(epoxide with } O^-) \xrightarrow[H_2O]{HCl} \text{(diol product)}$$

35. **a.** $\text{cyclohexanol} \xrightarrow[\text{pyridine}]{PBr_3} \text{bromocyclohexane} \xrightarrow[\text{2. CuI}]{\text{1. Li}} (\text{cyclohexyl})_2CuLi \xrightarrow[\text{2. HCl}]{\text{1. epoxide}} \text{cyclohexyl-}CH_2CH_2OH$

b. $\text{(2-propanol)} \xrightarrow[\text{pyridine}]{PBr_3} \text{(2-bromopropane)} \xrightarrow[Et_2O]{Mg} \text{(isopropyl-MgBr)} \xrightarrow{D_2O} \text{(isopropyl-D)}$

36. Remember that the double bond that comes from the alkenyl-organoboron compound always has the E (trans) configuration.

a. **C** cannot be prepared by a Heck reaction, because the double bond in the alkenyl-organoboron compound reactant of a Heck reaction always has the E configuration in the product. Both double bonds in **C** have the Z configuration.

A cannot be *directly* synthesized by a Heck reaction. A Heck reaction couples a vinylic halide and an alkene, so the product will contain at least two double bonds. However, **A** can be made by a Heck reaction followed by reduction of the double bonds.

b. The starting materials for the synthesis of **A** are shown above. Those for the synthesis of **B** are shown here.

37. $CH_3CH{=}CHCH_2CH_3$
                         2-pentene

A = $CH_3CH{=}$ **B** = $CH_3CH_2CH{=}$

A = a piece containing two carbons and **B** = a piece containing three carbons. The two pieces can be put together in four different ways.

$$\textbf{A—A} \quad \textbf{B—B} \quad \textbf{A—B} \quad \textbf{B—A}$$

Two of the four pieces (**A—B** and **B—A**) represent 2-pentene. Therefore, the maximum yield of 2-pentene is 50%.

38. Bombykol can be prepared using a Suzuki reaction. First, make the organoboron compound. Draw the vinylic bromide so you can see what product will be obtained from the coupling reaction.

Bombykol can also be prepared using a Heck reaction.

39. **a.** 2-butene and 3-hexene

b. Two symmetrical alkenes. The only alkenes that will be synthesized other than the target alkene are the reactant alkenes.

40. There are three possibilities **A**, **B**, and **C**. Only the bromine bonded to the benzylic carbon can be replaced by hydroxide ion via a nucleophilic substitution reaction. Both bromines can react with magnesium to form an organomagnesium group that can react with acid.

Compound **A**, for example, will react as follows:

41. **a.**   **b.**   **c.**

42. Some alkenes can undergo ring-opening metathesis to form high-molecular-weight polymers. Relieving the strain of a ring that contains a double bond (recall that sp^2 carbons have 120° bond angles) causes the reaction to go to completion.

a. First draw the product obtained by breaking the double bond and forming new double bonds. Seeing that six-membered rings are formed allows you to draw a reasonable structure for the polymer.

$-CH=CHCHCH_2CH_2CHCH=CHCHCH_2CH_2CHCH=CH-$

$-CH=CH$ ... $CH=CH-$

$CH=CH$

b.

$-CH=CHCHCH_2CH_2CHCH=CHCHCH_2CH_2CHCH=CH-$

$-CH=CH$ $CH=CH$ $CH=CH-$

Chapter 12 Practice Test

1.   Which is more reactive, an organocadmium compound (electronegativity of Cd = 1.5) or an organolead compound (electronegativity of lead = 1.6)?

2.   Which of the following alkyl halides could be used to form a Grignard reagent?

3.   What are the products of the following reactions?

   a.   + $\xrightarrow[\text{(CH}_3\text{CH}_2\text{)}_3\text{N}]{\text{PdL}_2}$

   b.   + $(\text{CH}_3\text{CH}_2\text{CH}_2)_2\text{CuLi}$ $\longrightarrow$

   c.   +   ⌐⌐⌐⌐B(OR)$_2$ $\xrightarrow[\text{HO}^-]{\text{PdL}_2}$

4.   What are the products of the following reactions?

   a.   Br   + $\xrightarrow[\text{(CH}_3\text{CH}_2\text{)}_3\text{N}]{\text{PdL}_2}$

   b.   Br   + $(\text{CH}_3\text{CH}_2\text{CH}_2)_2\text{CuLi}$ $\longrightarrow$

   c.   Br   +   ⌐⌐⌐⌐B(OR)$_2$ $\xrightarrow[\text{HO}^-]{\text{PdL}_2}$

5.   Draw the products of each of the following reactions:

   a. $\xrightarrow{\text{Grubbs catalyst}}$

   b. $\xrightarrow{\text{Grubbs catalyst}}$

   c. $\xrightarrow{\text{Grubbs catalyst}}$

6.   Indicate how each of the following compounds could be prepared using the given starting material:

a.

b.

7.   What alcohols would be formed from the reaction of ethylene oxide with the following organocuprates?

a. $(CH_3CH_2CH_2)_2CuLi$

b.

8.   Which of the following could be used as a reactant in a Heck or a Suzuki reaction?

9.   Show how the following compounds can be prepared by a:

a.   Heck reaction.   b.   Suzuki reaction

1.   2.

10.   What are the products of the following reactions?

a.

b.

11.   Using any needed reagents, show how the following synthesis could be carried out.

Important Terms

alkane	a hydrocarbon that contains only single bonds.
alkoxy radical	an alkoxy group with an unpaired electron on oxygen.
allylic radical	a species with an unpaired electron on an allylic carbon.
benzylic radical	a species with an unpaired electron on a benzylic carbon.
combustion	a reaction with oxygen that takes place at high temperatures and converts alkanes (or other organic compounds) to carbon dioxide and water.
free radical **(radical)**	an atom or a molecule with an unpaired electron.
halogenation reaction	the reaction of an alkane (or other organic compounds) with a halogen.
heterolytic bond cleavage **(heterolysis)**	breaking a bond with the result that both bonding electrons stay with one of the previously bonded atoms.
homolytic bond cleavage **(homolysis)**	breaking a bond with the result that each of the atoms that formed the bond gets one of the bonding electrons.
initiation step	the step in which radicals are created and/or the step in which the radical needed for the first propagating step is created.
peroxide	a compound with an O—O bond.
peroxide effect	peroxide causes the initial step in the addition of HBr to an alkene to be a bromine radical instead of a proton.
primary alkyl radical	an alkyl radical with the unpaired electron on a primary carbon.
propagation step	in the first of a pair of propagation steps, a radical reacts to produce another radical that reacts in the second propagation step to produce the radical that was the reactant in the first propagation step.
radical (often called a free radical)	an atom or a molecule with an unpaired electron.
radical addition reaction	an addition reaction in which the first species that adds is a radical.
radical chain reaction	a reaction in which radicals are formed and react in repeating propagating steps.
radical inhibitor	a compound that traps radicals.

radical initiator	a compound that creates radicals.
radical substitution reaction	a substitution reaction that has a radical intermediate.
reactivity-selectivity principle	the greater the reactivity of a species, the less selective it will be.
saturated hydrocarbon	a hydrocarbon that contains only single bonds (it is saturated with hydrogen).
secondary alkyl radical	an alkyl radical with the unpaired electron on a secondary carbon.
termination step	two radicals combine to produce a molecule in which all the electrons are paired.
tertiary alkyl radical	an alkyl radical with the unpaired electron on a tertiary carbon.

Solutions to Problems

1. $Cl_2 \xrightarrow{h\nu} 2\ Cl\cdot$ } initiation

2.
$$Cl_2 \xrightarrow{h\nu} 2\ Cl\cdot$$

$$Cl\cdot + CH_4 \longrightarrow \cdot CH_3 + HCl$$

$$\cdot CH_3 + Cl_2 \longrightarrow CH_3Cl + Cl\cdot$$

$$Cl\cdot + CH_3Cl \longrightarrow \cdot CH_2Cl + HCl$$

$$\cdot CH_2Cl + Cl_2 \longrightarrow CH_2Cl_2 + Cl\cdot$$

$$Cl\cdot + CH_2Cl_2 \longrightarrow \cdot CHCl_2 + HCl$$

$$\cdot CHCl_2 + Cl_2 \longrightarrow CHCl_3 + Cl\cdot$$

$$Cl\cdot + CHCl_3 \longrightarrow \cdot CCl_3 + HCl$$

$$\cdot CCl_3 + Cl_2 \longrightarrow CCl_4 + Cl\cdot$$

$$Cl\cdot + \cdot CCl_3 \longrightarrow CCl_4$$

3.   **a.**
$$\underset{\underset{CH_3}{|}}{CH_3CH_2CHCH_2CCH_2CH_3}$$

with CH_3 groups as shown

b.   Six secondary hydrogens

4.   **a.** 3   **c.** 5   **e.** 5   **g.** 2   **i.** 4

       b. 3   **d.** 1   **f.** 4   **h.** 4

5.   Solved in the text.

6.   Solved in the text.

7.   **a.**   The major product would be the one obtained from the removal of a tertiary hydrogen.

$$
\underset{\underset{Br}{|}}{\overset{\overset{CH_3}{|}}{CH_3CCH_2CH_3}}
$$

   b.   The relative amounts of the four possible products would be:

   $6 \times 1 = 6$ $1 \times 1600 = 1600$ $2 \times 82 = 164$ $3 \times 1 = 3$

   percentage of major product $= \dfrac{1600}{1773} \times 100 = 90.2\%$

8.   **a.**   Chlorination, because the halogen is substituting for a primary hydrogen.
   b.   Bromination, because the halogen is substituting for a tertiary hydrogen.
   c.   Because the molecule has only one kind of hydrogen, both chlorination and bromination will form only one monohalogenated product.

9.   Solved in the text.

10.   **a.** $\underset{}{\overset{\overset{CH_3}{|}}{CH_3CHCH_3}} \xrightarrow[hv]{Br_2} \underset{\underset{Br}{|}}{\overset{\overset{CH_3}{|}}{CH_3CCH_3}}$

   b. $\underset{}{\overset{\overset{CH_3}{|}}{CH_3CHCH_3}} \xrightarrow[hv]{Br_2} \underset{\underset{Br}{|}}{\overset{\overset{CH_3}{|}}{CH_3CCH_3}} \xrightarrow{HO^-} \underset{\underset{OH}{|}}{\overset{\overset{CH_3}{|}}{CH_3CCH_3}}$

   c. $\underset{}{\overset{\overset{CH_3}{|}}{CH_3CHCH_3}} \xrightarrow[hv]{Br_2} \underset{\underset{Br}{|}}{\overset{\overset{CH_3}{|}}{CH_3CCH_3}} \xrightarrow{HO^-} \underset{\underset{OH}{|}}{\overset{\overset{CH_3}{|}}{CH_3CCH_3}} \xrightarrow[\Delta]{H_2SO_4} \overset{\overset{CH_3}{|}}{CH_2{=}CCH_3}$

11.   To start peroxide formation, the chain-initiating radical removes a hydrogen atom from an α-carbon of the ether.

   a.   **D** is most apt to form a peroxide, because removal of a hydrogen from an α-carbon forms a secondary radical.

   b.   **B** is least apt to form a peroxide, because it does not have any hydrogens bonded to its α-carbons.

12.

13.   **a.**

b.

c.

d.

14.   **a.**

b.

15.   Bromination can occur at either of the two tertiary positions. Because the radical intermediate is sp^2 hybridized and therefore planar, the incoming bromine can add to a tertiary radical from either the top or the bottom of the plane. (Notice that there are four stereoisomers: *R,R*, *R,S*, *S,S*, and *S,R*.)

16.   **a.**   Two stereoisomers are formed because the reaction forms a compound with an asymmetric center.

b. The reaction forms two constitutional isomers, and each constitutional isomer has two stereoisomers.

17.

18. Solved in the text.

19. The product of the reaction in Problem 18 has one asymmetric center. Therefore, two allylic substituted bromoalkenes are obtained, one with the *R* configuration at the asymmetric center and one with the *S* configuration.

(*R*)-4-bromo-2-pentene   (*S*)-4-bromo-2-pentene

20. **a.**

b.

c.

d.

21.   **a.**   **1.**   There are two sets of secondary allylic hydrogens, **a** and **b** (plus a less reactive set of primary allylic hydrogens on the methyl group).
Removing one of the **a** allylic hydrogens forms an intermediate in which the unpaired electron is shared by two secondary allylic carbons. Removing one of the **b** allylic hydrogens forms an intermediate in which the unpaired electron is shared by a tertiary allylic carbon and a secondary allylic carbon.

Therefore, the major products are obtained by removing a **b** allylic hydrogen.

b.   **1.**   Each of the products has one asymmetric center.
The *R* and the *S* stereoisomers will be obtained for each product.

2.   The product has two asymmetric centers.
Because addition of Br$_2$ is anti, only the stereoisomers with the bromine atoms on opposite sides of the ring are obtained.

3.   The product does not have an asymmetric center, so it does not have stereoisomers.

4.   The product has two asymmetric centers.
Because radical addition of HBr can be either syn or anti, four stereoisomers are obtained as products.

22.   **a.**

b.

c.

d.

23.   Four atoms (three carbons and one oxygen) share the unpaired electrons.

24. Antioxidants are radical inhibitors. That is, they react with radicals and, thereby, prevent radical chain reactions. There are several OH groups in a catechin that, upon losing a hydrogen atom as a result of reacting with a reactive radical, form a radical that is stabilized by electron delocalization. This highly stabilized radical is sufficiently unreactive that it cannot damage cells by reacting with them. An example of one of the stabilized radicals is shown below.

25. **a.** $CH_2{=}CHCHCH_2CH_3 \longleftrightarrow CH_2CH{=}CHCH_2CH_3$

$\downarrow Br_2$ $\qquad\qquad \downarrow Br_2$

$CH_2{=}CHCHCH_2CH_3$ + $BrCH_2CH{=}CHCH_2CH_3$
$\qquad\quad |$
$\qquad\quad Br$

b. In the radical intermediate that leads to the major products, the unpaired electron is shared by a primary carbon and a tertiary carbocation. In the radical intermediate that leads to the minor products, the unpaired electron is shared by a primary and a secondary carbocation. It, therefore, is not as stable as the intermediate that leads to the major products.

$$CH_3\!\!-\!\!C{=}CH{=}CH_2$$
$$\quad\;\;\delta\!\cdot \qquad \delta\!\cdot$$

$$CH_3C{=}CHCH_2Br \;+\; CH_3CCH{=}CH_2$$
$$\qquad\qquad\qquad\qquad\qquad\quad |$$
$$\qquad\qquad\qquad\qquad\qquad\; Br$$
major products

$$CH_2{=}C{=}CHCH_3$$
$$\delta\!\cdot \qquad\quad \delta\!\cdot$$

$$BrCH_2C{=}CHCH_3 \;+\; CH_2{=}CCHCH_3$$
$$\qquad\qquad\qquad\qquad\qquad\qquad |$$
$$\qquad\qquad\qquad\qquad\qquad\; Br$$
minor products

In bromination, selectivity is much more important than probability. Therefore, even though twice as many hydrogens are available for removal by a bromine radical that leads to the minor products, they will still be minor products because the easier-to-remove hydrogens lead to the major products.

c.

Br

major product

d.

Cl

e. no reaction
(There is no light or heat, so radicals cannot be formed.)

f.

+   +   +

26. a.

dimethylpropane

b. CH₃CH₂CHCH₂CH₂CH₃

3-methylhexane

27. a.

Four secondary hydrogens
can be removed to form
this product

b.

Eight secondary
hydrogens can be
removed to form
this product

c.

Six secondary
hydrogens can be
removed to form
this product

28. Because a bromine radical is much more selective than a chlorine radical, the bromine radical will remove a tertiary hydrogen in order to form a tertiary radical.

a.

b.

c.

29. a.

d.

b.

e.

+

c.

f.

+

In **e** and **f**, because one product is under kinetic control and the other is under thermodynamic control, the major product will depend on the conditions under which the reaction is carried out.

30.

$$\begin{array}{c} \text{CH}_3 \\ | \\ \text{CH}_3\text{CCH}_3 \\ | \\ \text{Cl} \end{array}$$

$$1 \times x = x$$

$$\begin{array}{c} \text{CH}_3 \\ | \\ \text{CH}_3\text{CHCH}_2\text{Cl} \end{array}$$

$$9 \times 1 = 9$$

$$\frac{\text{fraction of the total that is}}{\text{1-chloro-2-methylpropane}} = \frac{\text{1-chloro-2-methylpropane}}{\text{1-chloro-2-methylpropane} + \text{2-chloro-2-methylpropane}} = 0.64$$

$$\frac{9}{9 + x} = 0.64$$

$$9 = 0.64\,(9 + x)$$

$$9 = 5.76 + 0.64x$$

$$3.24 = 0.64x$$

$$x = 5$$

We have found that it is five times easier for a chlorine radical to remove a hydrogen atom from a tertiary carbon than from a primary carbon.

31.   Removing a hydrogen atom from ethane by an iodine radical is a highly endothermic reaction ($\Delta H^\circ = 101 - 71 = 30 \text{ kcal/mol}$; see Table 5.1 on page 210 of the text), so the iodine radicals will reform I_2 rather than remove a hydrogen atom.

$$\text{CH}_3\text{CH}_3 \;+\; \text{I}\cdot \;\;\xrightarrow{\;\;\big/\!\!\!\!\!\!\!\!\!\times\;\;}\;\; \text{CH}_3\overset{\bullet}{\text{C}}\text{H}_2 \;+\; \text{HI}$$
$$\text{101 kcal/mol} \qquad\qquad\qquad\qquad \text{71 kcal/mol}$$

32.

33.   It is easier to break a C—H bond than a C—D bond.
Because a bromine radical is less reactive (and more selective) than a chlorine radical, a bromine radical has a greater preference for the more easily broken C—H bond. Bromination, therefore, would have a greater deuterium kinetic isotope effect than chlorination.

34.   **a.** Five monochlorination products are possible.

b. The two compounds below will be obtained in the greatest yield because each one can be formed by removing one of four hydrogens to form a secondary radical. More of the compound on the left will be obtained because the methyl group in the compound on the right will provide some steric hindrance to the approach of the chlorine atom.

4 × 3.8 = 15.2   4 × 3.8 = 15.2

c. The number of possible stereoisomers for each compound is indicated below each structure.

   1   1   4   4   2

These have 0   These have 2
asymmetric centers.   asymmetric centers.   This does not have
an asymmetric center,
but has cis-trans isomers.

A total of 12 stereoisomers can be obtained.

35.   **a.** CH_3CHCH_2Cl (with CH_3 above)

c. $CH_3CHCHCH_3$ (with CH_3 above, Br below)

e. (seven-membered ring with Br and CH_3 substituents)

b. CH_3CCH_3 (with CH_3 above, Br below)

d. $CH_2=CHCH_2CH_2CHCH_2Br$ (with CH_3 above)

f. (six-membered ring with $CH=CH_2$, Br, and CH_3 substituents)

36. a.

b.

c.

d.

e.

37.   a. One difference between this reaction and the monochlorination of ethane is the source of the chlorine radical. This reaction has two sets of propagation steps because two different radicals are generated in the initiation step. This reaction also has several more termination steps because of the two different radicals generated in the initiation step. (Bond dissociation energies can be found in Table 5.1 on page 210 of the text.)

propagation

$$Cl\cdot \ + \ CH_3CH_3 \ \longrightarrow \ HCl \ + \ CH_3\dot{C}H_2$$

$$CH_3\dot{C}H_2 \ + \ CH_3\underset{\underset{CH_3}{|}}{\overset{\overset{CH_3}{|}}{C}}OCl \ \longrightarrow \ CH_3CH_2Cl \ + \ CH_3\underset{\underset{CH_3}{|}}{\overset{\overset{CH_3}{|}}{C}}O\cdot$$

this species enters the first propagation cycle

termination

$$CH_3\dot{C}H_2 \ + \ Cl\cdot \ \longrightarrow \ CH_3CH_2Cl$$

$$CH_3\underset{\underset{CH_3}{|}}{\overset{\overset{CH_3}{|}}{C}}O\cdot \ + \ CH_3\underset{\underset{CH_3}{|}}{\overset{\overset{CH_3}{|}}{C}}O\cdot \ \longrightarrow \ CH_3\underset{\underset{CH_3}{|}}{\overset{\overset{CH_3}{|}}{C}}O - O\underset{\underset{CH_3}{|}}{\overset{\overset{CH_3}{|}}{C}}CH_3$$

$$CH_3\underset{\underset{CH_3}{|}}{\overset{\overset{CH_3}{|}}{C}}O\cdot \ + \ Cl\cdot \ \longrightarrow \ CH_3\underset{\underset{CH_3}{|}}{\overset{\overset{CH_3}{|}}{C}}OCl$$

$$CH_3\underset{\underset{CH_3}{|}}{\overset{\overset{CH_3}{|}}{C}}O\cdot \ + \ CH_3\dot{C}H_2 \ \longrightarrow \ CH_3\underset{\underset{CH_3}{|}}{\overset{\overset{CH_3}{|}}{C}}OCH_2CH_3$$

$$Cl\cdot \ + \ Cl\cdot \ \longrightarrow \ Cl_2$$

$$CH_3\dot{C}H_2 \ + \ CH_3\dot{C}H_2 \ \longrightarrow \ CH_3CH_2CH_2CH_3$$

b. $\Delta H° = $ bonds broken $-$ bonds formed

Let $x = $ the bond dissociation energy of the O—Cl bond.

$$-42\,kcal/mol = [\,101\,kcal/mol + x\,kcal/mol\,] - [\,85\,kcal/mol + 105\,kcal/mol\,]$$

$$-143\,kcal/mol = x\,kcal/mol - 190\,kcal/mol$$

$$x = 47\,kcal/mol$$

38.   The reaction forms a product with two new asymmetric centers.

$$CH_3CH_2\underset{}{\overset{\overset{CH_3}{|}}{C}}H - \underset{\underset{Br}{|}}{\overset{\overset{CH_3}{|}}{C}}CH_2CH_3$$

Because the reaction involves both syn and anti addition, four stereoisomers are formed.

39. **a.** HO⌒⌒ (alkene) $\xrightarrow[\text{peroxide}]{\text{HBr}}$ HO⌒⌒⌒Br $\xrightarrow{\text{NaH}}$ ⁻O⌒⌒⌒Br ⟶ (tetrahydrofuran ring with O)

b. ⌒⌒ $\xrightarrow[\text{peroxide}]{\text{HBr}}$ ⌒⌒Br $\xrightarrow{CH_3O^-}$ ⌒⌒OCH_3

c. (structure with OH) $\xrightarrow[\Delta]{H_2SO_4}$ (alkene) $\xrightarrow[\text{peroxide}]{\text{HBr}}$ (structure with Br)

d. ⌒⌒ $\xrightarrow{Br_2}$ (structure with Br, Br) $\xrightarrow{HO^-}$ (alkenyl Br)

$\xrightarrow[(CH_3CH_2)_3N]{PdL_2}$ (diene)

(straight chain alkane) $\xleftarrow[Pd/C]{H_2}$ (diene)

40.

$$CH_3CHCH_2CH_3$$
$$|$$
$$CH_3$$

2-methylbutane

2-Methylbutane forms two primary alkyl halides, one secondary alkyl halide, and one tertiary alkyl halide. First we need to calculate how much alkyl halide is formed from each primary hydrogen available, from each secondary hydrogen available, and from each tertiary hydrogen available.

a primary alkyl halide

$ClCH_2CHCH_2CH_3$
  |
 CH_3

Substitution of any one of six hydrogens leads to this product.
percentage of this product that is formed = 36%
percentage formed per hydrogen available = 36/6 = 6%

$CH_3CHCH_2CH_2Cl$
  |
 CH_3

Substitution of any one of three hydrogens leads to this product.
percentage of this product that is formed = 18%
percentage formed per hydrogen available = 18/3 = 6%

a secondary alkyl halide

       Cl
       |
$CH_3CHCHCH_3$
       |
      CH_3

Substitution of any one of two hydrogens leads to this product.
percentage of this product that is formed = 28%
percentage formed per hydrogen available = 28/2 = 14%

a tertiary alkyl halide

       Cl
       |
$CH_3CCH_2CH_3$
       |
      CH_3

Substitution of the one tertiary hydrogen leads to this product.
percentage of this product that is formed = 18%
percentage formed per hydrogen available = 18/1 = 18%

From the above calculations, we see that at 300 °C, the relative rates of removal of a hydrogen atom from a tertiary, secondary, and primary carbocation are:

$$18 : 14 : 6 = 3.0 : 2.3 : 1$$

In Section 13.4, we saw that at room temperature the relative rates are:

$$5.0 : 3.8 : 1$$

Thus, we can conclude that at higher temperatures, there is less discrimination by the radical between a tertiary, secondary, and primary hydrogen atom.

41.   The chlorine radical is more reactive at the higher temperature $(600\,°C)$, so it is less selective.

42.   **a.**

b.

c.

d.

e.

f.

43.   For an alkene to form the same product whether it reacts with HBr in the presence or absence of peroxide, it must be symmetrical about the double bond.

a.   **b.** $CH_3CH_2CH{=}CHCH_2CH_3$ $CH_3C{=}CCH_3$ (with two CH_3 groups)

cis or trans

44.   **a.** $CH_3{-}H$ + $Cl{-}Cl$ $\longrightarrow$ $CH_3{-}Cl$ + $H{-}Cl$

          105   58   84   103

$\Delta H° = $ bonds broken $-$ bonds formed

$\Delta H° = [\,105\,kcal/mol + 58\,kcal/mol\,] - [\,84\,kcal/mol + 103\,kcal/mol\,]$

$\Delta H° = 163 - 187 = -24\,kcal/mol$

b. $CH_3—H$ + $\overset{\cdot}{Cl}$ $\longrightarrow$ $\overset{\cdot}{CH_3}$ + $H—Cl$ $\Delta H° = 105 - 103 = 2\,\text{kcal/mol}$

$\overset{\cdot}{CH_3}$ + $Cl—Cl$ $\longrightarrow$ $CH_3—Cl$ + $\overset{\cdot}{Cl}$ $\Delta H° = 58 - 84 = -26\,\text{kcal/mol}$

Overall $\Delta H° = 2 + (-26) = -24\,\text{kcal/mol}$

c. If you cancel the things that are the same on opposite sides of the equations in part **b** and then add the two equations, you are left with the equation in part **a**.

$CH_3—H$ + $\overset{\cdot}{\cancel{Cl}}$ $\longrightarrow$ $\overset{\cdot}{\cancel{CH_3}}$ + $H—Cl$

$\cancel{CH_3}$ + $Cl—Cl$ $\longrightarrow$ $CH_3—Cl$ + $\overset{\cdot}{\cancel{Cl}}$

CH_4 + Cl_2 $\longrightarrow$ CH_3Cl + HCl

45.   The overall $\Delta H°$ value is the same for both mechanisms. However, the mechanism that occurs is controlled by the first propagation step. The first propagation step of the alternative mechanism is very endothermic, so it would not be able to compete with the first propagation step of the mechanism shown in Problem 44b.

$CH_3—H$ + $\cdot Cl$ $\longrightarrow$ $CH_3—Cl$ + $\cdot H$ $\Delta H° = 105 - 84 = 21\,\text{kcal/mol}$

$H\cdot$ + $Cl—Cl$ $\longrightarrow$ $H—Cl$ + $\cdot Cl$ $\Delta H° = 58 - 103 = -45\,\text{kcal/mol}$

Overall $\Delta H° = 21 + (-45) = -24\,\text{kcal/mol}$

46.

47.   The methyl radical that is created in the first propagation step reacts with Br_2, forming bromomethane.

$\cdot Br$ + CH_4 $\longrightarrow$ $\cdot CH_3$ + HBr

$\cdot CH_3$ + Br_2 $\longrightarrow$ CH_3Br + $\cdot Br$

If HBr is added to the reaction mixture, the methyl radical that is created in the first propagation step can react with Br_2 or with the added HBr. Because only reaction with Br_2 forms bromomethane (reaction with HBr reforms methane), the overall rate of formation of bromomethane is decreased.

$$\cdot Br \ + \ CH_4 \ \longrightarrow \ \cdot CH_3 \ + \ HBr$$

$$\cdot CH_3 \ + \ Br_2 \ \longrightarrow \ CH_3Br \ + \ \cdot Br$$

$$\cdot CH_3 \ + \ HBr \ \longrightarrow \ CH_4 \ + \ \cdot Br$$

48.

Chapter 13 Practice Test

1. How many monochlorinated products would be obtained from the monochlorination of the following alkanes? (Ignore stereoisomers.)

   a.

   b.

   c.

   d.

   e.

2. What is the first propagation step in the monochlorination of ethane?

3. When (S)-2-bromopentane is brominated, 2,3-dibromopentanes are formed. Which of the following compounds are **not** formed?

4. Determine the $\Delta H°$ of the two propagation steps in the monochlorination of ethane, using Table 5.1 on page 210 of the text.

5. Label the radicals in order from most stable to least stable. (Label the most stable #1.)

   $\dot{C}H_2CH_2CH_2CH=CH_2$ $CH_3CH_2\dot{C}HCH=CH_2$

   $CH_3CH_2CH_2CH=\dot{C}H$ $CH_3\dot{C}HCH_2CH=CH_2$

6. Draw the product(s) of each of the following reactions, ignoring stereoisomers.

   a. + NBS $\xrightarrow[\text{peroxide}]{\Delta}$

   b. $CH_3CH_2CH=CH_2$ + NBS $\xrightarrow[\text{peroxide}]{\Delta}$

7.   **a.**   Draw the products that will be obtained from the monochlorination of the following alkane at room temperature. (Ignore stereoisomers.)

$$
\begin{array}{ccc}
CH_3 & & CH_3 \\
| & & | \\
CH_3CCH_2CH_2CHCH_3 & + \ Cl_2 & \xrightarrow{\ h\nu\ } \\
| & & \\
CH_3 & &
\end{array}
$$

   b.   What product will be obtained in greatest yield?

   c.   What product would be obtained in greatest yield if the alkane were brominated instead of chlorinated?

8.   Calculate the yield of 2-chloro-2-methylbutane (as a percentage of all the monochlorinated products obtained) when 2-methylbutane is chlorinated in the presence of light at room temperature.

9.   What is the major product of each of the following reactions? Ignore stereoisomers.

   a. $CH_3CH{=}\overset{\overset{\displaystyle CH_3}{|}}{C}CH_2CH_3 \ + \ HBr \longrightarrow$

   c. $CH_3CH{=}\overset{\overset{\displaystyle CH_3}{|}}{C}CH_2CH_3 \ + \ HCl \longrightarrow$

   b. $CH_3CH{=}\overset{\overset{\displaystyle CH_3}{|}}{C}CH_2CH_3 \ + \ HBr \xrightarrow{\ peroxide\ }$

   d. $CH_3CH{=}\overset{\overset{\displaystyle CH_3}{|}}{C}CH_2CH_3 \ + \ HCl \xrightarrow{\ peroxide\ }$

10.   Which of the above reactions forms more than one stereoisomer?

CHAPTER 14
Mass Spectrometry, Infrared Spectroscopy, and Ultraviolet/Visible Spectroscopy

Important Terms

absorption band	a peak in a spectrum that occurs as a result of absorption of energy.
auxochrome	a substituent that when attached to a chromophore alters the λ_{max} and intensity of absorption of UV/Vis radiation.
base peak	the peak in a mass spectrum with the greatest intensity.
Beer-Lambert Law	an equation that states the relationship between the absorbance of UV/Vis light, the concentration of the sample, the length of the light path, and the molar absorptivity.
bending vibration	a vibration that does not occur along the line of the bond.
chromophore	the part of a molecule responsible for a UV or visible spectrum.
α-cleavage	homolytic cleavage of an alpha substituent.
electromagnetic radiation	radiant energy that displays wave properties.
electronic transition	promotion of an electron from its HOMO to its LUMO.
fingerprint region	the right-hand third of an IR spectrum ($1400-600$ cm^{-1}), where the absorption bands are characteristic of the compound as a whole.
fragment ion peak	a positively charged fragment of a molecular ion.
frequency	the velocity of a wave divided by its wavelength.
functional group region	the left-hand two-thirds of an IR spectrum ($4000-1400$ cm^{-1}), where most functional groups show absorption bands.
highest occupied molecular orbital (HOMO)	the highest energy molecular orbital that contains electrons.
Hooke's law	an equation that describes the motion of a vibrating spring.
λ_{max}	the wavelength at which there is maximum UV/Vis absorbance.
infrared radiation	electromagnetic radiation familiar to us as heat.
infrared spectroscopy	spectroscopy that uses infrared energy to provide a knowledge of the functional groups in a compound.
infrared spectrum (IR spectrum)	a plot of relative absorption versus wavenumber (or wavelength) of absorbed infrared radiation.

lowest unoccupied molecular orbital (LUMO)	the lowest energy molecular orbital that does not contain electrons.
mass spectrometry	an instrumental technique that provides a knowledge of the molecular weight and certain structural features of a compound.
mass spectrum	a plot of the relative abundance of the positively charged fragments produced in a mass spectrometer versus their m/z values.
McLafferty rearrangement	rearrangement of the molecular ion of a ketone that contains a γ-hydrogen; the bond between the α- and β-carbons breaks, and a γ-hydrogen migrates to the oxygen.
molar absorptivity	the absorbance obtained from a 1.00 M solution in a cell with a 1.00 cm light path.
molecular ion (M)	the radical cation formed by removing one electron from a molecule.
nominal molecular mass	mass to the nearest whole number.
radical cation	a species with a positive charge and an unpaired electron.
rule of 13	a rule that allows possible molecular formulas to be determined from the m/z value of the molecular ion.
spectroscopy	study of the interaction of matter and electromagnetic radiation.
stretching vibration	a vibration occurring along the line of the bond.
ultraviolet light	electromagnetic radiation with wavelengths ranging from 180 to 400 nm.
UV/Vis spectroscopy	the absorption of electromagnetic radiation that is useful in determining information about conjugated systems.
visible light	electromagnetic radiation with wavelengths ranging from 400 to 780 nm.
wavelength	distance from any point on one wave to the corresponding point on the next wave.
wavenumber	the number of waves in 1 cm.

Solutions to Problems

1.   Only positively charged fragments are accelerated through the analyzer tube.

$$CH_3CH_2\overset{+}{C}H_2 \qquad [CH_3CH_2CH_3]\overset{+}{\cdot} \qquad \overset{+}{C}H_2CH{=}CH_2$$

2.   The peak at $m/z = 57$ will be more intense for 2,2-dimethylpropane than for 2-methylbutane or pentane. The peak at $m/z = 57$ is due to loss of a methyl radical: loss of a methyl radical from 2,2-dimethylpropane forms a tertiary carbocation, whereas loss of a methyl radical from 2-methylbutane and pentane forms a less stable secondary and primary carbocation, respectively.

Notice that the mass spectrum of 2-methylbutane can be distinguished from those of the other isomers by the peak at $m/z = 43$. The peak at $m/z = 43$ will be most intense for 2-methylbutane because such a peak is due to loss of an ethyl radical, which forms a secondary carbocation. Pentane gives a less intense peak at $m/z = 43$ because loss of an ethyl radical from pentane forms a primary carbocation. 2,2-Dimethylpropane will not show a peak at $m/z = 43$, because it does not have an ethyl group.

3. Intense peaks should occur at $m/z = 57$ for loss of an ethyl radical ($86 - 29$), and at $m/z = 71$ for loss of a methyl radical ($86 - 15$).

$$\left[\begin{array}{c} CH_3 \\ | \\ CH_3CH_2CHCH_2CH_3 \end{array}\right]^{+\cdot} \longrightarrow \begin{array}{c} CH_3 \\ | \\ CH_3CH_2\overset{+}{C}H \end{array} + \dot{C}H_2CH_3$$
$$m/z = 57$$

$$\left[\begin{array}{c} CH_3 \\ | \\ CH_3CH_2CHCH_2CH_3 \end{array}\right]^{+\cdot} \longrightarrow CH_3CH_2\overset{+}{C}HCH_2CH_3 + \dot{C}H_3$$
$$m/z = 71$$

A secondary carbocation is formed in both cases. Because an ethyl radical is more stable than a methyl radical, the base peak will most likely be at $m/z = 57$.

4. Solved in the text.

5.   **a.** Dividing 72 by 13 gives 5 with 7 left over. Thus, the base value is C_5H_{12}. Because the compound contains only carbons and hydrogens, we know that the base value is also the molecular formula of the compound.

   b. Dividing 100 by 13 gives 7 with 9 left over. Thus, the base value is C_7H_{16}. Because the compound contains one oxygen, an O must be added to the base value and one C and four Hs must be subtracted. Therefore, the molecular formula is $C_6H_{12}O$.

   c. Dividing 102 by 13 gives 7 with 11 left over. Thus, the base value is C_7H_{18}. Because the compound contains two oxygens, two Os must be added to the base value and two Cs and eight Hs must be subtracted. Therefore, the molecular formula is $C_5H_{10}O_2$.

   d. Dividing 115 by 13 gives 8 with 11 left over. Thus, the base value is C_8H_{19}. Because the compound contains one oxygen, an O must be added to the base value and one C and four Hs must be subtracted. Because it contains an N, an N must be added to the base value and one C and three Hs must be subtracted. Therefore, the molecular formula is $C_6H_{12}NO$.

6.   **a.** **1.** $15 + (3 \times 14) + 16 = 73$ **2.** $16 + (3 \times 14) + 16 = 74$
   b. An alkane has an even-mass molecular ion. If a CH_2 group (14) of an alkane is replaced by an NH group (15), or if a CH_3 group (15) of an alkane is replaced by an NH_2 group (16), the molecular ion will have an odd mass.

   A second nitrogen in the molecular ion will cause it to have an even mass.

   Thus, for a molecular ion to have an odd mass, it must have an odd number of nitrogens.

   c. An even-mass molecular ion either has no nitrogens or has an even number of nitrogens.

7.   **a.** Dividing 86 by 13 gives 6 with 8 left over. Thus, the base value is C_6H_{14}.

   If the compound contains only carbons and hydrogens, the base value is also the molecular formula of the compound. Some possible structures would be:

If the compound contains one oxygen, the molecular formula would be $C_5H_{10}O$. Some possible structures would be:

If the compound contains two oxygens, the molecular formula would be $C_4H_6O_2$. Some possible structures would be:

b. Because the compound has an even-numbered mass, we know that it does not contain a nitrogen atom.

8. A hydrocarbon with molecular formula C_9H_{20} has a molecular mass of 128.
Since $C_9H_{20} = C_nH_{2n+2}$, we know that the hydrocarbon has no rings and no π bonds. The hydrocarbon is **2,6-dimethylheptane**.

$$CH_3\overset{CH_3}{\underset{}{C}HCH_2CH_2CH_2\overset{CH_3}{\underset{}{C}HCH_3} \longrightarrow CH_3\overset{CH_3}{\underset{+}{C}H \;+\; \overset{\cdot}{C}H_2CH_2CH_2\overset{CH_3}{\underset{}{C}HCH_3}$$

$m/z = 43$

$$CH_3\overset{CH_3}{\underset{\cdot}{C}H \;+\; CH_2CH_2CH_2\overset{CH_3}{\underset{+}{C}HCH_3}$$

$m/z = 85$

$$CH_3\overset{CH_3}{\underset{+}{C}HCH_2 \;+\; \overset{\cdot}{C}H_2CH_2\overset{CH_3}{\underset{}{C}HCH_3}$$

$m/z = 57$

$$CH_3\overset{CH_3}{\underset{\cdot}{C}HCH_2 \;+\; CH_2CH_2\overset{CH_3}{\underset{+}{C}HCH_3}$$

$m/z = 71$

2-Methyloctane would also be expected to give a base peak of $m/z = 43$ because it, too, will form a secondary (isopropyl carbocation) together with a primary radical. All other cleavages that form primary radicals form primary carbocations. However, we would expect fragments with $m/z = 29$ and 99 to be present to the same extent as those with $m/z = 57, 85$, and 71. Because fragments with $m/z = 29$ and 99 are not mentioned, we can conclude that the hydrocarbon shown below is less likely than the one shown above.

$$CH_3\overset{CH_3}{\underset{}{C}HCH_2CH_2CH_2CH_2CH_2CH_3 \longrightarrow CH_3\overset{CH_3}{\underset{}{C}HCH_2CH_2CH_2\overset{\cdot}{C}H_2 \;+\; \overset{+}{C}H_2CH_3$$

$m/z = 29$

$$CH_3\overset{CH_3}{\underset{}{C}HCH_2CH_2CH_2\overset{+}{C}H_2 \;+\; \overset{\cdot}{C}H_2CH_3$$

$m/z = 99$

9.   The ratio is 1:2:1.
        To get the M peak, both Br atoms must be ^{79}Br. To get the M+4 peak, they both must be ^{81}Br. To get the M + 2 peak, the first Br atom can be ^{79}Br and the second ^{81}Br, or the first can be ^{81}Br and the second ^{81}Br. So the relative intensity of the M+2 peak will be twice that of the others.

<u>M</u>	<u>M+2</u>	<u>M+4</u>
	^{79}Br ^{81}Br	
^{79}Br ^{79}Br	^{81}Br ^{79}Br	^{81}Br ^{81}Br

10.   The calculated exact masses show that only C_6H_{14} has an exact mass of 86.10955.

C_6H_{14} $6(12.00000) = 72.00000$ $C_4H_6O_2$ $4(12.00000) = 48.00000$
        $14(1.007825) = \underline{14.10955}$ $6(1.007825) = 6.04695$
        86.10955 $2(15.9949) = \underline{31.9898}$
                                                86.03675

$C_4H_{10}N_2$ $4(12.00000) = 48.00000$
        $10(1.007825) = 10.07825$
        $2(14.0031) = \underline{28.0064}$
        86.08465

11.   **a.** A low-resolution spectrometer cannot distinguish between them because they both have the same molecular mass (29).
        b. A high-resolution spectrometer can distinguish between these ions, because they have different exact molecular masses; one has an exact molecular mass of 29.039125, and the other an exact molecular mass of 29.002725.

12.   Because the compound contains chlorine, the M+2 peak is one-third the size of the M peak. Breaking the weak C—Cl bond heterolytically and therefore losing a chlorine atom from either the M+2 peak $(80 - 37)$ or the M peak $(78 - 35)$ gives the base peak with $m/z = 43$ ($CH_3CH_2CH_2^+$). It is more difficult to predict the other peaks.

Mass spectrum of 1-chloropropane

13.   The dominant fragmentation pathway in each case is loss of an alkyl radical via α-cleavage to form a cation in which the positive charge is shared by carbon and oxygen.

The base peak at $m/z = 73$ $(88 - 15)$, due to loss of a methyl radical, indicates that Figure 14.7a is the mass spectrum of **2-methoxy-2-methylpropane**.

The base peak at $m/z = 59$ $(88 - 29)$, due to loss of an ethyl radical, indicates that Figure 14.7b is the mass spectrum of **2-methoxybutane**.

The base peak at $m/z = 45$ $(88 - 43)$, due to loss of a propyl radical, indicates that Figure 14.7c is the mass spectrum of **1-methoxybutane**.

$$
\begin{array}{ccc}
\overset{\displaystyle CH_3}{\underset{\displaystyle OCH_3}{CH_3\overset{|}{\underset{|}{C}}CH_3}} & CH_3CH_2\overset{\displaystyle CHCH_3}{\underset{\displaystyle OCH_3}{|}} & CH_3CH_2CH_2CH_2OCH_3 \\[2em]
\text{2-methoxy-2-methylpropane} & \text{2-methoxybutane} & \text{1-methoxybutane}
\end{array}
$$

14. $R-CH_2-\ddot{O}H \xrightarrow{-e^-} R\overset{\frown}{}CH_2\overset{\frown}{}\overset{+}{\ddot{O}}H \longrightarrow R\cdot + CH_2=\overset{+}{\underset{\cdot\cdot}{O}}H$

$m/z = 31$

15.   The molecular ions with $m/z = 86$ indicate that both ketones have the molecular formula $C_5H_{10}O$. Figure 14.9a shows a base peak at $m/z = 43$ for loss of a propyl (or isopropyl) radical $(86 - 43)$, indicating that it is the mass spectrum of either 2-pentanone or 3-methyl-2-butanone, since each of these has a propyl or isopropyl group. The fact that the spectrum has a peak at $m/z = 58$, indicating loss of ethene $(86 - 28)$, indicates that the compound has a γ-hydrogen that enables it to undergo a McLafferty rearrangement. Therefore, the ketone must be 2-pentanone, since 3-methyl-2-butanone does not have a γ-hydrogen.

The fact that Figure 14.9b shows a base peak at $m/z = 57$ for loss of an ethyl radical $(86 - 29)$ indicates that it is the mass spectrum of 3-pentanone.

16.   All three ketones will have a molecular ion with $m/z = 86$.

3-pentanone   2-pentanone   3-methyl-2-butanone

$86 - 29 = 57$ $86 - 41 = 43$ $86 - 41 = 43$

3-Pentanone will have a base peak at $m/z = 57$, whereas the other two ketones will have base peaks at $m/z = 43$.

2-Pentanone will have a peak at $m/z = 58$ due to a McLafferty rearrangement.

3-Methyl-2-butanone does not have any γ-hydrogens. Therefore, it cannot undergo a McLafferty rearrangement, so it will not have a peak at $m/z = 58$.

17.   **a.** $CH_3CH_2CH_2CH_2CH_2\ddot{O}H$

$$\downarrow -e^-$$

$CH_3CH_2CH_2CH_2 \!-\! CH_2 \!-\! \overset{+}{\underset{\cdot\cdot}{O}}H \xrightarrow{\alpha\text{-cleavage}} CH_3CH_2CH_2\dot{C}H_2 + CH_2 \!=\! \overset{+}{\underset{\cdot\cdot}{O}}H$

$m/z = 88$ $\qquad\qquad\qquad\qquad\qquad\qquad\qquad\qquad\quad m/z = 31$

$\overset{H}{}$

$CH_3CH_2CHCH_2CH_2 \!-\! \overset{+}{\underset{\cdot\cdot}{O}}H \longrightarrow CH_3CH_2\dot{C}HCH_2\overset{+}{C}H_2 + H_2O$

$\qquad\qquad\qquad\qquad\qquad\qquad\qquad\qquad m/z = 70$

b. $CH_3CH_2CHCH_2CH_2CH_2CH_3$

$\underset{\cdot\cdot}{:}OH$

$$\downarrow -e^-$$

$CH_3CH_2 \!-\! CHCH_2CH_2CH_2CH_3 \xrightarrow{\alpha\text{-cleavage}} CH_3CH_2CH_2CH_2CH \!=\! \overset{+}{O}H + CH_3\dot{C}H_2$

$\overset{|}{\underset{+\,\cdot\cdot}{\cdot}OH}$

$m/z = 116$ $\qquad\qquad\qquad\qquad\qquad\qquad m/z = 87$

$CH_3CH_2CH \!-\! CH_2CH_2CH_2CH_3 \xrightarrow{\alpha\text{-cleavage}} CH_3CH_2CH \!=\! \overset{+}{O}H + CH_3CH_2CH_2\dot{C}H_2$

$\underset{+\,\cdot\cdot}{\cdot}OH$

$m/z = 59$

$CH_2CH_2CHCH_2CH_2CH_2CH_3 \longrightarrow \dot{C}H_2CH_2\overset{+}{C}HCH_2CH_2CH_2CH_3 + H_2O$

$\overset{|}{H}\underset{+\,\cdot\cdot}{\cdot}OH$

$\qquad\qquad\qquad\qquad\qquad\qquad\qquad m/z = 98$

$CH_3CH_2CHCH_2CHCH_2CH_3 \longrightarrow CH_3CH_2\overset{+}{C}HCH_2\dot{C}HCH_2CH_3 + H_2O$

$\underset{+\,\cdot}{:}OH H$

$\qquad\qquad\qquad\qquad\qquad\qquad\qquad m/z = 98$

c.

$$CH_3CH_2 - \ddot{O} - \underset{\underset{CH_3}{|}}{\overset{\overset{CH_2CH_3}{|}}{C}} CH_2CH_2CH_3$$

$$\downarrow -e^-$$

$$CH_3CH_2 - \overset{+}{\underset{..}{\ddot{O}}} - \underset{\underset{CH_3}{|}}{\overset{\overset{CH_2CH_3}{|}}{C}} CH_2CH_2CH_3 \longrightarrow CH_3CH_2 - \ddot{\underset{..}{O}}: + \underset{\underset{CH_3}{|}}{\overset{\overset{CH_2CH_3}{|}}{+C}} CH_2CH_2CH_3$$

$$m/z = 144 \qquad\qquad\qquad m/z = 99$$

$$CH_3CH_2 - \overset{+}{\underset{..}{\ddot{O}}} - \underset{\underset{CH_3}{|}}{\overset{\overset{CH_2CH_3}{|}}{C}} CH_2CH_2CH_3 \xrightarrow{\alpha\text{-cleavage}} CH_3CH_2 - \overset{+}{\underset{..}{\ddot{O}}} = \underset{}{C} CH_2CH_2CH_3 + \dot{C}H_3$$

$$m/z = 129$$

$$CH_3CH_2 - \overset{+}{\underset{..}{\ddot{O}}} - \underset{\underset{CH_3}{|}}{\overset{\overset{CH_2CH_3}{|}}{C}} CH_2CH_2CH_3 \xrightarrow{\alpha\text{-cleavage}} CH_3CH_2 - \overset{+}{\underset{..}{\ddot{O}}} = \underset{\underset{CH_3}{|}}{C} CH_2CH_2CH_3 + CH_3\dot{C}H_2$$

$$m/z = 115$$

$$CH_3CH_2 - \overset{+}{\underset{..}{\ddot{O}}} - \underset{\underset{CH_3}{|}}{\overset{\overset{CH_2CH_3}{|}}{C}} CH_2CH_2CH_3 \xrightarrow{\alpha\text{-cleavage}} CH_3CH_2 - \overset{+}{\underset{..}{\ddot{O}}} = \underset{\underset{CH_3}{|}}{C} CH_2CH_3 + CH_3CH_2\dot{C}H_2$$

$$m/z = 101$$

$$CH_3 - CH_2 - \overset{\dot{+}}{\underset{..}{\ddot{O}}} - \underset{\underset{CH_3}{|}}{\overset{\overset{CH_2CH_3}{|}}{C}} CH_2CH_2CH_3 \xrightarrow{\alpha\text{-cleavage}} CH_2 = \overset{+}{\underset{..}{\ddot{O}}} - \underset{\underset{CH_3}{|}}{\overset{\overset{CH_2CH_3}{|}}{C}} CH_2CH_2CH_3 + \dot{C}H_3$$

$$m/z = 129$$

d.

$$CH_3CCH_2CH_2CH_2CH_3$$

(O:, double bond above C)

$\downarrow -e^-$

α-cleavage

$CH_3\overset{|}{-}CCH_2CH_2CH_2CH_3$ $\longrightarrow$ $CH_3CH_2CH_2CH_2C\equiv\overset{+}{\ddot{O}}$ + $\dot{C}H_3$

$m/z = 100$ $m/z = 85$

$CH_3C\overset{|}{-}CH_2CH_2CH_2CH_3$ $\xrightarrow{\alpha\text{-cleavage}}$ $CH_3C\equiv\overset{+}{\ddot{O}}$ + $CH_3CH_2CH_2\dot{C}H_2$

$m/z = 43$

$CH_3CCH_2-CH_2-CHCH_3$ $\xrightarrow[\text{rearrangement}]{\text{McLafferty}}$ $CH_3\overset{:\overset{+}{O}H}{\underset{||}{C}}CH_2$ + $CH_2=CHCH_3$

$m/z = 58$

e. $CH_3CH_2CH\overset{..}{\ddot{C}l}:$ $\xrightarrow{-e^-}$ $CH_3CH_2CH-\overset{\cdot+}{\ddot{C}l}:$ $\longrightarrow$ $CH_3CH_2\overset{+}{C}H$ + $:\dot{C}l:$

$\overset{|}{C}H_3$ $\overset{|}{C}H_3$ $\overset{|}{C}H_3$

$m/z = 92$ and 94 $m/z = 57$

$CH_3CH_2CH\overset{\cdot+}{\ddot{C}l}:$ $\xrightarrow{\alpha\text{-cleavage}}$ $CH_3CH_2CH=\overset{+}{\ddot{C}l}:$ + $\dot{C}H_3$

$\overset{|}{C}H_3$ $m/z = 77$ and 79

$CH_3CH_2-CH-\overset{\cdot+}{\ddot{C}l}:$ $\xrightarrow{\alpha\text{-cleavage}}$ $CH_3CH=\overset{+}{\ddot{C}l}:$ + $CH_3\dot{C}H_2$

$\overset{|}{C}H_3$ $m/z = 63$ and 65

f. $\overset{CH_3}{\underset{CH_3}{CH_3C-\ddot{B}r:}}$ $\xrightarrow{-e^-}$ $\overset{CH_3}{\underset{CH_3}{CH_3C-\overset{\cdot+}{\ddot{B}r:}}}$ $\longrightarrow$ $\overset{CH_3}{\underset{CH_3}{CH_3\overset{+}{C}}}$ + $:\ddot{B}r:$

$m/z = 136$ and 138 $m/z = 57$

18.   We know from Section 6.5 that when (Z)-2-pentene reacts with water and an acid catalyst, 3-pentanol and 2-pentanol are formed. Both alcohols have a molecular weight of 88. (Notice that the first step in solving this problem is to use chemical knowledge to identify the products.) The absence of a molecular ion peak is consistent with the fact that the compounds are alcohols.

$\overset{H_3C}{\underset{H}{}}C=C\overset{CH_2CH_3}{\underset{H}{}}$ + H_2O $\xrightarrow{H_2SO_4}$ $CH_3CH_2\underset{\underset{OH}{|}}{C}HCH_2CH_3$ + $CH_3\underset{\underset{OH}{|}}{C}HCH_2CH_2CH_3$

(Z)-2-pentene   3-pentanol   2-pentanol

Figure 14.10a shows a base peak at $m/z = 59$ due to loss of an ethyl radical ($88 - 29$), indicating that it is the spectrum of 3-pentanol.

Figure 14.10b shows a base peak at $m/z = 45$ due to loss of a propyl radical ($88 - 43$), indicating that it is the spectrum of 2-pentanol.

19. The wavelength is the distance from the top of one wave to the top of the next wave. We see that **A** has a longer wavelength than **B**.
Infrared radiation has longer wavelengths than ultraviolet light because infrared radiation is lower in energy. Therefore, **A** depicts infrared radiation and **B** depicts ultraviolet light.

20. **a.** $2000 \ \text{cm}^{-1}$ (The larger the wavenumber, the higher the energy.)
  b. $8 \ \mu\text{m}$ (The shorter the wavelength, the higher the energy.)
  c. $2 \ \mu\text{m}$, because $2 \ \mu\text{m} = 5000 \ \text{cm}^{-1}$ ($5000 \ \text{cm}^{-1}$ is a larger wavenumber than $2000 \ \text{cm}^{-1}$.)

21. **a.** $\tilde{\nu} \ (\text{cm}^{-1}) = \dfrac{10{,}000}{4 \ (\mu\text{m})} = 2500 \ \text{cm}^{-1}$

  b. $\tilde{\nu} \ (\text{cm}^{-1}) = \dfrac{10{,}000}{\lambda(\mu\text{m})} = 200 \ \text{cm}^{-1}$

  $\lambda = 50 \ \mu\text{m}$

22. **a.** **1.** C≡C stretch   A triple bond is stronger than a double bond, so it takes more energy to stretch a triple bond.
      2. C—H stretch   It requires more energy to stretch a given bond than to bend it.
      3. C=N stretch   A double bond is stronger than a single bond, so it takes more energy to stretch a double bond.
      4. C=O stretch   A double bond is stronger than a single bond, so it takes more energy to stretch a double bond.

  b. **1.** C—O   Vibrations of lighter atoms occur at larger wavenumbers.
      2. C—C   Vibrations of lighter atoms occur at larger wavenumbers.

23. **a.** The carbon-oxygen stretch of phenol, because it has partial double-bond character as a result of electron delocalization.

  b. The carbon-oxygen double-bond stretch of a ketone, because it has more double-bond character. The double-bond character of the carbonyl group of an amide is reduced by electron delocalization.

c.   The C—N stretch of aniline, because it has partial double-bond character.

24.   A carbonyl group bonded to an sp^3 carbon will exhibit an absorption band at a larger wavenumber because a carbonyl group bonded to an sp^2 carbon of an alkene will have greater single-bond character as a result of electron delocalization.

25.   The C—O bond of the alcohol is a pure single bond. In contrast, the C—O bond of the carboxylic acid has double-bond character, so it is a stronger bond and therefore takes more energy to stretch it.

26.   a.   The C=O absorption band of an ester occurs at the largest wavenumber because the carbonyl group of an ester has the most double-bond character, since the predominant effect of the ester oxygen atom is inductive electron withdrawal.

The C=O absorption band of an amide occurs at the smallest wavenumber because the carbonyl group of an amide has the least double-bond character, since the predominant effect of the amide nitrogen atom is electron donation by resonance.

b.   The C=O absorption of the four-membered ring lactone occurs at the highest wavenumber because it is the least able to accommodate double-bond character in the ring, which is required if the carbonyl group is to have any single-bond character.

The C=O absorption of the six-membered ring lactone occurs at the lowest wavenumber because it is the most able to accommodate double-bond character in the ring.

27.   Ethanol dissolved in carbon disulfide will show the oxygen-hydrogen stretch at a larger wavenumber. There is extensive hydrogen bonding in the undiluted alcohol, and an oxygen-hydrogen bond is easier to stretch if it is hydrogen bonded, so the O—H stretch will be at a smaller wavenumber.

28.   The absorption band at 1100 cm^{-1} would be less intense if it were due to a C—N bond, because a smaller change in dipole moment is associated with the stretch of a C—N bond compared to the change in dipole moment associated with the stretch of a C—O bond.

29.   a.   The absorption band at 1700 cm^{-1} indicates that the compound has a carbonyl group.

The absence of an absorption band at 3300 cm^{-1} indicates that the compound is not a carboxylic acid.

The absence of an absorption band at 2700 cm^{-1} indicates that the compound is not an aldehyde.

The absence of an absorption band at 1100 cm^{-1} indicates that the compound is not an ester or an amide. The compound, therefore, must be a **ketone**.

b.   The absence of an absorption band at 3400 cm^{-1} indicates that the compound does not have an N—H bond.

The absence of a carbonyl absorption band between 1700 cm^{-1} and 1600 cm^{-1} indicates that the compound is not an amide. The compound, therefore, must be a **tertiary amine**.

30.   a.   An aldehyde would show absorption bands at 2820 and 2720 cm^{-1}. A ketone would not have these absorption bands.

b.   An open-chain ketone would have a methyl substituent and therefore an absorption band at 1385–1365 cm^{-1} that a cyclic ketone would not have.

c.   Cyclohexene would show an sp^3 C—H stretch slightly to the right of 3000 cm^{-1}. Benzene would not show an absorption band in this region.

d.   The cis isomer would show a carbon-hydrogen bending vibration at 730–675 cm^{-1}, whereas the trans isomer would show a carbon-hydrogen bending vibration at 960–980 cm^{-1}.

e.   Cyclohexene would show a carbon-carbon double-bond stretching vibration at 1680–1600 cm^{-1} and an sp^2 carbon-hydrogen stretching vibration at 3100–3020 cm^{-1}. Cyclohexane would not show these absorption bands.

f.   A primary amine would show a nitrogen-hydrogen stretch at 3500–3300 cm^{-1}, and a tertiary amine would not have this absorption band.

31.   a.   An absorption band at 1150–1050 cm^{-1} due to a C—O stretching vibration would be present for the ether and absent for the alkane.

b.   An absorption band at 3300–2500 cm^{-1} due to an O—H stretching vibration would be present for the carboxylic acid and absent for the ester.

c. An absorption band at 1385–1365 cm^{-1} due to C—H bending vibrations of a methyl group would be present for methylcyclohexane and absent for cyclohexane.

d. Only the terminal alkyne would show an absorption band at 3300 cm^{-1} due to an sp C—H stretching vibration.

e. An absorption band at 1780–1650 cm^{-1} due to a C=O stretching vibration would be present for the carboxylic acid and absent for the alcohol.

f. An absorption band at 2960–2850 cm^{-1} due to a sp^3 C—H stretching vibration would be present for the compound with the methyl group and absent for benzene.

32. 2-butyne, H_2, Cl_2, and ethene because they are symmetrical molecules.

33. The absorption bands in the vicinity of 3000 cm^{-1} indicate that the compound has hydrogens attached to both sp^2 and sp^3 carbons. The absence of absorptions at 1600–1800 cm^{-1} and the absence of broad absorptions between 2500 and 3650 cm^{-1} rules out compounds containing C=O, N—H, and O—H groups. The lack of absorption at 1600 cm^{-1} and 1500 cm^{-1} indicates that the compound does not have a benzene ring. The sp^2 hydrogens, therefore, must be those of an **alkene**.

The lack of significant absorption at 1600 cm^{-1} indicates that the compound must be an alkene with a relatively small (if any) dipole moment change when the vibration occurs. The absorption band at 965 cm^{-1} indicates that the compound is a ***trans*-alkene**.

The molecular ion with $m/z = 84$ suggests that the compound has a molecular formula of C_6H_{12}. The base peak with $m/z = 55$ indicates that the molecular ion most easily loses an ethyl radical ($84 - 29 = 55$). Because it most easily loses an ethyl radical, the ethyl group must be attached to an allylic carbon. The compound, therefore, is ***trans*-2-hexene**.

trans-2-hexene

34. The absorption band at ~1700 cm^{-1} indicates that the compound has a carbonyl group, and the absorption band at ~1600 cm^{-1} indicates that the compound has a carbon-carbon double bond. The absorption bands in the vicinity of 3000 cm^{-1} indicate that the compound has hydrogens attached to both sp^2 and sp^3 carbons. The absorption band at ~1380 cm^{-1} indicates that the compound has a methyl group. Because the compound has only four carbons and one oxygen, it must be **methyl vinyl ketone**. Notice that the carbonyl stretch is at a lower frequency (1700 cm^{-1}) than expected for a ketone (1720 cm^{-1}), because the carbonyl group has partial single-bond character due to electron delocalization.

35. $A = c l \varepsilon$

$$c = \frac{A}{l \varepsilon}$$

$$c = \frac{0.52}{12{,}600} = 4.1 \times 10^{-5} \, \text{M}$$

36. $A = c l \varepsilon$

$$c = \frac{A}{l \varepsilon}$$

$$\varepsilon = \frac{0.40}{4.0 \times 10^{-5}} = 10{,}000 \, \text{M}^{-1} \, \text{cm}^{-1}$$

37.   The compound has the same chromophore as methyl vinyl ketone. So it must have approximately the same value for its λ_{max}, which is 219 nm.

38.   **a.**

  b.

39.   **a.**   Blue results from absorption of light that has a longer wavelength than the light that produces purple when absorbed. The compound on the right has two $N(CH_3)_2$ auxochromes that will cause it to absorb at a longer wavelength than the compound on the left that has only one $N(CH_3)_2$ auxochrome. Therefore, the compound on the right is the blue compound.

  b.   They will be the same color at pH = 3 because the $N(CH_3)_2$ groups will be protonated and, therefore, will not possess the lone pair that causes the compound to absorb light of a longer wavelength.

40.   The two absorption bands that produce an observed green color correspond to the absorption bands that produce yellow and blue. So mixing yellow and blue produce green.

41.   NADH is formed as a product; it absorbs light at 340 nm. Therefore, the rate of the oxidation reaction can be determined by monitoring the increase in absorbance at 340 nm as a function of time.

42.   The Henderson-Hasselbalch equation (Section 2.10) shows that when the pH equals the pK_a, the concentration of the species in the acidic form is the same as the concentration of the species in the basic form.

From the data given, we see that the absorbance of the acid is 0. We also see that the absorbance ceases to change with increasing pH after the absorbance reaches 1.60. That means that all of the compound is in the basic form when the absorbance is 1.60. Therefore, when the absorbance is half of 1.60 (or 0.80), half of the compound is in the basic form and half is in the acidic form; in other words, the concentration in the acid form is the same as the concentration in the basic form. We see that the absorbance is 0.80 at pH = 5.0. Therefore, the pK_a of the compound is 5.0.

43. The molecular ion peak for these compounds is $m/z = 86$; the peak at $m/z = 57$ is due to loss of an ethyl radical ($86 - 29$), and the peak at $m/z = 71$ is due to loss of a methyl radical ($86 - 15$).

a. 3-Methylpentane will be more apt to lose an ethyl radical (forming a secondary carbocation and a primary radical) than a methyl radical (forming a secondary carbocation and a methyl radical). In addition, 3-methylpentane has two pathways to lose an ethyl radical. Therefore, the peak at $m/z = 57$ would be more intense than the peak at $m/z = 71$.

$$CH_3CH_2CHCH_2CH_3$$
$$|$$
$$CH_3$$

3-methylpentane

b. 2-Methylpentane has two pathways to lose a methyl radical (forming a secondary carbocation and a methyl radical in each pathway), and it cannot form a secondary carbocation by losing an ethyl radical. (Loss of an ethyl radical would form a primary carbocation and a primary radical.) Therefore, it will be more apt to lose a methyl radical than an ethyl radical, so the peak at $m/z = 71$ would be more intense than the peak at $m/z = 57$.

$$CH_3CHCH_2CH_2CH_3$$
$$|$$
$$CH_3$$

2-methylpentane

44.
1. the change in the dipole moment when the bond stretches or bends
2. the number of bonds that cause the absorption band
3. the concentration of the sample

45.
a. An absorption band at ~1250 cm^{-1} due to a C—O stretching vibration would be present for the ester and absent for the ketone.
b. An absorption band at 720 cm^{-1} due to in-phase rocking of the five adjacent methylene groups would be present for heptane and absent for methylcyclohexane.
c. An absorption band at 3650–3200 cm^{-1} due to an O—H stretching vibration would be present for the alcohol and absent for the ether.
d. An absorption band at 3500–3300 cm^{-1} due to an N—H stretching vibration would be present for the amide and absent for the ester.
e. The secondary alcohol would have an absorption band at 1385–1365 cm^{-1} for the methyl group. The primary alcohol does not have a methyl group, so it would not have this absorption band.
f. The trans isomer would have a C—H bending absorption band at 980–960 cm^{-1}, whereas the cis isomer would have the absorption band at 730–675 cm^{-1}. In addition, a weak C=C absorption band at 1680–1600 cm^{-1} would be present for the cis isomer and absent for the trans isomer, since only the trans isomer is symmetrical.
g. The C=O absorption band will be at a larger wavenumber for the ester (1740 cm^{-1}) than for the ketone (1720 cm^{-1}).

h. The C=O absorption band will be at a larger wavenumber for the β,γ-unsaturated ketone (1720 cm^{-1}) than for the α,β-unsaturated ketone (1680 cm^{-1}), since the double bonds in the latter are conjugated.

i. The alkene would have absorption bands at 1680–1600 cm^{-1} due to a C=C stretching vibration and at 3100–3020 cm^{-1} due to an sp^2 C—H stretching vibration that the alkyne would not have. The alkyne would have an absorption band at 2260–2100 cm^{-1} that the alkene would not have.

j. An absorption band at ~2820 and ~2720 cm^{-1} due to the aldehyde C—H stretching vibration would be present for the aldehyde and absent for the ketone.

k. Absorption bands at 1600 cm^{-1} and 1500 cm^{-1} (aromatic ring stretching vibrations) and at 3100–3020 cm^{-1} (sp^2 C—H stretching vibration) would be present for the compound with the benzene ring and absent for the compound with the cyclohexane ring. An absorption band at 2960–2850 cm^{-1} due to an sp^3 C—H stretching vibration would be present for the compound with the cyclohexane ring and absent for the compound with the benzene ring.

l. Absorption bands at 990 cm^{-1} and 910 cm^{-1} due to an sp^2 C—H bending vibration would be present for the terminal alkene and absent for the internal alkene.

46. Dividing 128 by 13 gives 9 with 11 left over. Thus, the base value is C_9H_{20}. Because the compound is a saturated hydrocarbon, we know that the base value is also the molecular formula of the compound. Some possible structures are:

47. **a.** If the reaction had occurred, the intensity of the absorption bands at ~1700 cm^{-1} (due to the carbonyl group) and at ~2700 cm^{-1} (due to the aldehyde C—H bond) of the reactant would have decreased. If all the aldehyde had reacted, these absorption bands would have disappeared.

b. If all the NH$_2$NH$_2$ had been removed, there would be no N—H absorption at ~3400 cm^{-1}.

48. If the force constants are approximately the same, the lighter atoms will absorb at higher frequencies.

$$C—C \quad > \quad C—N \quad > \quad C—O$$

49.

3600 cm^{-1}		3000		1800		1400		1000
OH	3300–3000	sp^3 CH	2950	C=O	1700	C—O	1250–1050
NH	3600–3200	O‖RCH	2700	C=C	1600	C—N	1230–1030
sp CH	3300	C≡C	2100	⬡	1500		
sp^2 CH	3050	C≡N	2250				

50.   The molecular weight of each of the alcohols is 158. The peak at $m/z = 140$ is due to loss of water $(158 - 18)$; each of the alcohols could show such a peak. The peaks at $m/z = 87$, 115, and 143 are due to loss of a group with five carbons (C_5H_{11}), a group with three carbons (C_3H_7), and a methyl group, respectively. Only 2,2,4-trimethyl-4-heptanol could lose all three groups via α-cleavage.

51. $CH_2{=}CHCH_2CH_2CH{=}CH_2$ $CH_3CH{=}CHCH{=}CHCH_3$
           1,5-hexadiene   2,4-hexadiene

One way to distinguish the two compounds is by the presence or absence of an absorption band at $\sim 1370 \ cm^{-1}$ due to the methyl group that 2,4-hexadiene has but that 1,5-hexadiene does not have. In addition, 2,4-hexadiene has conjugated double bonds; therefore its double bonds have some single-bond character due to electron delocalization. Consequently, they are easier to stretch than the isolated double bonds of 1,5-hexadiene. Thus, the carbon-carbon double-bond stretch of 2,4-hexadiene will be at a smaller wavenumber than the carbon-carbon double-bond stretch of 1,5-hexadiene.

52.   The fact that the abundance of the M+2 peak is 30% of the abundance of the M peak indicates that the compound has one chlorine atom. The peak at $m/z = 77$ is due to loss of the chlorine atom $(112 - 35 = 77)$. The fact that the peak at $m/z = 77$ does not fragment indicates it is a phenyl cation. Therefore, the compound is chlorobenzene.

53.   Dividing 112 by 13 gives 8 with 8 left over. Thus, the base value is C_8H_{16}, and because the compound is a hydrocarbon, this is also its molecular formula. The molecular formula indicates that it has one degree of unsaturation, which is accounted for by the fact that we know it has a six-membered ring. Possible structures are shown here. Possible stereoisomers are not shown: the second and third structures have three stereoisomers and the fourth structure has two stereoisomers.

54.   a.   and

The λ_{max} will be at a longer wavelength, because there are three conjugated double bonds.

b. and

The λ_{max} of the phenolate ion is at a longer wavelength than the λ_{max} of phenol. Because the pK_a of phenol is ~10, the λ_{max} will be at a longer wavelength at pH $= 11$ than at pH $= 7$.

The λ_{max} is pH independent, so it will be the same at pH 7 and 11.

c. and

The λ_{max} will be at a longer wavelength because the carbonyl group is conjugated with the benzene ring.

55. Hydrogens are more electron-withdrawing than alkyl groups. Therefore, the carbonyl group bonded to two relatively electron-withdrawing hydrogens has the largest wavenumber for its $C=O$ absorption band, whereas the carbonyl group bonded to two alkyl groups has the lowest wavenumber.

$$\underset{H\quad\quad H}{\overset{O}{\underset{}{\parallel}}C} \;>\; \underset{CH_3\quad H}{\overset{O}{\underset{}{\parallel}}C} \;>\; \underset{CH_3\quad CH_3}{\overset{O}{\underset{}{\parallel}}C}$$

56. The $C=O$ absorption band of the three compounds decreases in the following order.

$$>\qquad>$$

The first compound has the $C=O$ absorption band at the largest wavenumber, because a lone pair on the ring oxygen can be delocalized onto two different atoms; thus it is less apt than the lone pair in the other compounds to be delocalized onto the carbonyl oxygen atom.

or

The third compound has the $C = O$ absorption band at the smallest wavenumber because its carbonyl group has more single-bond character due to contributions from two other resonance contributors.

57.   4-Methyl-2-pentanone would show peaks at $m/z = 85$ (loss of a methyl group) and at $m/z = 43$ (loss of an isobutyl group) and a peak at $m/z = 58$ due to a McLafferty rearrangement.

4-methyl-2-pentanone

2-Methyl-3-pentanone would show peaks at $m/z = 71$ (loss of an ethyl group) and at $m/z = 57$ (loss of an isopropyl group). Because it does not have a γ-hydrogen, it cannot undergo a McLafferty rearrangement, so it does not have a peak at $m/z = 58$.

2-methyl-3-pentanone

58.   **a.** The absorption band at $\sim 2100\ \text{cm}^{-1}$ indicates a carbon-carbon triple bond, and the absorption band at $\sim 3300\ \text{cm}^{-1}$ indicates a hydrogen bonded to an *sp* carbon.

$$CH_3CH_2CH_2CH_2C \equiv CH$$

b. The absence of an absorption band at $\sim 2700\ \text{cm}^{-1}$ indicates that the compound is not an aldehyde, and the absence of a broad absorption band in the vicinity of $3000\ \text{cm}^{-1}$ indicates that the compound is not a carboxylic acid. The ester and the ketone can be distinguished by the absorption band at $\sim 1200\ \text{cm}^{-1}$ that indicates the carbon-oxygen single bond of an ester.

c. The absorption band at $\sim 1360\ \text{cm}^{-1}$ indicates the presence of a methyl group.

59. **a.** The broad absorption band at ~3300 cm^{-1} is characteristic of the oxygen-hydrogen stretch of an alcohol, and the absence of absorption bands at ~1600 cm^{-1} and ~3100 cm^{-1} indicates that it is not the alcohol with a carbon-carbon double bond.

b. The absorption band at ~1685 cm^{-1} indicates a carbon-oxygen double bond. The absence of a strong and broad absorption band at ~3000 cm^{-1} rules out the carboxylic acid, and the absence of an absorption band at ~2700 cm^{-1} rules out the aldehyde. Thus, it must be one of the ketones. The ketone with the conjugated carbonyl group would be expected to show a C=O stretch at ~1685 cm^{-1}, whereas the ketone with the isolated carbonyl group would show a C=O stretch at ~1720 cm^{-1}. Thus, the compound is the ketone with the conjugated carbonyl group.

c. The absorption band at ~1700 cm^{-1} indicates a carbon-oxygen double bond. The absence of an absorption band at ~1600 cm^{-1} rules out the ketones with the benzene or cyclohexene rings. The absence of absorption bands at ~2100 cm^{-1} and ~3300 cm^{-1} rules out the ketone with the carbon-carbon triple bond. Thus, it must be 4-ethylcyclohexanone.

60. The absorption bands at ~2700 cm^{-1} for the aldehyde hydrogen and at ~1380 cm^{-1} for the methyl group would distinguish the compounds.

   A would have the band at ~2700 cm^{-1} but not the one at ~1380 cm^{-1}.
   B would have the band at ~1380 cm^{-1} but not the one at ~2700 cm^{-1}.
   C would have both the band at ~2700 cm^{-1} and the one at ~1380 cm^{-1}.

61. 1-Hexyne will show absorption bands at ~3300 cm^{-1} for a hydrogen bonded to an *sp* carbon and at ~2100 cm^{-1} for the triple bond.

$$CH_3CH_2CH_2CH_2C\equiv CH$$
1-hexyne

2-Hexyne will show the absorption band at ~2100 cm^{-1} but not the one at ~3300 cm^{-1}.

$$CH_3CH_2CH_2C\equiv CCH_3$$
2-hexyne

3-Hexyne will show neither the absorption band at ~3300 cm^{-1} nor the one at ~2100 cm^{-1} (there is no change in dipole moment when the C≡C stretches).

$$CH_2CH_2C\equiv CCH_2CH_3$$
3-hexyne

62.   Dividing 116 by 13 gives 8 with 12 left over. Thus, the base value is C_9H_{20}. Because the compound contains two oxygens, two Os must be added to the base value and two Cs and eight Hs must be subtracted. Therefore, the molecular formula is $C_6H_{12}O_2$. Some possible structures are:

63.   Since only benzene absorbs light of 260 nm, the concentration of benzene can be determined by measuring the absorbance (A) at 260 nm, using the Beer-Lambert law, since the length of the light path of the cell (1.0 cm) is known and the molar absorptivity of benzene at 260 nm (ε) is also known.

$$\text{Beer-Lambert law: } A = c\,l\,\varepsilon$$

$$\text{Therefore, the concentration of benzene} = \frac{A}{l\varepsilon}$$

64.   **a.**

2700 cm^{-1}

1700 cm^{-1}

C=C   1600 cm^{-1}

c. C—O   ~1050 cm^{-1}
     O—H   3600–3200 cm^{-1} (broad)

e. *sp* CH   3300 cm^{-1} (narrow)
     C≡C   2100 cm^{-1}

b.

1700 cm^{-1}

1500 cm^{-1}, 1600 cm^{-1}

C—O   ~1250 cm^{-1}, ~1050 cm^{-1}

d.

1700 cm^{-1}

C—N   ~1030 cm^{-1}

N—H   3500–3300 cm^{-1}

f.

1700 cm^{-1}

~3000 cm^{-1} (broad)

C—O   ~1250 cm^{-1}

65.   Calculating the term in Hooke's law that depends on the masses of the atoms joined by a bond for a C—H bond and for a C—C bond shows why stretching vibrations for smaller atoms occur at larger wavenumbers.

for a carbon-hydrogen bond

$$\frac{m_1 + m_2}{m_1 \times m_2} = \frac{12 + 1}{12 \times 1}$$

$$= \frac{13}{12}$$

$$= 1.08$$

for a carbon-hydrogen bond

$$\frac{m_1 + m_2}{m_1 \times m_2} = \frac{12 + 12}{12 \times 12}$$

$$= \frac{24}{144}$$

$$= 0.17$$

66. Before the addition of acid, the compound is colorless because the benzene rings are not conjugated with each other, since they are separated by two single bonds. In the presence of acid, a carbocation is formed in which the three benzene rings are conjugated with each other as a result of electron delocalization. The conjugated carbocation is highly colored.

67. **a.** The absorption bands at 1720 cm^{-1} and ~2700 cm^{-1} indicate that the compound is an aldehyde (carbon-hydrogen stretch of an aldehyde hydrogen). The absence of an absorption band at ~1600 cm^{-1} rules out the aldehyde with the benzene ring. Thus, it must be the other aldehyde.

b. The absorption bands at ~3350 cm^{-1} and ~3200 cm^{-1} indicate that the compound is an amide (nitrogen-hydrogen stretch). The absence of an absorption band at ~3050 cm^{-1} indicates that the compound does not have hydrogens bonded to sp^2 carbons. Therefore, it is not the amide that has a benzene ring. Thus, it must be the other amide.

c. The absence of absorption bands at ~1600 cm^{-1} and ~1500 cm^{-1} indicates that the compound does not have a benzene ring. Thus, it must be the ketone. This is confirmed by the absence of an absorption band at ~1380 cm^{-1}, indicating that the compound does not have a methyl group.

68. The absorption band at ~1740 cm^{-1} indicates that the compound has a carbonyl group, and the absence of an absorption band at ~1380 cm^{-1} indicates that it has no methyl groups. The absence of an absorption band at ~1600 cm^{-1} indicates the compound does not have a carbon-carbon double bond, and the absence of an absorption band at ~3050 cm^{-1} indicates the compound does not have hydrogens bonded to sp^2 carbons.

From the molecular formula you can deduce that the compound is **cyclopentanone**.

69.   The broad absorption band at $\sim 3300 \text{ cm}^{-1}$ indicates that the compound has an OH group. The absorption bands at $\sim 2900 \text{ cm}^{-1}$ indicate that the compound has hydrogens attached to an sp^3 carbon. The compound, therefore, is **benzyl alcohol**.

70. $$\widetilde{v} = \frac{1}{2\pi c}\left[\frac{f(m_1 + m_2)}{m_1 m_2}\right]^{1/2}$$

$$\widetilde{v} = \frac{1}{2 \times 3.1416 \times 3 \times 10^{10} \text{ cm s}^{-1}}\sqrt{\frac{10 \times 10^5 \text{ g s}^{-2}\left(\dfrac{12}{6.02} + \dfrac{12}{6.02}\right) \times 10^{-23} \text{ g}}{\dfrac{12}{6.02} \times 10^{-23} \text{ g} \times \dfrac{12}{6.02} \times 10^{-23} \text{ g}}}$$

$$\widetilde{v} = \frac{1}{18.85 \times 10^{10}}\sqrt{10.0 \times 10^{28}}$$

$$\widetilde{v} = \frac{1}{18.85 \times 10^{10}} \times 3.16 \times 10^{14}$$

$$\widetilde{v} = 1676 \text{ cm}^{-1}$$

71.   In acidic solutions, the three benzene rings are isolated from one another, so phenolphthalein is colorless. In basic solutions, loss of the proton from one of the OH groups causes the five-membered ring to open. As a result, the number of conjugated double bonds increases, which causes the solution to become colored.

72.   The broad absorption band at $\sim 3300 \text{ cm}^{-1}$ indicates that the compound has an OH group. The absence of absorbance at $\sim 1700 \text{ cm}^{-1}$ shows that the compound does not have a carbonyl group. The absence of absorption at $\sim 2950 \text{ cm}^{-1}$ indicates the compound does not have any hydrogens bonded to sp^3 carbons. Therefore, the compound is **phenol**.

73. **a.** The IR spectrum indicates that the compound is an aliphatic ketone with at least one methyl group. The M peak at $m/z = 100$ indicates that the ketone is a hexanone. The peak at 43 ($100 - 57$) for loss of a butyl radical and the peak at 85 for loss of a methyl radical ($100 - 15$) suggest that the compound is **2-hexanone**.

$$CH_3-\overset{\overset{\textstyle O}{\textstyle \|}}{C}-CH_2CH_2CH_2CH_3$$

This is confirmed by the peak at 58 for loss of propene ($100 - 42$) as a result of a McLafferty rearrangement.

b. The equal heights of the M and M+2 peaks at 162 and 164 indicate that the compound contains bromine. The peak at $m/z = 83$ ($162 - 79$) is for the carbocation that is formed when the bromine atom is eliminated. The IR spectrum does not indicate the presence of any functional groups, and it shows that there are no methyl groups present. The m/z peak $= 83$ indicates a carbocation with a formula of C_6H_{11}. The fact that the compound does not contain a methyl group indicates that the compound is **bromocyclohexane**.

bromocyclohexane

c. The absorption bands at $\sim 1700 \text{ cm}^{-1}$ and $\sim 2700 \text{ cm}^{-1}$ indicate that the compound is an aldehyde. The molecular ion peak at $m/z = 72$ indicates that the aldehyde contains four carbons (C_4H_8O). The peak at $m/z = 44$ for loss of a group with molecular weight 28 indicates that ethene has been lost as a result of a McLafferty rearrangement.

A McLafferty rearrangement can occur only if the aldehyde has a γ-hydrogen. The only four-carbon aldehyde that has a γ-hydrogen is **butanal**.

$$CH_3CH_2CH_2-\overset{\overset{\textstyle O}{\textstyle \|}}{C}-H$$

Chapter 14 Practice Test

1.   Give one IR absorption band that could be used to distinguish each of the following pairs of compounds. Indicate the compound for which the band would be present.

   a.   and

   b.   and

   c. $CH_3CH_2CH_2CH_2OH$ and $CH_3CH_2CH_2OCH_3$

   d.   CHCH$_3$ and   CH$_2$CH$_2$OH
        OH

   e.   OCH$_3$ and

   f. $CH_3CH_2CH\!=\!CHCH_3$ and $CH_3CH_2C\!\equiv\!CCH_3$

   g. $CH_3CH_2C\!\equiv\!CH$ and $CH_3CH_2C\!\equiv\!CCH_3$

2.   Indicate whether each of the following is true or false:

   a.   The O—H stretch of a concentrated solution of an alcohol occurs at a higher frequency than the O—H stretch of a dilute solution.   T   F

   b.   Light of 2 μm is of higher energy than light of 3 μm.   T   F

   c.   It takes more energy for a bending vibration than for a stretching vibration.   T   F

   d.   Propyne will not have an absorption band at 3100 cm^{-1}, because there is no change in the dipole moment.   T   F

   e.   Light of 8 μm has the same energy as light of 1250 cm^{-1}.   T   F

   f.   The M+2 peak of an alkyl chloride is half the height of the M peak.   T   F

3.   The major peaks shown in the mass spectrum of a tertiary alcohol are at $m/z = 73, 87, 98$, and 101. Identify the alcohol.

4.   A 3.8×10^{-4} M solution of cyclohexanone shows an absorbance of 0.75 at 280 nm in a 1.00 cm cell. What is the molar absorptivity of cyclohexanone at 280 nm?

5.   How could you distinguish between the IR spectra of the following compounds?

a.
and

d. an aryl ketone and benzyl methyl ether structures with OCH₃

b.
and

e. styrene and ethylbenzene structures

c.
and

6.   Which compound has the greater λ_{max}? (300 nm is a greater λ_{max} than 250 nm.)

a. cyclohexene and cyclohexadiene structures   and

c. aniline (NH_2) and anilinium ($\overset{+}{N}H_3$) structures   and

b. N-methylaniline and benzylamine (NH_2) structures   and

7.   A solution of a compound with a molar absorptivity of 1200 $M^{-1}cm^{-1}$ at 297 nm gives an absorbance of 0.76 at that wavelength in a 1.0 cm quartz cell. What is the concentration of the solution?

8.   Draw possible structures for an alcohol that has a molecular ion with an m/z value of 60.

9.   A bond between a carbon and an atom of similar electronegativity breaks _____, whereas a bond between a carbon and a more electronegative atom breaks _____.

CHAPTER 15
NMR Spectroscopy

Important Terms

applied magnetic field	the externally applied magnetic field.
chemically equivalent protons	protons with the same connectivity relationship to the rest of the molecule.
chemical shift	location of a signal occurring in an NMR spectrum. It is measured downfield from a reference compound (most often TMS).
^{13}**C NMR**	nuclear magnetic resonance from carbon (^{13}C) nuclei.
COSY spectrum	a 2-D NMR spectrum showing ^{1}H-^{1}H correlations.
coupled protons	protons that split each other. Coupled protons have the same coupling constant.
coupling constant	the distance (in hertz) between two adjacent peaks of a split NMR signal.
DEPT ^{13}C NMR spectrum	a group of four ^{13}C NMR spectra that distinguish CH_3, CH_2, and CH groups.
diamagnetic anisotropy	the term used to describe the greater freedom of π electrons to move in response to a magnetic field as a consequence of their greater polarizability compared with σ electrons.
diamagnetic shielding	shielding by the local magnetic field that opposes the applied magnetic field.
diastereotopic hydrogens	two hydrogens bonded to the same carbon that will result in a pair of diastereomers when each of them is replaced in turn with deuterium.
2-D NMR	two-dimensional nuclear magnetic resonance.
doublet	an NMR signal that is split into two peaks.
doublet of doublets	an NMR signal that is split into four peaks of approximately equal height. A doublet of doublets is caused by splitting a signal into a doublet by one hydrogen and into another doublet by another (nonequivalent) hydrogen.
downfield	farther to the left-hand side of the spectrum.
effective magnetic field	the magnetic field that a nucleus "senses" through the surrounding cloud of electrons.
enantiotopic hydrogens	two hydrogens bonded to a carbon that is bonded to two other groups that are nonidentical.
Fourier transform NMR (FT-NMR)	a technique in which all the nuclei are excited simultaneously by an rf pulse, their relaxation monitored, and the data mathematically converted to a frequency spectrum.

434

geminal coupling	the mutual splitting of two nonidentical protons bonded to the same carbon.
gyromagnetic ratio	the ratio of the magnetic moment of a rotating charged particle to its angular momentum.
HETCOR spectrum	a 2-D NMR spectrum showing ^{13}C-^{1}H correlations.
high-resolution NMR spectroscopy	NMR spectroscopy that uses a spectrometer with a high operating frequency.
^{1}H NMR	nuclear magnetic resonance from hydrogen nuclei.
long-range coupling	splitting by a proton more than three σ bonds away.
magnetic resonance imaging (MRI)	NMR used in medicine. The difference in the way water is bound in different tissues produces the signal variation between organs as well as between healthy and diseased states.
methine hydrogen	a tertiary hydrogen.
MRI scanner	an NMR spectrometer used in medicine for whole-body NMR.
multiplet	an NMR signal split by two (or more) nonequivalent sets of protons.
multiplicity	the number of peaks in an NMR signal.
$N + 1$ rule	a rule that states that an ^{1}H NMR signal for a hydrogen with N equivalent hydrogens bonded to an adjacent carbon is split into $N + 1$ peaks; a proton-coupled ^{13}C NMR signal for a carbon bonded to N hydrogens is split into $N + 1$ peaks.
NMR spectroscopy	the absorption of rf radiation by nuclei in an applied magnetic field to determine the structural features of an organic compound. In the case of ^{1}H NMR spectroscopy, it reveals the carbon-hydrogen framework.
operating frequency	the frequency at which an NMR spectrometer operates.
prochiral carbon	a carbon (bonded to two hydrogens) that will become an asymmetric center if one of the hydrogens is replaced by deuterium.
pro-R-hydrogen	replacing this hydrogen with deuterium creates an asymmetric center with the R configuration.
pro-S-hydrogen	replacing this hydrogen with deuterium creates an asymmetric center with the S configuration.
proton-coupled ^{13}C NMR spectrum	a ^{13}C NMR spectrum in which each signal for a carbon is split by the hydrogens bonded to that carbon.
proton exchange	the transfer of a proton from one molecule to another.

quartet   an NMR signal that is split into four equally spaced peaks with an integral ratio of 1:3:3:1.

reference compound   a compound added to the sample whose NMR spectrum is to be taken. The position of the signals in the NMR spectrum are measured from the position of the signal given by the reference compound.

rf radiation   radiation in the radiofrequency region of the electromagnetic spectrum.

shielding   the electrons around a proton shield the proton from the full effect of the applied magnetic field. The more a proton is shielded, the farther to the right its signal appears in an NMR spectrum.

singlet   an unsplit NMR signal.

spin-spin coupling   the splitting of a signal in an NMR spectrum described by the $N + 1$ rule.

α-spin state nuclei in this spin state have their magnetic moments oriented in the same direction as the applied magnetic field.

β-spin state   nuclei in this spin state have their magnetic moments oriented opposite to the direction of the applied magnetic field.

splitting diagram (splitting tree)   a diagram that describes the splitting of a set of protons.

triplet   an NMR signal that is split into three equally spaced peaks with an integral ratio of 1:2:1.

upfield   farther to the right-hand side of the spectrum.

X-ray crystallography   a technique used to determine the arrangement of atoms within a crystal.

X-ray diffraction   a technique used to obtain images that are used to determine the electron density within a crystal.

Solutions to Problems

1. $\nu = \dfrac{\gamma}{2\pi}B_0$

$$= \frac{26.75 \times 10^7 \text{ rad } \text{T}^{-1} \text{ sec}^{-1} \times 1.0 \text{ T}}{2(3.1416) \text{ rad}}$$

$$= 43 \times 10^6 \text{ sec}^{-1}$$

$$= 43 \times 10^6 \text{ Hz} = 43 \text{ MHz}$$

2.   **a.** $\nu = \dfrac{\gamma}{2\pi}B_0$

$$B_0 = \frac{\nu \times 2\pi}{\gamma}$$

$$B_0 = \frac{360 \times 10^6 \times 2(3.1416)}{26.75 \times 10^7}$$

$$B_0 = \frac{226.2}{26.75}$$

$$B_0 = 8.46 \text{ T}$$

b. $B_0 = \dfrac{\nu \times 2\pi}{\gamma}$

$$B_0 = \frac{500 \times 10^6 \times 2(3.1416)}{26.75 \times 10^7}$$

$$B_0 = \frac{314.2}{26.75}$$

$$B_0 = 11.75 \text{ T}$$

From these calculations, you can see that the greater the operating frequency of the instrument (360 MHz versus 500 MHz), the more powerful the magnet (8.46 T versus 11.75 T) required to operate it.

3.

3   5   4   3   2

4.   **a.** 2   **e.** 1   **i.** 5   **m.** 3
       b. 1   **f.** 4   **j.** 4   **n.** 4
       c. 1   **g.** 3   **k.** 2   **o.** 3
       d. 3   **h.** 3   **l.** 3

5.   **A** would give two signals, **B** would give one signal, and **C** would give three signals.

6.   **a.**   **b.**   **c.**

All the Hs are equivalent.

The Hs attached to the front of the molecule are equivalent, and the methylene Hs are equivalent.

The Hs attached to the front of the molecule are equivalent, and the methylene Hs are not equivalent.

17.   Solved in the text.

18.   The heights of the integrals for the signals in the spectrum shown in Figure 15.10 are about 3.5 and 5.2. The ratio of the integrals, therefore, is $5.2/3.5 = 1.5$. This matches the ratio of the integrals calculated for **B**. (Later we will see that a signal at ~7 ppm is characteristic of a benzene ring.)

HC≡C—⟨benzene ring⟩—C≡CH

$$\frac{4}{2} = 2$$

CH₃—⟨benzene ring⟩—CH₃

$$\frac{6}{4} = 1.5$$

ClCH₂—⟨benzene ring⟩—CH₂Cl

$$\frac{4}{4} = 1.0$$

Br₂CH—⟨benzene ring⟩—CHBr₂

$$\frac{4}{2} = 2$$

19.   The signal farthest downfield in both spectra is the signal for the hydrogens bonded to the carbon that is also bonded to the halogen. Because chlorine is more electronegative than iodine, the farthest downfield signal should be farther downfield in the ¹H NMR spectrum for 1-chloropropane than in the ¹H NMR spectrum for 1-iodopropane. Therefore, the **first spectrum** in Figure 15.2 is the ¹H NMR spectrum for **1-iodopropane**, and the **second spectrum** in Figure 15.12 is the ¹H NMR spectrum for **1-chloropropane**.

20.   **C** is easiest to distinguish because it will have **two** signals, whereas **A** and **B** will each have **three** signals.

A and **B** can be distinguished by looking at the splitting of their signals.

The signals in the ¹H NMR spectrum of **A** will be (left to right across the spectrum): **triplet, triplet, multiplet**.

The signals in the ¹H NMR spectrum of **B** will be (left to right across the spectrum): **doublet, multiplet, doublet**.

21.   The signal at 12.2 ppm indicates that the compound is a carboxylic acid. From the molecular formula and the splitting patterns of the signals, the spectra can be identified as the ¹H NMR spectrum of:

a. $CH_3\underset{\underset{Cl}{|}}{C}H{-}\overset{\overset{O}{\|}}{C}{-}OH$ **b.** $ClCH_2CH_2{-}\overset{\overset{O}{\|}}{C}{-}OH$

22.   **a.** A triplet is caused by coupling to two equivalent protons on an adjacent carbon. The two protons can be aligned in three different ways: both with the field, one with the field and one against the field, or both against the field. That is why the signal is a triplet. There is only one way to align two protons that are with the field or two protons that are against the field. However, there are two ways to align two protons if one is with and one is against the field: with and against or against and with. Consequently, the peaks in a triplet have relative intensities of 1:2:1.

b. A quintet is caused by coupling to four equivalent protons on adjacent carbons. The four protons can be aligned in five different ways. The following possible arrangements for the alignment of four protons explain why the relative intensities of a quintet are 1:4:6:4:1.

23. **a.** three signals

| triplet | | triplet |
| | multiplet | |

b. three signals

$ICH_2CH_2CH_2Br$

| triplet | | triplet |
| | multiplet | |

c. two signals

$ClCH_2CH_2CH_2Cl$

| triplet | |
| | quintet |

d. three signals

| triplet | | triplet |
| | multiplet | |

24. Each compound will have two doublets. In addition, **A** will have two doublet of doublets, **B** will have one doublet of doublets and a singlet, and **C** will have no other signals.

A	
four signals	
(two doublets and	
two doublets of	
doublets)	

B

four signals
(two doublets, one
singlet, and one
doublet of doublets)

C

two signals
(two doublets)

25.

one signal

a singlet

two signals

a quintet (a) and
a triplet (b)

three signals

two triplets (a and c)
and a multiplet (b)

three signals

two doublets (a and b)
and a multiplet (c)

26.   **a.**   The signal at ~7.2 ppm indicates the presence of a benzene ring. The integrations shows that the ring has a single substituent; thus the ring has a formula of C_6H_5. Subtracting C_6H_5 from the molecular formula of the compound $(C_9H_{12} - C_6H_5)$ gives a substituent with a formula of C_3H_7. The triplet at ~0.9 ppm that integrates to 3 protons indicates a methyl group adjacent to a methylene group. The identical integration of the signals at 1.6 ppm and 2.6 ppm indicates two methylene groups. Thus, the compound is propylbenzene.

     b.   The triplet that integrates to 3 protons and the quartet that integrates to 2 protons indicate a CH_3CH_2 group bonded to an atom that is not bonded to any hydrogens. The molecular formula of $C_5H_{10}O$ indicates that the compound is diethyl ketone.

     c.   The signals between ~7 and 8 ppm that integrate to 5 protons indicate the presence of a monosubstituted benzene ring. Subtracting C_6H_5 from the molecular formula of the compound $(C_9H_{10}O_2 - C_6H_5)$ shows that the substituent has a formula of $C_3H_5O_2$.

The triplet that integrates to 3 protons and the quartet that integrates to 2 protons suggest a CH_3CH_2 group bonded to an atom that is not bonded to any hydrogens. Therefore, we can conclude that the compound is one of the following:

The substituent attached to the benzene ring in compound **A** withdraws electrons from the ring, whereas the substituent attached to the benzene ring in compound **B** donates electrons into the benzene ring. That one of the peaks of the benzene ring signal is at a higher frequency (farther downfield) than usual (>8 ppm) suggests that an electron-withdrawing substituent is present on the benzene ring. This is confirmed by the signal for the CH_2 group at 4.3 ppm, indicating that the methylene group is adjacent to an oxygen. Thus, the spectrum is that of compound **A**.

27.

a. (t) (q) $CH_3CH_2CH_2CH_3$

b. (s) $BrCH_2CH_2Br$

c.
(Remember that equivalent hydrogens do not split each other's signal.)

d. (m) (d) (d→)
The H on C-2 will not be split by the H on C-3, because they are equivalent. The Hs on C-6 will not be split by the Hs on C-5, because the Hs on C-5 and C-6 are equivalent.

e.
(s→H)

f. (t) (m) (t) (O) (s) $CH_3CH_2CH_2CCH_3$

g. (quin) (m) (t) $CH_3CH_2CHCH_2CH_3$ | Cl

h. (d) (m) (t) $CH_3CHCH_2CHCH_3$ | CH_3 CH_3

i. (d) (q) (d of d) (←t) (d) $CH_3CH—$ | Br (benzene ring)

j. (s) (d) (d) (s) CH_3—(ring)—OCH_3

k. (s) (s) CH_3—(ring)—CH_3

l. (d) (t→) (←s) Br ... Br (ring)

m. (d of d) (d) (t→) (ring)—NO_2

28.

a. $CH_3OCH_2CH_2OCH_3$

b. There are three possibilities:

$$CH_3CCH_2CH_2CCH_3 \ (O,O)$$
$$CH_3OCH_2C\equiv CCH_2OCH_3$$
(dioxane ring with CH_3, CH_3)

c. HC—(ring)—CH (with O, O)

29. Each spectrum is described going from left to right.

a. $BrCH_2CH_2CH_2CH_2Br$

   triplet   triplet (This triplet will not be split by the adjacent equivalent protons.)

b. two triplets (close to each other)   singlet   multiplet
(The table "^{1}H NMR Chemical Shifts" in the Study Section of Mastering indicates that a methylene group adjacent to an RO and a methylene group adjacent to a Br appear at about the same place.)

c. singlet (Equivalent Hs do not split each other's signals.)

    d. quartet singlet triplet   **k.** doublet doublet of septets doublet
    e. three singlets   **l.** triplet singlet quintet
    f. three doublets of doublets   **m.** singlet
    g. quartet triplet   **n.** singlet (Equivalent Hs do not split each other's signals.)
    h. singlet quartet triplet   **o.** singlet
    i. doublet multiplet doublet   **p.** quintet, multiplet, multiplet
    j. triplet quintet

30. There is no coupling between H_a and H_c, because they are separated by four σ bonds. (We will see in Section 15.17 that unless the sample is pure and dry, a proton bonded to an oxygen is not split and does not split other protons.)

There is no coupling between H_b and H_c, because they are separated by five σ bonds.

31. The two singlets in the ^1H NMR spectrum that each integrate to 3 protons suggest two methyl groups, one of which is adjacent to an oxygen. That the benzene ring protons (6.7–7.1 ppm) consist of two doublets (each integrating to two hydrogens) indicates a 1,4-disubstituted benzene ring. The IR spectrum also indicates a benzene ring (1600 and 1500 cm^{-1}) with hydrogens bonded to sp^2 carbons and no carbonyl group. The absorption bands in the 1250–1000 cm^{-1} region suggest there are two C-O single bonds, one with no double-bond character and one with some double-bond character.

$$CH_3O \underset{}{\longleftarrow}\hspace{-0.5em}\bigcirc\hspace{-0.5em}\longrightarrow CH_3$$

32.   **a.**   **b.**

33.   **a.** A   **b.** B and D

34. Solved in the text.

35. Solved in the text.

36. The OH proton of a carboxylic acid is more deshielded than the OH proton of an alcohol as a result of electron delocalization that decreases the electron density around the OH proton. In addition, the extent of hydrogen bonding affects the chemical shift of a proton bonded to an oxygen. Carboxylic acids exist as tightly hydrogen-bonded dimers (see page 728 in the text); this strongly deshields the OH proton.

37. The greater the extent of hydrogen bonding, the greater the chemical shift. Therefore, the ^{1}H NMR spectrum of pure ethanol would have the signal for the OH proton at a greater chemical shift because it would be hydrogen bonded to a greater extent.

38.

$$R\ddot{O}{-}H \ + \ {:}\ddot{O}H \ \rightleftharpoons \ R\ddot{O}{:}^- \ + \ H{-}\ddot{O}H$$

39. Figure 15.31 is the ^{1}H NMR spectrum of propanamide. Notice that the signals for the N-H protons are unusually broad. Because of the partial double-bond character of the C-N bond, there is no free rotation about the C-N bond, so the two N-H protons are not chemically equivalent. The quartet and triplet are characteristic of an ethyl group.

40. **a.**

1. 3	**4.** 4	**7.** 3
2. 3	**5.** 3	**8.** 3
3. 2	**6.** 4	**9.** 2

b. An arrow is drawn to the carbon that gives the signal at the lowest frequency.

1. $CH_3CH_2CH_2Br$
2.
3. CH_3CHCH_3 with Br
4. CH_3CH_2C(=O)OCH_3
5. $CH_3CH(CH_3)$C(=O)H
6.
7.
8.
9.

41. Each spectrum is described going from left to right:

1. triplet triplet quartet   **3.** doublet quartet   **5.** doublet doublet quartet

42.

		first	second	third
a.	^{1}H NMR	two signals	three signals	one signal
b.	^{13}C NMR	three signals	four signals	two signals

43.   **a.** The signal at 210 ppm is for the carbonyl carbon of a ketone. There are 10 other carbons in the compound but only 5 other signals. That suggests the compound is a symmetrical ketone with identical five-carbon alkyl groups.

$$CH_3CH_2CH_2CH_2CH_2 \overset{\overset{\displaystyle O}{\|}}{C} CH_2CH_2CH_2CH_2CH_3$$

b. Because there are only four signals for the six carbons of the benzene ring (the signals between 110 and 117 ppm), the compound must be a 1,4-disubstituted benzene with two different substituents.

$$Br\text{—}\bigcirc\text{—}CH_2CH_3$$

c. The signal at 212 ppm is for the carbonyl carbon of a ketone. There are five other carbons in the compound but only three other signals. This suggests the compound is a symmetrical cyclic ketone.

d. The molecular formula indicates the compound has one double bond or one ring. The presence of a signal at 130 ppm shows the presence of sp^2 carbons; thus, the compound must have a double bond. Each of the two sp^2 carbons must be bonded to the same groups, because the six carbons exhibit only three signals. Whether the compound is *cis*-3-hexene or *trans*-3-hexene cannot be determined from the spectrum. An ^{1}H NMR spectrum could distinguish the compounds because the coupling constant for trans protons is greater than the coupling constant for cis protons.

$$CH_3CH_2CH=CHCH_2CH_3$$

44.

If the triangles shown above are drawn on the spectrum in Figure 15.41 of the text, you will be able to easily identify the coupled protons.

Point **A** shows that the **a** protons are split by the **c** protons.
Point **B** shows that the **b** protons are split by the **d** protons.

45.   Cross peak X tells you that the hydrogens that produce the signal at ~ 1.0 ppm in the ^{1}H NMR spectrum are bonded to the carbon that produces the signal at ~ 19 ppm in the ^{13}C NMR spectrum.

46. **a.** **1.** 5   **4.** 2   **b.** **1.** 7   **4.** 2
             2. 5   **5.** 3   **2.** 7   **5.** 2
             3. 4   **6.** 3   **3.** 5   **6.** 4

47.   The H$_b$ proton will be split into a quartet by the H$_a$ protons and into a doublet by the H$_c$ protons.

a.

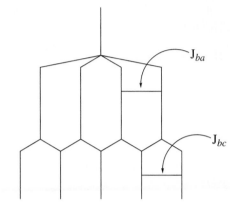

five peaks (a quintet)

b.

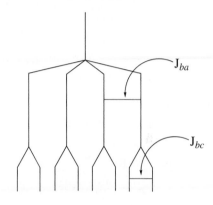

eight peaks (a doublet of quartets)

48. **a.** CH$_3$CHNO$_2$ ...

49. **a.** The spectrum must be that of **2-bromopropane**, because the NMR spectrum has two signals and the lowest frequency (farthest upfield) signal is a doublet and the other signal is given by a single hydrogen.

    b. The spectrum must be that of **1-nitropropane**, because the NMR spectrum has three signals and both the lowest frequency (farthest upfield) and highest frequency (farthest downfield) signals are triplets.

    c. The spectrum must be that of **ethyl methyl ketone**, because the NMR spectrum has three signals and the signals are a triplet, a singlet, and a quartet.

50.   By dividing the value of the integration steps by the smallest one, the ratios of the hydrogens are found to be 3:2:1:9.

$$\frac{40.5}{13} = 3.1 \quad \frac{27}{13} = 2.1 \quad \frac{13}{13} = 1 \quad \frac{118}{13} = 9.1$$

Because the ratios are given in the highest frequency to lowest frequency direction, a possible compound could be the following ester:

$$\begin{array}{cc} CH_3 & O \\ | & \| \\ CH_3C\!-\!CHCOCH_3 \\ | & | \\ CH_3 & CH_2Cl \end{array}$$

51.   **a.**   and

four signals   two signals

b.   Br⁓Br   and   Br⁓NO$_2$

two signals   three signals

c. $\begin{array}{cc} CH_3 & CH_3 \\ | & | \\ CH_3CH\!-\!CHCH_3 \end{array}$ and $\begin{array}{c} CH_3 \\ | \\ CH_3CCH_2CH_3 \\ | \\ CH_3 \end{array}$

two signals   three signals

d. $\begin{array}{cc} CH_3 & O \\ | & \| \\ CH_3C\!-\!COCH_3 \\ | \\ CH_3 \end{array}$ and $\begin{array}{c} OCH_3 \\ | \\ CH_3CCH_3 \\ | \\ OCH_3 \end{array}$

two signals with   two signals with
integration 3:1   integration 1:1

e. CH_3O⎯⟨⟩⎯CH_2CH_3 and CH_3⎯⟨⟩⎯OCH_2CH_3

the singlet would be at a higher   the quartet would be at a higher
frequency than the quartet   frequency than the singlet

f.   and

two signals   three signals

g. $\begin{array}{c} CH_3CHCl \\ | \\ CH_3 \end{array}$ and $\begin{array}{c} CH_3CDCl \\ | \\ CH_3 \end{array}$

two signals   one signal

h.

the lowest frequency signal would be a doublet   the lowest frequency signal would be a singlet

i.

three signals   two signals

j.

the signals for the benzene ring protons   the signals for the benzene ring protons
plus two signals with integration 3:1   plus two signals with integration 4.5:1

52. **a.** Chemical shift in ppm is independent of the operating frequency.
b. Chemical shift in hertz is proportional to the operating frequency.
c. The coupling constant in hertz is independent of the operating frequency.
d. The frequency required for NMR spectroscopy is lower than that required for IR spectroscopy, which is lower than that required for UV/Vis spectroscopy.

53. **a.** CH_3CH_2CHBr **b.** $CH_3CH_2CH_2CH_2Br$ **c.** CH_3CHCH_2Br
              | |
              CH_3 CH_3

54. **a.** **1.** 5   **2.** 3   **3.** 4   **4.** 3   **5.** 4   **6.** 4

      b. **1.** 5   **2.** 4   **3.** 3   **4.** 3   **5.** 3   **6.** 3

55. **a.**

CH_3
|
CH_3CCH_2Br
|
Br

b.

c. $CH_3CH_2\overset{\displaystyle O}{\overset{\displaystyle \|}{C}}OCH_2CH_3$

56. The singlet at 210 ppm indicates the carbonyl group of a ketone. The splitting of the other two signals indicates an isopropyl group. The molecular formula indicates that the compound must have two isopropyl groups.

57.

CH₃CCH₃ with CH₃ above and Cl below
tert-butyl chloride
Compound A

CH₃CHCH₂CH₃ with Cl below
sec-butyl chloride
Compound B

58. **a.** CH_3—C(=O)—CH_2CCH_3 (with CH₃ above and CH₃ below)

b. CH_3CH—C(=O)—$CH_2CH_2CH_3$ (with CH₃ below)

c. CH_3CH—C(=O)—$CHCH_3$ (with CH₃ below each)

59. CH_3CH_2CH=CH_2

1-butene

cis-2-butene

CH₃C=CH₂ (with CH₃ above)

2-methylpropene

It would be better to use ^{13}C NMR, because you would have to look only at the number of signals in each spectrum: 1-butene will show four signals, cis-2-butene will show two signals, and 2-methylpropene will show three signals. (In the ^1H NMR spectrum, 1-butene will show five signals, and cis-2-butene and 2-methylpropene will both show two signals.)

60.

CH_3CH_2—C(=O)—OCH_3

A

three signals

singlet, quartet, triplet
(singlet farthest downfield)

CH_3—C(=O)—OCH_2CH_3

B

three signals

singlet, quartet, triplet
(quartet farthest downfield)

H—C(=O)—$OCH_2CH_2CH_3$

C

four signals

singlet, triplet multiplet, triplet

H—C(=O)—$OCHCH_3$ (with CH₃ below)

D

three signals

singlet, septet doublet

C can be distinguished from **A, B,** and **D**, because **C** has four signals and the others have three signals.

D can be distinguished from **A** and **B**, because the three signals of **D** are a singlet, a septet, and a doublet, whereas the three signals of **A** and **B** are a singlet, a quartet, and a triplet.

A and **B** can be distinguished, because the highest frequency signal in **A** is a singlet, whereas the highest frequency signal in **B** is a quartet.

61.   It is the ^1H NMR spectrum of tert-butyl methyl ether.

CH_3Br + CH₃C—O⁻ (with CH₃ above and CH₃ below) ⟶ CH₃C—OCH₃ (with CH₃ above and CH₃ below)

tert-butyl methyl ether

62. **a.** $CH_3CH_2CH_2CH_2OCH_3$

c.
CH_3CH with Cl below, attached to $C=O$ and OCH_3

b.
structure with CH_3, C, CH_3, CH_3C, OCH_3, O

d. $CH_3CH_2CH_2$ attached to $C=O$ and OH

63. **a.**
(numbers 19, 19, 30, 70, OH)

b.
(numbers 26, 36, 24, 71—OH)

The numbers indicate the ppm value of the signal given by that carbon.

64. The broad signal at ~2.9 ppm is for the H_a proton that is bonded to the oxygen. The signal for the H_b protons at ~4.2 ppm is split into a doublet by the H_e proton. The H_c proton and the H_d proton are each split by the H_e proton; the coupling constant is greater for the trans protons. Thus, the H_c proton is at ~5.1 ppm and the H_d proton is at ~5.3 ppm. The H_e proton at ~6.0 ppm is split by the H_b protons and by the H_c proton and the H_d proton.

65. In the 6.5–8.1 ppm region, each compound will have two doublets. In addition, **A** will have two doublet of doublets, **B** will have one doublet of doublets and a second singlet, and **C** will have no other signals.

In the 6.5–8.1 ppm region:
	A	**B**	**C**
	four signals (two doublets and two doublets of doublets)	four signals (one singlet, two doublets, and one doublet of doublets)	two signals (two doublets)

66. The signals at ~7.2 ppm indicate that the compounds whose spectra are shown in parts **a** and **b** each contains a benzene ring. From the molecular formula, you know that each compound has five additional carbons.

a. The integration of the benzene ring protons indicate that the ring is monosubstiuted. The hydrogens bonded to the five additional carbons in part **a** must all be accounted for by two singlets with integral ratios of ~2:9, indicating that the compound is **2,2-dimethyl-1-phenylpropane**.

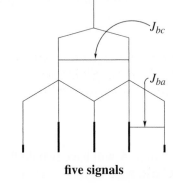

2,2-dimethyl-1-phenylpropane

b. The two singlets in the spectrum in part **b** have integral ratios of ~1:3, indicating a methyl substituent and a *tert*-butyl substituent. The integration of the benzene ring protons indicate that the ring is 1,4-disubstituted. Thus, the compound is **1-*tert*-butyl-4-methylbenzene**.

1-*tert*-butyl-4-methylbenzene

67.

five signals

68.   Oxygen is more electronegative than chlorine. (See Table 1.3 on page 11 of the text.)

a.

                                                            b   a   c

                                                 $ClCH_2CH_2OH$

      c   a   b

triplet   doublet   triplet

of

triplets

⟵ frequency

b.

     c   a   b

singlet   triplet   triplet

⟵ frequency

c.

a   b
ClCH$_2$CH$_2$OH

← frequency

d.

← frequency

e.

CH$_3$ carbons

CH$_2$ carbons

CH carbons

all carbons

69. If addition of HBr to propene follows the rule that says that the electrophile adds to the sp^2 carbon that is bonded to the most hydrogens, the product of the reaction will give an NMR spectrum with two signals (a doublet and a septet). If addition of HBr does not follow the rule, the product will give an NMR spectrum with three signals (two triplets and a multiplet).

$$\begin{array}{cc} & \text{Br} \\ & | \\ \text{CH}_3\text{CH}{=}\text{CH}_2 \ + \ \text{HBr} \longrightarrow \ \ \text{CH}_3\text{CHCH}_3 & \text{CH}_3\text{CH}_2\text{CH}_2\text{Br} \end{array}$$

follows the rule   does not follow the rule
two signals   **three signals**

70. a.   b.   c.   d.

71. Bromomethane gives a signal at 2.7 ppm for its three methyl protons, and 2-bromo-2-methylpropane gives a signal at 1.8 ppm for its nine methyl protons. If equal amounts of each were present in a solution, the ratio of the hydrogens (and, therefore, the ratio of the relative integrals) would be 3:1. Because the relative integrals are 1:6, there must be an 18-fold greater concentration of bromomethane in the mixture.

$$3:x = 1:6$$
$$x = 18$$

72.   Using the formula given on page 651 of the text:

$$\Delta E = h\nu = \frac{h\gamma}{2\pi}B_0$$

Given: Planck's constant $= h = 6.626 \times 10^{-34}$ Js   (page 224 of the text)

1 cal $= 4.184$ J   (page 207 of the text)

300 MHz $= 7.046$ T   (page 651 of the text)

γ for ^{1}H $= 2.675 \times 10^8$ T^{-1}s^{-1}

$$\Delta E = \frac{h\gamma}{2\pi}B_0$$

$$= \frac{6.626 \times 10^{-34} \text{ Js} \times 2.675 \times 10^8 \text{ T}^{-1}\text{s}^{-1}}{2(3.1416)} \times 1.4092 \text{ T} \times \frac{1 \text{ cal}}{4.184 \text{ J}}$$

$$= 9.50 \times 10^{-26} \text{ cal}$$

73.   All four spectra show a singlet at ~2.0 ppm, suggesting they are all esters with a methyl group attached to the carbonyl group. Now the problem becomes determining the nature of the group attached to the oxygen.

$$\underset{CH_3}{\overset{\displaystyle \overset{O}{\underset{\|}{C}}}{}}\!\!\!\!-O-R$$

a.   The highest frequency signal in the first spectrum is a triplet that integrates to 2 protons, indicating that a CH_2 group is attached to the oxygen and is bonded to another CH_2 group. The lowest frequency signal is a triplet that integrates to 3 protons, indicating a methyl group that is attached to a CH_2 group. The presence of two multiplets that each integrate to 2 protons confirms the structure.

$$\underset{CH_3}{\overset{\displaystyle \overset{O}{\underset{\|}{C}}}{}}\!\!\!\!-O-CH_2CH_2CH_2CH_3$$

b.   The highest frequency signal in the second spectrum is a multiplet that integrates to 1 proton, indicating that the carbon attached to the oxygen is attached to one proton and two nonequivalent carbons bonded to hydrogens. The lowest frequency signal is a triplet that integrates to 3 protons, indicating a methyl group attached to a CH_2 group. The doublet at ~1.2 ppm that integrates to 3 protons is due to a methyl group attached to a carbon bonded to one hydrogen.

$$\underset{CH_3}{\overset{\displaystyle \overset{O}{\underset{\|}{C}}}{}}\!\!\!\!-O-\underset{\underset{CH_3}{|}}{C}HCH_2CH_3$$

c.   The highest frequency signal is a doublet that integrates to 2 protons, indicating that the methylene group that is attached to the oxygen is attached to a carbon bonded to one hydrogen. The lowest frequency signal that integrates to 6 protons indicates two equivalent methyl groups that are attached to a carbon bonded to one hydrogen.

$$\underset{CH_3}{\overset{\displaystyle \overset{O}{\underset{\|}{C}}}{}}\!\!\!\!-O-CH_2\underset{\underset{CH_3}{|}}{C}HCH_3$$

d. The group attached to the oxygen in the fourth spectrum has only one kind of hydrogen. The carbon attached to the oxygen is not bonded to any hydrogens, because there is no signal at ~4.0.

$$\underset{CH_3}{\overset{\overset{\displaystyle O}{\|}}{C}}\;\; O{-}\underset{CH_3}{\overset{CH_3}{\underset{|}{\overset{|}{C}CH_3}}}$$

74.

$$\underset{\overset{|}{OH}}{\underset{|}{CH_3}}CH_3CHCHCH_3 \;\;\xrightarrow[\Delta]{HBr}\;\; \underset{\overset{|}{Br}}{\underset{|}{CH_3}}CH_3CH_2CCH_3$$

compound A   compound B

75. **a.** The IR spectrum indicates that the compound is a ketone. The molecular formula shows there are 12 hydrogens in the compound. Therefore, the signals in the NMR spectrum are due to 2, 4, and 6 protons. The doublet in the NMR spectrum at ~0.9 ppm that integrates to 6 protons suggests an isopropyl group. There is a singlet at ~2.1 ppm on top of a multiplet, which is due to the methine proton of an isopropyl group plus a three-hydrogen singlet for a methyl group adjacent to the carbonyl carbon. The doublet at ~2.2 ppm that integrates to 2 protons suggests a CH_2 group adjacent to the carbonyl group that couples to an adjacent methine proton. Knowing that the compound contains six carbons helps in the identification.

b. The IR spectrum indicates that this oxygen-containing compound is not a carbonyl compound or an alcohol; the absorption band at ~1000 cm^{-1} suggests that it is an ether. The doublet at ~1.1 ppm in the ^{1}H NMR spectrum that integrates to 6 protons and the septet that integrates to 1 proton indicate an isopropyl group. Since there are no other signals in the NMR spectrum, the compound must be a symmetrical ether. The compound is **diisopropyl ether**.

c. The IR absorption band at ~3400 cm^{-1} indicates the compound is an alcohol. The two doublets in the ^{1}H NMR spectrum at 7.3 and 8.1 ppm that integrate to 4 protons indicate a 1,4-disubstituted benzene ring with a strongly electron-withdrawing substituent. The two triplets that each integrate to 2 protons and the two multiplets that each integrate to 2 protons indicate that the four-carbon substituent is not branched. The broad signal at 2.1 ppm is due to the OH proton.

d. The IR absorption bands at ~1700 and 2700 cm^{-1} indicate the compound is an aldehyde. The two doublets at ~7.0 and 7.8 ppm in the ^{1}H NMR spectrum indicate a 1,4-disubstituted benzene ring. That none of the remaining NMR signals is a doublet suggests that the aldehyde group is attached directly to the benzene ring. The two triplets and two multiplets indicate an unbranched substituent. The triplet at ~4.0 ppm indicates that the group giving this signal is next to an electron-withdrawing group.

76. **a.** The IR absorption band at ~1730 cm^{-1} and the absence of an aldehyde C-H stretch suggest the compound is a ketone. The signals in the ^{1}H NMR spectrum are consistent with a $CH_3CH_2CH_2(C{=}O)$ group. That there are no other signals in the ^{1}H NMR spectrum suggests it is a symmetrical ketone. The mass spectrum shows a molecular ion with $m/z = 114$; this is consistent with the ketone shown below. Also, the large fragments at $m/z = 71$ (M − propyl) and at $m/z = 43$ (a propyl cation) are consistent with the expected α-cleavage fragmentation. The fact that there are only three signals in the NMR spectrum suggests it is a symmetrical ketone. The splitting pattern confirms the structure.

b. The M+2 peak at $m/z = 156$ in the mass spectrum indicates the compound contains chlorine; the IR spectrum indicates it is a ketone; the NMR spectrum indicates it has a monosubstituted benzene ring. The singlet at ~4.7 ppm indicates that the CH_2 group giving this signal is in a strongly electron-withdrawing environment. The major fragment ions at $m/z = 105(M − CH_2Cl)$ and at $m/z = 77$ ($C_6H_5^+$) are consistent with the structure shown below.

77. The DEPT ^{13}C NMR spectrum indicates that the compound has six carbons. The molecular formula shows that the compound has six carbons and an oxygen. The signal at 220 ppm suggests a ketone. The following ketones have the single CH_3 group, the single CH group, and the three CH_2 groups indicated by the DEPT spectrum. The highest frequency signals are a CH_2 group and a CH group, indicating that these are the groups attached to the carbonyl carbon. Thus, the compound on the right is responsible for the spectrum.

or   **or**

78. There is a singlet for the three methyl groups that are attached to the quaternary carbon; the singlet is on top of the triplet for the other methyl group. The methylene group shows the expected quartet.

Chapter 15 Practice Test

1. How many signals would you expect to see in the ^{1}H NMR spectrum of each of the following compounds?

2. Indicate the multiplicity of each of the indicated sets of protons. (That is, indicate whether it is a singlet, doublet, triplet, quartet, quintet, multiplet, or doublet of doublets.)

3. How could you distinguish the following compounds using ^{1}H NMR spectroscopy?

4. Indicate whether each of the following statements is true or false:

  a. The signals on the right of an NMR spectrum are deshielded compared to the signals on the left.   T   F

  b. Dimethyl ketone has the same number of signals in its ^{1}H NMR spectrum as in its ^{13}C NMR spectrum.   T   F

c. In the ^{1}H NMR spectrum of the compound shown below, the lowest frequency signal (the one farthest upfield) is a singlet and the highest frequency signal (the one farthest downfield) is a doublet.   T   F

$$O_2N \text{—} \bigcirc \text{—} CH_3$$

d. The greater the frequency of the signal, the greater its chemical shift in ppm.   T   F

5.   For each compound:

a. Indicate the number of signals you would expect to see in its ^{1}H NMR spectrum.
b. Indicate the hydrogen or set of hydrogens that would give the highest frequency (farthest downfield) signal.
c. Indicate the multiplicity of that signal.

$$
\begin{array}{ccc}
 & \overset{\displaystyle O}{\underset{\displaystyle \|}{}} & \\
\textbf{1. } CH_3CH_2CH_2Cl & \textbf{2. } CH_3CH_2COCH_3 & \textbf{3. } CH_3CHCH_3 \\
 & & | \\
 & & Br
\end{array}
$$

6.   For each compound of Problem 5:

a. Indicate the number of signals you would expect to see in its ^{13}C NMR spectrum.
b. Indicate the carbon that would give the highest frequency (farthest downfield) signal.
c. Indicate the multiplicity of that signal in a proton-coupled ^{13}C NMR spectrum.

CHAPTER 16

Reactions of Carboxylic Acids and Carboxylic Acid Derivatives

Important Terms

acid anhydride	
acyl adenylate	a carboxylic acid derivative with AMP as the leaving group.
acyl group	
acyl halide	
acyl phosphate	a carboxylic acid derivative with a phosphate leaving group.
acyl transfer reaction	a reaction that transfers an acyl substituent from one group to another.
alcoholysis	a reaction with an alcohol that converts one compound into two compounds.
amide	
amino acid	an α-amino carboxylic acid.
aminolysis	a reaction with an amine that converts one compound into two compounds.
biosynthesis	synthesis that occurs in a biological system.
carbonyl carbon	the carbon of a C=O group.
carbonyl compound	a compound that contains a C=O group.
carbonyl group	a carbon doubly bonded to an oxygen (C=O).
carbonyl oxygen	the oxygen of a C=O group.
carboxyl group	or —COOH or —CO$_2$H

carboxylic acid	$\underset{R}{\overset{\displaystyle O}{\overset{\|}{\underset{\displaystyle \quad OH}{C}}}}$
carboxylic acid derivative	a compound that is hydrolyzed to a carboxylic acid.
carboxyl oxygen	the single-bonded oxygen of a carboxylic acid or ester.
catalyst	a species that increases the rate of a reaction without being consumed in the reaction.
ester	$\underset{R}{\overset{\displaystyle O}{\overset{\|}{\underset{\displaystyle \quad OR'}{C}}}}$
fat	a triester of glycerol that exists as a solid at room temperature.
fatty acid	a long-chain carboxylic acid.
Fischer esterification reaction	a reaction of a carboxylic acid with excess alcohol and an acid catalyst.
Gabriel synthesis	a method used to convert an alkyl halide into a primary amine, involving S_N2 attack of phthalimide ion on an alkyl halide followed by hydrolysis.
hydrolysis	a reaction with water that converts one compound into two compounds.
hydrophobic interactions	the attractive forces between hydrocarbon groups in water.
imide	a compound with two acyl groups bonded to a nitrogen.
lactam	a cyclic amide.
lactone	a cyclic ester.
lipid	a water-insoluble compound found in a living system.
lipid bilayer	two layers of phosphoacylglycerols arranged so that their polar heads are on the outside and their nonpolar fatty acid chains are on the inside.
membrane	the material that surrounds the cell in order to isolate its contents.
micelle	a spherical aggregation of molecules, each with a long hydrophobic tail and a polar head, arranged so the polar head points to the outside of the sphere.
mixed anhydride	an acid anhydride with two different R groups.
mixed triglyceride	a triglyceride in which the fatty acid components are different.

nitrile   a compound that contains a carbon-nitrogen triple bond.

$$R—C\equiv N$$

nucleophilic acyl   a reaction in which a group bonded to an acyl group is substituted by another group.
substitution reaction

nucleophilic addition-   another name for a nucleophilic acyl substitution reaction, which emphasizes the
elimination reaction   two-step nature of the reaction: a nucleophile adds to the carbonyl carbon in the
                               first step, and a group is eliminated in the second step.

oil   a triester of glycerol that exists as a liquid at room temperature.

phosphoglyceride   formed when two OH groups of glycerol form esters with fatty acids and the terminal
(phosphoacylglycerol)   OH group is part of a phosphodiester.

phosphoanhydride bond   the bond holding two phosphate groups together within a diphosphate or triphosphate.

phospholipid   a lipid that contains a phosphate group.

polyunsaturated fatty acid a fatty acid with more than one double bond.

saponification   hydrolysis of a fat under basic conditions.

simple triglyceride   a triglyceride in which the fatty acid components are the same.

soap   a sodium or potassium salt of a fatty acid.

symmetrical anhydride   an acid anhydride with identical R groups.

tetrahedral intermediate   the intermediate formed in an addition-elimination (or a nucleophilic acyl substitution)
                               reaction.

thioester   the sulfur analog of an ester.

$$\underset{R}{\overset{\overset{\displaystyle O}{\|}}{\diagdown}}\overset{C}{\underset{\diagup}{}}\underset{SR'}{}$$

transesterification reaction   the reaction of an ester with an alcohol to form a different ester.

wax   an ester formed from a long straight-chain carboxylic acid and a long straight-
                               chain alcohol.

Solutions to Problems

1.   **a.** benzyl acetate   **b.** isopentyl acetate   **c.** methyl butyrate

2.   The lactone would be a three-membered ring, which is too strained to form.

3.   **a.** potassium butanoate   **f.** propanamide
           potassium butyrate   propionamide

        b. isobutyl butanoate   **g.** γ-butyrolactam or
           isobutyl butyrate   2-azacyclopentanone

        c. *N,N*-dimethylhexanamide   **h.** cyclopentanecarboxylic acid

        d. pentanoyl chloride   **i.** β-methyl-δ-valerolactone or
           valeryl chloride   5-methyl-2-oxacyclohexanone

        e. 5-methylhexanoic acid
           δ-methylcaproic acid

4.   **a.** $CH_3-\overset{\displaystyle O}{\overset{\|}{C}}-O-C_6H_5$

        d. $CH_3CH_2CH(CH_3)CH_2CH_2-\overset{\displaystyle O}{\overset{\|}{C}}-OH$

        g. cyclohexyl–$\overset{\displaystyle O}{\overset{\|}{C}}$–Cl

        b. 5-ethyl-2-pyrrolidinone (ring with NH, C=O, CH₂CH₃ substituent)

        e. $CH_3CH_2CH_2CH(Cl)-\overset{\displaystyle O}{\overset{\|}{C}}-OCH_2CH_3$

        h. $CH_3CH_2CH_2CH(Cl)-\overset{\displaystyle O}{\overset{\|}{C}}-OH$

        c. $CH_3-\overset{\displaystyle O}{\overset{\|}{C}}-NHCH_2-C_6H_5$

        f. $CH_3CH(Br)CH_2-\overset{\displaystyle O}{\overset{\|}{C}}-NH_2$

5.   **B** is a correct statement. The delocalization energy (resonance energy) is greater for the amide than for the ester because the second resonance contributor of the amide has a greater predicted stability and so contributes more to the overall structure of the amide; nitrogen is less electronegative than oxygen, so it is more stable with a positive charge.

6.   The carbon-oxygen single bond in an alcohol is longer because, as a result of electron delocalization, the carbon-oxygen single bond in a carboxylic acid has some double-bond character.

$$RCH_2-OH \qquad R-\overset{\displaystyle O}{\overset{\|}{C}}-\overset{..}{\underset{..}{O}}H \longleftrightarrow R-\overset{\displaystyle O^-}{\underset{}{C}}=\overset{+}{\underset{..}{O}}H$$

longer   shorter

7.   **a.**   The bond between oxygen and the methyl group is the longest, because it is a pure single bond, whereas the other two carbon-oxygen bonds have some double-bond character.

more stable   less stable

3 = longest   1 = shortest

The bond between carbon and the carbonyl oxygen is the shortest, because it has the most double-bond character.

b.   Notice that the longer the bond, the lower its IR stretching frequency.

1 = highest frequency
3 = lowest frequency

8.   The key factor is the extent to which the lone pair on Y can be delocalized onto the carbonyl oxygen. The smaller the contribution from the resonance contributor on the right, the greater will be the double-bond character of the $C=O$ and the greater will be its IR stretching frequency.

When Y = Cl, delocalization of chlorine's lone pair onto the carbonyl oxygen is minimal, because of the poor overlap of the large $3p$ orbital of chlorine with the small $2p$ orbital of carbon. Thus, the $C=O$ group of the acyl chloride has the highest IR stretching frequency.

acyl chloride	$\sim 1800 \ cm^{-1}$
ester	$\sim 1730 \ cm^{-1}$
amide	$\sim 1640 \ cm^{-1}$

When Y = NH_2, the resonance contributor on the right is more stable than the analogous resonance contributors for the other carboxylic acid derivatives, because nitrogen can better accommodate a positive charge than can oxygen. Therefore, the $C=O$ bond of the amide has the least double-bond character, so it stretches at the lowest frequency.

9.   **a.**   Stearic acid has the higher melting point, because it has two more methylene groups (giving it a greater surface area and therefore more intermolecular van der Waals interactions) than palmitic acid.

b.   Palmitic acid has the higher melting point, because it does not have any carbon-carbon double bonds, whereas palmitoleic acid has a cis double bond that prevents the molecules from packing closely together.

10.   CH$_3$CH$_2$CH$_2$CH$_2$CH$_2$CH=CHCH$_2$CH=CHCH$_2$CH=CHCH$_2$CH=CHCH$_2$CH$_2$CH$_2$COOH

1. excess O$_3$
2. (CH$_3$)$_2$S

CH$_3$CH$_2$CH$_2$CH$_2$CH$_2$CH   3 HCCH$_2$CH   HCCH$_2$CH$_2$CH$_2$CH

hexanal   3 malonaldehyde units   gluteraldehyde

11.   **a.** Because HCl is a stronger acid than H$_2$O, Cl⁻ is a weaker base than HO⁻.
Therefore, Cl⁻ will be eliminated from the tetrahedral intermediate, so the product of the reaction will be acetic acid. Because the solution is basic, acetic acid will be in its basic form as a result of losing a proton.

acetyl chloride   acetic acid   + Cl⁻

b. Because H$_2$O is a stronger acid than NH$_3$, HO⁻ is a weaker base than ⁻NH$_2$.
Therefore, HO⁻ will be eliminated from the tetrahedral intermediate, so the reactant will reform. In other words, no reaction will take place.

acetamide

12.   **a.** Acetyl chloride will have the stretching vibration for its carbonyl group at the highest frequency because it has the most C=O double-bond character, since it has the smallest contribution from the resonance contributor with a positive charge on Y.

b. Acetamide will have the stretching vibration for its carbonyl group at the lowest frequency because it has the least C=O double-bond character, since it has the largest contribution from the resonance contributor with a positive charge on Y.

13.   **a.** CH$_3$ OCH$_3$ + NaCl $\longrightarrow$ no reaction, because Cl⁻ is weaker base than CH$_3$O⁻

b. CH$_3$ Cl + NaOH $\longrightarrow$ CH$_3$ OH + NaCl $\xrightarrow{\text{HO}^-}$ CH$_3$ O⁻ + H$_2$O

c. CH$_3$—C(=O)—NH$_2$ +   NaCl   ⟶   no reaction, because Cl$^-$ is a weaker base than $^-$NH$_2$

d. CH$_3$—C(=O)—NH$_2$ +   NaOH   ⟶   no reaction, because HO$^-$ is a weaker base than $^-$NH$_2$

14.   It is a true statement.

If the nucleophile is the stronger base, it will be harder to eliminate the nucleophile from the tetrahedral intermediate (**B**) than to eliminate the group attached to the acyl group in the reactant. In other words, the hill that has to be climbed from the intermediate back to the reactants (**B** to **A**) is higher than the hill that has to be climbed from the intermediate to the products (**B** to **C**). Since the transition state with the highest energy is the transition state of the rate-limiting step, the first step is the rate-limiting step.

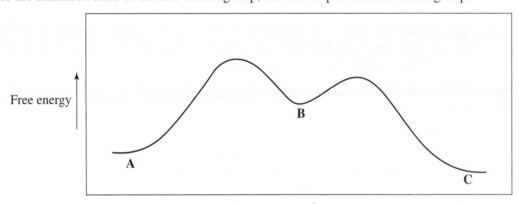

15.   **a.**   a new carboxylic acid derivative

b.   no reaction

c.   a mixture of two carboxylic acid derivatives

16.   **a.**   CH$_3$CH$_2$CH$_2$OH   **c.**   (CH$_3$)$_2$NH   **e.**   HO—⬡

b.   NH$_3$ **d.**   H$_2$O   **f.**   HO—⬡—NO$_2$

17.   Solved in the text.

18.   **a.**

CH$_3$—C(=O:)—Cl   +   H$_2$O: ⇌   CH$_3$—C(—O:$^-$)(—$^+$OH)—Cl   ⇌   CH$_3$—C(—O:$^-$)—Cl   +   HB$^+$

B: H

⟶

CH$_3$—C(=Ö)—OH   +   Cl$^-$

b.

19. A protonated amine has a pK_a ~ 11. Therefore, the amine will be protonated by the acid that is produced in the reaction, and a protonated amine is not a nucleophile. Excess amine is used in order to have some unprotonated amine available to react as a nucleophile.

A protonated alcohol has a pK_a ~ −2. Therefore, the alcohol will not be protonated by the acid that is produced in the reaction.

20.

a.

CH_3O^- (path a) and HO^- (path b) have the same leaving propensity so either one can be eliminated.

b.

21.   **a.** **1.** The carbonyl group of an ester is a weak electrophile.

           2. Water is a weak nucleophile.

           3. $^{-}OCH_3$ is a strong base and, therefore, a poor leaving group.

    b. Aminolysis is faster, because an amine is a better nucleophile than water.

22.   Solved in the text.

23.   **a.**

              c. The mechanism for the hydrolysis of a cyclic ester is the same as that for an acyclic ester.

    b.

24.   The mechanism for the acid-catalyzed reaction of acetic acid and methanol is the exact reverse of the mechanism for the acid-catalyzed hydrolysis of methyl acetate.

25.   **a.** Any species with an acidic proton can be represented by HB^+.

    b. Any species with a lone pair can be represented by :B.

c. H_3O^+

d. H_3O^+ if excess water is used, $CH_3\overset{+}{O}H_2$ if excess water is not used.

26.   Solved in the text.

27.   These are transesterification reactions.

a.

b.

28.

29.   The mechanism is the same as that on the top of page 745 of the text except that the nucleophile is methanol rather than water.

30.   **a.** The conjugate base ($CH_3CH_2CH_2O^-$) of the reactant alcohol ($CH_3CH_2CH_2OH$) can be used to increase the rate of the reaction because it is the nucleophile that we want to become attached to the carbonyl carbon.

       b. If H^+ is used as a catalyst, the amine will be protonated in the acidic solution and therefore will not be able to react as a nucleophile.

           If HO^- is used as a catalyst, HO^- will be the best nucleophile in the solution, so it will add to the ester, and the product of the reaction will be a carboxylate ion rather than an amide.

           If RO^- is used as a catalyst, RO^- will be the nucleophile, and the product of the reaction will be an ester rather than an amide.

31.

$$CH_3CH_2CH_2\overset{\displaystyle O}{\underset{\displaystyle }{\overset{\displaystyle \|}{C}}}\!\!-\!\!\overset{18}{O}CH_3 \quad \text{and} \quad CH_3\overset{18}{O}H$$

32.   **a.** The alcohol (CH_3CH_2OH) contained the ^{18}O label.

       b. The carboxylic acid would have contained the ^{18}O label.

33.   Solved in the text.

34.   When the oil is saponified, it forms glycerol and three equivalents of carboxylate ion. In losing glycerol, the oil loses three carbons and five hydrogens.

       Therefore, the three equivalents of carboxylate ion have a combined molecular formula of $C_{42}H_{81}O_6$.

       Dividing by three gives a molecular formula of $C_{14}H_{27}O_2$ for the carboxylate ion.

35.   Glyceryl tripalmitate has a higher melting point because the carboxylic acid components are saturated and can therefore pack more closely together than the cis-unsaturated carboxylic acid components of glyceryl tripalmitoleate.

36.   To be optically inactive, the fat must have a plane of symmetry. In other words, the fatty acids at C-1 and C-3 must be identical. Therefore, stearic acid must be at C-1 and C-3.

$$
\begin{array}{l}
CH_2\!-\!O\!-\!\overset{\displaystyle O}{\overset{\displaystyle \|}{C}}\!-\!(CH_2)_{16}CH_3 \\
\;\;|\qquad\qquad\;\; O \\
\;\;\;\;\;\;\;\;\;\;\;\;\;\; \| \\
CH\!-\!O\!-\!C\!-\!(CH_2)_{10}CH_3 \\
\;\;|\qquad\qquad\;\; O \\
\;\;\;\;\;\;\;\;\;\;\;\;\;\; \| \\
CH_2\!-\!O\!-\!C\!-\!(CH_2)_{16}CH_3
\end{array}
$$

37.   To be optically active, the fat must not have a plane of symmetry. Therefore, the two stearic acid groups must be attached to adjacent alcohol groups.

$$
\begin{array}{l}
CH_2\!-\!O\!-\!\overset{\displaystyle O}{\overset{\displaystyle \|}{C}}\!-\!(CH_2)_{16}CH_3 \\
\;\;|\qquad\qquad\;\; O \\
\;\;\;\;\;\;\;\;\;\;\;\;\;\; \| \\
CH\!-\!O\!-\!C\!-\!(CH_2)_{16}CH_3 \\
\;\;|\qquad\qquad\;\; O \\
\;\;\;\;\;\;\;\;\;\;\;\;\;\; \| \\
CH_2\!-\!O\!-\!C\!-\!(CH_2)_{10}CH_3
\end{array}
$$

38.

a.

or

or

b.

or

or

39.   In the first step, protonation occurs on the carbon that results in formation of the most stable carbocation (a tertiary carbocation). In the second step, the pair of electrons of the second π bond provides the nucleophile that adds to the carbocation because, as a result of that reaction, a stable six-membered ring is formed.

40. a. CH$_3$CH$_2$CH$_2$—C(=O)—Cl + 2 CH$_3$CH$_2$NH$_2$ **b.** C$_6$H$_5$—C(=O)—Cl + 2 CH$_3$NHCH$_3$

41.   Only #2 and #5 will form amides. #1 will form an amide if it is heated to about 225 °C.

42.

43.   The relative reactivities of the amides depend on the basicities of their leaving groups: the weaker the base, the more reactive the amide. (*para*-Nitroanilinium ion has a pK_a value of 0.98 and the anilinium ion has a pK_a value of 4.58; page 940 in the text.)

44. a. pentyl bromide   **b.** isohexyl bromide   **c.** benzyl bromide   **d.** cyclohexyl bromide

45.   The reaction of an alkyl halide with ammonia leads to primary, secondary, tertiary amines, and even quaternary ammonium ions. Thus, the yield of primary amine would be relatively low.

In contrast, the Gabriel synthesis forms only primary amines. The reaction of an alkyl halide with azide ion also forms only primary amines, because the compound formed from the initial reaction of the two reagents is not nucleophilic, so polyalkylation does not occur.

46.   **a.**   butanenitrile
            propyl cyanide

b.   4-methylpentanenitrile
        isopentyl cyanide

47.   Notice that the alkyl halide has one less carbon than the target carboxylic acid, because the alkyl halide will obtain a carbon from the cyanide ion.

a. $CH_3CH_2CH_2Br$

b. CH_3CHCH_2Br
        |
        CH_3

c.   —Br

49.   **a.**

b.   The mechanisms are exactly the same.

50.

51.   When an acid anhydride reacts with an amine, the tetrahedral intermediate does not have to lose a proton before it eliminates the carboxylate ion, because the carboxylate ion is a weaker base (pK_a of its conjugate acid is ~5) than the amine (pK_a of its conjugate acid is ~10), so the carboxylate ion is the better leaving group.

52.   a. At neutral pH, both carboxyl groups of succinic acid will be in their basic form.

b. Without acetic anhydride, the leaving group would be hydroxide ion. Acetic anhydride causes the reaction to take place via two successive addition-elimination reactions. In both reactions the leaving group is acetate ion, which is much less basic than hydroxide ion and therefore a better leaving group.

53. PDS lasts longer, because its ester linkage will hydrolyze more slowly since its alkoxide ion leaving group is a stronger base. You can determine which of the alkoxide ions is the stronger base by comparing how close the negatively charged oxygen atom is to an electron-withdrawing group. An electron-withdrawing group stabilizes the base, making it a weaker base and therefore a better leaving group. In Dexon, there is a carbonyl group attached to the CH$_2$ group that is next to the negatively charged oxygen. In PDS, there are two hydrogens in place of the double-bonded oxygen in Dexon. Thus, the alkoxide leaving group of Dexon is a weaker base and therefore a better leaving group.

$$\overset{\quad\;\; O}{\underset{\quad\;\; \parallel}{}}$$
$$^-OCH_2C-O-\qquad\qquad ^-OCH_2CH_2OCH_2C-O-$$
$$\text{Dexon}\qquad\qquad\qquad\quad \text{PDS}$$

54.   a. CH$_3$CH$_2$—COOH $\xrightarrow{\text{SOCl}_2}$ CH$_3$CH$_2$—COCl $\xrightarrow{\text{C}_6\text{H}_5\text{—OH}}$ CH$_3$CH$_2$—COO—C$_6$H$_5$

b.

Benzoic acid $\xrightarrow{\text{SOCl}_2}$ benzoyl chloride $\xrightarrow{\text{2 CH}_3\text{CH}_2\text{NH}_2}$ N-ethylbenzamide

55.

a. $CH_3CH_2CH_2CH_2CH_2\overset{\displaystyle O}{\overset{\|}{C}}\!\!-\!\!N\overset{\displaystyle CH_3}{\underset{\displaystyle CH_3}{}}$

d. $CH_3CH_2C\equiv N$

h.

CH_2CH_3

b. $CH_3CH_2CH_2\underset{\displaystyle CH_3}{\overset{\displaystyle CH_3}{C}}CH_2\overset{\displaystyle O}{\overset{\|}{C}}\!\!-\!\!NH_2$

e. $CH_3CH_2\overset{\displaystyle O}{\overset{\|}{C}}\!\!-\!\!NH_2$

i.

c.

f. $CH_3\overset{\displaystyle O}{\overset{\|}{C}}\!\!-\!\!O^- \; Na^+$

j.

g.

k.

56.

a. pentanoyl chloride
valeryl chloride

e. acetic benzoic anhydride

b. *N,N*-dimethylbutanamide
N,N-dimethylbutyramide

f. propanoic anhydride
propionic anhydride

c. (*S*)-3-methylpentanoic acid
(*S*)-*β*-methylvaleric acid

g. 5-ethylheptanoic acid

d. propyl propanoate
propyl propionate

h. (*R*)-3-methylhexanenitrile

57.

a.

b.

c.

d.

e.

a carboxylic acid will
be in its basic form in
a basic solution

f.

g.

h.

i.

j.

k.

58.   a. CH_3CH_2—C(=O)—OH + CH_3OH

c.

+ $CH_3CH_2\overset{+}{N}H_3$ Cl^-

b.

+ HCl

d. CH_3CH_2—C(=O)—OH + CO_2 + CH_3OH

59. **a.** The weaker the base attached to the acyl group, the stronger its electron-withdrawing ability; therefore, the easier it is to form the tetrahedral intermediate. *para*-Chlorophenol is a stronger acid than phenol so the conjugate base of *para*-chlorophenol is a weaker base than the conjugate base of phenol.

b. The tetrahedral intermediate collapses by eliminating the OR group. The weaker the OR group is as a base, the easier it is to eliminate it.

Thus, the rate of both formation of the tetrahedral intermediate and collapse of the tetrahedral intermediate is decreased by increasing the basicity of the OR group.

60. Because acetate ion is a weak base, the S_N2 reaction will form only a substitution product (see page 468 of the text). The product of the S_N2 reaction is an ester which, when hydrolyzed, forms cyclohexanol (the target molecule) and acetic acid.

61. **a.** Methyl acetate has a resonance contributor that butanone does not have. This resonance contributor causes methyl acetate to be more polar than butanone. Because methyl acetate is more polar, it has the greater dipole moment.

b. Because it is more polar, the intermolecular forces holding methyl acetate molecules together are stronger, so we would expect methyl acetate to have a higher boiling point.

62. Propyl formate is easy to distinguish because it is the only ester that will show four signals. The other three esters each show three signals.

Isopropyl formate can be distinguished by its unique splitting pattern: a singlet, a doublet, and a septet. The splitting patterns of the other two esters are the same: a singlet, a triplet, and a quartet. They can be distinguished because the highest frequency signal is due to the protons attached to the carbon that is next to the oxygen. For ethyl acetate it will be a quartet, whereas for methyl propionate it will be a singlet.

4 signals

3 signals

singlet, doublet, septet

3 signals

singlet, triplet, quartet

The signal farthest downfield is a quartet.

3 signals

singlet, triplet, quartet

The signal farthest downfield is a singlet.

63. The reaction of methylamine with propionyl chloride generates a proton that will protonate unreacted amine, thereby destroying the amine's nucleophilicity. If two equivalents of CH_3NH_2 are used, one equivalent will remain unprotonated and be able to react as a nucleophile with propionyl chloride to form *N*-methylpropanamide.

64.   a.

b. The carboxyl oxygen will be labeled. Only one isotopically labeled oxygen can be incorporated into the ester because the bond between the methyl group and the labeled oxygen does not break, so there is no way for the carbonyl oxygen to become labeled.

c. In the presence of an acid catalyst, the ester will be hydrolyzed to a carboxylic acid and an alcohol. Both oxygen atoms of the carboxylic acid will be labeled for the same reason that both oxygen atoms of the carboxylic acid are labeled in **a**. The alcohol will not contain any label, because the bond between the methyl group and the oxygen does not break.

65.   The carbonyl IR absorption band decreases in the order:

$$\underset{CH_3\quad Cl}{\text{C}=O} \;>\; \underset{CH_3\quad OCH_3}{\text{C}=O} \;>\; \underset{CH_3\quad H}{\text{C}=O} \;>\; \underset{CH_3\quad NH_2}{\text{C}=O}$$

When Y = Cl, delocalization of chlorine's lone pair onto the carbonyl oxygen is minimal, because of the poor overlap of the large $3p$ orbital of chlorine with the small $2p$ orbital of carbon. Thus, the C=O group of the acyl chloride has the highest IR stretching frequency.

The predominant effect of the oxygen of an ester is inductive electron withdrawal, so its carbonyl group stretches at a higher frequency than the carbonyl group of the aldehyde (see page 620 of the text).

The carbonyl group of the amide stretches at the lowest frequency, because the nitrogen is less electronegative than oxygen so it is better able to accommodate a positive charge. Therefore, the resonance contributor on the right makes a larger contribution to the overall structure. The larger the contribution from the resonance contributor on the right, the greater the single bond character of the C=O group.

66.   **a.** $CH_3COOH \xrightarrow{SOCl_2} CH_3COCl \xrightarrow{CH_3CH_2CH_2OH} CH_3COOCH_2CH_2CH_3$

$CH_3COO^- \xrightarrow{CH_3CH_2CH_2Br} CH_3COOCH_2CH_2CH_3$

b. $CH_3COOH \xrightarrow{SOCl_2} CH_3COCl \xrightarrow{(CH_3)_2CH(CH_2)_2OH} CH_3COO(CH_2)_2CH(CH_3)_2$

$CH_3COO^- \xrightarrow{(CH_3)_2CH(CH_2)_2Br} CH_3COO(CH_2)_2CH(CH_3)_2$

c. $CH_3CH_2CH_2COOH \xrightarrow{SOCl_2} CH_3CH_2CH_2COCl \xrightarrow{CH_3CH_2OH} CH_3CH_2CH_2COOCH_2CH_3$

$CH_3CH_2CH_2COO^- \xrightarrow{CH_3CH_2Br} CH_3CH_2CH_2COOCH_2CH_3$

d.

67. **a.** isopropyl alcohol and HCl   **c.** ethylamine

b. aqueous sodium hydroxide   **d.** water and HCl

68. **a.**

b.   **d.** CH₃CH₂

c. no reaction; an amide
will not react with water
without a catalyst

69. The offset in the NMR spectrum shows that there is a broad signal at ~ 10 ppm, which is characteristic of a COOH group. The two triplets and the multiplet are characteristic of a propyl group.

$$-CH_2-CH_2-CH_2-$$
$$t \quad\quad m \quad\quad t$$

Therefore, the compound is butanoic acid.

The molecular formula shows that the unknown reactant has one more carbon than butanoic acid; the IR spectrum shows it is an ester. Since butanoic acid is formed from acid hydrolysis of the ester, the compound must be the methyl ester of butanoic acid.

70. Aspartame has an amide group and an ester group that will be hydrolyzed in an aqueous solution of HCl. Because the hydrolysis is carried out in an acidic solution, the carboxylic acid groups and the amino groups in the products will be in their acidic forms.

71.   **a.**   1, 3, 4, 6, 7, 9 will not form the indicated products under the given conditions.
b.   9 will form the product shown in the presence of an acid catalyst.

72. The tertiary amine is a better nucleophile than the alcohol, so formation of the amide with a positively charged nitrogen will be faster than formation of the new ester would have been. The positively charged amide is more reactive than an ester, so formation of the new ester by reaction of the alcohol with the charged amide will be faster than formation of the ester by reaction of the alcohol with the starting ester would have been. In other words, both reactions that occur in the presence of the tertiary amine are faster than the single reaction that occurs in the absence of the tertiary amine.

73. 1-Bromobutane undergoes an S_N2 reaction with NH_3 to form butylamine (**A**). Butylamine can then react with 1-bromobutane to form dibutylamine (**B**). The amines each form an amide upon reaction with acetyl chloride. The IR spectrum of **C** exhibits an NH stretch at about ~ 3300 cm^{-1}, whereas the IR spectrum of **D** does not exhibit this absorption band, because the nitrogen in **D** is not bonded to a hydrogen.

7.   **a.** Two stereoisomers are obtained, because the reaction creates an asymmetric center in the product.

CH$_3$COCH$_2$CH$_2$CH$_3$ $\xrightarrow[\text{2. H}_3\text{O}^+]{\text{1. CH}_3\text{CH}_2\text{MgBr}}$

(S)-3-methyl-3-hexanol   (R)-3-methyl-3-hexanol

b. Only one compound is obtained, because the product does not have an asymmetric center.

CH$_3$COCH$_2$CH$_2$CH$_3$ $\xrightarrow[\text{2. H}_3\text{O}^+]{\text{1. CH}_3\text{MgBr}}$ CH$_3$CCH$_2$CH$_2$CH$_3$ with OH and CH$_3$

8.   **a.** Solved in the text.

b.   **1.** Solved in the text.

2. CH$_3$COOR + 2 CH$_3$MgBr

4. CH$_3$COOR + 2 CH$_3$CH$_2$MgBr

6. CH$_3$COOR + 2 ⬡—MgBr

9.   If a secondary alcohol is formed from the reaction of a formate ester with excess Grignard reagent, the two alkyl substituents of the alcohol will be identical because they both come from the Grignard reagent. Therefore, only the following two alcohols (**B** and **D**) can be prepared that way.

CH$_3$CHCH$_3$ with OH   CH$_3$CH$_2$CHCH$_2$CH$_3$ with OH

10.

11.   **A** and **C** will not undergo nucleophilic addition with a Grignard reagent. **A** has an H bonded to a nitrogen and **C** has an H bonded to an oxygen; these acidic hydrogens will react with the Grignard reagent, converting it into an alkane, before the Grignard reagent has a chance to react as a nucleophile.

B will undergo a nucleophilic addition-elimination reaction with the Grignard reagent to form a ketone. This will be followed by a nucleophilic addition reaction of the ketone with another equivalent of the Grignard reagent. The product will be a tertiary alcohol.

12.   **a.**

b.

c.

13.   **a.** $HC\equiv CH \xrightarrow{NaNH_2} HC\equiv C^- \xrightarrow{CH_3CH_2CH} HC\equiv CCHCH_2CH_3 \xrightarrow{HCl} HC\equiv CCHCH_2CH_3$

b. $HC\equiv CH \xrightarrow{NaNH_2} HC\equiv C^- \xrightarrow{CH_3Br} HC\equiv CCH_3 \xrightarrow{NaNH_2} {}^-C\equiv CCH_3$

c. HC≡CH $\xrightarrow{\text{NaNH}_2}$ HC≡C⁻ $\xrightarrow{\text{CH}_3\text{CH}_2\text{Br}}$ HC≡CCH₂CH₃ $\xrightarrow{\text{NaNH}_2}$ ⁻C≡CCH₂CH₃

$$\text{CH}_3\overset{\overset{\displaystyle O}{\|}}{\text{C}}\text{CH}_3 \downarrow$$

$$\underset{\underset{\displaystyle \text{CH}_3}{|}}{\overset{\overset{\displaystyle \text{OH}}{|}}{\text{CH}_3\text{CC}}}\text{≡CCH}_2\text{CH}_3 \quad \xleftarrow{\text{HCl}} \quad \underset{\underset{\displaystyle \text{CH}_3}{|}}{\overset{\overset{\displaystyle \text{O}^-}{|}}{\text{CH}_3\text{CC}}}\text{≡CCH}_2\text{CH}_3$$

Notice that ethyne must be alkylated before it reacts as a nucleophile with the carbonyl compound. The reaction that follows shows that if ethyne is alkylated after nucleophilic addition to the ketone, alkylation on oxygen will compete with alkylation on carbon.

$$^-\text{C}≡\text{CH} \xrightarrow{\text{CH}_3\overset{\overset{\displaystyle O}{\|}}{\text{C}}\text{CH}_3} \underset{\underset{\displaystyle \text{CH}_3}{|}}{\overset{\overset{\displaystyle \text{O}^-}{|}}{\text{CH}_3\text{CC}}}\text{≡CH} \xrightarrow{\text{NaNH}_2} \underset{\underset{\displaystyle \text{CH}_3}{|}}{\overset{\overset{\displaystyle \text{O}^-}{|}}{\text{CH}_3\text{CC}}}\text{≡C}^- \xrightarrow{\text{CH}_3\text{CH}_2\text{Br}} \underset{\underset{\displaystyle \text{CH}_3}{|}}{\overset{\overset{\displaystyle \text{OCH}_2\text{CH}_3}{|}}{\text{CH}_3\text{CC}}}\text{≡CCH}_2\text{CH}_3$$

14. Similar to the way a Grignard reagent reacts with an ester, an acetylide ion first undergoes a nucleophilic addition-elimination reaction with the ester to form a ketone; this is followed by a nucleophilic addition reaction to form a tertiary alcohol.

$$\underset{\text{R}}{\overset{\overset{\displaystyle O}{\|}}{\text{C}}}\text{OR} \xrightarrow{\text{RC}≡\text{C}^-} \text{R}-\underset{\underset{\displaystyle \underset{\displaystyle \text{R}}{\overset{\displaystyle |||}{\underset{\displaystyle \text{C}}{\text{C}}}}}{\overset{\overset{\displaystyle \text{O}^-}{|}}{\text{C}}}-\text{OR} \longrightarrow \underset{\text{R}}{\overset{\overset{\displaystyle O}{\|}}{\text{C}}}\text{C}≡\text{CR} \xrightarrow[\text{2. pyridinium chloride}]{\text{1. RC}≡\text{C}^-} \text{R}-\underset{\underset{\displaystyle \underset{\displaystyle \text{R}}{\overset{\displaystyle |||}{\underset{\displaystyle \text{C}}{\text{C}}}}}{\overset{\overset{\displaystyle \text{OH}}{|}}{\text{C}}}-\text{C}≡\text{CR}$$

$$\text{+ RO}^- \qquad\qquad\qquad\qquad\qquad\qquad\qquad \text{a tertiary alcohol}$$
$$\text{a ketone}$$

Notice that the alkoxide ion is protonated by pyridinium chloride rather than by HCl or H_3O^+ that would add to the triple bonds.

15. The reaction is carried out with excess cyanide ion in order to have some unprotonated cyanide ion to act as a nucleophile. HCl is a strong acid, so it dissociates completely and protonates the cyanide ion. Thus, there is no HCl in the reaction mixture. So HCN is the only acid available to protonate the alkoxide ion.

$$\text{HCl} + {}^-\text{C}≡\text{N} \longrightarrow \text{HC}≡\text{N} + \text{Cl}^-$$

16. No, an acid must be present in the reaction mixture in order to protonate the oxygen of the cyanohydrin. Otherwise the cyano group will be eliminated and the reactants will reform.

17. Strong acids like HCl and H_2SO_4 have very weak conjugate bases (Cl^- and HSO_4^-), which are excellent leaving groups. When these bases add to the carbonyl group, they are readily eliminated, reforming the starting materials. Cyanide ion is a strong enough base to not be eliminated unless the oxygen in the product is negatively charged.

18. Solved in the text.

19.

1. $CH_3CH_2Br \xrightarrow[Et_2O]{Mg} CH_3CH_2MgBr \xrightarrow[2.\ H_3O^+]{1.\ CO_2} CH_3CH_2\overset{\displaystyle O}{\underset{}{C}}-OH$

2. $CH_3CH_2Br \xrightarrow{^-C\equiv N} CH_3CH_2C\equiv N \xrightarrow[\Delta]{HCl,\ H_2O} CH_3CH_2\overset{\displaystyle O}{\underset{}{C}}-OH$

20.   **a.** $CH_3\overset{\displaystyle}{\underset{\displaystyle CH_3}{CH}}CH_2OH$ **b.** cyclohexyl—OH   **c.** $(CH_3)_3C$—cyclohexyl—OH   **d.** phenyl—$\overset{\displaystyle OH}{\underset{}{CH}}CH_3$

21.   **a.** $CH_3CH_2CH_2CH_2OH \ + \ CH_3CH_2OH$ **c.** benzyl—$CH_2OH \ + \ CH_3OH$

b. benzyl—CH_2OH **d.** $CH_3CH_2CH_2CH_2CH_2OH$

22.   **a.** $H\overset{\displaystyle O}{\underset{}{C}}NHCH_2$—phenyl   **or**   phenyl—$\overset{\displaystyle O}{\underset{}{C}}NHCH_3$ **c.** $CH_3\overset{\displaystyle O}{\underset{}{C}}NHCH_2CH_3$

b. $CH_3\overset{\displaystyle O}{\underset{}{C}}NH_2$ **d.** $CH_3\overset{\displaystyle O}{\underset{\displaystyle CH_2CH_3}{C}}NCH_2CH_3$

23.   **a.** **1.** $LiAlH_4$ **2.** H_2O

b. HCl, H_2O, Δ

c. **1.** HCl, H_2O, Δ **2.** $LiAlH_4$ **3.** HCl

24.   **a.** phenyl—CH_2OH **c.** cyclohexyl—OH

b. $CH_3CH_2CH_2CH_2NH_2$ **d.** no reaction

25.

26.   The multiply bonded carbons of alkenes and alkynes do not possess a partial positive charge so they do not react with the negatively charged hydride ion. The multiply bonded carbons of imines and nitriles do react with hydride ion because the carbons possess a partial positive charge since they are bonded to a relatively electron-withdrawing nitrogen.

27.   **a.**   **c.**

   b.   **d.**

   $+$ CH_3OH

28.   The nitrile reacts with the Grignard reagent to form an imine, which can then be hydrolyzed to a ketone.

29.   The OH group withdraws electrons inductively and this makes the ammonium ion a stronger acid.

30.   Figure 17.2 for imine formation when the amine is hydroxylamine (pK_a of its conjugate acid $= 6.0$) shows that the maximum rate is obtained when the pH is about 1.5 units lower than the pK_a of the amine's conjugate acid. Thus for an amine such as ethylamine whose conjugate acid has a $pK_a \sim 11$, imine formation should be carried out at about pH $= 9.5$.

31. For the derivations of the equations used to calculate the amount of a compound that is present in either its acidic or its basic form, refer to Special Topic I in this study guide.

a. fraction present in the acidic form $= \dfrac{[H^+]}{K_a + [H^+]}$

$$\dfrac{[H^+]}{K_a + [H^+]} = \dfrac{3.2 \times 10^{-5}}{3.2 \times 10^7 + 3.2 \times 10^{-5}}$$

$$= \dfrac{3.2 \times 10^{-5}}{3.2 \times 10^7}$$

$$= 1 \times 10^{-12}$$

b. fraction present in the acidic form $= \dfrac{[H^+]}{K_a + [H^+]}$

$$\dfrac{[H^+]}{K_a + [H^+]} = \dfrac{3.2 \times 10^{-2}}{3.2 \times 10^7 + 3.2 \times 10^{-2}}$$

$$= \dfrac{3.2 \times 10^{-2}}{3.2 \times 10^7}$$

$$= 1 \times 10^{-9}$$

c. fraction present in the basic form $= \dfrac{K_a}{K_a + [H^+]}$

$$\dfrac{K_a}{K_a + [H^+]} = \dfrac{1.0 \times 10^{-6}}{1.0 \times 10^{-6} + 3.2 \times 10^{-2}}$$

$$= \dfrac{1.0 \times 10^{-6}}{3.2 \times 10^{-2}}$$

$$= 3.1 \times 10^{-3}$$

32. **a.**   **b.**

33.   **a.**   Notice that both of these mechanisms have a feature that we have seen in many mechanisms—formation of three tetrahedral intermediates: first a protonated tetrahedral intermediate, then a neutral tetrahedral intermediate, and then a second protonated tetrahedral intermediate.

1.

a protonated tetrahedral intermediate

neutral tetrahedral intermediate

protonated tetrahedral intermediate

2.

β-carbon

b.   The only difference is the first step of the mechanism: in imine hydrolysis, the acid protonates the nitrogen; in enamine hydrolysis, the acid protonates the β-carbon.

34.   **a.**

cyclopentanone + $CH_3CH_2NH_2$ $\underset{}{\overset{\text{trace } H^+}{\rightleftharpoons}}$ =NCH_2CH_3 + H_2O

b.

cyclopentanone + CH_3CH_2NH (with CH_2CH_3) $\underset{}{\overset{\text{trace } H^+}{\rightleftharpoons}}$ enamine—N(with CH_2CH_3, CH_2CH_3) + H_2O

c.

Ph—C(=O)—CH_3 + $CH_3(CH_2)_5NH_2$ $\underset{}{\overset{\text{trace } H^+}{\rightleftharpoons}}$ $\text{Ph—C(CH}_3\text{)=N(CH}_2)_5CH_3$ + H_2O

d.

Ph—C(=O)—CH_3 + cyclohexyl—NH_2 $\underset{}{\overset{\text{trace } H^+}{\rightleftharpoons}}$ $\text{Ph—C(CH}_3\text{)=N—cyclohexyl}$ + H_2O

35.   A tertiary amine will be obtained, because the primary amine synthesized in the first part of the reaction by reductive animation will react with the excess carbonyl compound, forming an imine that will be reduced to a secondary amine. The secondary amine will then react with the carbonyl compound, forming an enamine that will be reduced to a tertiary amine.

$$R_2C{=}O + NH_3 \xrightarrow{NaBH_3CN} R_2CHNH_2 \xrightarrow[NaBH_3CN]{R_2C=O} R_2CHNHCHR_2 \xrightarrow[NaBH_3CN]{R_2C=O} (R_2CH)_3N$$

primary amine   secondary amine   tertiary amine

36.   **a.**

b.

37.

38.   Because an electron-withdrawing substituent decreases the stability of a ketone and increases the stability of a hydrate, the compound with the electron-withdrawing nitro substituents has the largest equilibrium constant for addition of water.

39. The electron-withdrawing chlorines decrease the stability of the aldehyde (and therefore decrease its concentration), increasing the equilibrium constant for hydrate formation compared to, for instance, acetaldehyde.

$$K_{eq} = \frac{[\text{hydrate}]}{[\text{aldehyde}][H_2O]}$$

the concentration ⟶ the concentration is decreased

40. **a.** hemiacetals: 1, 7, 8

b. acetals: 2, 3, 5

c. hydrates: 4, 6

41. **a.** Hemiacetals are unstable in basic solution, because the base can remove a proton from an OH group, thereby providing an oxyanion that can expel the OR group (since the charge that develops on the oxygen in the transition state is negative) and form a stable aldehyde.

b. In order for an acetal to form, the CH_3O group in the hemiacetal must eliminate an OH group. Hydroxide ion is too basic to be eliminated by a CH_3O group (since the charge that develops on the oxygen in the transition state is positive) but water, a much weaker base, can be eliminated by a CH_3O group. In other words, the OH group must be protonated before it can be eliminated by a CH_3O group. Therefore, acetal formation must be carried out in an acidic solution.

c. Hydrate formation can be catalyzed by hydroxide ion because a group does not have to be eliminated after hydroxide ion attacks the aldehyde or ketone.

42. When a tetrahedral intermediate collapses, the intermediate that is formed is very unstable because of the positive charge on the sp^2 oxygen atom. In the case of an acetal, the only way to form a neutral species is to reform the acetal.

acetal   unstable

In the case of a hydrate, a neutral species can be formed by loss of a proton.

hydrate   unstable

43. a.

both groups would
be reduced

b. $NaBH_4$ because it is not strong enough a reducing agent to reduce the less reactive ester

44. An acetal has a very poor leaving group (e.g., CH_3O^-).

45. If the OH group of the carboxylic acid is not protected, both OH groups will be converted to COOH groups.

46.   If the yield of each step is 80%, the yield of **B** is 80%; the yield of **C** from **B** is 80% times the yield of **B**; the yield of **D** from **C** is 80% times the yield of **C**; and so on.

a.   A $\longrightarrow$ B $\longrightarrow$ C $\longrightarrow$ D $\longrightarrow$ E $\longrightarrow$ F $\longrightarrow$ G

   80%   0.80 × 0.80   0.80 × 0.64   0.80 × 0.51   0.80 × 0.41   0.80 × 0.33

            64%   51%   41%   33%   26%

b.   G $\longrightarrow$ H $\longrightarrow$ I

   26%   0.80 × 0.26   0.80 × 0.21

              21%   17%

47.   **a.**

b.

c. The carbonyl group has to be protected to prevent it from reacting with the Grignard reagent. This synthesis is easier to understand if the CO_2 needed in the third step is written as $O=C=O$.

48.   a.

c.

b.

d.

49.

a.

1.   Solved in the text.

3. $(C_6H_5)_2C=O$ + $CH_3CH=P(C_6H_5)_3$

or $(C_6H_5)_2C=P(C_6H_5)_3$ + $CH_3CH=O$

2. $=O$ + $CH_3CH_2CH=P(C_6H_5)_3$

4. $-CH=O$ + $CH_2=P(C_6H_5)_3$

or $=P(C_6H_5)_3$ + $CH_3CH_2CH=O$

or $-CH=P(C_6H_5)_3$ + $CH_2=O$

b.

1. Solved in the text.   2. $CH_3CH_2CH_2Br$ 3. CH_3CH_2Br 4. CH_3Br

or   or   or

—Br $(C_6H_5)_2CHBr$ —CH_2Br

c. The preferred pair of reagents is the pair that has the ylide synthesized from the less sterically hindered alkyl halide.

1. Solved in the text.   **2.** $CH_3CH_2CH=P(C_6H_5)_3$ **3.** $CH_3CH=P(C_6H_5)_3$ **4.** $CH_2=P(C_6H_5)_3$

\+   \+   \+

=O $(C_6H_5)_2C=O$ —CH=O

50.   a.

The above synthesis of cyclohexanol will also form an elimination product (cyclohexene).
The following procedure will give a greater yield cyclohexanol, because the substitution product is favored over the elimination product by a weak base (see page 468 of the text). The substitution product can be hydrolyzed to the target molecule.

b.

c.

or

d.

e.

$$(C_6H_5)_3P{=}O \quad + $$

f.

For a review of why 1-ethylcyclohexene is the final product of the dehydration, see page 495 in the text.

51.

a.

weak base favors
conjugate addition

b.

steric hindrance
at the β-carbon
favors direct addition

c. $CH_3C{=}CHCCH_3$ (with OH, CH₃, CH₃ substituents)

strong base favors
direct addition

d. $CH_3CH{=}CHCH_2OH$

strong base favors
direct addition

52.   Conjugate addition occurs in **a** and **c** because the nucleophile is a relatively weak base. Nucleophilic addition-elimination occurs in **b** because the cabonyl group is very reactive. Because excess base is used in part **d**, both nucleophilic addition-elimination and conjugate addition occur.

a.
$$CH_3CHCH_2 \overset{\overset{\displaystyle O}{\displaystyle \|}}{\underset{}{C}} OCH_3$$
$$\underset{Br}{|}$$

c.
$$CH_3CHCH_2 \overset{\overset{\displaystyle O}{\displaystyle \|}}{\underset{}{C}} OCH_3$$
$$\underset{NH_2}{|}$$

b.
$$CH_3CH{=}CH \overset{\overset{\displaystyle O}{\displaystyle \|}}{\underset{}{C}} OCH_3$$

d.
$$CH_3CHCH_2 \overset{\overset{\displaystyle O}{\displaystyle \|}}{\underset{}{C}} NH_2$$
$$\underset{NH_2}{|}$$

53.

a.
$$CH_3CH \overset{\overset{\displaystyle O}{\displaystyle \|}}{\underset{}{C}} H$$
$$\underset{CH_3}{|}$$

d.

g.
$$CH_3CH_2CHCH_2CH_2 \overset{\overset{\displaystyle O}{\displaystyle \|}}{\underset{}{C}} H$$
$$\underset{Br}{|}$$

b.
$$CH_3CH{=}CHCH_2CH_2 \overset{\overset{\displaystyle O}{\displaystyle \|}}{\underset{}{C}} H$$

e.
$$CH_3 \overset{\overset{\displaystyle O}{\displaystyle \|}}{\underset{}{C}} CH_2 \overset{\overset{\displaystyle O}{\displaystyle \|}}{\underset{}{C}} CH_3$$

h.

c.
$$CH_3CHCH_2CH_2 \overset{\overset{\displaystyle O}{\displaystyle \|}}{\underset{}{C}} CH_2CH_2CHCH_3$$
$$\underset{CH_3}{|} \qquad\qquad \underset{CH_3}{|}$$

f.
$$CH_3CH_2 \overset{\overset{\displaystyle O}{\displaystyle \|}}{\underset{}{C}} CHCH_2CH_2CH_3$$
$$\underset{Br}{|}$$

i.
$$CH_3 \overset{\overset{\displaystyle O}{\displaystyle \|}}{\underset{}{C}} CHCH_2CH_2 \overset{\overset{\displaystyle O}{\displaystyle \|}}{\underset{}{C}} H$$
$$\underset{CH_3}{|}$$

54.   **a.**
$$CH_3CH_2O \underset{CH_3CH_2}{\overset{OCH_2CH_3}{\underset{|}{\overset{|}{C}}}} H$$

e. $CH_3CH_2CH_2CH_2OH \quad + \quad CH_3CH_2OH$

b.

f.

c.
$$CH_3CH_2\underset{\overset{|}{OH}}{CH}CH_3$$

d.
$$CH_3CH_2\underset{\underset{\underset{N}{\|}}{\underset{C}{|}}}{\overset{OH}{\overset{|}{C}}}CH_2CH_3$$

g.

55. The greater the steric hindrance at the site of nucleophilic addition, the less reactive the carbonyl compound.

56. a.

b. The reaction of an organocuprate with ethylene oxide and the hydroboration/oxidation of an alkene both lead to $R'CH_2CH_2OH$. Because this alcohol does not have a methyl group on the β-carbon, these two methods cannot be used to synthesize isobutyl alcohol.

$$R'CH_2CH_2OH \text{ cannot lead to } CH_3\underset{\underset{CH_3}{|}}{C}HCH_2OH$$

57.

a. CH$_3$CH$_2$CH$_2$ C=C H / CH$_3$, H
reduction

b. CH$_3$CH$_2$CH$_2$NHCH$_3$ reduction

c. [benzene ring]—CH$_2$CH$_2$CH$_3$
reduction

d. [benzene ring]—CH$_2$OH   reduction

e. [benzene ring]—CH$_2$OH   +   CH$_3$OH
reduction

f. [benzene ring]—C(=O)—OH   oxidation

58.

a. [cyclohexanone] $\xrightarrow[\text{2. H}_3\text{O}^+]{\text{1. NaBH}_4}$ [cyclohexanol]

b. [cyclohexanone] $\xrightarrow[\text{2. H}_3\text{O}^+]{\text{1. NaBH}_4}$ [cyclohexanol] $\xrightarrow[\Delta]{\text{H}_2\text{SO}_4}$ [cyclohexene]

c. [cyclohexanone] $\xrightarrow[\text{2. H}_3\text{O}^+]{\text{1. NaBH}_4}$ [cyclohexanol] $\xrightarrow[\text{pyridine}]{\text{PBr}_3}$ [cyclohexyl bromide]

d. [cyclohexanone] + NH$_3$ (excess) $\xrightarrow[\text{Pd/C}]{\text{H}_2}$ [cyclohexylamine NH$_2$]

e. [cyclohexyl]—Br $\xrightarrow{^-\text{C}\equiv\text{N}}$ [cyclohexyl]—C≡N $\xrightarrow[\text{Raney Ni}]{\text{H}_2}$ [cyclohexyl]—CH$_2$NH$_2$

product of **c**

f. [cyclohexanone] + CH$_3$NH—CH$_3$ $\xrightarrow[\text{acid}]{\text{trace}}$ [cyclohexene]—N(CH$_3$)$_2$ $\xrightarrow{\text{NaBH}_3\text{CN}}$ [cyclohexyl]—N(CH$_3$)$_2$

g.

product of **c**

or

product of **c**

h.

1. HSCH$_2$CH$_2$CH$_2$SH/HCl
2. H$_2$, Raney Ni

or

1. NaBH$_4$
2. H$_3$O$^+$

H$_2$SO$_4$
Δ

H$_2$
Pd/C

i.

1. CH$_3$CH$_2$MgBr
2. H$_3$O$^+$

H$_2$SO$_4$
Δ

H$_2$
Pd/C

or

+ CH$_3$CH=P(C$_6$H$_5$)$_3$

H$_2$
Pd/C

59.

HO ... H—B⁺ ⇌ HÖ ... ⁺ÖH H ⇌ OH O⁺—H :B

HB⁺

OCH₃ ⇌ B: H OCH₃ CH₃ÖH O⁺ ⇌ H +OH Ö: ⇌ :ÖH H—B⁺ OH O

+ H₂O

60.

a.

$$\xrightarrow{H_2CrO_4}$$

$$\xrightarrow{SOCl_2 \atop pyridine}$$

1. NaBH₄ 2. H₃O⁺

CH₃CH₂CH₂OH

b.

OH $$\xrightarrow[\substack{CH_3COOH \\ 0\ °C}]{NaOCl}$$

1. CH₃CH₂MgBr 2. H₃O⁺

OH

NaOCl | CH₃COOH 0 °C

c. OH $$\xrightarrow[\substack{CH_3COOH \\ 0\ °C}]{NaOCl}$$ O

1. CH₃NH₂ 2. NaBH₃CN

NHCH₃

d. O

1. NaBH₄ 2. H₃O⁺

OH $$\xrightarrow[\Delta]{H_2SO_4}$$

1. O₃, −78 °C 2. (CH₃)₂S

H ... H (O, O)

H₂CrO₄

HO ... OH (O, O)

e.

61. **a.** $CH_3OH \xrightarrow[\text{pyridine}]{PBr_3} CH_3Br \xrightarrow[\text{2. CuI}]{\text{1. Li}} (CH_3)_2CuLi \xrightarrow[\text{2. HCl}]{\text{1. H}_2C=O} CH_3CH_2OH$

b. $CH_4 \atop \text{excess} \xrightarrow[hv]{Br_2} CH_3Br \xrightarrow[\text{2. CuI}]{\text{1. Li}} (CH_3)_2CuLi \xrightarrow[\text{2. HCl}]{\text{1. ethylene oxide}} CH_3CH_2CH_2OH$

62. **a.**

$+ \ CH_3CH_2\overset{+}{N}H_3$

c.

e.

b. $CH_3CH_2\overset{\displaystyle OH}{\underset{\displaystyle CH_2CH_3}{C}}CH_3$

d. $CH_3CH_2\overset{\displaystyle OH}{\underset{\displaystyle CH_2CH_3}{C}}CH_2CH_3$

f.

g.

i. CH_3NH

h.

j. CH_3O

63.

64.

a.

b.   CH₃CH₂CH₂CH₂CH₂SH

65. The offset shows there is a signal at ~10.0 ppm, indicating an aldehyde. The ^{1}H NMR spectrum is that of benzaldehyde. Phenylmagnesium bromide, therefore, must react with a compound with one carbon to form an alcohol that can be oxidized to benzaldehyde. Therefore, Compound **Z** must be formaldehyde.

66. **a.**

b.

c.

67.   **a.**   three signals in its ^{1}H NMR spectrum   **b.**   three signals in its ^{13}C NMR spectrum

68.

69.   **a.**

b.

c.

d.

70. In parts **a**, **b**, and **d**, a new asymmetric center is formed in the product, so a racemic mixture is formed. In part **c**, both *E* and *Z* stereoisomers are formed.

a.

b. HO   CH$_3$ HO   CH$_3$

c.

d.

71. a. CH$_3$CH$_2$ C CH$_2$CH$_2$CH$_2$CH$_3$ + —MgBr

CH$_2$CH$_3$ + CH$_3$CH$_2$CH$_2$CH$_2$MgBr

CH$_2$CH$_2$CH$_2$CH$_3$ + CH$_3$CH$_2$MgBr

b.

$$\underset{\substack{\text{O} \\ \| \\ \text{C}}}{CH_3CH_2 \quad CH_2CH_2CH_3} \quad + \quad CH_3CH_2MgBr$$

$$\underset{\substack{\text{O} \\ \| \\ \text{C}}}{CH_3CH_2 \quad CH_2CH_3} \quad + \quad CH_3CH_2CH_2MgBr$$

$$\underset{\substack{\text{O} \\ \| \\ \text{C}}}{CH_3CH_2CH_2 \quad OCH_2CH_3} \quad + \quad 2\ CH_3CH_2MgBr$$

72.   **a.**   **b.**   **c.**   **d.**

73.   Enovid would have its carbonyl stretch at a higher frequency. The carbonyl group in Norlutin has some single-bond character because of electron delocalization as a result of having conjugated double bonds. The single-bond character causes the carbon-oxygen bond to be easier to stretch than the carbon-oxygen bond in Enovid, which has isolated double bonds and therefore no electron delocalization.

Norlutin

74.   Notice that ketal formation can be used to protect either ketones or 1,2-diols. Thus in part **a**, the 1,2-diol is protected using acetone, whereas in part **b**, 1,2-ethanediol is used to protect a ketone.

a.   **b.**

75.   **a.**   **b.**

c.

d.

76.   a.

b.

77.

dimethylchlorosulfonium ion

78.   The absorption bands at 1600 cm^{-1}, 1500 cm^{-1}, and >3000 cm^{-1} in the IR spectrum indicate that the compound has a benzene ring. The absorption band at 1720 cm^{-1} suggests that it is an aldehyde or a ketone and the carbonyl group is not conjugated with the benzene ring. The absorption bands at 2720 cm^{-1} and 2820 cm^{-1} indicates that the compound is an aldehyde.

The broad signal at 1.8 ppm in the ^{1}H NMR spectrum of the product indicates an OH group. We also see there are three different kinds of hydrogens (other than the OH group and the benzene ring hydrogens). The doublet at ~1.2 ppm that integrates to 3 protons indicates a methyl group adjacent to a carbon bonded to one hydrogen. The doublet at 3.6 ppm that integrates to 2 protons is due a methylene group bonded to a CH group. The chemical shift of this signal suggests that the electron-withdrawing OH group is also bonded to this methylene group.

The IR spectrum is the spectrum of 2-phenylpropanal and the ^{1}H NMR spectrum is the spectrum of 2-phenyl-1-propanol.

79.   In the first step of each reaction, cyanide is eliminated because it is a weak base. In the second step, cyanide ion adds to the β-carbon, again, because it is a weak base.

80.   We have seen that the Wittig reaction proceeds via a concerted $[2+2]$ cycloaddition reaction.

In contrast, the sulfonium ylide adds to the carbonyl group to form the intermediate shown here. The intermediate has a nucleophilic (an oxyanion) that attacks the methylenecarbon because $(CH_3)_2S$ is a weak base and, therefore, a very good leaving group.

81.   **a.**

b.

The synthesis can also be carried out in one step.

c.

or

d.

e.

82.   **a.**

b.

83.   **a.**   The OH group on C-5 of glucose reacts with the aldehyde group in an intramolecular reaction, forming a cyclic hemiacetal. Because the reaction creates a new asymmetric center, two cyclic hemiacetals can form, one with the *R* configuration at the new asymmetric center and one with the *S* configuration.

b.   The two products can be drawn in their chair conformations by putting the largest group (CH_2OH) in the equatorial position and then putting the other groups in axial or equatorial positions depending on whether they are cis or trans to one another. For example, the OH attached to the carbon that is next to the carbon attached to the CH_2OH group is trans to the CH_2OH group. The hemiacetal on the left has all but one of its OH groups in the equatorial position, whereas the hemiacetal on the right has all its OH groups in the equatorial position. Since a compound can avoid unfavorable 1,3-diaxial interactions by having its substituents in equatorial positions, the hemiacetal on the right is more stable.

less stable   more stable

84.   The alkyl bromide is 1-bromo-2-phenylethane. The molecular formula of the product of the Wittig reaction indicates that the ketone that reacts with the phosphonium ylide has three carbons (that is, the ketone is acetone).

85.

86. **a.** —CH=O + $(C_6H_5)_3P$=CHCH$_2$CH$_2$CH$_3$ **or**

—CH=P(C$_6$H$_5$)$_3$ +   O=CHCH$_2$CH$_2$CH$_3$

b. + $(C_6H_5)_3P$=CHCH$_2$CH$_3$ **or**   + O=CHCH$_2$CH$_3$

c. —CH=O + $(C_6H_5)_3P$=CH—

d. =O + $(C_6H_5)_3P$=CH$_2$ **or** =P(C$_6$H$_5$)$_3$ + O=CH$_2$

87. The first step is conjugate addition of an organocuprate and the second step is direct addition of a Grignard reagent.

88.   **a.**

b.

89.   The compound that gives the ^{1}H NMR spectrum is 2-phenyl-2-butanol.

Therefore, the compound that reacts with methylmagnesium bromide is 1-phenyl-1-propanone.

90.   a.

b.

c.

91.

92.

93.   **a.**   This carboxyl group is more acidic, because its conjugate base is more stable since it is closest to the electron-withdrawing keto group. (Recall that the more stable the base, the stronger its conjugate acid.)

b.   The data show that the amount of hydrate decreases with increasing pH until about pH = 6, and that increasing the pH beyond 6 has no effect on the amount of hydrate.

A hydrate is stabilized by electron-withdrawing groups. A COOH group is electron withdrawing, but a COO⁻ group is less so. In acidic solutions, where both carboxylic acid groups are in their acidic (COOH) forms, the compound exists as essentially all hydrate. As the pH of the solution increases and the COOH groups become COO⁻ groups, the amount of hydrate decreases. Above pH = 6, where both carboxyl groups are fully in their basic (COO⁻) forms, there is no change in the amount of hydrate.

oxaloacetic acid   hydrate

94.

95. **a.** The negative ρ value obtained when hydrolysis is carried out in a basic solution indicates that electron-donating substituents increase the rate of the reaction. This means that the rate-determining step must be protonation of the carbon (the first step), because the more electron donating the substituent, the greater the partial negative charge on the carbon and the easier it will be to protonate.

b. The positive ρ value obtained when hydrolysis is carried out in an acidic solution indicates that electron-withdrawing substituents increase the rate of the reaction. This means that the rate-determining step must be nucleophilic addition of water to the iminium carbon to form the tetrahedral intermediate (the second step), because electron withdrawal increases the electrophilicity of the iminium carbon, making it more susceptible to nucleophilic addition.

96. Notice that in the mechanisms in part **a**, there is an equilibrium between a protonated intermediate, a neutral intermediate, and a second protonated intermediate, just as we saw in many previous acid-catalyzed mechanisms in this chapter and Chapter 16.

a.

b.

Chapter 17 Practice Test

1.   What is the product of each of the following reactions?

a.   (phenyl ketone) $-CCH_2CH_3$ + NH_2OH $\xrightarrow{\text{trace acid}}$

b.   (cyclopentanone)   +   (piperidine, N–H) $\xrightarrow{\text{trace acid}}$

c. CO_2 $\xrightarrow[\text{2. HCl, H}_2\text{O}]{\text{1. CH}_3\text{CH}_2\text{CH}_2\text{MgBr}}$

d.   (cyclohexanone)   + CH_3CH_2OH excess $\xrightarrow{\text{HCl}}$

e. $CH_3CH_2\overset{\displaystyle O}{\overset{\|}{C}}CH_2CH_3$ $\xrightarrow[\text{2. HCl}]{\text{1. CH}_3\text{MgBr}}$

f.   (cyclohexanone)   + NH_3 excess $\xrightarrow[\text{Pd/C}]{\text{H}_2}$

g. $CH_3CH_2\overset{\displaystyle O}{\overset{\|}{C}}CH_2CH_3$ $\xrightarrow[\text{HCl}]{^-C\equiv N}$

h. $CH_3CH=CH\overset{\displaystyle O}{\overset{\|}{C}}CH_3$ + CH_3SH $\longrightarrow$

i.   (phenyl ester) $-\overset{\displaystyle O}{\overset{\|}{C}}OCH_2CH_3$ $\xrightarrow[\text{2. HCl}]{\text{1. 2 CH}_3\text{CH}_2\text{CH}_2\text{MgBr}}$

2.   Which of the following alcohols cannot be prepared by the reaction of an ester with excess Grignard reagent?

$$\underset{\underset{CH_3}{|}}{CH_3CH_2\overset{\overset{OH}{|}}{C}CH_2CH_2CH_3} \qquad \underset{\underset{CH_3}{|}}{CH_3\overset{\overset{OH}{|}}{C}CH_2CH_3} \qquad \underset{\underset{CH_3}{|}}{CH_3\overset{\overset{OH}{|}}{C}CH_3}$$

3.   Name the following compounds:

a.

b.

c.

4.   Give an example of each the following:

   a.   an enamine

   b.   an acetal

   c.   an imine

   d.   a hemiacetal

   e.   a phenylhydrazone

5.   Which is more reactive toward nucleophilic addition?

   a.   butanal or methyl propyl ketone

   b.   2-heptanone or 3-pentanone

6.   Indicate how the following compounds could be prepared using the given starting material:

   a. $CH_3CH_2CH_2Br \longrightarrow CH_3CH_2CH_2\overset{\overset{O}{||}}{C}OCH_2CH_3$

   b. $CH_3\overset{\overset{O}{||}}{C}OCH_3 \longrightarrow \underset{\underset{OH}{|}}{CH_3CHCH_3}$

   c. $CO_2 \longrightarrow CH_3CH_2OH$

   d.

CHAPTER 18
Reactions at the α-Carbon of Carbonyl Compounds

Important Terms

acetoacetic ester synthesis	synthesis of a methyl ketone using ethyl acetoacetate as the starting material.
aldol addition	a reaction between two molecules of an aldehyde (or two molecules of a ketone) that connects the α-carbon of one with the carbonyl carbon of the other.
aldol condensation	an aldol addition followed by elimination of water.
annulation reaction	a ring-forming reaction.
α-carbon	a carbon adjacent to a carbonyl carbon.
carbon acid	a compound that contains a carbon bonded to a relatively acidic hydrogen.
Claisen condensation	a reaction between two molecules of an ester that connects the α-carbon of one with the carbonyl carbon of the other and eliminates an alkoxide ion.
condensation reaction	a reaction combining two molecules while removing a small molecule (usually water or an alcohol).
crossed aldol addition	an aldol addition using two different aldehydes or ketones.
crossed Claisen condensation	a Claisen condensation using two different esters.
decarboxylation	loss of carbon dioxide.
Dieckmann condensation	an intramolecular Claisen condensation.
β-diketone	a ketone with a second ketone carbonyl group at the β-position.
E1cB	a two-step elimination reaction that proceeds via a carbanion intermediate.
enolization	keto-enol interconversion.
gluconeogenesis	the synthesis of D-glucose from pyruvate.
glycolysis	the breakdown of D-glucose into two molecules of pyruvate.
haloform reaction	the conversion of a methyl ketone to a carboxylic acid and haloform using Cl_2 (or Br_2 or I_2) and HO^-.
Hell-Volhard-Zelinski (HVZ) reaction	the conversion of a carboxylic acid into an α-bromocarboxylic acid using $PBr_3 + Br_2$.
α-hydrogen	a hydrogen bonded to the carbon adjacent to a carbonyl carbon.

keto-enol tautomerism (keto-enol interconversion)	an interconversion of keto and enol tautomers.
β-keto ester	an ester with a ketone carbonyl group at the β-position.
Kolbe-Schmidt carboxylation reaction	a reaction of a phenolate ion with carbon dioxide under pressure.
malonic ester synthesis	synthesis of a carboxylic acid using diethyl malonate as the starting material.
Michael reaction	the addition of an α-carbanion to the β-carbon of an α,β-unsaturated carbonyl compound.
mixed aldol addition	an aldol addition using two different aldehydes or ketones (see crossed aldol addition).
mixed Claisen condensation	a Claisen condensation in which two different esters are used (see crossed Claisen condensation).
Robinson annulation	a Michael reaction followed by an intramolecular aldol condensation.
α-substitution reaction	a reaction that puts a substituent on an α-carbon in place of an α-hydrogen.
tautomers	constitutional isomers that are in rapid equilibrium; for example, keto and enol tautomers. The keto and enol tautomers differ only in the location of a double bond and a hydrogen.

b.

15.   The compound formed in 14a has no stereoisomers.
The compound formed in 14b has a new asymmetric center, so both the *R* and *S* stereoisomers will be obtained.

16.   a.

b.

c.

17.   a.

b.

$$\text{cyclohexanone} + \text{pyrrolidine} \xrightarrow[\text{acid}]{\text{trace}} \text{enamine} \xrightarrow{CH_3CH_2CCl, \ O} \text{iminium } Cl^-$$

$$\xrightarrow{HCl \mid H_2O}$$

pyrrolidinium Cl^- + 2-propanoylcyclohexanone (CCH_2CH_3)

c.

$$\text{cyclohexanone} + \text{pyrrolidine} \xrightarrow[\text{acid}]{\text{trace}} \text{enamine} \xrightarrow{CH_2=CHCCH_3, \ O} \text{iminium } Cl^-$$

$$\xrightarrow{HCl \mid H_2O}$$

pyrrolidinium Cl^- + $CH_2CH_2CCH_3$ substituted cyclohexanone

18.

a.

$$CH_3CHCH_2 \quad \underset{\|}{C} \quad NH_2$$
$$CH_3CH_2 \quad \underset{\|}{CH} \quad \underset{\|}{C} \quad OCH_3$$
$$O \qquad O$$

b.

$$CH_3CH_2CHCH_2 \quad \underset{\|}{C} \quad OCH_3$$
$$CH_3 \quad CHC\equiv N$$
$$\underset{\|}{C}$$
$$O$$

19.   a.

HO⁻

b.

CH₃CH₂O⁻

20.   a.

b.

c.

21.   a. CH₃CH₂CH₂ — C(=O) — H   **b.** CH₃ — C(=O) — CH₃   **c.**

d. CH₃CH₂ — C(=O) — CH₂CH₃

22.

aldol addition   aldol condensation

A bond has formed between the α-carbon of one
molecule and the carbonyl carbon of another.

23.   a. Solved in the text.

b. CH₃ — C(=O) — CH₃ $\xrightarrow{\text{LDA/THF}}$ ⁻:CH₂ — C(=O) — CH₃ $\xrightarrow{\text{CH}_3\text{CH}_2\text{Br}}$ CH₃CH₂CH₂ — C(=O) — CH₃

24.

a.

LDA / THF

1. add slowly

2. H$_2$O

b.

1. CH$_3$CH$_2$—CO—CH$_2$CH$_3$

add slowly

2. H$_2$O

HOCH$_2$CH(CH$_3$)—CO—CH$_2$CH$_3$

c.

LDA / THF

1.

add slowly

2. H$_2$O

Δ | H$_3$O$^+$ or HO$^-$

d.

1.

add slowly

2. H$_2$O

+ H$_2$O

25.

+

26.

27. **a.**

b.

28.   **A**, **B**, and **D** cannot undergo a Claisen condensation.
A cannot, because a proton cannot be removed from an sp^2 carbon.
B and **D** cannot, because they do not have an α-hydrogen.

29.   a.

b.

30.   a. Only one carbonyl compound has α-hydrogens, so the one with α-hydrogens is added slowly to a basic solution of the one without α-hydrogens.

b. Both carbonyl compounds have α-hydrogens, so LDA is used to form the enolate and the other carbonyl compound is added slowly.

c. Only one carbonyl compound has α-hydrogens, so the one with α-hydrogens is added slowly to a basic solution of the one without α-hydrogens.

31.

a 1,7-diester

32.   a.

b.

33.   No, because an intramolecular reaction would lead to a strained four-membered ring. Therefore, an intermolecular reaction will be preferred; the reaction will involve removal of a proton from the more acidic central carbon.

intramolecular product   intermolecular product

34.   **a.**   **b.**   **c.**   **d.**

35.   **a.**

b.

36.   **a.**   retrosynthetic analysis

synthesis

b.

retrosynthetic analysis

synthesis

$$CH_2{=}CH{-}\overset{\overset{\displaystyle O}{\|}}{C}{-}CH_2CH_3 \ + \ CH_3{-}\overset{\overset{\displaystyle O}{\|}}{C}{-}H \ \xrightarrow[\Delta]{HO^-}$$

c.

retrosynthetic analysis

synthesis

$$CH_2{=}\overset{\overset{\displaystyle CH_3}{|}}{\underset{\underset{\displaystyle CH_3}{|}}{C}}{-}\overset{\overset{\displaystyle O}{\|}}{C}{-}CH_3 \ + \ CH_3CH_2{-}\overset{\overset{\displaystyle O}{\|}}{C}{-}C_6H_5 \ \xrightarrow[\Delta]{HO^-}$$

d.

retrosynthetic analysis

$$CH_2{=}CH{-}\overset{\overset{\displaystyle O}{\|}}{C}{-}CH_2CH_3 \ + \ \ \xrightarrow[\Delta]{HO^-}$$

37.   **A** and **D** are β-keto acids so they can be decarboxylated on heating.
B cannot be decarboxylated because it does not have a carboxyl group.
The electrons left behind if **C** were decarboxylated cannot be delocalized onto an oxygen.

38.   **a.** methyl bromide   **c.** benzyl bromide
b. methyl bromide (twice)   **d.** isobutyl bromide

39.   **a.** An S_N2 reaction cannot be done on bromobenzene. (See Section 9.5 of the text.)
b. An S_N2 reaction cannot be done on vinyl bromide. (See Section 9.5 of the text.)
c. An S_N2 reaction cannot be done on a tertiary alkyl halide. (Only elimination occurs; see Table 10.6 on page 470 of the text.)

40.   **a.** ethyl bromide   **b.** pentyl bromide   **c.** benzyl bromide

41.   Solved in the text.

42.   **a.**

b.

c.

d.

43. Because the catalyst is a hydroxide ion rather than an enzyme, four stereoisomers will be formed since two asymmetric centers are created in the product.

44. **Seven** moles. The first two carbons in the fatty acid come from acetyl CoA.
Each subsequent two-carbon piece comes from malonyl CoA.
Because this amounts to 14 carbons for the synthesis of the 16-carbon fatty acid, seven moles of malonyl CoA are required.

45. **a.** **Three** deuteriums would be incorporated into palmitic acid, because only one CD₃COSR is used in the synthesis.

b. **Seven** deuteriums would be incorporated into palmitic acid, because seven ⁻OOCCD₂COSR are used in the synthesis (for a total of 14 Ds), and each ⁻OOCCD₂COSR loses one deuterium in the dehydration step $(14\,\text{Ds} - 7\,\text{Ds} = 7\,\text{Ds})$.

46. It tells you that an imine is formed as an intermediate. Because an imine is formed, the only source of oxygen for acetone formation is $H_2^{18}O$.

If decarboxylation occurred without imine formation, most of the acetone would contain ^{16}O. There would be some ^{18}O incorporated into acetone because acetone would react with $H_2^{18}O$ and form a hydrate. The ^{18}O would become incorporated into acetone when the hydrate reforms acetone.

47. **a.** CH₃—C(=O)—CH₂—C(=O)—OCH₂CH₃

b. HO—C(=O)—CH(CH₃)—C(=O)—OH

c. R—C(=O)—CH₂—C(=O)—OR'

d. 1-hydroxycyclopentene (cyclopentene with OH)

e. CH₃CH₂CH₂CH₂—C(=O)—OH

48.   **a.**   **c.**   **e.** $CH_3CHCH_2CH_2$

b. $CH_3CH_2CH_2CH$ **d.**   **f.**

49.

50.   The electron-withdrawing nitro group will cause the signal to occur at a higher frequency (will have a greater chemical shift). The more acidic the hydrogen, the greater the chemical shift.

$$CH_3NO_2 \quad CH_2(NO_2)_2 \quad CH(NO_2)_3$$
$$\delta\,4.33 \qquad \delta\,6.10 \qquad \delta\,7.52$$

51.   The compound on the far right loses CO_2 at the lowest temperature because it is a β-keto acid. Thus, the electrons left behind when CO_2 is removed can be delocalized onto oxygen.

52.   **a.**   **c.**

b. H_3C **d.**

53.   When the ketone enolizes, the asymmetric center is lost. When the enol reforms the ketone, an asymmetric center is created; the *R* and *S* enantiomers can be formed equally easily, so a racemic mixture is obtained.

(*R*)-2-methyl-1-
phenyl-1-butanone

You need a ketone with an α-carbon that is an asymmetric center. Racemization will occur when an α-hydrogen is removed from the asymmetric center.

54.   Because water is a stronger acid than the ketone, formation of the enolate ion will be reversible (and will favor the reacants) when hydroxide is used to remove the α-hydrogen (Section 18.6). Under reversible conditions, the thermodynamic enolate will predominate. Therefore, the deuterium will substitute for a hydrogen attached to the more substituted α-carbon.

thermodynamic enolate

Because the ketone is a much stronger acid than diisopropylamine, formation of the enolate ion will be irreversible when LDA is used to remove the α-hydrogen. Under irreversible condition, the kinetic enolate will predominate. Therefore, the deuterium will substitute for a hydrogen attached to the less substituted α-carbon.

kinetic enolate

55. The initially formed addition product loses water immediately, because the new double bonds are particularly stable (and therefore easy to form) since each is conjugated with a benzene ring.

$$+ \quad 2 H_2O$$

56.

57. The haloform reaction requires that a group be created that is a weaker base than hydroxide ion so that hydroxide ion is not the group eliminated from the tetrahedral intermediate. For an alkyl group to be the weaker base, the α-carbon must be bonded to three halogen atoms. The only alkyl group that can become bonded to three halogen atoms at the α-carbon is a methyl group (because it is bonded to three hydrogens).

58.

A
$C_5H_8O_3$

$\xrightarrow[\Delta]{HCl, H_2O}$ B $\xrightarrow[H_2O]{HO^-}$ C $\xrightarrow[\Delta]{HCl}$ D

1. CH_3O^-
2. CH_3Br

E $\xrightarrow[\Delta]{HCl, H_2O}$ H $\xrightarrow[\substack{I_2 \\ excess}]{\substack{HO^- \\ excess}}$ I $\xrightarrow[pyridine]{SOCl_2}$ J

$\xrightarrow{CH_3OH}$ K

1. CH_3O^-
2. CH_3Br

F $\xrightarrow[\Delta]{HCl, H_2O}$ G

1. CH_3O^-
2. HCl

L

59.   **a.**   **b.**   **c.**

60.

Notice that the β-substituted compounds are prepared via an α,β-unsaturated ketone, which can be prepared by dehydration of α-hydroxycyclohexanone.

61. Remember that there are no positively charged organic reactants, intermediates, or products in a basic solution, and no negatively charged organic reactants, intermediates, or products in an acidic solution.

62.   a.

b.

c.

d.

63. The Reformatsky reaction requires an organozinc reagent that is obtained by treating an α-bromo ester with zinc.

a.

$$CH_3CH_2CH_2\overset{\overset{\displaystyle O}{\|}}{C}H \;+\; BrCH_2\overset{\overset{\displaystyle O}{\|}}{C}OCH_3 \;\xrightarrow[\text{2. H}_2\text{O}]{\text{1. Zn}}\; CH_3CH_2CH_2\overset{\displaystyle OH}{\underset{}{C}}HCH_2\overset{\overset{\displaystyle O}{\|}}{C}OCH_3$$

b.

$$CH_3CH_2\overset{\overset{\displaystyle O}{\|}}{C}H \;+\; CH_3CH_2\underset{\underset{\displaystyle Br}{|}}{C}H\overset{\overset{\displaystyle O}{\|}}{C}OCH_3 \;\xrightarrow[\text{2. H}_2\text{O}]{\text{1. Zn}}\; CH_3CH_2\overset{\displaystyle OH}{\underset{\underset{\displaystyle CH_2CH_3}{|}}{C}}HCH\overset{\overset{\displaystyle O}{\|}}{C}OCH_3$$

$$\xrightarrow[\text{excess}]{\text{HCl} \;|\; \text{H}_2\text{O}}\; CH_3CH_2\overset{\displaystyle OH}{\underset{\underset{\displaystyle CH_2CH_3}{|}}{C}}HCH\overset{\overset{\displaystyle O}{\|}}{C}OH$$

c.

$$CH_3CH_2\overset{\overset{\displaystyle O}{\|}}{C}H \;+\; CH_3\underset{\underset{\displaystyle Br}{|}}{C}H\overset{\overset{\displaystyle O}{\|}}{C}OCH_3 \;\xrightarrow[\text{2. H}_2\text{O}]{\text{1. Zn}}\; CH_3CH_2\overset{\displaystyle OH}{\underset{\underset{\displaystyle CH_3}{|}}{C}}HCH\overset{\overset{\displaystyle O}{\|}}{C}OCH_3$$

$$\xrightarrow[\text{excess}]{\text{HCl} \;|\; \text{H}_2\text{O}}\; CH_3CH_2\overset{\displaystyle OH}{\underset{\underset{\displaystyle CH_3}{|}}{C}}HCH\overset{\overset{\displaystyle O}{\|}}{C}OH$$

$$CH_3CH_2CH=\underset{\underset{\displaystyle CH_3}{|}}{C}\overset{\overset{\displaystyle O}{\|}}{C}OH \;\xleftarrow[\Delta]{\text{H}_2\text{SO}_4}\;$$

d.

$$CH_3CH_2\overset{\overset{\displaystyle O}{\|}}{C}CH_2CH_3 \;+\; BrCH_2\overset{\overset{\displaystyle O}{\|}}{C}OCH_3 \;\xrightarrow[\text{2. H}_2\text{O}]{\text{1. Zn}}\; CH_3CH_2\overset{\displaystyle OH}{\underset{\underset{\displaystyle CH_2CH_3}{|}}{C}}CH_2\overset{\overset{\displaystyle O}{\|}}{C}OCH_3$$

64. The ketone is 2-hexanone. Therefore, the alkyl halide is a propyl halide (propyl bromide, propyl chloride, or propyl iodide).

$$CH_3\overset{\overset{\displaystyle O}{\|}}{C}\underbrace{CH_2CH_2CH_2CH_3}$$

This part comes from acetoacetic ester.

65.

a.

1. LDA
2. CH₃CH₂CH₂CH₂Br

b. 1.

1. [pyrrolidine] trace acid

2. CH₃CH₂CH₂CCl
3. HCl, H₂O

2.

1. LDA

2. CH₃CH₂CH₂CCl
   add slowly

c. CH₃CH₂O—C(=O)—OCH₂CH₃ + CH₃CH₂O⁻

1. [cyclohexanone]
   add slowly
2. HCl

d. H—C(=O)—OCH₂CH₃ + CH₃CH₂O⁻

1. [cyclohexanone]
   add slowly
2. HCl

e.

+ CH₂=CHCCH₃

HO⁻
a Robinson annulation

H₂SO₄
Δ

H₂ | Pd/C

66.   The positive iodoform test indicates that compound **B** is a methyl ketone. The two singlets with a ratio of 3 : 1 indicate that if the methyl group represents the "3", there is only one other hydrogen in the compound or, if the methyl group represent the "1", there are nine other hydrogens. The former does not agree with the molecular formula. Therefore, the compound must have three identical methyl groups that are attached to a carbon that is not attached to a hydrogen.

3,3-Dimethyl-1-butyne will show two singlets with a ratio of 9 : 1. When it adds water, it will form a methyl ketone with two singlets with a ratio of 3 : 1.

two singlets (9 : 1)   two singlets (3 : 1)

67.   **a.**

b.

c.

d.

68. **a.** The first reaction is an intermolecular Claisen condensation. An intramolecular condensation reaction forms the final product.

b. The reaction involves two successive aldol condensations.

c. The base removes a proton from the carbon that is flanked by two carbonyl groups; this is followed by an intramolecular S$_N$2 reaction.

d. A proton is removed from the α-carbon that is flanked by a carbonyl group and a nitrile; then an intramolecular S_N2 reaction occurs.

69.

how the reagents become connected

70.   a.

alanine

b.

glycine

71. **A** can be prepared by removing an α-hydrogen from acetone with LDA and then adding formaldehyde, followed by dehydration.

B can be prepared by removing an α-hydrogen from propionaldehyde with LDA and then slowly adding acetaldehyde, followed by dehydration.

C can be prepared by removing an α-hydrogen from 3-hexanone with LDA and then adding formaldehyde, followed by dehydration. The yield is poor because 3-hexanone is an asymmetrical ketone; therefore, two different α-carbanions are formed that lead to two different α,β-unsaturated ketones.

D can be prepared by removing an α-hydrogen from acetaldehyde with LDA and then slowly adding acetone, followed by dehydration.

E can be prepared via an intramolecular aldol condensation using 5-oxooctanal.

F can be prepared via an intramolecular aldol condensation using 1,7-heptanedial.

A small amount of **G** can be formed via an intramolecular aldol condensation using 7-oxooctanal, but this compound can form two different α-carbanions that can react with a carbonyl group to form a six-membered ring. Because an aldehyde is more reactive than a ketone toward nucleophilic addition, the target molecule will be a minor product.

minor product   **major product**

72.   a.

b.

c.

73.   At low temperatures ($-78\,°C$), the proton will be more apt to be removed from the methyl group, because its hydrogens are the most accessible and are slightly more acidic (that is, the kinetically controlled product will be formed).

At higher temperatures ($25\,°C$), the proton will be more apt to be removed from the more substituted α-carbon, because in that way the more stable enolate is formed (the one with the more stable double bond). Thus, the thermodynamically controlled product is formed.

74.

75. The compound that gives the ^{1}H NMR spectrum is 4-phenyl-3-buten-2-one. The singlet at 2.3 ppm that integrates to 3 protons is the methyl group. Because the benzene ring protons that show signals between 7 and 8 ppm apparently integrate to six protons, we know that the signals include one of the vinylic protons of the double bond. The other vinyl proton is the doublet (6.7 ppm) that integrates to one proton. The compounds that would form this compound (via an aldol condensation) are benzaldehyde and acetone.

4-phenyl-3-buten-2-one

76. The middle carbonyl group is hydrated, because the hydrate is stabilized by the electron-withdrawing carbonyl groups on either side of it.

77.

Release of strain
in the three-membered
ring causes a ring bond
to be the most likely
bond to break

78.

79. **a.**

b.

80.

81.   **a.** The mechanism starts with an isomerization that converts isolated double bonds to conjugated double bonds.

b. The carbonyl group on the left is protonated because electron delocalization causes it to be more basic than the other carbonyl group.

82. The benzoin condensation requires the aldehydic hydrogen to be removed. This hydrogen cannot be removed unless the electrons left behind can be delocalized onto an electronegative atom. The nitrogen of the cyano group serves that purpose. Therefore, the reaction will not occur if hydroxide ion is used instead of cyanide ion, because the electrons left behind if the hydrogen were to be removed cannot be delocalized.

83.   To arrive at the final product, three equivalents of malonyl thioester are needed. (See page 892 in the text.)
Because only one acetate ion was used, only one carbon is labeled in the product.

84.

85.

86.

Notice that in the last step of the reaction, a diester is converted into a diamide.

87.

Chapter 18 Practice Test

1. Rank the following compounds in order from most acidic to least acidic:

2. Draw a structure for each of the following:

    a. the most stable enol tautomer of 2,4-pentanedione

    b. a β-keto ester

3. Draw the product of each of the following reactions:

a.

b.

c.

d.

e.

f.

g.

4.   Give an example of each of the following:

    a. an aldol addition

    b. an aldol condensation

    c. a Claisen condensation

    d. a Dieckmann condensation

    e. a malonic ester synthesis

    f. an acetoacetic ester synthesis

5.   Draw the products of the following crossed aldol addition:

6.   What ester would be required to prepare each β-ketoester?

CHAPTER 19
Reactions of Benzene and Substituted Benzenes

Important Terms

activating substituent	a substituent that increases the reactivity of an aromatic ring. Electron-donating substituents activate aromatic rings toward electrophilic attack, and electron-withdrawing substituents activate aromatic rings toward nucleophilic attack.
arenediazonium salt	$Ar \overset{+}{N}\equiv N \ \ X^-$
azo linkage	a $-N=N-$ bond.
aromatic compound	a cyclic and planar compound with an uninterrupted cloud of electrons containing an odd number of pairs of π electrons.
benzyl group	$\langle\!\!\!\!\!\bigcirc\!\!\!\!\!\rangle\!-CH_2-$
Clemmensen reduction	a reaction that reduces the carbonyl group of a ketone to a methylene group using $Zn(Hg)/HCl$, Δ.
deactivating substituent	a substituent that decreases the reactivity of an aromatic ring. Electron-withdrawing substituents deactivate aromatic rings toward electrophilic attack, and electron-donating substituents deactivate aromatic rings toward nucleophilic attack.
donate electrons by resonance (resonance electron donation)	donation of electrons through p orbital overlap with neighboring π bonds.
electrophilic aromatic substitution reaction	a reaction in which an electrophile substitutes for a hydrogen of an aromatic ring.
Friedel-Crafts acylation	an electrophilic aromatic substitution reaction that puts an acyl group on an aromatic ring.
Friedel-Crafts alkylation	an electrophilic aromatic substitution reaction that puts an alkyl group on an aromatic ring.
Gatterman-Koch reaction	a reaction that uses a high-pressure mixture of carbon monoxide and HCl to form benzaldehyde.
halogenation	reaction with a halogen (Br_2, Cl_2, I_2).
inductive electron withdrawal	withdrawal of electrons through a σ bond.
meta director	a substituent that directs an incoming substituent meta to an existing substituent.

Meisenheimer complex   a resonance-stabilized complex formed by addition of a nucleophile to a benzene ring

nitration   substitution of a nitro group (NO_2) for a hydrogen of an aromatic ring.

nitrosamine
(*N*-nitroso compound)   an amine with a nitroso (N=O) substituent bonded to its nitrogen atom.

nucleophilic aromatic
substitution (S_NAr) reaction   a reaction in which a nucleophile substitutes for a halo-substituent on a benzene ring

ortho-para director   a substituent that directs an incoming substituent ortho and para to an existing substituent.

phenyl group C_6H_5—

principle of microscopic
reversibility   a principle that states that the mechanism for a reaction in the forward direction has the same intermediates and the same transition states as the mechanism for the reaction in the reverse direction.

Sandmeyer reaction   the reaction of an arenediazonium ion with a cuprous salt.

Schiemann reaction   the reaction of an arenediazonium ion with HBF_4.

S_NAr reaction   a nucleophilic aromatic substitution reaction.

sulfonation   substitution of a hydrogen of an aromatic ring with a sulfonic acid group (SO_3H).

Suzuki reaction   a reaction that will replace a bromine substituent on an aromatic ring with an alkyl, alkenyl, or aryl substituent.

withdraw electrons by
resonance (resonance
electron withdrawal)   withdrawal of electrons through *p* orbital overlap with neighboring π bonds.

Wolff-Kishner reduction   a reaction that reduces the carbonyl group of a ketone to a methylene group using NH_2NH_2/HO^-, Δ.

Solutions to Problems

1.   **a.** $CH_3CHCH_2CH_2CH_2CH_3$ **b.** [benzene ring] CH_2OH **c.** $CH_3CH_2CHCH_2CH_3$ **d.** [benzene ring] CH_2Br
          [benzene ring below] CH_2
                                                                                                 [benzene ring below]

2.   Electrophilic addition to benzene is an endergonic reaction because benzene is more stable than the product of the reaction. (See Figure 19.2 on page 912 of the text.)

      An alkene is much less stable than benzene because it is not aromatic. Electrophilic addition to an alkene is an exergonic reaction because the alkene is less stable than the product of the reaction. (The overall reaction converts a σ bond and a π bond into two σ bonds; a σ bond is stronger and, therefore, of lower energy than a π bond.)

3.   Ferric bromide activates Br_2 for nucleophilic attack by accepting a pair of electrons from it. Hydrated ferric bromide cannot do this because it has already accepted a pair of electrons from water.

$$H-\overset{\overset{\displaystyle +}{\cdot\cdot}}{O}-\overset{\overset{\displaystyle Br}{|}}{\underset{\underset{\displaystyle Br}{|}}{Fe}}-Br$$
$$\underset{\displaystyle H}{|}$$

4.   Solved in the text.

5.   In sulfonation (the forward reaction), **A** to **B** has a smaller rate constant (a higher energy barrier and, therefore, a slower reaction) than **B** to **C**; **A** to **B** is the rate-determining step.

      In desulfonation (the reverse reaction), **C** to **B** has a smaller rate constant than **B** to **A**, but **B** to **A** is the rate-determining step, because once **B** is formed, it is easier for it to go back to **C** than to proceed to **A**. Therefore, **B** to **A** is the bottleneck (or rate-determining) step.

      This example shows that the rate-determining step of a reaction is not necessarily the step with the smallest rate constant; it is the step that has the transition state with the highest energy on the reaction coordinate.

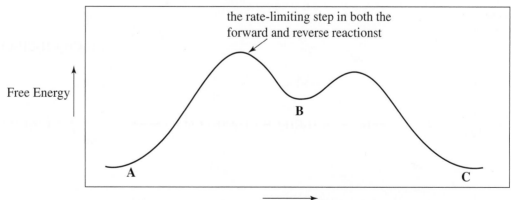

6.

7.   A carbocation rearrangement occurs in **b.** and **e.**

a.

CH$_2$CH$_3$

c.

CH$_3$
CHCH$_2$CH$_3$

e.

CH$_3$
CCH$_3$
CH$_3$

b.

CH$_3$
CHCH$_3$

d.

CH$_3$
CCH$_3$
CH$_3$

f.

CH$_2$CH=CH$_2$

8.   **a.**

You could also use 1-chloropentane as the alkyl halide because it would rearrange to form the desired alkyl group.

b.

or

or

9. **a.** COOH

c. CH₂OCH₃

b. COOH, COOH

d. CH₂CH₂NH₂

10. **a.** Solved in the text.

b.

$$\text{benzene} \xrightarrow[\text{AlCl}_3]{\text{CH}_3\text{CH}_2\text{Cl}} \text{CH}_2\text{CH}_3 \xrightarrow[\text{hv}]{\text{Br}_2} \text{CHCH}_3 (\text{Br})$$

or Br₂ | FeBr₃

Br-benzene

$$\xrightarrow[\substack{\text{PdL}_2 \\ (\text{CH}_3\text{CH}_2)_3\text{N}}]{\text{CH}_2=\text{CH}_2} \text{CH}=\text{CH}_2$$

tert-BuO⁻

CH=CH₂

c. CH=CH₂ (product of **b**) $\xrightarrow[\text{peroxide}]{\text{HBr}}$ CH₂CH₂Br

d. CH=CH₂ (product of **b**) $\xrightarrow[\text{2. HO}^-, \text{H}_2\text{O}_2, \text{H}_2\text{O}]{\text{1. BH}_3/\text{THF}}$ CH₂CH₂OH

e. benzene $\xrightarrow[\text{H}_2\text{SO}_4]{\text{HNO}_3}$ NO₂ $\xrightarrow[\text{Pd/C}]{\text{H}_2}$ NH₂

f. benzene $\xrightarrow[\text{AlCl}_3]{\text{CH}_3\text{Cl}}$ CH₃ $\xrightarrow[\Delta]{\text{H}_2\text{CrO}_4}$ COOH

11. **a.** *ortho*-ethylphenol   or   2-ethylphenol

b. *meta*-bromochlorobenzene   or   1-bromo-3-chlorobenzene

c. *meta*-bromobenzaldehyde   or   3-bromobenzaldehyde

d. *ortho*-ethylmethylbenzene   or   2-ethylmethylbenzene

12.   **a.**

    b.

    c.

    d.

13.   **a.**

    c.

    e.

    g.

    b.

    d.

    f.

    h.

    i.

14.   **a.**   1,3,5-tribromobenzene   **c.**   *para*-bromomethylbenzene or 4-bromomethylbenzene
    b.   *meta*-nitrophenol or 3-nitrophenol   **d.**   *ortho*-dichlorobenzene or 1,2-dichlorobenzene

15.   **a.**   donates electrons by resonance and withdraws electrons inductively
    b.   donates electrons by hyperconjugation
    c.   withdraws electrons by resonance and withdraws electrons inductively
    d.   donates electrons by resonance and withdraws electrons inductively
    e.   donates electrons by resonance and withdraws electrons inductively
    f.   withdraws electrons inductively

16.   **a.**   phenol $>$ toluene $>$ benzene $>$ bromobenzene $>$ nitrobenzene

       A chloro substituent attached to a methyl group withdraws electrons inductively and, therefore, causes the substituent to be less activating. Do not confuse this with a chloro substituent attached to a benzene ring that withdraws electrons inductively **and** donates electrons by resonance.

    b.   toluene $>$ chloromethylbenzene $>$ dichloromethylbenzene $>$ difluoromethylbenzene

17.   Solved in the text.

18. **a.**

b.

19. **a.** $CH_2CH_2CH_3$

$+$

$CH_2CH_2CH_3$

NO_2

c.

e.

b. Br

$+$

Br

NO_2

d.

f.

$+$

20. They are all meta directors:

a. this group withdraws electrons by resonance from the ring. The relatively electronegative nitrogen atom causes it to also withdraw electrons inductively from the ring.

b. NO_2 withdraws electrons inductively and withdraws electrons by resonance.

c. CH_2OH withdraws electrons inductively from the ring.

d. COOH withdraws electrons inductively and withdraws electrons by resonance.

e. CF_3 withdraws electrons inductively from the ring.

f. $N{=}O$ withdraws electrons inductively and withdraws electrons by resonance.

You could draw resonance contributors for electron donation into the ring by resonance. However, the most stable resonance contributors are obtained by electron flow out of the benzene ring toward oxygen, the most electronegative atom in the compound.

resonance electron donation
into the ring

resonance electron withdrawal
out of the ring

21. For each compound, determine which benzene ring is more highly activated. The more highly activated ring is the one that undergoes electrophilic aromatic substitution.

a. Solved in the text.

b.

c. Because the right-hand ring is highly activated and a catalyst is employed, monobrominated, dibrominated, and tribrominated compounds can be obtained. However, since only one equivalent of Br_2 is available, there is not enough bromine to brominate all the reactive positions. Therefore, a mixture of unreacted starting material and the four products shown below will be obtained.

d. Because the left-hand ring is highly activated and a catalyst is employed, both ortho positions can be brominated. Because there is not enough Br_2 available to brominate both reactive positions, a mixture of unreacted starting material and the two products shown below will be obtained.

22.   For **a**, **b**, and **c**: the closer the electron-withdrawing substituent is to the COOH group, the stronger the acid.

a. ClCH$_2$COOH

d.   COOH

The predominant effect of the CH$_3$O group is to donate electrons by resonance, so it decreases the acidity of the carboxylic acid.

f. HCOOH

A hydrogen is electron withdrawing compared to a methyl group, since a methyl group can donate electrons by hyperconjugation.

b. O$_2$NCH$_2$COOH

c. H$_3$$\overset{+}{\text{N}}CH_2$COOH

e. HOOCCH$_2$COOH

It is easier to lose the first proton from a dicarboxylic acid than the second proton. (See Table 16.3 on page 769.)

g. FCH$_2$COOH

Fluorine is more eletronegative than chlorine.

h.

COOH

Cl

We know that F withdraws electrons inductively more than Cl does, because F is more electronegative than Cl.

We know that F donates electrons by resonance better than Cl does, because F donation involves 2p–2p overlap, whereas Cl donation involves 3p–2p overlap.

Because Table 19.1 on page 932 of the text shows that F is less deactivating than Cl toward electrophilic aromatic substitution, we know that overall F donates electrons better than Cl. Thus, resonance electron donation is more important than inductive electron withdrawal. (See the answer to Problem 17.)

23.   When *para*-nitrophenol loses a proton, the negative charge in the conjugate base can be delocalized onto the nitro substituent. Therefore, the *para*–nitro substituent decreases the pK$_a$ both by resonance electron withdrawal and by inductive electron withdrawal.

When *meta*-nitrophenol loses a proton, the negative charge in the conjugate base cannot be delocalized onto the nitro substituent. Therefore, the *meta*-nitro substituent can decrease the pK$_a$ only by inductive electron withdrawal. Therefore, the para isomer has a lower pK$_a$.

24. **a.** Because fluorine and chlorine both deactivate the benzene ring toward electrophilic aromatic substitution, they increase acidity compared to unsubstituted benzoic acid. Problem 17 explains that fluorine is the weakest deactivator of the halogens, because electron donation by resonance is more important than inductive electron withdrawal. Therefore, *o*-chlorobenzoic acid is the strongest acid of the three.

b. Benzoic acid is the weakest acid of the three, because it does not have a deactivating (overall electron-withdrawing) substituent on the ring.

25. Notice that in all three syntheses, the Friedel-Crafts reaction has to be done first. Both NO_2 and SO_3H are meta directors, and a Friedel-Crafts reaction cannot be carried out if there is a meta director on the ring.

a.

b.

c.

26. No reaction will occur in **a** and **c**, because a Friedel-Crafts reaction cannot be carried out on a ring that possesses a meta director.

a. no reaction

c. no reaction

b.

+

d.

+

27.   **a.**   Unlike *m*-dipropylbenzene that requires a coupling reaction to be used in its synthesis, a coupling reaction does not have to be used in the synthesis of *p*-dipropylbenzene. The propyl group is an ortho/para director, so propylbenzene can undergo a Friedel-Crafts reaction.

b.   A coupling reaction can be used in the synthesis of *p*-dipropylbenzene. Notice that bromination occurs after the reduction of the carbonyl group. In contrast, bromination occurs before the reduction reaction in the synthesis of *m*-dipropylbenzene.

28.   In Problems **c**, **g**, and **h**, the carbonyl group can be converted to a methylene group by H_2, Pd/C (as is used in **c**), or by a Wolff-Kishner reduction (as is used in **g**), or by a Clemmensen reduction (as is used in **h**).

a.

b.

c. Note that benzaldehyde is formed by the Gatterman-Koch reaction (page 919 in the text).

d.

e.

f.

g.

h.

i. Notice that in the second step of the synthesis, epoxidation followed by the addition of hydride ion (page 508) is used to add water to the double bond in order to avoid the carbocation rearrangement that would occur with the acid-catalyzed addition of water.

29. **a.**

COOH is a meta director and Br is an ortho-para director, so both direct to the same position.

b.

COOH directs to the meta position. The same product will be obtained regardless of which COOH is the director.

Less of this compound will be obtained because of steric hindrance.

c.

COOH directs to its meta position and Cl directs to its ortho position, so they both direct to the same position on the ring.

d.

A methoxy substituent is strongly activating and a fluorine substituent is deactivating, so the methoxy substituent will do the directing.

e.

The aldehyde group is a meta director and the methoxy group is an ortho-para director, so both direct to the same position.

f.

Less of this product will be obtained because of steric hindrance.

30.   Yes, the advice is sound.

The para isomer will form one product, because the formyl and ethyl groups both direct to the same positions and both positions result in the same product.

The ortho isomer will form two products, because the formyl and ethyl groups both direct to the same positions but different products are obtained from each position.

The meta isomer will form four products because the formyl and ethyl groups direct to four different positions and a different product is obtained from each position.

low yield

31.   Solved in the text.

32.   FeBr$_3$ would complex with the amino group, converting it into a meta director. The NH$_2$ group is a strongly activating substituent, so a Lewis acid is not needed.

33.   Because a diazonium ion is electron withdrawing, it deactivates the benzene ring. A deactivated benzene ring would be too unreactive to undergo an electrophilic substitution reaction at the cold temperature necessary to keep the benzenediazonium ion from decomposing.

34.

35.   **a.**   Solved in the text.

          b.   The reaction steps are the same as those in **a** (answered on page 950 of the text) except for the last step. In the last step, the meta-bromo-substituted diazonium ion should undergo the reaction shown here rather than the reaction with CuBr.

c. The first approach is longer, but it will generate a higher yield of the ortho isomer.

or

d. The nitro group cannot be placed on the benzene ring first, because a Friedel-Crafts reaction cannot be carried out on a ring with a meta director. Because formyl chloride is too unstable to be purchased, benzaldehyde is prepared by the Gatterman-Koch reaction (page 919 in the text).

e.

f.

or

36. Because the para position is occupied, the electrophile will add to the ortho position. (Attack at the ortho position is slower than attack at the para position because of steric hindrance, but since the para position is not available, the slower reaction will prevail.)

37.   You can see why nucleophilic attack occurs on the neutral nitrogen if you compare the products of nucleophilic attack on the two nitrogens. Nucleophilic attack on the neutral nitrogen forms a stable product, whereas nucleophilic attack on the positively charged nitrogen would form an unstable compound with two charged nitrogen atoms.

compared to

The terminal nitrogen is electrophilic because of electron withdrawal by the positively charged nitrogen. If you draw the resonance contributors, you can see that the "neutral" nitrogen is electron deficient.

38.   **a.**

activated ring   **diazonium ion**

b.

activated ring   **diazonium ion**

39.   Immediately after hydrolysis of the amide bond, carbon dioxide is lost. We know then that the indicated amide bond is the one that is hydrolyzed because carbon dioxide can then be lost since the electrons left behind can be delocalized onto the carbonyl oxygen.

40. The nitrosamine formed from a secondary amine cannot form a diazonium ion when the nitrogen-nitrogen double bond is formed because the nitrogen cannot lose its positive charge by losing a proton. Therefore, the reaction stops at the nitrosamine.

nitrosamine

41. Note: Diazomethane is both explosive and toxic, so it should be synthesized only in small amounts by experienced laboratory workers.

The first step of the reaction is formation of the methyldiazonium ion as a result of removal of a proton from the carboxylic acid by diazomethane.

In the second step of the reaction, the carboxylate ion displaces nitrogen gas (N_2) from the methyldiazonium ion in an S_N2 reaction. High yields are obtained because the only side product is a gas.

diazomethane   methyldiazonium
                                        ion

42. From the resonance contributors, you can see that the reason that *meta*-chloronitrobenzene does not react with hydroxide ion is because the negative charge that is generated on the benzene ring cannot be delocalized onto the nitro substituent.

Electron delocalization onto the nitro substituent can occur only if the nitro substituent is ortho or para to the site of nucleophilic attack.

43. **a.** The more nitro substituents ortho and para to the halogen, the faster the rate of nucleophilic aromatic substitution.

$$1\text{-chloro-2,4-dinitrobenzene} \ > \ p\text{-chloronitrobenzene} \ > \ \text{chlorobenzene}$$

b. The fewer nitro substituents attached to the benzene ring, the faster the rate of electrophilic aromatic substitution.

$$\text{chlorobenzene} \ > \ p\text{-chloronitrobenzene} \ > \ 1\text{-chloro-2,4-dinitrobenzene}$$

44. **a.**

b.

c.

d. This synthesis would be carried out exactly as in **c** (above) except for the last step, which would be:

45. **a.** BrCH$_2$CH$_2$CH$_2$CHCH$_2$CH$_3$ $\xrightarrow{\text{NaH}}$
              |
             OH

b.

c. In this synthesis and in the synthesis of **d**, a primary alkyl halide cannot be used in the first step because the carbocation will rearrange to a secondary carbocation and will result in the formation of a methyl-substituted five-membered ring. In contrast, a primary allyl halide can be used because the initially formed allyl cation is stabilized by electron delocalization and is not prone to rearrangement.

Br

CH$_2$CH=CHCH$_2$Cl   →   AlCl$_3$ →   NBS, Δ / peroxide   →   Br

H$_2$, Pd/C

tert-BuO$^-$ ←   Br

d.

CH$_2$CH=CHCH$_2$Cl   →   AlCl$_3$ →   NBS, Δ / peroxide

Br

H$_2$, Pd/C

COOH   ← H$_2$O, HCl / Δ ← C≡N (N) ← $^-$C≡N ← Br → Mg / Et$_2$O → MgBr → 1. CO$_2$ 2. HCl → COOH

e.   BrCH$_2$CH$_2$CH$_2$CHCH$_2$CH=CH$_2$ → NaH → O⟩CH$_2$CH=CH$_2$
                        |
                        OH

f.   O⟩CH$_2$CH=CH$_2$ → 1. BH$_3$/THF   2. HO$^-$, H$_2$O$_2$, H$_2$O → O⟩CH$_2$CH$_2$CH$_2$OH

product of **e**

46.   **a.** OH   **d.** O=C–H (benzaldehyde)   **g.** CH$_3$

b. CH$_2$O (benzyl phenyl ether)   **e.** OCH$_3$ **h.** C(CH$_3$)$_3$

c. C≡N   **f.** CH=CH$_2$ **i.** CH$_2$Cl

47.
  a. *m*-bromobenzoic acid
  b. 1,2,4-tribromobenzene
  c. 2,6-dimethylphenol

  d. *p*-nitrostyrene
  e. *m*-ethylanisole
  f. 3,5-dichlorobenzenesulfonic acid

  g. *o*-bromomethylbenzene
  h. *p*-cyclohexylmethylbenzene
  i. 2-chloro-4-ethylazobenzene

48.

49.

a.

b.

c.

d.

e.

f.

g.

h.

i.

50.
   a. CH_2CH_3 donates electrons by hyperconjugation and does not donate or withdraw electrons by resonance.
   b. NO_2 withdraws electrons inductively and withdraws electrons by resonance.
   c. Br deactivates the ring and directs ortho/para.
   d. OH withdraws electrons inductively, donates electrons by resonance, and activates the ring.
   e. $^+NH_3$ withdraws electrons inductively and does not donate or withdraw electrons by resonance.

51.   a.   b.   c.

52.   a.   c.

   b.   +   d.

e.

g.

f.

h.

53.   The more activating the substituent, the more basic is the NH_2 group.

$$CH_3O-\underset{}{\bigcirc}-NH_2 \;>\; CH_3-\underset{}{\bigcirc}-NH_2 \;>\; Br-\underset{}{\bigcirc}-NH_2 \;>\; CH_3\overset{O}{\underset{}{C}}-\underset{}{\bigcirc}-NH_2$$

54.   The least reactive compound will be the one that yields the highest percentage of meta product.

1.

most reactive   least reactive
                                        highest % meta product

2.

least reactive   most reactive
highest % meta product

3.

most reactive   least reactive
                          highest % meta product

55. CH₃O— ⟨benzene ring⟩ —OCH₃

56. These are intramolecular Friedel-Crafts alkylation reactions. Notice that a carbocation rearrangement occurs in part **a**.

a.

HB⁺

b.

57.   a.

b.

c.

d.

e.

f.

g.

h.

i.

58. **a.** Two products will be obtained. Less of the product on the right will be obtained because of steric hindrance.

b. One product will be obtained. (Notice that the four positions are equivalent.)

c. Two products could be obtained, but little, if any, of one of the products will be obtained because of steric hindrance.

59.   **a.**   anisole > ethylbenzene > benzene > chlorobenzene > nitrobenzene

b.   2,4-dinitrophenol > 1-methyl-2,4-dinitrobenzene > 1-chloro-2,4-dinitrobenzene

c.   *p*-cresol > *p*-xylene > toluene > benzene

d.   phenol > propylbenzene > benzene > benzoic acid

e.   *p*-chloromethylbenzene > *p*-methylnitrobenzene > 2-chloro-1-methyl-4-nitrotoluene > 1-methyl-2,4-dinitrobenzene

f.   fluorobenzene > chlorobenzene > bromobenzene > iodobenzene

60.   **a.**

b.

c.

d.

e.

f.

61.   Three ways to synthesize anisole from benzene are shown below.

1.

2.

3.

62.

a. NO$_2$ ... COOH

c. OCH$_3$ ... Cl

e. O / CCH$_3$ ... Br ... minor

g. CH$_3$ ... NO$_2$

b. O / OCCH$_3$ ... CH$_3$OC ... O

d. OH ... COOH

f. OH ... minor ... CH$_3$

h. Cl ... minor ... Cl

63.

1. (benzoyl chloride) + (toluene) $\xrightarrow[\text{2. H}_2\text{O}]{\text{1. AlCl}_3}$ (product)

2. (benzene) + (4-methylbenzoyl chloride) $\xrightarrow[\text{2. H}_2\text{O}]{\text{1. AlCl}_3}$ (product)

64. Because the 2p orbital of oxygen overlaps better than the 3p orbital of sulfur with the 2p orbital of carbon, oxygen is better than sulfur at donating electrons by resonance. Therefore, the benzene ring of anisole is more activated toward electrophilic aromatic substitution than is the benzene ring of thioanisole.

65. Anisole will undergo electrophilic aromatic substitution (D$^+$ is the electrophile) primarily at the ortho and para positions.

(anisole) + D$_2$O $\xrightarrow{\text{D}_2\text{SO}_4}$ (2,4,6-trideuterated anisole)

66. The compound with the methoxy substituent is the more reactive because it forms the more stable carbocation intermediate. The carbocation intermediate is stabilized by resonance electron donation.

CH$_3\ddot{\text{O}}$ ... $\overset{+}{\text{C}}$H—CH$_3$

67.   The signal at ~7 ppm that integrates to 5H indicates a monosubstituted benzene ring. A monosubstituted benzene ring contains six carbons and five hydrogens. Subtracting these atoms from the molecular formula of $C_{13}H_{20}$ gives us a substituent with 7 carbons and 15 hydrogens. The compound shown below has the correct number of carbons and hydrogens that will give two singlets with one (9H) having 1.5 times the area of the second (6H).

68.   The rate-determining step in the S_N1 reaction is the formation of the tertiary carbocation. An electron-donating substituent will stabilize the carbocation and cause it to be more easily formed. An electron-withdrawing substituent will destabilize the carbocation and cause it to be less easily formed.

69.   **a.**

b.

c.

d.

e.

f.

g.

h.

i.

70.   **a.**   The halogens withdraw electrons inductively and donate electrons by resonance. Because they all deactivate the benzene ring toward electrophilic aromatic substitution, we know that their electron-withdrawing effect is stronger than their electron-donating effect. Therefore, an *ortho*-halo-substituted benzoic acid will be a stronger acid than benzoic acid.

b.   Because fluorine is the weakest deactivator of the halogens (Table 19.1), we know that overall it donates electrons better than the other halogens. Therefore, *ortho*-fluorobenzoic acid is the weakest of the *ortho*-halo-substituted benzoic acids.

c.   The smaller the halogen, the more electronegative it is and therefore the better it is at withdrawing electrons inductively. The smaller the halogen, the better it is at donating electrons by resonance because a $2p$ orbital of carbon overlaps better with a $2p$ orbital of a halogen than with a $3p$ orbital of a halogen and overlaps better with a $3p$ orbital than with a $4p$ orbital. Therefore, Br does not withdraw electrons as well as Cl, and Br does not donate electrons as well as Cl, so their pK_a values are similar.

71.   **a.**   The weaker the base attached to the acyl group, the stronger its electron-withdrawing ability; therefore, the easier it is to form the tetrahedral intermediate. (*para*-Chlorophenol is a stronger acid than phenol so the conjugate base of *para*-chlorophenol is a weaker base than the conjugate base of phenol, etc.)

b.   The tetrahedral intermediate collapses by eliminating the OR group. The lower the basicity of the OR group, the easier it is to eliminate it.

Thus, the rate of both formation of the tetrahedral intermediate and collapse of the tetrahedral intermediate is decreased by increasing the basicity of the OR group.

72.   **a.**

b.   The major product is 1,4-dimethyl-2-nitrobenzene
para-Xylene is more reactive than benzene toward electrophilic aromatic substitution. The methyl groups activate the ortho positions, so all four positions on the ring are activated. Attack on any one of the four leads to the same compound.

73.   The reactions is part **a** and part **b** are intramolecular Friedel-Crafts acylation reactions.

a.

b.

c.

74.   The ^{1}H NMR spectrum is the spectrum of butylbenzene.
The benzene ring protons at ~7.2 ppm that integrate to 5 protons indicate a monosubstitutied benzene ring.
The two triplets and two multiplets indicate a straight-chain butyl group.

Therefore, the acyl chloride has a straight-chain propyl group and a carbonyl group that will be reduced to a methylene group.

75.   *meta*-Xylene will react more rapidly. In *meta*-xylene both methyl groups activate the same position, whereas in *para*-xylene each methyl group activates a different position.

76.   **a.**   **b.**   **c.**

77.   Recall that the highest frequency signal is the one with the most deshielded proton, that is, the one with the least electron density. (The signal farthest to the left on the spectrum.)

   a.   The oxygen withdraws electrons inductively from the carbon bonded to the hydrogens being compared, so the oxygen-containing compound has the signal at the highest frequency.

$$\overleftarrow{CH_3O} - CH_2C\underline{H}_3$$

   b.   The oxygen donates electrons by resonance, so the oxygen-containing compound has the signal at the lowest frequency.

$$CH_3\ddot{\underset{..}{O}} - CH = CH_2 \longleftrightarrow CH_3\overset{+}{\underset{..}{O}} = CH - \overset{-}{\ddot{C}}H_2$$

   Therefore, the non-oxygen-containing compound has the signal at the highest frequency.

$$CH_3CH = C\underline{H}_2$$

78.   <u>generation of the electrophile</u>

<u>reaction of benzene with the electrophile</u>

79.   **a.**

   b.

80.   The original compound was 1,3-dibromobenzene. 1,4-Dibromobenzene would form only one product, and 1,2-dibromobenzene would form only two.

1,3-dibromobenzene

1,4-dibromobenzene

1,2-dibromobenzene

81.   The greater the electron-withdrawing ability of the para substituent, the greater the K_{eq} for hydrate formation (see Section 17.11 in the text).

82.   **a.**

b.

The reason is that there are four ways to carry out the reaction in part **a**, but only three ways to carry it out in part **b** is because only a carbonyl group adjacent to a benzene ring can be reduced to a methylene group by H_2, Pd/C.

83.   a. HCl adds to the alkene, forming a secondary carbocation that undergoes a 1,2-hydride shift to form a tertiary carbocation. The tertiary carbocation is an electrophile that can either add to the double bond in a second molecule of the reactant (in an intermolecular reaction) or add to the benzene ring in the same molecule of the reactant (in an intramolecular reaction).

The intramolecular reaction is favored because it forms a stable five-membered ring. (See Section 9.8 in the text.) After the electrophile adds to the benzene ring, a base (B:) in the reaction mixture removes a proton and the aromaticity of the benzene ring is restored.

an intramolecular reaction

b. As in part **a**, an electrophile is formed that can react in either an intermolecular reaction or an intramolecular reaction. Seeing that the product of the reaction has two benzene rings and that there are twice as many carbons in the product as in the reactant indicates that two reactant molecules react in an intermolecular reaction. In this case, the intermolecular reaction is favored, because the intramolecular reaction would lead to a highly strained three-membered ring. The electrophile that is formed in the intermolecular reaction can add to the benzene ring in an intramolecular reaction to form a stable five-membered ring.

84.   a.

b.

85.

86.

A

B

C

D

E

F

G

H

I

J

87. The carbocation formed by putting an electrophile at the ortho or para positions can be stabilized by resonance electron donation from the phenyl substituent.

The carbocation formed by putting an electrophile at the meta position cannot be stabilized by resonance electron donation from the phenyl substituent.

88. **a.** The chloro substituent primarily withdraws electrons inductively. (It only minimally donates electrons by resonance.) The closer it is to the COOH group, the more it can withdraw electrons from the OH bond and the stronger the acid. Therefore, the ortho isomer is the strongest acid and the para isomer is the weakest acid.

b. The nitro substituent withdraws electrons inductively. It also withdraws electrons by resonance if it is ortho or para to the COOH group. Therefore, the ortho and para isomers are the strongest acids, and the ortho isomer is a stronger acid than the para isomer because of the greater inductive electron withdrawal from the closer position.

c. The amino substituent primarily donates electrons by resonance, but it can donate electrons by resonance to the COOH group only if it is ortho or para to it. Therefore, the meta isomer is the strongest acid. Seeing that the ortho isomer is the weakest acid tells us that resonance electron donation to the COOH group is more efficient from the ortho position.

89. The signal at 7.1-7.3 ppm that integrates to 5 protons indicates a monosubstituted benzene ring. The doublet at 1.3 ppm that integrates to 6 protons and the septet at 2.9 ppm that integrates to 1 proton indicate an isopropyl group. Therefore, compound **A** is isopropylbenzene.

90. **1.**

2.

91. **a.** The hydroxy-substituted carbocation intermediate is more stable, because the positive charge can be stabilized by resonance electron donation from the OH group.

b. The carbanion with the negative charge meta to the nitro group is more stable, because a negative charge in this position can be delocalized onto the nitro group but a negative charge ortho to the nitro group position cannot.

92. Synthesizing benzaldehyde from benzene would be easy if formyl chloride could be used, but this compound is unstable and must be generated in situ via a Gatterman-Koch reaction (see page 919 in the text). The synthesis of benzaldehyde shown here is done via a Friedel-Crafts alkylation reaction.

93. **a.**

CH$_3$Cl / AlCl$_3$ → (toluene)

Br$_2$ / FeBr$_3$ → (4-bromotoluene, Br)

$\underset{\text{HO}^-}{\overset{\text{PdL}_2}{\longrightarrow}}$ with B(OR)$_2$ reagent → (CH$_3$, CH$_3$C=CH$_2$)

or

$\underset{\text{AlCl}_3}{\overset{\text{CH}_3\text{CHCH}_3,\ \text{Cl}}{\longrightarrow}}$ → CH$_3$CHCH$_3$

$\overset{\text{Br}_2}{\underset{hv}{\longrightarrow}}$ → CH$_3$CCH$_3$, Br

$\xrightarrow{\textit{tert}\text{-BuO}^-}$ → CH$_3$C=CH$_2$

$\underset{\text{AlCl}_3}{\overset{\text{CH}_3\text{Cl}}{\longrightarrow}}$ → CH$_3$, CH$_3$C=CH$_2$

b.

$\overset{\text{Br}_2}{\underset{\text{FeBr}_3}{\longrightarrow}}$ → (bromobenzene, Br)

$\overset{\text{HNO}_3}{\underset{\text{H}_2\text{SO}_4}{\longrightarrow}}$ → Br, NO$_2$

$\overset{\text{Br}_2}{\underset{\text{FeBr}_3}{\longrightarrow}}$ → Br, Br, NO$_2$

$\overset{\Delta}{\underset{\text{CH}_3\text{O}^-}{\longrightarrow}}$ → OCH$_3$, Br, NO$_2$

c.

$\underset{\text{2. H}_2\text{O}}{\overset{\text{1. (CH}_3)_2\text{CHCCl, AlCl}_3}{\longrightarrow}}$ → CH(CH$_3$)$_2$, O

$\overset{\text{Zn(Hg)}}{\underset{\substack{\text{HCl, }\Delta\ \textbf{or}\ \text{H}_2\text{, Pd/C} \\ \textbf{or} \\ \text{H}_2\text{NNH}_2 \\ \text{HO}^-, \Delta}}{\longrightarrow}}$ → CH$_2$CH(CH$_3$)$_2$

$\underset{\text{2. H}_2\text{O}}{\overset{\text{1. CH}_3\text{CCl + AlCl}_3}{\longrightarrow}}$ → CH$_3$, O, CH$_2$CH(CH$_3$)$_2$

$\overset{\text{H}_2\text{SO}_4}{\longleftarrow}$ → O, CH$_3$, SO$_3$H, CH$_2$CH(CH$_3$)$_2$

d.

$\underset{\text{2. H}_2\text{O}}{\overset{\text{1. CH}_3\text{CCl, AlCl}_3}{\longrightarrow}}$ → CH$_3$, C, O

$\overset{\text{Br}_2}{\underset{\text{FeBr}_3}{\longrightarrow}}$ → CH$_3$, C, O, Br

$\overset{\text{H}_2}{\underset{\text{Pd/C}}{\downarrow}}$ → CH$_3$CH$_2$, Br

$\overset{(\text{CH}_3)_3\text{CCl}}{\underset{\text{AlCl}_3}{\longleftarrow}}$ → C(CH$_3$)$_3$, CH$_3$CH$_2$, Br

94. The unsplit signal at ~7.1 ppm suggests that all the benzene ring hydrogens in the product are chemically equivalent. The triplet at 1.2 ppm and quartet at 2.6 ppm indicate an ethyl group.

$$CH_3CH_2-\!\!\!\bigcirc\!\!\! \xrightarrow[\text{AlCl}_3]{\text{CH}_3\text{CH}_2\text{Cl}} CH_3CH_2-\!\!\!\bigcirc\!\!\!-CH_2CH_3$$

95. a. The first three compounds will not show a carbonyl stretch at 1700 cm^{-1}, and the bottom four will show this absorption band. The first three can be distinguished from one another by the presence or absence of the indicated absorption bands.

CH₂OH   CH₂OH   CH₂OCH₃

band at 3300 cm^{-1} band at 3300 cm^{-1} no band at 3300 cm^{-1}

no band at 1600 cm^{-1} band at 1600 cm^{-1} band at 1600 cm^{-1}

The last four compounds all have an absorption band at 1700 cm^{-1}.
They can be distinguished by the presence or absence of the indicated absorption bands.

O‖C–OH	O‖C–H	O‖C–OCH₃	O‖C–OCH₃

large broad band at $\sim 2500\text{–}3500 \text{ cm}^{-1}$

band at 1250 cm^{-1}

no band at 2700 cm^{-1}

no band at $\sim 2500\text{–}3500 \text{ cm}^{-1}$

no band at 1250 cm^{-1}

band at 2700 cm^{-1}

no band at $\sim 2500\text{–}3500 \text{ cm}^{-1}$

band at 1250 cm^{-1}

no band at 2700 cm^{-1}

no band at $\sim 2500\text{–}3500 \text{ cm}^{-1}$

no band at 1250 cm^{-1}

no band at 2700 cm^{-1}

b. This is the only compound without the characteristic benzene ring hydrogens at ~7−8 ppm.

CH₂OH

Only two compounds will have two signals other than the signals for the benzene ring hydrogens. They can be distinguished by integration (2:3 versus 2:1), or by the two sharp singlets for the ether versus the somewhat broader singlet for the hydrogen bonded to oxygen.

CH₂OCH₃   CH₂OH

Each of the following four compounds has only one signal (a singlet) in addition to the benzene ring hydrogens. The four can be distinguished by the position of the singlet.

~2 ppm   ~4 ppm   ~9–10 ppm   ~10–12 ppm

96. A fluoro substituent is more electronegative than a chloro substituent. Therefore, nucleophilic attack on the carbon bearing the fluoro substituent will be easier than nucleophilic attack on the carbon bearing the chloro substituent. In addition, the smaller fluoro substituent provides less steric hindrance to attack by the nucleophile.

A fluoro substituent is a stronger base than a chloro substituent, so elimination of the halogen in the second step of the reaction will be harder for a fluoro-substituted benzene than for a chloro-substituted benzene.

The fact that the fluoro-substituted compound is more reactive tells you that attack of the nucleophile on the aromatic ring is the rate-determining step of the reaction.

97. **a.** The alkyl diazonium ion is very unstable. Loss of N_2 and a 1,2-hydride shift forms a *tert*-butyl carbo-cation, which can undergo either substitution or elimination.

$$CH_3CHCH_2NH_2 \xrightarrow[\text{HCl}]{\text{NaNO}_2} CH_3CCH_2-N\equiv N \longrightarrow CH_3\overset{+}{C}CH_3 \longrightarrow$$

products: $CH_3C=CH_2 + H_3O^+$ (with CH_3) via H_2O; and $CH_3CCH_3 + H_3O^+$ (with OH and CH_3) via H_2O

b. The cation formed from the diazonium ion will undergo a pinacol-like rearrangement. (See Chapter 11, Problem 84.)

$$CH_3-\overset{OH}{\underset{CH_3}{C}}-\overset{NH_2}{\underset{CH_3}{C}}-CH_3 \xrightarrow[\text{HCl}]{\text{NaNO}_2} CH_3-\overset{OH}{\underset{CH_3}{C}}-\overset{\overset{+}{N}\equiv N}{\underset{CH_3}{C}}-CH_3 \longrightarrow CH_3-\overset{:\overset{..}{O}H}{\underset{CH_3}{C}}-\overset{+}{\underset{CH_3}{C}}-CH_3 + N_2$$

$$H^+ + CH_3-\overset{O}{C}-\overset{CH_3}{\underset{CH_3}{C}}-CH_3 \longleftarrow CH_3-\overset{+OH}{C}-\overset{CH_3}{\underset{CH_3}{C}}-CH_3$$

98.

For the last reaction, see Problem 16 on page 416.

99.   The configuration of the asymmetric center in the reactant will be retained only if the asymmetric center undergoes two successive S_N2 reactions. In the first step of the reaction, the NH_2 group is converted to a good leaving group. The first S_N2 reaction involves the carboxylate ion that is closest to the carbon attached to the leaving group. The product of the reaction is a strained three-membered ring that reacts with the other carboxylate ion in a second S_N2 reaction, forming a more stable five-membered ring.

100.   A chloro group is a better leaving group than the ammonium group, so the product is formed without hydroxide ion catalysis.

A methoxy group is a poorer leaving group than the ammonium group, so the ammonium group is eliminated, reforming starting materials. If hydroxide ion is added to the solution, hydroxide ion converts the ammonium group into an amino group. Since the amino group is a poorer leaving group than the methoxy group, the methoxy group will be eliminated.

101.   a.

b.

102.

103. **a.**

$CH_2CH_2NH_2$ (on benzene ring)

$\xrightarrow{CH_3CCl \text{ (O)}}$

$CH_2CH_2NHCCH_3$ (on benzene ring, amide C=O)

$\xrightarrow[H_2SO_4]{HNO_3}$

$CH_2CH_2NHCCH_3$ (on benzene ring with para NO_2, amide C=O)

$\xdownarrow{H_2, Pd/C}$

$CH_2CH_2NHCCH_3$ (on benzene ring with para NH_2, amide C=O)

$\xleftarrow[0\,°C]{HCl, NaNO_2}$

$CH_2CH_2NHCCH_3$ (on benzene ring with para $\overset{+}{N}\!\equiv\!N$ diazonium, amide C=O)

$\xleftarrow[\substack{Cu(NO_3)_2,\\ H_2O\\ \textbf{or}\\ H_3O^+,\ \Delta}]{Cu_2O}$

$CH_2CH_2NHCCH_3$ (on benzene ring with para OH, amide C=O)

$\xdownarrow{\substack{1.\ HCl,\ H_2O,\ \Delta\\ 2.\ HO^-}}$

$CH_2CH_2NH_2$ (on benzene ring with para OH)

b.

$CH_2CH_2NH_2$ (on benzene ring with para OH)

$\xrightarrow[H_2SO_4]{HNO_3}$

$CH_2CH_2\overset{+}{N}H_3$ (on benzene ring with NO_2 ortho to OH, para OH)

$\xrightarrow[2.\ HO^-]{1.\ H_2,\ Pd/C}$

$CH_2CH_2NH_2$ (on benzene ring with NH_2 ortho to OH, para OH)

$\xdownarrow[0\,°C]{\substack{HCl\\ NaNO_2}}$

$CH_2CH_2\overset{+}{N}\!\equiv\!N$ (on benzene ring with $\overset{+}{N}\!\equiv\!N$ ortho to OH)

$\xleftarrow[\substack{Cu(NO_3)_2,\\ H_2O\\ \textbf{or}\\ H_3O^+,\ \Delta}]{Cu_2O}$

CH_2CH_2OH (on benzene ring with ortho OH and para OH)

$\xleftarrow[\substack{2.\ \bar{N}_3\\ 3.\ H_2,\ Pd/C}]{1.\ PCl_3}$

$CH_2CH_2NH_2$ (on benzene ring with ortho OH and para OH)

c.

CH₂CH₂Cl

$\xrightarrow[\text{2. H}_2\text{, Pd/C}]{\text{1. }\bar{\text{N}}_3}$

CH₂CH₂NH₂

CH₂CH₂Cl

$\xrightarrow[\substack{\text{2. HCl, H}_2\text{O, }\Delta \\ \text{3. HO}^-}]{\text{1. phthalimide anion}}$

CH₂CH₂NH₂

d.

CH₂Cl

$\xrightarrow{\text{}^-\text{C}\equiv\text{N}}$

CH₂C≡N

$\xrightarrow[\text{Raney Ni}]{\text{H}_2}$

CH₂CH₂NH₂

e.

O
‖
CH

$\xrightarrow[\text{2. H}_3\text{O}^+]{\text{1. NaBH}_4}$

CH₂OH

$\xrightarrow{\text{SOCl}_2}$

CH₂Cl

$\xrightarrow{\text{}^-\text{C}\equiv\text{N}}$

CH₂C≡N

$\xrightarrow[\text{Raney Ni}]{\text{H}_2}$

CH₂CH₂NH₂

104.

105.   a.

b.

c.

106. **a.**

b. The substituent in ibuprofen is placed on the ring by a Friedel-Crafts acylation reaction. A Friedel-Crafts reaction cannot be done in the synthesis of ketoprofen, because the benzene rings are deactivated and deactivated rings cannot undergo Friedel-Crafts reactions.

107.

108.   Notice that the Friedel-Crafts reaction can be done on a ring with a sulfonic acid group, because its deactivation is offset by the activating amide group. The sulfonic acid group is put on the ring to block the para position and thereby force the methyl groups to go to the ortho positions.

109.

Chapter 19 Practice Test

1.   Name each of the following:

a. [structure: benzene ring with CH₃ and NO₂ (meta)]

b. [structure: benzene ring with Br, Br, Br]

c. [structure: benzene ring with COOH and CH₂CH₃]

d. [structure: benzene ring with Cl and OH (para)]

2.   Rank the following compounds in order from most reactive to least reactive toward reaction with $Br_2/FeBr_3$:

[structures: benzene-Br; benzene-NH-C(=O)-CH₃; benzene-CH₂CH₃; benzene-C(=O)-NHCH₃; benzene]

3.   For each of the following pairs of compounds, circle the one that is the stronger acid:

a. [benzene with COOH and Cl (para)] or [benzene with COOH and CH₃ (para)]

c. [benzene with ⁺NH₃ and CH₃ (para)] or [benzene with ⁺NH₃ and OCH₃ (para)]

b. [benzene with OH and NO₂ (para)] or [benzene with OH and Br (para)]

d. [benzene with COOH and CH₃ (para)] or [benzene with COOH]

4.   a.   Which is more reactive in a nucleophilic substitution reaction, *para*-bromonitrobenzene or *para*-bromoethylbenzene?

b.   Which is more reactive in an electrophilic substitution reaction, *para*-bromonitrobenzene or *para*-bromoethylbenzene?

5.   What acid anhydride would you use in the synthesis of propylbenzene?

6.   Draw the mechanism for the formation of the nitronium ion from nitric acid and sulfuric acid.

7.   Draw the major product(s) of each of the following reactions:

a.
NO$_2$ (benzene ring)   +   H$_2$SO$_4$ $\longrightarrow$

b.
OCH$_3$ (benzene ring)   +   CH$_3$Cl $\xrightarrow{\text{AlCl}_3}$

c.
CH$_2$CH$_3$ (benzene ring) $\xrightarrow[\Delta]{\text{H}_2\text{CrO}_4}$

d.
Cl / NO$_2$ (benzene ring)   +   CH$_3$O$^-$ $\xrightarrow{\Delta}$

e.
Cl / OCH$_3$ (benzene ring)   +   HNO$_3$ $\xrightarrow{\text{H}_2\text{SO}_4}$

f.
(benzene ring with C(=O)CH$_3$)   +   Cl$_2$ $\xrightarrow{\text{FeCl}_3}$

8.   Indicate whether each of the following statements is true or false:

a. Benzoic acid is more reactive than benzene toward electrophilic aromatic substitution.   T   F

b. *para*-Chlorobenzoic acid is more acidic than *para*-methoxybenzoic acid.   T   F

c. A CH=CH$_2$ group is a meta director.   T   F

d. *para*-Nitroaniline is more basic than *para*-chloroaniline.   T   F

9.   Draw the resonance contributors for the carbocation intermediate that is formed when benzene reacts with an electrophile (Y$^+$).

<u>**CHAPTER 20**</u>
More About Amines • Reactions of Heterocyclic Compounds

<u>**Important Terms**</u>

furan	a five-membered ring aromatic compound containing an oxygen heteroatom.
heteroatom	an atom other than a carbon or a hydrogen.
heterocyclic compound (heterocycle)	a cyclic compound in which one or more of the ring atoms are heteroatoms.
imidazole	a five-membered ring aromatic compound containing two nitrogen heteroatoms.
ligation	sharing of nonbonded electrons with a metal.
porphyrin ring system	a compound that consists of four pyrrole rings joined by one-carbon bridges.
purine	a pyrimidine ring fused to an imidazole ring.
pyrimidine	a benzene ring with nitrogens at the 1- and 3-positions.
pyrrole	a five-membered ring aromatic compound containing a nitrogen heteroatom.
thiophene	a five-membered ring aromatic compound containing a sulfur heteroatom.

Solutions to Problems

1.
    a. 2,2-dimethylazacyclopropane or
2,2-dimethylaziridine

    b. 4-ethylazacyclohexane
4-ethylpiperidine

    c. 2-methylthiacyclopropane or
2-methylthiirane

    d. 3-methylazacyclobutane or
3-methylazetidine

    e. 2,3-dimethyloxacyclopentane or
2,3-dimethyltetrahydrofuran

    f. 2-ethyloxacyclobutane or
2-ethyloxetane

2. The oxygen in morpholine withdraws electrons inductively, which make protonated morpholine the stronger acid. Recall that inductive electron withdrawal increases acidity (Section 2.7).

$pK_a = 9.28$ morpholine $pK_a = 11.12$ piperidine

3.   **a.**

    b. We would predict that the conjugate acid of 3-quinuclidinone has a lower pK_a than the conjugate acid of morpholine because the sp^2 oxygen of 3-quinucldinone is more electronegative than the sp^3 oxygen of morpholine. 3-quinuclidinone is closer to the nitrogen.

conjugate acid
of 3-quinuclidinone

$pK_a = 7.46$

conjugate acid
of morpholine

$pK_a = 9.28$

    c. The conjugate acid of 3-chloroquinuclidine has a lower pK_a than the conjugate acid of 3-bromoquinuclidine because chlorine is more electronegative than bromine, so it is better at withdrawing electrons inductively.

conjugate acid of
3-chloroquinuclidine

conjugate acid of
3-bromoquinuclidine

4.   Solved in the text.

5.   a.

c.

b.

d.

6.

7.

2-deuteriopyrrole

8.   Pyrrole and its conjugate base are both aromatic. Cyclopentadiene does not become aromatic until it loses a proton. It is the drive for the nonaromatic compound to become a stable aromatic compound that causes cyclopentadiene to be a stronger acid than pyrrole.

aromatic   **aromatic**

nonaromatic   **aromatic**

9.   Solved in the text.

10.   Pyridine will act as an amine with the alkyl bromide, forming a quaternary ammonium salt.

11. The first reaction is a two-step nucleophilic aromatic substitution reaction (S_NAr).

The second reaction is a one-step nucleophilic substitution reaction (S_N2).

12. a.

the hydride ion
is a better leaving
group than O^{2-}

2-pyridone

b. 4-Pyridone is also formed because nucleophilic addition of hydroxide ion can take place at the 4-position as well as at the 2-position. It proceeds by the same mechanism as the one shown in part **a** for the formation of 2-pyridone.

4-pyridone

13. It is easiest to remove a proton from the methyl group of the *N*-alkylated pyridine because the electrons left behind when the proton is removed can be delocalized onto the positively charged nitrogen. (A positively charged nitrogen more readily accepts the delocalized electrons than does a neutral nitrogen.)

It is easier to remove a proton from 4-methylpyridine than from 3-methylpyridine because in the former, the electrons left behind when the proton is removed can be delocalized onto the electronegative nitrogen. In contrast, in 3-methylpyridine the electrons can be delocalized onto only carbons.

14.   There are three possible sites for electrophilic substitution: C-2, C-4, and C-5. To determine the major product, compare the relative stabilities of the carbocations formed in the first step of the reaction.

Substitution at C-2

Substitution at C-2 forms an intermediate with three resonance contributors; all the atoms in one contributor have complete octets and a positive charge on N; one contributor has a carbon with an incomplete octet, and one has a nitrogen with an incomplete octet.

Substitution at C-4

Substitution at C-4 forms an intermediate with two resonance contributors; all the atoms in one contributor have complete octets and a positive charge on N; one contributor has a carbon with an incomplete octet.

Substitution at C-5

Substitution at C-5 forms an intermediate with three resonance contributors; all the atoms in one contributor have complete octets and a positive charge on N; two contributors have a carbon with an incomplete octet.

Substitution at C-4 forms the least stable intermediate because it has only two of the three resonance contributors that the others have. Substitution at C-5 forms the most stable intermediate because a positively charged carbon with an incomplete octet is more stable than a positively charged nitrogen with an incomplete octet.

Therefore, the major product of the reaction is 5-bromo-*N*-methylimidazole.

5-bromo-*N*-methylimidazole

15.   Pyrrole and imidazole are both more reactive than benzene, because each reacts with an electrophile to form a carbocation intermediate that can be stabilized by resonance electron donation into the ring by a nitrogen atom.

Pyrrole is more reactive than imidazole, because the second nitrogen atom of imidazole cannot donate electrons into the ring by resonance but can withdraw electrons from the ring inductively.

pyrrole   imidazole   benzene

16.   Imidazole forms intermolecular hydrogen bonds, whereas *N*-methylimidazole cannot form hydrogen bonds because it does not have a hydrogen bonded to a nitrogen.

Because the hydrogen bonds have to be broken in order for the compound to boil, imidazole has a higher boiling point.

17.   The second nitrogen in imidazole, onto which the electrons left behind when the proton is removed can be delocalized, causes imidazole to be a stronger acid than pyrrole because nitrogen is more electronegative than carbon.

imidazole   pyrrole

$pK_a = 14.4$ $pK_a = 17$

18.   For the derivation of the equation used in this problem, see Special Topic I, page 46.

$$\text{fraction of imidazole in the acidic form} = \frac{[H^+]}{K_a + [H^+]}$$

$$pH = 7.4; [H^+] = 4.0 \times 10^{-8}$$

$$pK_a = 6.8; K_a = 1.58 \times 10^{-7} \text{ (see Table 20.1)}$$

$$\text{fraction of imidazole in the acidic form} = \frac{4.0 \times 10^{-8}}{1.58 \times 10^{-7} + 4.0 \times 10^{-8}}$$

$$= \frac{4.0 \times 10^{-8}}{1.98 \times 10^{-7}}$$

$$= 0.20$$

$$\text{percent of imidazole in the acidic form} = 20\%$$

19.

enol form of guanine   enol form of cytosine

20.   The second nitrogen in protonated pyrimidine (that withdraws electrons inductively) causes it to be a stronger acid than protonated pyridine.

protonated pyrimidine   protonated pyridine
$pK_a = 1.0$ $pK_a = 5.2$

21.   **a.** 2-methylazacyclobutane or 2-methylazetidine
b. 2,3-dimethylazacyclohexane or 2,3-dimethylpiperidine
c. 3-chloropyrrole
d. 2-ethyl-5-methylazacyclohexane or 2-ethyl-5-methylpiperidine

22.   **a.**

b.

c. $CH_3CH_2CH_2CH_2CH_2NH_2$

d.

e.

f.

$\overset{+}{N}H_2CH_2CH_2CH_3$ Br^-

$\overset{+}{N}H(CH_2CH_2CH_3)_2$ Br^-

$\overset{+}{N}(CH_2CH_2CH_3)_3$ Br^-

g.

h.

$CH_2CH_2CH_2CH_3$

i.

COOH

23.

A is the most acidic, because it becomes aromatic when it loses a proton.

B, **C**, and **D** are the next most acidic, because in all three, the proton is bonded to a positively charged nitrogen.

B and **C** are more acidic than **D**, because in **B** and **C** the proton to be lost is bonded to an sp^2 nitrogen, which is more electronegative than the sp^3 nitrogen in **D**.

B is more acidic than **C**, because the uncharged nitrogen in **C** can donate electrons by resonance to the positively charged nitrogen.

Neutral compounds **E**, **F**, and **G** are the least acidic. **E** and **F** are more acidic than **G**, because **E** and **F** lose a proton from an sp^2 nitrogen, whereas a proton is lost from a less electronegative sp^3 nitrogen in **G**.

E is more acidic than **F**, because the negative charge on **E**'s conjugate base can be delocalized onto the second nitrogen.

24.   The compound on the right (the compound shown below) is easier to decarboxylate, because the electrons left behind when CO_2 is removed can be delocalized onto nitrogen, forming a stable neutral species as a result of proton donation to nitrogen. The electrons left behind when the other compound loses CO_2 cannot be delocalized.

25.   The *N*-, *O*-, and *S*-substituted benzenes have the same relative reactivity toward electrophilic aromatic substitution as the *N*-, *O*-, and *S*-containing five-membered heterocyclic rings, and for the same reason.

The *N*-substituted benzene is more reactive than the *O*-substituted benzene, because nitrogen is more effective than oxygen at donating electrons to the benzene ring because it is less electronegative than oxygen. (Recall that electrophilic substitution is aided by electron donation to the ring because it stabilizes the carbocation formed in the rate-limiting step.)

The *S*-substituted benzene is the least reactive because the lone-pair electrons of sulfur are in its $3p$ orbital, whereas the lone-pair electrons of nitrogen and oxygen are in its $2p$ orbital. Electron delocalization by overlap of the $3p$ orbital of sulfur with the $2p$ orbital of carbon is less effective than electron delocalization by overlap of the $2p$ orbital of nitrogen with the $2p$ orbital of carbon.

26. By comparing the carbocation intermediates formed in the rate-determining step when the amino-substituted compound undergoes electrophilic substitution at C-3 and C-4, you can see that the carbocation formed in the case of substitution at C-3 is more stable; one of its resonance contributors is relatively stable because it is the only one that does not have an atom with an incomplete octet. As a result, the amino-substituted compound undergoes electrophilic substitution predominantly at C-3. Therefore, the keto-substituted compound is the one that undergoes electrophilic substitution predominantly at C-4.

relatively stable

27. a. The Lewis acid, AlCl₃, complexes with nitrogen, causing the aziridine ring to open when it is attacked by the nucleophilic benzene ring. The ring will open in the direction that puts the partial positive charge on the more substituted carbon because it is more stable than the transition state that would have the positive charge on the less substituted carbon.

Therefore, the major and minor products are those shown.

major   minor

b. Yes, epoxides can undergo similar reactions.

28.   Oxygen is the negative end of the dipole in both compounds. Tetrahydrofuran has the greater dipole moment because in furan the effect of the electron-withdrawing oxygen is mitigated by the ability of oxygen to donate electrons by resonance into the ring.

tetrahydrofuran   furan
1.73 D   0.70 D

29.   The products of the reaction show that the hydrogen is removed from the β-carbon that is bonded to the most hydrogens. Therefore, we can conclude that the transition state has a carbanion-like one. (See page 451 in the text.)

30.   **a.**   **c.**

b.   **d.**

The carbon of the methyl group is the β-carbon that is bonded to the most hydrogens.

31.   Pyrrolidine is a saturated nonaromatic compound, whereas pyrrole and pyridine are unsaturated aromatic compounds. The C-2 hydrogens of pyrrolidine are at δ 2.82 ppm, about where one would expect the signal for hydrogens bonded to an sp^3 carbon adjacent to an electron-withdrawing amino group.

The C-2 hydrogens of pyrrole and pyridine are expected to be at a higher frequency because of diamagnetic anisotropy (Section 15.8). Because the nitrogen of pyrrole donates electrons into the ring and the nitrogen of pyridine withdraws electrons from the ring, the C-2 hydrogens of pyrrole are in an environment with a greater electron density, so they should show a signal at a lower frequency relative to the C-2 hydrogens of pyridine. Thus, the C-2 hydrogens of pyrrole are at δ 6.42 ppm, and the C-2 hydrogens of pyridine are at δ 8.50 ppm.

32.   A UV spectrum results from the π electron system. The lone-pair electrons on the nitrogen in aniline are delocalized into the benzene ring and, thus, are part of the π system. Protonation of aniline removes two electrons from the π system. This has a significant effect on its UV spectrum.

The lone-pair electrons on the nitrogen atom in pyridine are sp^2 electrons and thus are not part of the π system. Protonation of pyridine, therefore, does not remove any electrons from the π system and has only a minor effect on the UV spectrum.

33.   When ammonia loses a proton, the electrons left behind remain on nitrogen.

When pyrrole loses a proton, the electrons left behind can be delocalized onto the four ring carbons. Electron delocalization stabilizes the anion and makes it easier to form. Recall that stabilizing the base increases the acidity of its conjugate acid.

34.   The electrophile adds preferentially to the 2-position.

35.   Before we can answer the questions, we must figure out the mechanism of the reaction. Once the mechanism is known, it will be relatively easy to determine how a change in a reactant will affect the product. The mechanism is shown below.

Propenal, an α,β-unsaturated aldehyde, undergoes a conjugate addition reaction with aniline. This is followed by an intramolecular electrophilic aromatic substitution reaction. Dehydration of the alcohol results in 1,2-dihydroquinoline, which is oxidized to quinoline by nitrobenzene.

a.

b.

c.

meta-ethylaniline   2-methyl-2-pentenal   2,7-diethyl-3-methylquinoline

36.   **a.**   The electrophile adds to the 3-position.

b.   The electrophile adds preferentially to the 2-position.

37.   **a.**

You can see why the nitro substituent goes to this position by examining the relative stabilities of the possible carbocation intermediates.

The carbocation has three resonance contributors.

The carbocation has two resonance contributors.

relatively unstable The carbocation has three resonance contributors, but the first one is relatively unstable because the positive charge is on the carbon attached to the electron-withdrawing substituent.

b.

You can understand why the bromo substituent goes to this position by examining the relative stabilities of the possible carbocation intermediates.

relatively unstable

relatively unstable

c.

The other nitrogen is not alkylated, because its lone pair is delocalized into the pyridine ring, so it is not available to react with the alkyl halide.

d.

PCl$_3$ substitutes a Cl for an OH, as it does in alcohols and carboxylic acids.

e. In the first step, hydroxide ion removes the most acidic hydrogen from the compound. A hydrogen bonded to the C-4 methyl group is the most acidic hydrogen, because the electrons left behind when the proton is removed can be delocalized onto the positively charged nitrogen of the pyridine ring.

f.

38.   The increase in the electron density of the ring as a result of resonance electron donation by oxygen causes pyridine-*N*-oxide to be more reactive toward electrophilic aromatic substitution than pyridine because the extra electron-density will stabilize the carbocation intermediate.

In Section 19.15, we saw that substituents that are able to donate electrons by resonance into the ring are ortho/para directors. Therefore, pyridine-*N*-oxide will undergo electrophilic aromatic substitution at the 2- and 4-positions. Because the 2-positions are somewhat sterically hindered, pyridine-*N*-oxide undergoes electrophilic substitution primarily at the 4-position.

39.

40.   The electrophile adds preferentially to the 2-position.

pyrrole   *para*-(*N,N*-dimethylamino)-
            benzaldehyde

Repeat the
previous series
of reactions
adding to the
5-position of
the pyrrole ring.

colored compound

41.

42. **a.**

b.

c.

43.

Repeat two more times.

1. Repeat with benzaldehyde.
2. Instead of using another molecule of pyrrole, do an intramolecular reaction with the pyrrole at the end of the chain.

Chapter 20 Practice Test

1. Give two names for each of the following compounds:

a.

b.

c.

d.

2. Draw the product of each of the following reactions:

a.   +   2 CH$_3$CH$_2$NH$_2$ $\longrightarrow$

b.   + $\xrightarrow{\text{trace acid}}$

c.   +   Br$_2$ $\xrightarrow[\text{300 °C}]{\text{FeBr}_3}$

d.   +   CH$_3$I $\longrightarrow$

e.   +   Cl$_2$ $\longrightarrow$

3. Which of the following is a stronger acid?

a.   or

b.   or

c.   or

d.   or

4. Indicate whether each of the following statements is true or false:

a. 4-Chloropyridine is more reactive toward nucleophilic aromatic substitution than is 3-chloropyrrole.   T   F

b. Pyrrole is more reactive toward electrophilic aromatic substitution than is furan.   T   F

c. Pyrrole is more reactive toward electrophilic aromatic substitution than is benzene.   T   F

d. Pyridine is more reactive toward electrophilic aromatic substitution than is benzene.   T   F

5. Show how the following compound could be prepared from the given starting material:

CHAPTER 21
The Organic Chemistry of Carbohydrates

Important Terms

aldaric acid	an aldose whose aldehyde and primary alcohol groups have been oxidized to carboxylic acids.
alditol	an aldose or a ketose whose carbonyl group has been reduced to an alcohol.
aldonic acid	an aldose whose aldehyde group has been oxidized to a carboxylic acid.
aldose	a polyhydroxy aldehyde.
amino sugar	a sugar in which one of the OH groups is replaced by an NH_2 group.
anomeric carbon	the carbon in a cyclic sugar that is the carbonyl carbon in the straight-chain form.
anomeric effect	preference for the axial position by certain substituents bonded to the anomeric carbon.
anomers	two cyclic sugars that differ in configuration only at the carbon that is the carbonyl carbon in the straight-chain form.
bioorganic compound	an organic compound that is found in a biological system.
carbohydrate	a sugar, a saccharide. Naturally occurring carbohydrates have the D-configuration.
complex carbohydrate	a carbohydrate that contains two or more sugar molecules linked together; it can be hydrolyzed to simple sugars.
deoxy sugar	a sugar in which one of the OH groups has been replaced by a hydrogen.
disaccharide	a compound containing two sugar molecules linked together.
enediol rearrangement	a base-catalyzed reaction that interconverts monosaccharides.
epimerization	changing the configuration of a carbon by removing a proton and then reprotonating it.
epimers	monosaccharides that differ in configuration at only one carbon.
furanose	a five-membered ring sugar.
furanoside	a five-membered ring glycoside.
glycoprotein	a protein that is covalently bonded to an oligosaccharide.
glycoside	the acetal of a sugar.

653

N-glycoside	a glycoside with a nitrogen instead of an oxygen at the glycosidic linkage.
glycosidic bond	the bond between the anomeric carbon of one sugar and an alcohol residue of a second sugar in a glycoside.
α-1,4$'$-glycosidic linkage	a glycosidic linkage between the C-1 of one sugar and the C-4 of a second sugar with the oxygen atom at C-1 in the axial position.
α-1,6$'$-glycosidic linkage	a glycosidic linkage between the C-1 of one sugar and the C-6 of a second sugar with the oxygen atom at C-1 in the axial position.
β-1,4$'$-glycosidic linkage	a glycosidic linkage between the C-1 of one sugar and the C-4 of a second sugar with the oxygen atom at C-1 in the equatorial position.
Haworth projection	a way to show the structure of a sugar in which the five- and six-membered rings are represented as being flat.
heptose	a monosaccharide with seven carbons.
hexose	a monosaccharide with six carbons.
ketose	a polyhydroxy ketone.
Kiliani-Fischer synthesis	a method used to increase the number of carbons in an aldose by one, resulting in the formation of a pair of C-2 epimers.
molecular recognition	the recognition of one molecule by another as a result of specific interactions.
monosaccharide	a single sugar molecule.
mutarotation	a slow change in optical rotation to an equilibrium value.
nonreducing sugar	a sugar that cannot be oxidized by reagents such as Ag^+ and Cu^+. Nonreducing sugars are not in equilibrium with the open-chain aldose or ketose.
oligosaccharide	three to ten sugar molecules linked by glycosidic bonds.
oxocarbenium ion	an ion in which the positive charge is shared by a carbon and an oxygen.
pentose	a monosaccharide with five carbons.
polysaccharide	a compound containing 10 or more sugar molecules linked together.
pyranose	a six-membered ring sugar.
pyranoside	a six-membered ring glycoside.

reducing sugar	a sugar that can be oxidized by reagents such as Ag^+ and Cu^+. Reducing sugars are in equilibrium with the open-chain aldose or ketose forms.
simple carbohydrate	a single sugar molecule.
tetrose	a monosaccharide with four carbons.
triose	a monosaccharide with three carbons.
Wohl degradation	a method used to shorten an aldose by one carbon.

Solutions to Problems

1.   D-Ribose is an aldopentose.
D-Sedoheptulose is a ketoheptose.
D-Mannose is an aldohexose.

2.   Notice that an L-sugar is the mirror image of a D-sugar.

L-glucose   L-fructose

3.   First determine whether the structure represents R-glyceraldehyde or S-glyceraldehyde. Then, since R-glyceraldehyde = D-glyceraldehyde and S-glyceraldehyde = L-glyceraldehyde, we can answer the question.

L-glyceraldehyde   L-glyceraldehyde   D-glyceraldehyde

4.   **a.** enantiomers because they are mirror images
b. diastereomers because the configuration of one asymmetric center is the same in both and the configuration of the other asymmetric center is the opposite in both

5.   **a.** D-ribose   **b.** L-talose   **c.** L-allose   **d.** L-ribose

6.   **a.** D-glucose = $(2R,3S,4R,5R)$-2,3,4,5,6-pentahydroxyhexanal
b. Because D-mannose is the C-2 epimer of D-glucose, the systematic name of D-mannose can be obtained just by changing the configuration of the C-2 carbon in the systematic name of D-glucose.

D-mannose = $(2S,3S,4R,5R)$-2,3,4,5,6-pentahydroxyhexanal

c. D-Galactose is the C-4 epimer of D-glucose. Therefore, each of its carbon atoms, except C-4, has the same configuration as it has in D-glucose.

D-galactose = $(2R,3S,4S,5R)$-2,3,4,5,6-pentahydroxyhexanal

d. L-Glucose is the mirror image of D-glucose, so each carbon in L-glucose has the opposite configuration to that in D-glucose.

L-glucose = $(2S,3R,4S,5S)$-2,3,4,5,6-pentahydroxyhexanal

7.   D-psicose

8.   **a.** A ketoheptose has four asymmetric centers ($2^4 = 16$ stereoisomers).

   b. An aldoheptose has five asymmetric centers ($2^5 = 32$ stereoisomers).

   c. A ketotriose does not have an asymmetric center; therefore, it has no stereoisomers.

9.

D-fructose   an enolate ion   an enediol

10.   Removal of an α-hydrogen creates an enol that can enolize back to the ketone (using the OH at C-2) or can enolize to an aldehyde (using the OH at C-1). The aldehyde has a new asymmetric center (indicated by an *); one of the epimers is D-glucose, and the other is D-mannose.

D-fructose

D-glucose
D-mannose

11.   D-talose and D-galactose in addition to D-tagatose and D-sorbose.

D-tagatose

D-talose   D-galactose   D-tagatose   D-sorbose

12.   a.   When D-idose is reduced, D-iditol is formed.

D-idose   D-iditol

b. When D-sorbose is reduced, C-2 becomes an asymmetric center, so both D-iditol and the C-2 epimer of D-iditol (D-gulitol) are formed.

D-sorbose   →(1. NaBH$_4$ 2. H$_3$O$^+$)→   D-iditol   +   D-gulitol

13. **a.** **1.** D-Altrose is reduced to the same alditol that D-talose is reduced to. The easiest way to answer this question is to draw D-talose and its alditol. Then draw the monosaccharide with the same configuration at C-2, C-3, C-4, and C-5 as D-talose, reversing the functional groups at C-1 and C-6. (Put the primary alcohol group at the top and the aldehyde group at the bottom.) When this structure is rotated 180° in the plane of the paper, the monosaccharide can be identified.

D-talose   →(1. NaBH$_4$ 2. H$_3$O$^+$)→   the alditol of D-talose   ←(1. NaBH$_4$ 2. H$_3$O$^+$)   rotate 180°   D-altrose

2. L-Gulose is reduced to the same alditol that D-glucose is reduced to.

D-glucose   →(1. NaBH$_4$ 2. H$_3$O$^+$)→   the alditol of D-glucose   ←(1. NaBH$_4$ 2. H$_3$O$^+$)   rotate 180°   L-gulose

3. L-Galactose is reduced to the same alditol that D-galactose is reduced to.

$$
\begin{array}{c}
\text{D-galactose} \\
\underset{\text{CH}_2\text{OH}}{\overset{\text{HC=O}}{\mid}}
\quad\quad\text{H—OH}\quad\text{HO—H}\quad\text{HO—H}\quad\text{H—OH}
\end{array}
$$

D-galactose   1. NaBH₄ / 2. H₃O⁺ → the alditol of D-galactose ← 1. NaBH₄ / 2. H₃O⁺   (rotate 180°)   L-galactose

b. 1. The ketohexose (D-tagatose) with the same configuration at C-3, C-4, and C-5 as D-talose will form the same alditol (D-talitol) that D-talose forms. The other alditol (D-galactitol) that is formed is the one with the opposite configuration at C-2.

D-tagatose   1. NaBH₄ / 2. H₃O⁺ → D-talitol + D-galactitol

2. Similarly, the ketohexose (D-psicose) with the same configuration at C-3, C-4, and C-5 as D-allose will form the same alditol that D-allose forms. The other alditol (D-altritol) is the one with the opposite configuration at C-2.

D-psicose   1. NaBH₄ / 2. H₃O⁺ → D-allitol + D-altritol

14.   **a.**   L-gulose (Note the similarity of this problem to Problem **13. a. 2.**)

D-glucose   D-glucaric acid   L-gulose

b.   L-Gularic acid, because it is also the oxidation product of L-gulose.

c.   D-allose and L-allose, D-altrose and D-talose, L-altrose and L-talose, D-galactose and L-galactose

15.   The monosaccharides are the aldoses with one additional carbon that are C-2 epimers and whose other asymmetric centers have the same configuration as the given aldose.

a.   D-gulose and D-idose   **b.**   L-xylose and L-lyxose

16.   The monosaccharides are the aldoses whose C-1 has been removed, whose C-2 carbon has been converted to an aldehyde, and whose other asymmetric centers have the same configuration as the given aldose.

a.   D-allose and D-altrose   **b.**   D-glucose and D-mannose   **c.**   L-allose and L-altrose

17.   Solved in the text.

18.

D-glucose   D-mannose   D-arabinose   D-erythrose

  A   B   C   D

19.   The hemiacetals in **a** and **b** have one asymmetric center; therefore, each has two stereoisomers. The hemiacetals in **c** and **b** have two asymmetric centers; therefore, each has four stereoisomers.

a.   Solved in the text.

b.

c.

d.

20.   **a.**

b.

c.

D-glucose   L-glucose
the mirror image of D-glucose

21.

Copyright © 2014 Pearson Education, Inc.

22.

D-erythrose   α-D-erythrofuranose   β-D-erythrofuranose

23.   First recall that, in the chair conformation, an α-anomer has the anomeric OH group in the axial position and a β-anomer has the anomeric OH group in the equatorial position. Then recall that glucose has all of its OH groups in equatorial positions. Now this question can be answered easily.

    a. Solved in the text.

    b. D-Idose differs in configuration from D-glucose at C-2, C-3, and C-4. Therefore, the OH groups at C-2, C-3, and C-4 in β-D-idose are in axial positions.

    c. D-Allose is a C-3 epimer of D-glucose. Therefore, the OH group at C-3 is in the axial position and, because it is the α-anomer, the OH group at C-1 (the anomeric carbon) is also in the axial position.

24.

β-D-galactose   ethyl β-D-galactoside   ethyl α-D-galactoside

25.   If more than a trace amount of acid is used, the amine that acts as a nucleophile when it forms the *N*-glycoside becomes protonated, and a protonated amine is not a nucleophile.

26.   **a.** Solved in the text.

    b. α-D-talose (or α-D-talopyranose)   (reducing; it is a hemiacetal)

    c. methyl α-D-galactoside (or methyl α-D-galactopyranoside)   (nonreducing; it is an acetal)

    d. ethyl β-D-psicoside (or ethyl β-D-psicofuranoside)   (nonreducing; it is an acetal)

27.   specific rotation of glucose + specific rotation of fructose = −22.0

$$+52.7 + \text{specific rotation of fructose} = -22.0$$
$$\text{specific rotation of fructose} = -22.0 + (-52.7)$$
$$\text{specific rotation of fructose} = -74.7$$

28.   **a.** Amylose has α-1,4′-glycosidic linkages, whereas cellulose has β-1,4′-glycosidic linkages.

    b. Amylose has α-1,4′-glycosidic linkages, whereas amylopectin has both α-1,4′-glycosidic linkages and α-1,6′-glycosidic linkages.

    c. Glycogen and amylopectin have the same kind of linkages, but glycogen has a higher frequency of α-1,6′-glycosidic linkages.

    d. Cellulose has a hydroxy group at C-2, whereas chitin has an *N*-acetyl amino group at that position.

29. A proton is more easily lost from the C-3 OH group because the electron that are left behind when the proton is removed are delocalized onto an oxygen. When a proton is removed from the C-2 OH group, the electrons that are left behind are delocalized onto a carbon. Because oxygen is more electronegative than carbon, a negatively charged oxygen is more stable than a negatively charged carbon. Recall that the more stable the base, the stronger its conjugate acid.

30. a. People with type O blood can receive blood only from other people with type O blood because types A, B, and AB blood have sugar components that type O blood does not have.

b. People with type AB blood can give blood only to other people with type AB blood because type AB blood has sugar components that types A, B, and O blood do not have.

31.

32. C-2 epimer = D-mannose   C-4 epimer = D-galactose
C-3 epimer = D-allose   C-5 epimer = L-idose

33. **a.** D-lyxose   **b.** D-talose   **c.** D-psicose

34.

D-threose   ---Kiliani-Fischer synthesis--->   sugar A (D-xylose)   +   sugar B (D-lyxose)   ---HNO₃--->   optically inactive   +   optically active

35. **a.** D-ribose and L-ribose, D-arabinose and L-arabinose, D-xylose and L-xylose, D-lyxose and L-lyxose

b. D-ribose and D-arabinose, D-xylose and D-lyxose, L-ribose and L-arabinose, L-xylose and L-lyxose

c. D-arabinose, L-arabinose, D-lyxose, and L-lyxose

36. **a.** 2R,3S,4R,5R   **b.** 2R,3S,4S,5R   **c.** 2R,3R,4R   **d.** 2R,3S,4R   **e.** 3R,4S,5R

37.

methyl α-D-ribofuranose   methyl β-D-ribofuranose

methyl α-D-ribopyranose   methyl β-D-ribopyranose

38. **a.** methyl β-D-sorboside   **b.** ethyl β-D-guloside   **c.** methyl α-D-idoside

39. A monosaccharide with a molecular weight of 150 must have five carbons (five C's = 60, five O's = 80, and 10 H's = 10 for a total of 150). All aldopentoses are optically active. Therefore, the compound must be a ketopentose. The following is the only ketopentose that would not be optically active.

40.

D-glucose

D-allose

41.

D-ribose   D-erythrose

   A   **B**

42.   The hexose is D-altrose.

1.   Knowing that $(+)$-glyceraldehyde is D-glyceraldehyde gives the configuration at C-5.

2.   Knowing that the second Wohl degradation gives D-erythrose gives the configuration at C-4.

3.   Knowing that the first Wohl degradation followed by oxidation gives an optically inactive aldaric acid gives the configuration at C-3.

4.   Knowing that the original hexose forms an optically active aldaric acid gives the configuration at C-2.

D-altrose

43.

44.   There are two possible structures for an optically active five-carbon alditol formed from a Wohl degradation of a D-hexose:

all OH groups will be in equatorial positions   only the starred OH group will be in an axial position

The unknown aldohexose has only one OH group in an axial position. Therefore the aldohexose that gives the alditol on the left needs to have the OH on C-2 pointing to the left (so it will be axial). The other aldohexose has an axial hydrogen, so needs to have the OH on C-2 pointing to the right (so it will not be axial). The two possible aldoses are D-mannose and D-galactose:

D-mannose

D-galactose

If the two aldoses are reduced to alditols, the alditol of D-mannose will be optically active and the alditol of D-galactose will be optically inactive.

alditol of D-mannose

alditol of D-galactose

45. The hydrogen that is bonded to the anomeric carbon will be the hydrogen that has its signal at the highest frequency, because it is the only hydrogen that is bonded to a carbon that is bonded to two oxygens. So the two anomeric hydrogen (one from the α-anomer and one from the β-anomer) are responsible for the two high-frequency doublets.

46.

the β-D-glucuronide

the α-D-glucuronide

47.   1. As Fischer did, we can narrow our search to eight aldohexoses because there are eight pairs of enantiomers. First we need to find an aldopentose that forms (+)-galactose as a product of a Kiliani-Fischer synthesis. That sugar is the one known as (−)-lyxose. The Kiliani-Fischer synthesis on (−)-lyxose yields two sugars with melting points that show them to be the sugars known as (+)-galactose and (+)-talose. Now we know that (+)-galactose and (+)-talose are C-2 epimers. The eight aldohexoses are sugars 1 and 2, 3 and 4, 5 and 6, or 7 and 8. (See page 1021 of the text.)

2. When (+)-galactose and (+)-talose react with HNO_3, (+)-galactose forms an optically inactive aldaric acid and (+)-talose forms an optically active aldaric acid. Thus, (+)-galactose and (+)-talose are sugars 1 and 2 or 7 and 8. Because (+)-galactose is the one that forms the optically inactive aldaric acid, it is either sugar 1 or 7.

3. To determine the structure of (+)-galactose, we can go back to (−)-lyxose, the sugar that forms sugars 7 and 8 by a Kiliani-Fischer synthesis, and oxidize it with HNO_3. Finding that the aldaric acid is optically active allows us to conclude that (+)-galactose is sugar 7, because the aldopentose that leads to sugars 1 and 2 would give an optically inactive aldaric acid.

48.   **D-Arabinose**. The only D-aldopentoses that are oxidized to optically active aldaric acids are D-arabinose and D-lyxose. A Wohl degradation of D-arabinose forms D-erythrose, whereas a Wohl degradation of D-lyxose forms D-threose. Because D-erythrose forms an optically inactive aldaric acid but D-threose does not, we can conclude that the D-aldopentose is D-arabinose.

49.

50.

hyaluronic acid

51. She can take a sample of one of the sugars and oxidize it with nitric acid to an aldaric acid or reduce it with sodium borohydride to an alditol. If the product is optically active, the sugar was D-lyxose; if the product is not optically active, the sugar was D-xylose.

52.

53.

aldaric acid of
D-glucose

rotate 180°

54.

β-maltose

+ Cl⁻

The β-anomer will also be formed when H_2O adds to the top of the plane of the oxocarbenium ion.

55.   **10** aldaric acids
Each of the following pairs forms the same aldaric acid:

D-allose and L-allose   L-altrose and L-talose

D-galactose and L-galactose   D-gulose and L-glucose

D-altrose and D-talose   L-gulose and D-glucose

Thus twelve aldohexoses form six aldaric acids. The other four aldohexoses each form a distinctive aldaric acid, and $6 + 4 = 10$.

56.   We know the hexose is a ketohexose because it does not react with Br_2; we know it is a 2-ketohexose because if it were a 3-ketohexose it would not be able to form a hemiacetal (because the hemiacetal would have an unstable four-membered ring), and therefore would not undergo mutarotation.

Knowing that the hexose is oxidized by Tollens reagent to the aldonic acids D-talonic acid and D-galactonic acid tells us that the aldonic acids and the ketohexose have the same configuration at C-3, C-4, and C-5. Therefore, the hexose is D-tagatose.

D-talonic acid   D-galactonic acid   D-tagatose

57.

D-fructose with one deuterium

D-fructose with two deuteriums

58.

a.   b.   c.   d.   e.   f.

59.   Let A = the fraction of D-glucose in the α-form and B = the fraction of D-glucose in the β-form.

$$A + B = 1$$
$$B = 1 - A$$

specific rotation of A = 112.2
specific rotation of B = 18.7
specific rotation of the equilibrium mixture = 52.7
specific rotation of the mixture = specific rotation of A × fraction of D-glucose in the α-form +
    specific rotation of B × fraction of D-glucose in the β-form

$$52.7 = 112.2\,A + (1 - A)\,18.7$$
$$52.7 = 112.2\,A + 18.7 - 18.7\,A$$
$$34.0 = 93.5\,A$$
$$A = 0.36$$
$$B = 0.64$$

This calculation shows that 36% is in the α-form and 64% is in the β-form. This agrees with the values given in Section 21.10.

60.   Let A = the fraction of D-galactose in the α-form and B = the fraction of D-galactose in the β-form.

$$A + B = 1$$
$$B = 1 - A$$

specific rotation of A = 150.7
specific rotation of B = 52.8
specific rotation of the equilibrium mixture = 80.2
specific rotation of the mixture = specific rotation of A × fraction of galactose in the α-form +
    specific rotation of B × fraction of galactose in the β-form

$$80.2 = 150.7\,A + 52.8\,(1 - A)$$
$$80.2 = 150.7\,A + 52.8 - 52.8\,A$$
$$27.4 = 203.5\,A$$
$$A = 0.135$$

Thus, 13.5% is α-D-galactose and 76.5% is β-D-galactose.

61.   Silver oxide increases the leaving tendency of the iodide ion from methyl iodide, thereby allowing the nucleophilic substitution reaction to take place with the poorly nucleophilic alcohol groups. Because mannose is missing the methyl substituent on the oxygen at C-6, the disaccharide must be formed using the C-6 OH group of mannose and the anomeric carbon of galactose.

62.   D-Altrose will most likely exist as a furanose because:

(1) the furanose is particularly stable since all the large substituents are trans to each other, and

(2) the pyranose has two of its OH groups in the unstable axial position.

D-altrofuranose   D-altropyranose

63.   From its molecular formula and the fact that only glucose is formed when trehalose is hydrolyzed, we know that it is a disaccharide. Trehalose can be a nonreducing sugar only if the anomeric carbon of one glucose is connected to the anomeric carbon of the other glucose. Because it can be hydrolyzed by maltase, its glycosidic linkage must have the same geometry as an α-1,4′-glycosidic linkage. Therefore, the OH group attached to the anomeric carbon of the glucose on the left must be axial, and the OH group attached to the anomeric carbon of the glucose on the right must be equatorial. (Notice that the glucose on the right is drawn upside down and is reversed horizontally.)

64.

65. Because all the glucose units have six-membered rings, the 5-position is never methylated.
2,3,4,6-tetra-*O*-methyl-D-glucose has only its 1-position in an acetal linkage.
2,4,6-tri-*O*-methyl-D-glucose has its 1- and 3-positions in an acetal linkage.
2,3,4-tri-*O*-methyl-D-glucose has its 1- and 6-positions in an acetal linkage.
2,4-di-*O*-methyl-D-glucose has its 1-, 3-, and 6-positions in an acetal linkage.

66. In the case of D-idose, the chair conformer with both the OH substituent at C-1 and the CH$_2$OH substituent in axial positions (which is necessary for the formation of the anhydro form) has the OH substituents at C-2, C-3, and C-4 in equatorial positions. Thus, this is a relatively stable conformer because three of the five large substituents are in the more stable equatorial position.

In the case of D-glucose, the chair conformer with both the OH substituent at C-1 and the CH$_2$OH substituent in axial positions has the OH substituents at C-2, C-3, and C-4 in axial positions. This is a relatively unstable conformer because all the large substituents are in less stable axial positions.

anhydro form of D-idose   anhydro form of D-glucose

Therefore, a large percentage of D-idose but only a small percentage of D-glucose exists in the anhydro form at 100 °C.

Chapter 21 Practice Test

1.   Draw the product(s) of each of the following reactions:

a.

$$\begin{array}{c} \text{H} \diagup \text{O} \\ \text{C} \\ \text{H} \rule{1em}{0.4pt} \text{OH} \\ \text{HO} \rule{1em}{0.4pt} \text{H} \\ \text{H} \rule{1em}{0.4pt} \text{OH} \\ \text{H} \rule{1em}{0.4pt} \text{OH} \\ \text{CH}_2\text{OH} \end{array}$$

$\xrightarrow[\Delta]{\text{HNO}_3}$

b.

$\xrightarrow[\text{CH}_3\text{OH}]{\text{HCl}}$

c.

$$\begin{array}{c} \text{H} \diagup \text{O} \\ \text{C} \\ \text{HO} \rule{1em}{0.4pt} \text{H} \\ \text{HO} \rule{1em}{0.4pt} \text{H} \\ \text{H} \rule{1em}{0.4pt} \text{OH} \\ \text{CH}_2\text{OH} \end{array}$$

1. $\text{HC}\equiv\text{N}$
2. $\text{H}_2/\text{Pd/BaSO}_4$
3. $\text{HCl, H}_2\text{O}$

$\longrightarrow$

d.

$$\begin{array}{c} \text{H} \diagup \text{O} \\ \text{C} \\ \text{H} \rule{1em}{0.4pt} \text{OH} \\ \text{H} \rule{1em}{0.4pt} \text{OH} \\ \text{H} \rule{1em}{0.4pt} \text{OH} \\ \text{CH}_2\text{OH} \end{array}$$

$+ \quad \text{Br}_2 \quad \xrightarrow{\text{H}_2\text{O}}$

2.   Indicate whether each of the following statements is true or false:

a.	Glycogen contains α-1,4′- and β-1,6′-glycosidic linkages.	T	F
b.	D-Mannose is a C-1 epimer of D-glucose.	T	F
c.	D-Glucose and L-glucose are anomers.	T	F
d.	D-Erythrose and D-threose are diastereomers.	T	F
e.	Wohl degradations of D-glucose and D-gulose form the same aldopentose.	T	F

3. Which of the following sugars will form an optically active aldaric acid?

4. When crystals of D-fructose are dissolved in a basic aqueous solution, two aldohexoses are obtained. Identify the aldohexoses.

5. Draw three tetroses that form the same enol.

6. D-Talose and _____ are reduced to the same alditol.

7. What is the main structural difference between amylose and cellulose?

8. What aldohexoses are formed from a Kiliani-Fischer synthesis starting with D-xylose?

9. What aldohexose is the C-3 epimer of D-glucose?

10. Draw the most stable chair conformer of β-D-allose, a C-3 epimer of β-D-glucose.

CHAPTER 22
The Organic Chemistry of Amino Acids, Peptides, and Proteins

Important Terms

amino acid	an α-amino carboxylic acid. Naturally occurring amino acids have the L-configuration.
D-amino acid	the configuration of an amino acid drawn in a Fischer projection with the carboxyl group on top, the hydrogen is on the left, and the amino group is on the right.
L-amino acid	the configuration of an amino acid drawn in a Fischer projection with the carboxyl group on top, the hydrogen on the right, the amino group on the left.
amino acid analyzer	an instrument that automates the ion-exchange separation of amino acids.
anion-exchange resin	a resin that binds anions.
antiparallel β-pleated sheet	a type of secondary structure in which the adjacent hydrogen-bonded peptide chains in a β-pleated sheet run in opposite directions.
automated solid-phase peptide synthesis	an automated technique that synthesizes a peptide (in the C-terminal to N-terminal direction) while its C-terminal amino acid is attached to a solid support.
cation-exchange resin	a resin that binds cations.
coil conformation (loop conformation)	the part of a protein that is highly ordered but not in an α-helix or a β-pleated sheet.
C-terminal amino acid	the terminal amino acid of a peptide (or protein) that has a free carboxyl group.
denaturation	the destruction of the highly organized secondary and tertiary structure of a protein.
dipeptide	two amino acids linked together by an amide bond.
disulfide	a compound with an S—S bond.
disulfide bridge	a disulfide (S—S) bond formed by two cysteine residues in a peptide or protein.
Edman's reagent	phenyl isothiocyanate; the reagent used to determine the N-terminal amino acid of a polypeptide.
electrophoresis	a technique that separates amino acids on the basis of their pI values.
endopeptidase	an enzyme that hydrolyzes a peptide bond that is not at the end of a peptide chain.
essential amino acid	an amino acid that humans must obtain from their diet because they either cannot synthesize it at all or cannot synthesize it in adequate amounts.

exopeptidase	an enzyme that hydrolyzes a peptide bond at the end of a peptide chain.
fibrous protein	a water-insoluble protein that has its polypeptide chains arranged in bundles.
globular protein	a water-soluble protein that tends to have a roughly spherical shape.
***α*-helix**	the backbone of a polypeptide coiled in a right-handed spiral with hydrogen bonding occurring within the helix.
hydrophobic interactions	interactions between nonpolar groups. These interactions increase stability by decreasing the amount of structured water (increasing entropy).
interchain disulfide bridge	a disulfide bridge between two cysteine residues in different peptide chains.
intrachain disulfide bridge	a disulfide bridge between two cysteine residues in the same peptide chain.
ion-exchange chromatography	a technique that uses a column packed with an insoluble resin to separate compounds on the basis of their charge and polarity.
isoelectric point (pI)	the pH at which there is no net charge on an amino acid.
kinetic resolution	separating enantiomers based on the difference in their rate of reaction with an enzyme.
loop conformation (coil conformation)	see coil conformation.
N-terminal amino acid	the terminal amino acid of a peptide (or protein) that has a free amino group.
oligomer	a protein with more than one peptide chain.
oligopeptide	three to ten amino acids linked by amide bonds.
paper chromatography	a technique that separates amino acids based on polarity.
parallel *β*-pleated sheet	a type of secondary structure in which the adjacent hydrogen-bonded peptide chains in a *β*-pleated sheet run in the same direction.
partial hydrolysis	a technique that hydrolyzes only some of the peptide bonds in a polypeptide.
peptidase	an enzyme that catalyzes the hydrolysis of a peptide bond.
peptide	a polymer of amino acids linked together by amide bonds.
peptide bond	the amide bond that links the amino acids in a peptide or protein.
***β*-pleated sheet**	a type of secondary structure in which the backbone of a polypeptide extends in a zigzag structure with hydrogen bonding between neighboring chains.

polypeptide   many amino acids linked by amide bonds.

primary structure   the sequence of amino acids and the location of the disulfide bridges in a protein.

protein   a naturally occurring polymer of 40 to 4000 amino acids linked together by amide bonds.

quaternary structure   a description of the way in which the individual polypeptide chains of an oligomeric protein are arranged with respect to one other.

random coil   the conformation of a totally denatured protein.

secondary structure   a description of the conformation of the backbone of a protein.

side chain   the substituent attached to the α-carbon of an amino acid.

structural protein   a protein that gives strength to a biological structure.

subunit   an individual chain of an oligomeric protein.

tertiary structure   a description of the three-dimensional arrangement of all the atoms in a protein.

thin-layer chromatography   a technique that separates compounds on the basis of their polarity.

tripeptide   three amino acids linked by amide bonds.

zwitterion   a compound with a negative charge and a positive charge on nonadjacent atoms.

Solutions to Problems

1.   **a.** When the imidazole ring is protonated, the double-bonded nitrogen is the one that accepts the proton. The lone-pair electrons on the single-bonded nitrogen are delocalized and therefore are not available to be protonated.

In contrast, the lone-pair electrons on the double-bonded nitrogen are not delocalized, because the constraints of the ring will not allow two adjacent double bonds.

Furthermore, if the lone pair on the single-bonded nitrogen were protonated, the compound would lose its aromaticity.

In contrast, when the double-bonded nitrogen is protonated, the protonated compound is still aromatic.

b. The lone-pair electrons on the double-bonded nitrogen are protonated, because the lone-pair electrons on the other nitrogens are delocalized and therefore cannot be protonated. In addition, protonation of the double-bonded nitrogen leads to a highly resonance-stabilized conjugate acid.

2.   **a.** D-alanine is (R)-alanine   **b.** D-asparate is (R)-aspartate

c. The α-carbons of all the D-amino acids except cysteine have the R-configuration. Similarly, the α-carbons of all the L-amino acids except cysteine have the S-configuration.

In all the amino acids except cysteine, the amino group has the highest priority and the carboxyl group has the second-highest priority. In cysteine, the thiomethyl group has a higher priority than the carboxylate group, because sulfur has a greater atomic number than oxygen, causing the counterclockwise arrow to be a clockwise arrow.

D-cysteine

(S)-cysteine

3. Solved in the text.

4. Isoleucine is the only other amino acid that has more than one asymmetric center. Like threonine, it has two asymmetric centers.

isoleucine

5. Alanine will exist predominately as a zwitterion in an aqueous solution with pH > 2.34 and pH < 9.69.

6. The electron-withdrawing $^+NH_3$ substituent on the α-carbon increases the acidity of the carboxyl group.

7. **a.** Solved in the text.   **c.** H_2N CH_2CH_2CH O^- **e.** H_2N $NHCH_2CH_2CH_2CH$ O^-
            $^+NH_3$ $^+NH_3$

b. CH_2CH O^- **d.** $H_3\overset{+}{N}CH_2CH_2CH_2CH_2CH$ O^- **f.** HO CH_2CH O^-
        $^+NH_3$ $^+NH_3$ $^+NH_3$

8. **a.** HO CH_2CH_2CH OH **c.** ^-O CH_2CH_2CH O^-
                    $^+NH_3$ $^+NH_3$

b. HO CH_2CH_2CH O^- **d.** ^-O CH_2CH_2CH O^-
            $^+NH_3$ NH_2

9.   **a.**   The carboxyl group of the aspartic acid side chain is a stronger acid than the carboxyl group of the glutamic acid side chain because the carboxyl group of the aspartic acid side chain is closer to the electron-withdrawing protonated amino group.

    b.   The protonated lysine side chain is a stronger acid than the protonated arginine side chain. The protonated arginine side chain has less of a tendency to lose a proton because its positive charge is delocalized over three nitrogens.

10.   In order for the amino acid to have no net charge, the two amino groups must have a +1 charge between them in order to cancel out the −1 charge of the carboxylate group. Because they are positively charged in their acidic forms and neutral in their basic forms, the sum of their charges will be +1 at the midpoint of their pK_a values.

11.   **a.**   asparagine $pI = \dfrac{2.02 + 8.84}{2} = \dfrac{10.86}{2} = 5.43$

    b.   arginine $pI = \dfrac{9.04 + 12.48}{2} = \dfrac{21.52}{2} = 10.76$

    c.   serine $pI = \dfrac{2.21 + 9.15}{2} = \dfrac{11.36}{2} = 5.68$

    d.   aspartate $pI = \dfrac{2.09 + 3.86}{2} = \dfrac{5.95}{2} = 2.98$

12.   **a.**   aspartate $(pI = 2.98)$

    b.   arginine $(pI = 10.76)$

    c.   Aspartate because its pI is lower than that of glutamate. Therefore, pH 6.20 is farther away from the pH at which it has no net charge. The farther an amino acid with no net charge has to "move" to get to the given pH value, the more charged it will become.

    d.   Methionine because at pH = 6.20 methionine is farther away from the pH at which it has no net charge (pI of methine = 5.75, pI of glycine = 5.97).

13.   Tyrosine and cysteine each have two groups that are neutral in their acidic forms and negatively charged in their basic forms. Unlike other amino acids that have similarly ionizing groups, the pK_a value of one of the two similarly ionizing groups in tyrosine and in cysteine is close to the pK_a value of the group that ionizes differently. Therefore, the group that ionizes differently cannot be ignored in calculating the pI.

14.

$$
\begin{array}{c}
\ \ \ \ \ \ \ \ \ \ \ \ \ \ \ \ \ \ O \\
\ \ \ \ \ \ \ \ \ \ \ \ \ \ \ \ \ \ \| \\
\ \ \ \ \ \ \ \ \ \ \ \ \ \ \ \ \ \ C \\
CH_3CH \diagup \ \ \ \diagdown H \\
\ \ \ \ \ | \\
\ \ \ \ \ CH_3
\end{array}
$$

the R group of the aldehyde is the same as the R group of the amino acid

15.   Leucine and isoleucine both have C_4H_9 side chains and, therefore, have the same polarity. Consequently, the spots for both amino acids appear at the same place on the chromatographic plate. Therefore, the chromatographic plate has one less spot than the number of amino acids.

16.   Because the amino acid analyzer contains a cation-exchange resin (it binds cations), the less positively charged the amino acid, the less tightly it is bound to the column. Using buffer solutions of increasingly higher pH to elute the column causes the amino acids bound to the column to become increasingly less positively charged, so they can be released from the column.

17.   Cation-exchange chromatography releases amino acids in order of their pI values. The amino acid with the lowest pI is released first because at a given pH it will be the amino acid with the highest concentration of negative charge, and negatively charged molecules are not bound by the negatively charged resin. The relatively nonpolar resin will release polar amino acids before nonpolar amino acids.

   a.   Asp (pI = 2.98) is more negative at pH = 4 than is Ser (pI = 5.68).

   b.   Gly is more polar than Ala.

   c.   Ser is more polar than Leu.

   d.   Tyr is more polar than Phe.

18.   A column containing an anion-exchange resin releases amino acids in reverse order of their pI values (the opposite of what happens in a cation-exchange resin). The amino acid with the highest pI is released first, because at a given pH it will be the amino acid with the highest concentration of positive charge.

$$His > Val > Ser > Asp$$

19.   The first equivalent of ammonia will react with the acidic proton of the carboxylic acid to form ammonium ion, which is not nucleophilic and therefore cannot substitute for Br. Thus, a second equivalent of ammonia is needed for the desired nucleophilic substitution reaction.

20.   **a.**   The following reactions show that pyruvic acid forms alanine, oxaloacetic acid forms aspartate, and α-ketoglutarate forms glutamate.

If reductive amination (page 795) is carried out in the cell, only the L-isomer of each amino acid will be formed.

   b.   Recall that imine formation is best carried at a pH about 1.5 units lower than the pK_a of the protonated ammonium ion (Section 18.8), that is, at about pH = 8. Therefore, the carboxyl groups will be in their basic forms. If reductive amination is carried out in the laboratory, both the D- and L-isomer (a racemic mixture) of each amino acid will be formed.

21. Notice that the R group attached to the Br is the same as the R group attached to the α-carbon of the amino acid.

$$R—Br \quad \text{corresponds to} \quad R—\underset{\underset{^+NH_3}{|}}{C}HCOO^-$$

a. leucine   **b.** methionine

22. As in Problem 21, the R group attached to the Br is the same as the R group attached to the α-carbon of the desired amino acid.

a. 4-bromo-1-butanamine   **b.** benzyl bromide

23. Notice that the R group attached to the carbonyl group of the aldehyde is the same as the R group attached to the α-carbon of the amino acid.

$$R—\overset{\overset{O}{\|}}{C}H \quad \text{corresponds to} \quad R—\underset{\underset{^+NH_3}{|}}{C}HCOO^-$$

a. alanine   **b.** isoleucine   **c.** leucine

24. Convert the amino acids into esters using $SOCl_2$ followed by ethanol. Then treat the esters with pig liver esterase. Because the enzyme hydrolyzes only esters of L-amino acids, the products will be the L-amino acid, ethanol, and the ester of the D-amino acid. These compounds can be readily separated. After they are separated, the D-amino acid can be obtained by acid-catalyzed hydrolysis of the ester of the D-amino acid. This separation technique is called a kinetic resolution, because the enantiomers are separated (resolved) as a result of reacting at different rates in the enzyme-catalyzed reaction.

25.

the peptide bonds are indicated by arrows

26.   **A-G-M   A-M-G   M-G-A   M-A-G   G-A-M   G-M-A**

27.

the less stable configuration has the R groups
on the same side of the double bond

28.   The bonds on either side of the α-carbon can freely rotate. In other words, the bond between the α-carbon and the carbonyl carbon and the bond between the α-carbon and the nitrogen (the bonds indicated by arrows) can freely rotate. The bond between the C and N (the peptide bond) cannot rotate because it has partial double-bond character.

29.

30.   **a.**   glutamate, cysteine, and glycine

b.   In forming the amide bond between glutamate and cysteine, the amino group of cysteine reacts with the γ-carboxyl group of glutamate rather than with its α-carboxyl group.

31. **Leu-Val** and **Val-Val** will be formed because the amino group of leucine is protected, so leucine cannot react with a carboxyl group (that is, leucine cannot be the C-terminal amino acid). The amino group of valine can react equally easily with the carboxyl group of leucine and the carboxyl group of valine.

32. If valine's carboxyl group is activated with thionyl chloride, the OH group of serine, as well as the NH$_2$ group of serine, would react readily with the very reactive acyl chloride, forming both an ester (with its OH group) and an amide (with its NH$_2$ group).

If valine's carboxyl group is activated with DCC, an imidate will be formed. Because an imidate is less reactive than an acyl chloride, the imidate will react with the more reactive NH$_2$ group in preference to the less reactive OH group.

(CH₃)₃CO   NHCH   O⁻ → DCC → (CH₃)₃CO   NHCH   O

N-protected valine

CH(CH₃)₂

an imidate

N-protected valine

H₂NCH   O⁻
CH₂OH

+ (CH₃)₃CO   NHCH   NHCH   O⁻

CH(CH₃)₂   CH₂OH

33.

(CH₃)₃CO   O   OC(CH₃)₃ + H₂NCH   O⁻ → (CH₃)₃CO   NHCH   O⁻

CH₂   CH₂

CH(CH₃)₂   CH(CH₃)₂

Leu   N-protected Leu

1. DCC
2. H₂NCH   O⁻
   CH₂

   Phe

(CH₃)₃CO   NHCH   NHCH   O⁻

CH₂   CH₂

CH(CH₃)₂

1. DCC
2. H₂NCH   O⁻
   CH₃   Ala

$(CH_3)_3CO$ — NHCH — NHCH — NHCH — O^- (Boc-tripeptide: side chains $CH_2CH(CH_3)_2$, $CH_2C_6H_5$, CH_3)

1. DCC

2. H_2NCH — O^- , $CH(CH_3)_2$ Val

$(CH_3)_3CO$ — NHCH — NHCH — NHCH — NHCH — O^- (side chains $CH_2CH(CH_3)_2$, $CH_2C_6H_5$, CH_3, $CH(CH_3)_2$)

CF_3CO_2H | CH_2Cl_2

$CH_3C(=CH_2)CH_3$ + CO_2 + $H_3\overset{+}{N}CH$ — NHCH — NHCH — NHCH — O^-

Leu   Phe   Ala   Val

34.   Notice that the number of steps is one less than the number of amino acids in the peptide.

a.   5.8%

2	3	4	5	6	7	8	9
70%	49%	34%	24%	17%	12%	8.2%	5.8%

b.   4.4%

2	3	4	5	6	7	8	9	10	11	12	13	14	15
80%	64%	51%	41%	33%	26%	21%	17%	13%	11%	8.6%	6.9%	5.5%	4.4%

35.

$$(CH_3)_3CO\underset{\underset{O}{\|}}{C}-O-\underset{\underset{O}{\|}}{C}OC(CH_3)_3 \;+\; H_2N\underset{\underset{CH(CH_3)_2}{|}}{CH}\underset{\underset{O}{\|}}{C}O^- \longrightarrow (CH_3)_3CO\underset{\underset{O}{\|}}{C}-NH\underset{\underset{CH(CH_3)_2}{|}}{CH}\underset{\underset{O}{\|}}{C}O^- \Big| \text{ClCH}_2\text{—}\bullet$$

$$(CH_3)_3CO\underset{\underset{O}{\|}}{C}-NH\underset{\underset{CH(CH_3)_2}{|}}{CH}\underset{\underset{O}{\|}}{C}OCH_2\text{—}\bullet$$

$$\Big| \begin{array}{c} CF_3COOH \\ CH_2Cl_2 \end{array}$$

$$\underset{\underset{CH_2}{\|}}{\overset{CH_3}{\underset{|}{CH_3C}}} \;+\; CO_2 \;+\; H_2N\underset{\underset{CH(CH_3)_2}{|}}{CH}\underset{\underset{O}{\|}}{C}OCH_2\text{—}\bullet$$

$$(CH_3)_3CO\underset{\underset{O}{\|}}{C}-NH\underset{\underset{CH_3}{|}}{CH}\underset{\underset{O}{\|}}{C}O^- \quad\xrightarrow{\text{DCC}}\quad (CH_3)_3CO\underset{\underset{O}{\|}}{C}-NH\underset{\underset{CH_3}{|}}{CH}\underset{\underset{O}{\|}}{C}O-DCC$$

$$(CH_3)_3CO\underset{\underset{O}{\|}}{C}-NH\underset{\underset{CH_3}{|}}{CH}\underset{\underset{O}{\|}}{C}-NH\underset{\underset{CH(CH_3)_2}{|}}{CH}\underset{\underset{O}{\|}}{C}OCH_2\text{—}\bullet$$

$$\Big| \begin{array}{c} CF_3COOH \\ CH_2Cl_2 \end{array}$$

$$\underset{\underset{CH_2}{\|}}{\overset{CH_3}{\underset{|}{CH_3C}}} \;+\; CO_2 \;+\; H_2N\underset{\underset{CH_3}{|}}{CH}\underset{\underset{O}{\|}}{C}-NH\underset{\underset{CH(CH_3)_2}{|}}{CH}\underset{\underset{O}{\|}}{C}OCH_2\text{—}\bullet$$

$$(CH_3)_3CO\underset{\underset{CH_2}{|}}{\underset{\underset{O}{\|}}{C}-NH}CH\underset{\underset{O}{\|}}{C}O^- \quad\xrightarrow{\text{DCC}}\quad (CH_3)_3CO\underset{\underset{CH_2}{|}}{\underset{\underset{O}{\|}}{C}-NH}CH\underset{\underset{O}{\|}}{C}O-DCC$$

(with benzyl side chains on the last two structures)

$(CH_3)_3O$ —NHCH— NHCH— NHCH—OCH_2— ⬤

CF$_3$COOH
CH$_2$Cl$_2$

CH_3C + CO_2 + H_2NCH —NHCH— NHCH—OCH_2— ⬤

$(CH_3)_3CO$—NHCH—O^- $\xrightarrow{\text{DCC}}$ $(CH_3)_3CO$—NHCH—O—DCC

$(CH_3)_3CO$—NHCH— NHCH— NHCH— NHCH—OCH_2— ⬤

CF$_3$COOH
CH$_2$Cl$_2$

CH_3C + CO_2 + H_2NCH —NHCH— NHCH— NHCH—OCH_2— ⬤

HF, H$_2$O

$H_3\overset{+}{N}CH$ —NHCH— NHCH— NHCH—OH + $HOCH_2$— ⬤

36.   It is an S_N2 reaction followed by dissociation of a proton.

37.   Because insulin has two peptide chains, treatment with Edman's reagent would release two PTH-amino acids in approximately equal amounts.

38.   Knowing that the N-terminal amino acid is Gly, look for a peptide fragment that contains Gly.

"Fragment 6" tells you that the second amino acid is Arg.

"Fragment 5" tells you that the next two are Ala-Trp or Trp-Ala.

"Fragment 4" tells you that Glu is next to Ala, so the third and fourth amino acids must be Trp-Ala and the fifth is Glu.

"Fragment 7" tells you that the sixth amino acid is Leu.

"Fragment 8" tells you that the next two are Met-Pro or Pro-Met.

"Fragment 3" tells you that Pro is next to Val, so the seventh and eighth amino acids must be Met-Pro and the ninth is Val.

"Fragment 2" tells you that the last amino acid is Asp.

Gly-Arg-Trp-Ala-Glu-Leu-Met-Pro-Val-Asp

39.   Cysteine can react with cyanogen bromide, but the sulfur would not be positively charged, so it would be a poor leaving group. In addition, the lactone will not be formed because it would have a strained four-membered ring. Without lactone formation, the imine would not be formed, so cleavage cannot occur.

40.   **a.** His-Lys   Leu-Val-Glu-Pro-Arg   Ala-Gly-Ala

          b. Leu-Gly-Ser-Met-Phe-Pro-Tyr   Gly-Val

41.   Solved in the text.

42. The data from treatment with Edman's reagent and carboxypeptidase A identify the first and last amino acids.

<u>Leu</u>   __   __   __   __   __   __   <u>Ser</u>

The data from cleavage with cyanogen bromide identify the position of Met and identify the other amino acids in the pentapeptide and tripeptide but not their order.

cleavage with cyanogen bromide

Arg, Lys, Tyr   Arg, Phe

<u>Leu</u>   __   __   __   <u>Met</u> | __   __   <u>Ser</u>

The data from treatment with trypsin put the remaining amino acids in the correct positions.

<u>Leu</u>   <u>Tyr</u>   <u>Lys</u> | <u>Arg</u> | <u>Met</u>   <u>Phe</u>   <u>Arg</u> | <u>Ser</u>

43. **a.** Trypsin cleaves at Arg and Lys. There are two possible primary structures:

Val-Gly-Asp-Lys-Leu-Glu-Pro-Ala-Arg-Ala-Leu-Gly-Asp

or

Leu-Glu-Pro-Ala-Arg-Val-Gly-Asp-Lys-Ala-Leu-Gly-Asp

The two possible primary structures can be distinguished by Edman's reagent. Edman's reagent would release Val in one case and Leu in the other.

b. Trypsin cleaves at Arg and Lys. There are two possible primary structures:

Ala-Glu-Pro-Arg-Ala-Met-Gly-Lys-Val-Leu-Gly-Glu

or

Ala-Met-Gly-Lys-Ala-Glu-Pro-Arg-Val-Leu-Gly-Glu

The two possible primary structures can be distinguished by treatment with cyanogen bromide. Cyanogen bromide would cleave one of the possible polypeptides into two hexamers and the other into a dimer and a decamer.

44. 74 amino acids/3.6 amino acids per turn of the helix = 20.6 turns of the helix

20.6×5.4 Å = **110 Å** in an α-helix (5.4 Å is the repeat distance of the α-helix.)

74 amino acids $\times$ 3.5 Å = **260 Å** in fully extended polypeptide chain

45. It would fold so that its nonpolar residues are on the outside of the protein in contact with the nonpolar membrane and its polar residues are on the inside of the protein.

46. A protein folds to maximize the number of polar groups on the surface of the protein and the number of nonpolar groups on the inside of the protein.

a. A cigar-shaped protein has the greatest surface area to volume ratio, so it has the highest percentage of polar amino acids.

b. A subunit of a hexamer would have the smallest percentage of polar amino acids, because part of the surface of the subunit can be on the inside of the hexamer and therefore have nonpolar amino acids on its surface.

a scheme of two possible quaternary structures of a hexamer

47.   An amino acid is insoluble in diethyl ether (a relatively nonpolar solvent), because an amino acid exists as a highly polar zwitterion at neutral pH. In contrast, carboxylic acids and amines are less polar because they either are neutral or have a single charge depending on the extent of dissociation in diethyl ether.

48.   **a.** pH = 9.60   **b.** pH = 5.97 (its pI)   **c.** pH = 2.34

49.   **a.** Val-Arg-Gly-Met-Arg-Ala   Ser

   b. Ser-Phe-Lys-Met   Pro-Ser-Ala-Asp

   c. Arg   Ser-Pro-Lys   Lys   Ser-Glu-Gly

50.   As an amino acid moves from a solution with a pH equal to its pI to a more basic solution, the amino acid becomes more and more negatively charged. Because asparagine has a lower pI than leucine, in a solution of pH = 7.3, asparagine has moved farther from its pI than has leucine. Asparagine therefore will have a higher percentage of negative charge at pH = 7.3.

51.

52.   **a.**

c.

   b.

d.

53.   The student is correct. At the pI, the total of the positive charges on the tripeptide's amino groups must be one to balance the one negative charge of the carboxylate group. When the pH of the solution is equal to the pK_a of a lysine residue, the three lysine groups each have one-half a positive charge for a total of one and one-half positive charges. Thus, the solution must be more basic than this in order to have just one positive charge.

54.   Since the mixture of amino acids is in a solution of pH = 5, **His** will have an overall positive charge and **Glu** will have an overall negative charge. **His**, therefore, will migrate to the cathode, and **Glu** will migrate to the anode.

Ser is more polar than **Thr** (both have OH groups, but **Thr** has an additional carbon).

Thr is more polar than **Met** (**Met** has SCH$_2$CH$_2$ instead of CHOH). **Met** is more polar than **Leu** (**Leu** has four carbons where **Met** has three carbons and a sulfur).

55. We know that "Fragment 3" (Leu-Pro-Phe) is at the C-terminal end of the polypeptide. Thus, the only question we need to answer is, which of the two fragments obtained by cleavage with cyanogen bromide, which begin with Gly and end with Met, is nearest the C-terminal end of the polypeptide? The answer is obtained from "Fragment 4" obtained from trypsin cleavage. The Met nearest to the C-terminal end must be preceded by Arg or Lys. Therefore, the polypeptide has the following sequence:

Gly-Leu-Tyr-Phe-Lys-Ser-Met-Gly-Leu-Tyr-Lys-Val-Ile-Arg-Met-Leu-Pro-Phe

56. You would (correctly) expect serine and cysteine to have lower pK_a values than alanine, since a hydroxymethyl and a thiomethyl group are more electron withdrawing than a methyl group. Because oxygen is more electronegative than sulfur, you would expect serine to have a lower pK_a than cysteine. The fact that cysteine has a lower pK_a than serine can be explained by stabilization of serine's carboxyl proton by hydrogen bonding to the β-OH group of serine, which causes it to have less of a tendency to be removed by a base.

57. Each compound has two groups that can act as a buffer, one amino group and one carboxyl group. Thus, the compound in higher concentration (0.2 M glycine) will be a more effective buffer.

58. Groups that are not fully charged at the given pH are shown in the form that predominates at that pH. For example, tyrosine has a pK_a = 9.11, so at pH = 7 it is shown in its acidic (neutral) form.

a.

b.

c.

d.

59.

60. **a.** When the polypeptide is treated with maleic anhydride, lysine reacts with maleic anhydride (see Problem 59), but the amino group of arginine is not sufficiently nucleophilic to react with maleic anhydride. Therefore, trypsin will cleave only at arginine residues because the enzyme will no longer recognize lysine residues.

b. Four fragments will be obtained from the polypetide. Remember that trypsin will not cleave the Arg-Pro bond.

c. The N-terminal end of each fragment will be positively charged because of the $^+NH_3$ group.
The C-terminal end will be negatively charged because of the COO^- group.
Arginine residues will be positively charged.
Aspartate and glutamate residues will be negatively charged.
Lysine residues will be negatively charged because they are attached to the maleic acid group.

Elution order: **A > D > C > B**

A Gly-Ala-Asp-Ala-Leu-Pro-Gly-Ile-Leu-Val-Arg   overall charge = 0

      +   −   +−

B Asp-Val-Gly-Lys-Val-Glu-Val-Phe-Glu-Ala-Gly-Arg   overall charge = −3

    +−   −   −   −   +−

C Ala-Glu-Phe-Lys-Glu-Pro-Arg   overall charge = −2

    +   −   −   −   +−

D Leu-Val-Met-Lys-Val-Glu-Gly-Arg-Pro-Val-Ala-Ala-Gly-Leu-Trp   overall charge = −1

    +   −   −   +   −

61. First, mark off where the chains would have been cleaved by chymotrypsin (C-side of Phe, Trp, Tyr).

Val-Met-Tyr │-Ala-Cys-Ser-Phe ┤-Ala-Glu-Ser

Ser-Cys-Phe│ -Lys-Cys-Trp │-Lys-Tyr ┤Cys-Phe ┤Arg-Cys-Ser

Then, from the fragments given, you can determine where the disulfide bridges are in the original intact peptide. For example, "Fragment 2" has two Phe, two Cys, and one Ser. Therefore, the first and fourth fragments of the second row must be connected by a disulfide bond.

62. The methyl ester of phenylalanine rather than phenylalanine itself should be added in the second peptide bond-forming step, because if esterification of phenylalanine is done after amide bond formation, both the carboxyl group of phenylalanine and the γ-carboxyl group of aspartate could be esterified. Both the carboxyl group of phenylalanine and the γ-carboxyl group of aspartate can be activated by DCC. Therefore, some product will be obtained in which the amide bond is formed with the γ-carboxyl group of aspartate rather than with the α-carboxyl group.

63. a.

intermediate I   intermediate II

b.

3-methylbutanal   leucine

c.

2-methylbutanal   isoleucine

64.   Ser-Glu-Leu-Trp-Lys-Ser-Val-Glu-His-Gly-Ala-Met

From the experiment with carboxypeptidase A, we know the C-terminal amino acid is Met.

"Fragment 12" tells us the amino acid adjacent to Met is Ala.

"Fragment 5" tells us the next amino acid is Gly.

"Fragment 2" tells us the next amino acid is His.

"Fragment 7" tells us the next amino acid is Glu.

"Fragment 10" tells us the next amino acid is Val.

"Fragment 3" tells us the next amino acid is Ser.

"Fragment 9" tells us the next amino acid is Lys.

"Fragment 1" tells us the next amino acid is Trp.

"Fragment 8" tells us the next amino acid is Leu.

"Fragment 11" tells us the next amino acid is Glu.

"Fragment 4" tells us the next (first) amino acid is Ser.

65. a. x^y x = the number of amino acids
               y = the number of spots in the peptide for each amino acid
               $20^8 = 25,600,000,000$

b. $20^{100} = 1.3 \times 10^{130}$

66.   The pK_a of the carboxylic acid group of the amino acid is lower than the pK_a of the carboxylic acid group of the dipeptide, because the positively charged ammonium group of the amino acid is more strongly electron withdrawing than the amide group of the peptide. This causes the amino acid to be a stronger acid and therefore have a lower pK_a.

The pK_a of the ammonium group of the dipeptide is lower than the pK_a of the ammonium group of the amino acid, because the amide group of the dipeptide is more strongly electron withdrawing than the carboxylate group of the amino acid.

67.   Finding that there is one less spot than the number of amino acids tells you that the spots for two of the amino acids superimpose. Since leucine and isoleucine have identical polarities, they are good candidates for being the amino acids that migrate to the same location.

Val (pI = 5.97), Trp (pI = 5.89), and Met (pI = 5.75) can be ordered based on their pI values, because the one with the greatest pI will be the one with the greatest amount of positive charge at pH = 5. (See Problem 50.)

68.

69. Oxidation of dithiothreitol is an intramolecular reaction, so it occurs with a larger rate constant than the oxidation of 2-mercaptoethanol, which is an intermolecular reaction. The reverse reduction reaction should occur with about the same rate constant in both cases. Increasing the rate of the oxidation reaction while keeping the rate of the reduction reaction constant is responsible for the greater equilibrium constant, because $K_{eq} = k_1/k_{-1}$, where k_1 is the rate constant for oxidation and k_{-1} is the rate constant for reduction.

$$2 \ HOCH_2CH_2SH \underset{k_{-1}}{\overset{k_1}{\rightleftharpoons}} HOCH_2CH_2S-SCH_2CH_2OH$$

70.   a.

b.

c. $(CH_3)_2CHC$ ... $\xrightarrow[\text{trace acid}]{\text{NH}_3 \text{ excess}}$ ... $(CH_3)_2CHC$ ... $\xrightarrow[\text{Pd/C}]{\text{H}_2}$ ... $(CH_3)_2CHCH$

valine

d. $C_2H_5OC-CH-COC_2H_5$ + potassium phthalimide $\longrightarrow$

Br

α-bromomalonic ester   potassium phthalimide

$\xrightarrow{CH_3CH_2O^-}$

CH_3CH_2OH

$\downarrow \begin{array}{c} CH_3CHBr \\ | \\ CH_3 \end{array}$

COOH

+ CO_2 + $H_3\overset{+}{N}-CH-\underset{OH}{\overset{O}{C}}$ $\xleftarrow[\Delta]{HCl, H_2O}$

COOH

$CH(CH_3)_2$ + 2 CH_3CH_2OH

phthalic acid   valine

CH_3CH

CH_3 Br^-

e. $C_2H_5OC-CH-COC_2H_5$ $\xrightarrow{CH_3CH_2O^-}$ $C_2H_5OC-\overset{..}{C}-COC_2H_5$ $\xrightarrow[CH_3]{CH_3CHBr}$ $C_2H_5OC-C-COC_2H_5$

N   N   N

H   CCH_3   H   CCH_3   H   CCH_3

O   O   O Br^-

acetamidomalonic ester CH_3CH_2OH

$\downarrow \begin{array}{c} HCl, H_2O, \Delta \end{array}$

2 CH_3CH_2OH + $CH_3\underset{OH}{\overset{O}{C}}$ + CO_2 + $H_3\overset{+}{N}-CH-\underset{OH}{\overset{O}{C}}$

acetic acid $CH(CH_3)_2$

valine

71.   Because the native enzyme has four disulfide bridges, we know that the denatured enzyme has eight cysteine residues. The first cysteine has a one in seven chance of forming a disulfide bridge with the correct cysteine. The first cysteine of the next pair has a one in five chance, and the first cysteine of the third pair has a one in three chance.

$$\frac{1}{7} \times \frac{1}{5} \times \frac{1}{3} = 0.0095$$

If disulfide bridge formation were entirely random, the recovered enzyme should have 0.95% of its original activity. The fact that the enzyme the chemist recovered had 80% of its original activity supports his hypothesis that disulfide bridges form after the minimum energy conformation of the protein has been achieved. In other words, disulfide bridge formation is not random, but is determined by the tertiary structure of the protein.

72.   Formation of the tetrahedral intermediate (the two arrows that represent electron flow to the right) and collapse of the tetrahedral intermediate (the two arrows that represent electron flow to the left) are shown together in the second step.

a thiazoline

a PTH-amino acid

73.   The spot marked with an **X** is the peptide that is different in the normal and mutant polypeptide. The spot is closer to the cathode and farther to the right, indicating that the substituted amino acid in the mutant has a greater pI and is less polar.

The fingerprints are those of hemoglobin (normal) and sickle-cell hemoglobin (mutant). In sickle-cell hemoglobin, a glutamate in the normal polypeptide is substituted with a valine. This agrees with our observation that the substituted amino acid is less negative and more nonpolar. (See the discussion of sickle-cell anemia on page 1224 of the text.)

74.   **a.**   Acid-catalyzed hydrolysis indicates the peptide contains 12 amino acids.

__ __ __ __ __ __ __ __ __ __ __ __

b.   Treatment with Edman's reagent indicates that Val is the N-terminal amino acid.

Val __ __ __ __ __ __ __ __ __ __ __

c.   Treatment with carboxypeptidase A indicates that Ala is the C-terminal amino acid.

Val __ __ __ __ __ __ __ __ __ __ Ala

d.   Treatment with cyanogen bromide indicates that Met is the fifth amino acid with Arg, Gly, Ser in an unknown order in positions 2, 3, and 4.

Val __ __ __ Met __ __ __ __ __ __ Ala
     Arg, Gly, Ser

e.   Treatment with trypsin indicates that Arg is the third amino acid, and Ser is second, Gly is fourth, Tyr is sixth, and Lys is seventh. Since Lys is in the terminal fragment, cleavage did not occur at Lys, so Pro must be at lysine's cleavage site, but we don't know whether Lys-Pro comes before of after Phe and Ser.

Val Ser Arg Gly Met Tyr Lys __ __ __ __ Ala
                              Lys-Pro, Phe, Ser

f.   Treatment with chymotrypsin indicates that Phe is the tenth amino acid and Ser is the eleventh.

Val Ser Arg Gly Met Tyr Lys Lys Pro Phe Ser Ala

Chapter 22 Practice Test

1. Draw the structure of the following amino acids at pH = 7:

    a. glutamic acid   b. lysine   c. isoleucine   d. arginine   e. asparagine

2. Draw the form of histidine that predominates at:

    a. pH = 1   b. pH = 4   c. pH = 8   d. pH = 11

3. Answer the following:

    a. Alanine has a pI = 6.02 and serine has a pI = 5.68. Which would have the highest concentration of positive charge at pH = 5.50?

    b. Which amino acid is the only one that does not have an asymmetric center?

    c. Which are the two most nonpolar amino acids?

    d. Which amino acid has the lowest pI?

4. Why does the carboxyl group of alanine have a lower pK_a than the carboxyl group of propanoic acid?

5. Indicate whether each of the following statements is true or false:

    a. A cigar-shaped protein has a greater percentage of polar residues than a spherical protein.   T   F

    b. Naturally occurring amino acids have the L-configuration.   T   F

    c. There is free rotation about a peptide bond.   T   F

6. What compound is obtained from mild oxidation of cysteine?

7. Define each of the following:

    a. the primary structure of a protein

    b. the tertiary structure of a protein

    c. the quaternary structure of a protein

8.   Identify the spots.

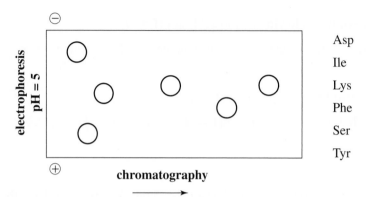

9.   Calculate the pI of each of the following amino acids:

    a.   phenylalanine $(pK_a s = 2.16, 9.18)$ **b.**   arginine $(pK_a s = 2.17, 9.04, 12.48)$

10.   From the following information, determine the primary sequence of the decapeptide:

    a.   Acid hydrolysis gives: Ala, 2 Arg, Gly, His, Ile, Lys, Met, Phe, Ser

    b.   Reaction with Edman's reagent liberated Ala

    c.   Reaction with carboxypeptidase A liberated Ile

    d.   Reaction with cyanogen bromide (cleaves on the C-side of Met)

        1.   Gly, 2 Arg, Ala, Met, Ser

        2.   Lys, Phe, Ile, His

    e.   Reaction with trypsin (cleaves on the C-side of Arg and Lys)

        1.   Arg, Gly

        2.   Ile

        3.   Phe, Lys, Met, His

        4.   Arg, Ser, Ala

    f.   Reaction with thermolysin (cleaves on the N-side of Leu, Ile, Phe, Trp, Tyr)

        1.   Lys, Phe

        2.   2 Arg, Ser, His, Gly, Ala, Met

        3.   Ile

11.   Describe how acetaldehyde can be converted to alanine.

12.   Draw the mechanism for the conversion of a thiol to a disulfide in a basic solution of Br_2.

CHAPTER 23
Catalysis in Organic Reactions and in Enzymatic Reactions

Important Terms

acid catalyst	a catalyst that increases the rate of a reaction by donating a proton.
active site	a pocket or cleft in an enzyme where the substrate is bound.
acyl-enzyme intermediate	an amino acid residue of an enzyme that has been acylated while catalyzing a reaction.
amino acid side chain	the substituent on the α-carbon of an amino acid.
base catalyst	a catalyst that increases the rate of a reaction by removing a proton.
catalyst	a substance that increases the rate of a reaction without itself being consumed in the overall reaction.
covalent catalysis (nucleophilic catalysis)	catalysis that occurs as a result of a nucleophile forming a covalent bond with one of the reactants.
effective molarity	the concentration of the reagent that would be required in an intermolecular reaction for it to have the same rate as an intramolecular reaction.
electrophilic catalyst	an electrophile that facilitates a reaction.
electrostatic catalysis	the stabilization of a charge by an opposite charge.
enzyme	a protein that is a catalyst.
general-acid catalysis	catalysis in which a proton is transferred to the reactant during the slow step of the reaction.
general-base catalysis	catalysis in which a proton is removed from the reactant during the slow step of the reaction.
induced fit model	a model that describes the specificity of an enzyme for its substrate: the shape of the active site does not become completely complementary to the shape of the substrate until after the enzyme has bound the substrate.
intramolecular catalysis	catalysis in which the catalyst that facilitates the reaction is part of the molecule undergoing reaction.
lock-and-key model	a model that describes the specificity of an enzyme for its substrate: the substrate fits the enzyme like a key fits into a lock.
metal-ion catalysis	catalysis in which the species that facilitates the reaction is a metal ion.

molecular recognition	the recognition of one molecule by another as a result of specific interactions; for example, the specificity of an enzyme for its substrate.
nucleophilic catalysis (covalent catalysis)	catalysis that occurs as a result of a nucleophile forming a covalent bond with one of the reactants.
nucleophilic catalyst	a catalyst that increases the rate of a reaction by acting as a nucleophile.
pH-activity profile or pH-rate profile	a plot of the activity of an enzyme as a function of the pH of the reaction mixture.
relative rate	the relative rate is obtained by dividing the actual rate constant by the rate constant of the slowest reaction in the group being compared.
site-specific mutagenesis	a technique that substitutes one amino acid of a protein for another.
specific-acid catalysis	catalysis in which the proton is fully transferred to the reactant before the slow step of the reaction.
specific-base catalysis	catalysis in which the proton is completely removed from the reactant before the slow step of the reaction.
substrate	the reactant of an enzyme-catalyzed reaction.

Solutions to Problems

1.   A catalyst increases the rate of a reaction by decreasing the energy between the reactant and the transition state of the rate-limiting step. That is, it decreases the height of the energy hill of the rate-limiting step. The following parameters would be different for a reaction carried out in the presence of a catalyst: $\Delta H^{\ddagger}$, E_a, $\Delta S^{\ddagger}$, $\Delta G^{\ddagger}$, k_{rate}. These are the parameters that reflect the difference in energy between the reactant and the transition state. The other factors do not change because they reflect the difference in energy between the reactant and product, which is not affected by catalysis.

2.   Notice that (1) and (2) have only the first part of the mechanism for the acid-catalyzed hydrolysis of an ester (because the final product of the reaction is a tetrahedral intermediate), (3) has only the second part (because the initial reactant is a tetrahedral intermediate), and (4) has both the first and second parts.

Mechanism for acid-catalyzed ester hydrolysis: first part

a. **Similarities:** the first step is protonation of the carbonyl compound, the second step is addition of a nucleophile to the protonated carbonyl compound, and the third step is loss of a proton.

b. **Differences:** the carbonyl compound that is used as the starting material; the nucleophile used in (2) is an alcohol rather than water.

Mechanism for acid-catalyzed ester hydrolysis: second part

a. **Similarities:** the first step is protonation of the tetrahedral intermediate, and the second step is elimination of a group from the tetrahedral intermediate. In two of the three reactions, the third step is loss of a proton.

b. **Differences:** the group that is protonated O in the first two and N in the third and the group that is eliminated from the tetrahedral intermediate. The third step in acetal formation is not loss of a proton, because the intermediate does not have a proton to lose. Instead, the third step is addition of a nucleophile, and the fourth step is loss of a proton.

3.   Both slow steps are specific-acid catalyzed; the compound is protonated before the slow step in each case.

4.   a. In specific-acid catalysis, the proton is added to the reactant before the O—C bond forms.

b. In general-acid catalysis, the proton is added to the reactant and the O—C bond forms in the same step.

5.   Solved in the text.

6.   **a.** In specific-base catalysis, the proton is removed from the reactant before the O—C bond forms.

b. In general-base catalysis, the proton is removed from the reactant and the O—C bond forms in the same step.

7.   The metal ion catalyzes the decarboxylation reaction by complexing with the negatively charged oxygen of the carboxylate group and carbonyl oxygen of the β-keto group, thereby making it easier for the carbonyl oxygen to accept the electrons that are left behind when CO_2 is eliminated.

Because acetoacetate and the monoethyl ester of dimethyloxaloacetate do not have a negatively charged oxygen on one carbon and a carbonyl group on an adjacent carbon with which to form a complex, a metal ion does not catalyze decarboxylation of these compounds.

8. Co^{2+} can catalyze the reaction in three different ways. It can complex with the reactant, increasing the susceptibility of the carbonyl group to nucleophilic addition. It can also complex with water, increasing the tendency of water to lose a proton, resulting in a better nucleophile for hydrolysis. And it can complex with the leaving group, decreasing its basicity and thereby making it a better leaving group.

9. Because the reacting groups in the trans isomer are pointed in opposite directions, they cannot react in an intramolecular reaction. Because they can react only via an intermolecular pathway, they will have approximately the same rate of reaction as they would have if the reacting groups were in separate molecules. Consequently, the relative rate would be expected to be close to one.

10. a. The nucleophile can attack the back side of either of the two ring carbons to which the sulfur is bonded in the intermediate, thereby forming two trans products. There are two nucleophiles (water and ethanol), so a total of four products will be formed.

    b.

11. Solved in the text.

12. The tetrahedral intermediate has two leaving groups, a carboxylate ion and a phenolate ion. When there are no nitro substituents on the aromatic ring, the carboxylate ion is a weaker base (a better leaving group) than the 2,4-dinitrophenolate ion, so the tetrahedral intermediate reforms the ester. Thus, the reaction proceeds through general-base catalyzed hydrolysis of the ester. In contrast, the 2,4-dinitrophenolate ion is a weaker base (better leaving group) than the carboxylate ion, so the tetrahedral intermediate forms the anhydride. Thus, the reaction proceeds through hydrolysis of the anhydride formed by nucleophilic catalysis.

13.   If the *ortho*-carboxyl substituent acts as an intramolecular base catalyst, ^{18}O would be incorporated into acetic acid and not into salicylic acid.

If the *ortho*-carboxyl substituent acts as an intramolecular nucleophilic catalyst, ^{18}O would be incorporated into both salicylic acid (if water adds to the carbonyl group attached to the benzene ring) and acetic acid (if water adds to the carbonyl group attached to the methyl group).

here water adds to the carbonyl group
attached to the benzene ring

Not all the salicylic acid would contain ^{18}O, because the anhydride intermediate has two different carbonyl groups. If water adds to the benzyl carbonyl group (above), salicylic acid will contain ^{18}O. If, however, water adds to the acetyl carbonyl group (below), acetic acid will contain ^{18}O.

here water adds to the carbonyl group
attached to the methyl group

14. Solved in the text.

15. **2**, **3**, and **4** are bases, so they can help remove a proton.

16. Ser-Ala-Leu would be more readily cleaved by carboxypeptidase A, because the nonpolar isobutyl substituent of phenylalanine would be more attracted to the hydrophobic pocket of the enzyme than would the negatively charged substituent of aspartate.

17. Glu 270 adds to the carbonyl group of the ester, forming a tetrahedral intermediate. Collapse of the tetrahedral intermediate is most likely catalyzed by an acid group of the enzyme in order to increase the leaving ability of the RO group. (Perhaps the HO substituent of tyrosine is close enough in the esterase to be the catalyst.) The group that donates the proton can then act as a base catalyst to remove a proton from water as it hydrolyzes the anhydride.

18.   Because arginine extends farther into the binding pocket, it must be the one that forms direct hydrogen bonds. Lysine, which is shorter, needs the mediation of a water molecule in order to engage in bond formation with aspartate.

19.   The side chains of D-Arg and D-Lys are not positioned to bind correctly at the active site. They would be able to bind at a mirror image of the active site.

20.   NAM would contain ^{18}O because it is the ring that would undergo nucleophilic attack by $H_2^{18}O$.

21.   Lemon juice contains citric acid. Some of the side chains of the enzyme will become protonated in an acidic solution. This will change the charge of the group (for example, a negatively charged aspartate, when protonated, becomes neutral; a neutral lysine, when protonated, becomes positively charged), and because the shape of an enzyme is determined by the interaction of the side chains, changing the charges of the side chains will cause the enzyme to undergo a conformational change that leads to denaturation. When the enzyme is denatured, it loses its ability to catalyze the reaction that causes apples to turn brown.

22.   In the absence of an enzyme, D-fructose is in equilibrium with D-glucose and D-mannose as a result of an enediol rearrangement (Section 21.5). Both C-2 epimers are formed, because a new asymmetric center is formed at C-2 and it can have either the *R* or the *S* configuration. Enzyme-catalyzed reactions are typically highly stereoselective—the enzyme catalyzes the formation of a single stereoisomer. Thus, D-fructose is in equilibrium with only D-glucose in the presence of the enzyme that catalyzes the enediol rearrangement.

23.   The ascending leg of the pH-rate profile is due to a group that is a general-base catalyst because the rate is at a maximum when the group is in its basic form. From the description of the mechanism, we know that amino acid is histidine.

The descending leg of the pH-rate profile is due to a group that is a general-acid catalyst because the rate is at a maximum when the group is in its acidic form. From the description of the mechanism, we know that amino acid is lysine.

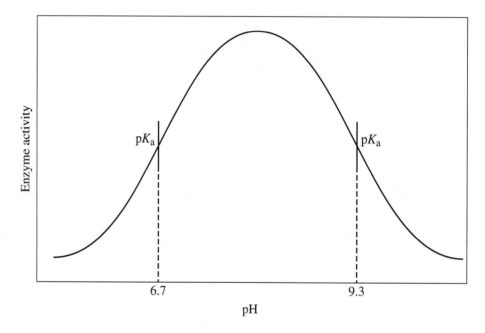

24.   **a.**   Because the catalytic group is an acid catalyst, it will be active in its acidic form and inactive in its basic form. The pH at the midpoint of the curve corresponds to the pK_a of the catalytic group because $pH = pK_a$ when $[HA] = [A^-]$ (see page 72 of the text).

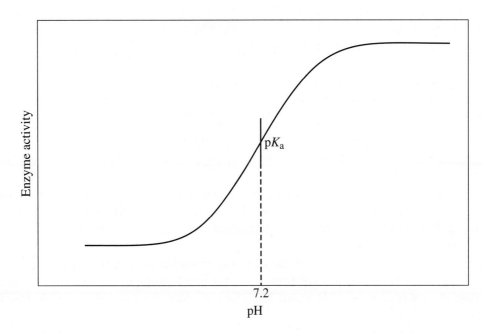

b. Because the catalytic group is a base catalyst, it will be inactive in its acidic form and active in its basic form.

25. Because **2** is a primary amine, it can form an imine. Notice that **1** cannot form an imine because the lone pair on the NH_2 group is delocalized onto the oxygen, so this NH_2 group is not a nucleophile. The N in **3** is not nucleophilic because its lone pair is delocalized (the lone pair is part of indole's π cloud).

26.

27.   The positively charged nitrogen of the protonated imine can accept the electrons that are left behind when the C3—C4 bond breaks.

In the absence of the protonated imine, the electrons would be delocalized onto a neutral oxygen (see Problem 26). The neutral oxygen is not as electron withdrawing as the positively charged nitrogen of the protonated imine. In other words, imine formation makes it easier to break the C3—C4 bond.

28.   In order to break the C3—C4 bond, the carbonyl group has to be at the 2-position as it is in fructose, so it can accept the electrons that result from C3—C4 bond cleavage; the carbonyl group at the 1-position in glucose cannot accept these electrons. Therefore, glucose must isomerize to a ketose before the reaction can occur.

glucose-6-phosphate   fructose-6-phosphate

29.   Cysteine residues react with iodoacetic acid because a thiol is a good nucleophile and iodine is a good leaving group. If a cysteine residue is at the active site of an enzyme, adding a substituent to the sulfur in this way could interfere with the enzyme's being able to bind the substrate or it could interfere with positioning the tyrosine residue that catalyzes the reaction. Adding a substituent to cysteine might also cause a conformational change in the enzyme that could destroy its activity.

30. The following compound will lose HBr more rapidly, because the negatively charged oxygen is in position to act as an intramolecular general-base catalyst.

31. The following compound will form an anhydride more rapidly because it forms a five-membered-ring anhydride, which is less strained and than the seven-membered-ring anhydride formed by the other compound. The greater stability of the five-membered-ring product causes the transition state leading to its formation to be more stable than the transition state leading to formation of the seven-membered-ring product.

32. In order to hydrolyze an amide, the NH_2 group in the tetrahedral intermediate has to leave in preference to the less basic OH group. This can happen if the NH_2 group is protonated because $^+NH_3$ is a weaker base and, therefore, easier to eliminate than OH. Of the four compounds, two (*ortho*-carboxybenzamide and *ortho*-hydroxybenzamide) have substituents that can protonate the NH_2 by acting as acid catalysts in a hydrolysis reaction carried out at pH = 3.5.

ortho-carboxybenzamide

Because the carboxy group withdraws electrons from the ring by resonance and the OH group donates electrons to the ring by resonance, formation of the tetrahedral intermediate will be faster for the carboxy-substituted compound.

The carboxy group is a stronger acid than the OH group (and with a $pK_a = 4.2$, 83% will be in the acidic form at pH = 3.5). Therefore, a larger fraction of the tetrahedral intermediate will be protonated than in the case of the hydroxy-substituted compound, so collapse of the tetrahedral intermediate will be faster. Therefore, the *ortho*-carboxybenzamide has the faster rate of hydrolysis.

33.   **a.** $CH_3CH_2\overset{..}{\underset{..}{S}}CH_2CH_2\overset{\frown}{—}\overset{\frown}{Cl}$ intramolecular nucleophilic catalysis

b.

intramolecular general-acid catalysis

The OH substituent is protonating the leaving group as it departs, causing it to be a weaker base and, therefore, a better leaving group.

34.   If the *ortho*-carboxylate ion is acting as a general-base catalyst, the kinetic isotope effect will be greater than 1.0 (see Section 10.8), because an O—H (or O—D) bond is broken in the slow step of the reaction, and an O—D bond is stronger than an O—H bond and therefore is harder to break.

If the *ortho*-carboxylate ion is acting as a nucleophilic catalyst, the kinetic isotope effect will be about 1.0, because an O—H (or O—D) bond is not broken in the slow step of the reaction.

35. $\dfrac{\text{rate constant for the catalyzed reaction}}{\text{rate constant for the uncatalyzed reaction}}$ $\dfrac{1.5 \times 10^6 \text{ M}^{-1}\text{ s}^{-1}}{0.6 \text{ M}^{-1}\text{ s}^{-1}} = 2.5 \times 10^6$

36. Co^{2+} can catalyze the hydrolysis reaction by complexing with three nitrogen atoms in the substrate as well as with water. Complexation increases the acidity of water, thereby providing a better nucleophile (metal-bound hydroxide) for the hydrolysis reaction. Complexation with three nitrogens locks the nucleophile into the correct position for addition to the carbonyl carbon.

37.   Class I aldolases use a protonated imine as a site to which electrons can be delocalized (see Problem 27). Class II aldolases use a metal ion complexed with the carbonyl group as a site to which electrons can be delocalized.

38.   Knowing that the reaction is much slower if the N is replaced by a CH tells us that the N must be acting as a catalyst. We see that the N is in position to be an intramolecular nucleophilic catalyst.

39.   In the absence of morpholine, the first step of the reaction is addition of water to the **ester**, because addition of water to the aldehyde would form a hydrate that would be in equilibrium with the aldehyde.

There are two possible mechanisms that can account for the observed catalysis by morpholine. In the presence of morpholine, the first step of the reaction (in both mechanisms) is addition of morpholine to the **aldehyde**. The reaction of morpholine with the aldehyde is a faster reaction than the reaction of water with the ester, because morpholine is a better nucleophile than water and an aldehyde is more susceptible to nucleophilic addition than an ester.

In one mechanism, the negatively charged aldehyde oxygen (which is a much better nucleophile than water) is the nucleophile that adds to the ester. The tetrahedral intermediate collapses to form a lactone. Iminium ion formation followed by imine hydrolysis gives the final product.

In the other mechanism, the reaction of morpholine with the aldehyde forms an iminium ion. The positively charged iminium ion makes it easier for water to add to the ester by stabilizing the negative charge that develops on the oxygen. Collapse of the tetrahedral intermediate followed by imine hydrolysis forms the final product.

40.

|H₂O   water exchanges for bicarbonate

41.   At pH = 12, the nucleophile is hydroxide ion. Addition of hydroxide ion to the carbonyl group is faster in **A**, because the negative charge on the oxygen that is created in the tetrahedral intermediate is stabilized by the positively charged nitrogen.

At pH = 8, the nucleophile is water because the concentration of hydroxide ion is only 1×10^{-6} M. Addition of water to the carbonyl group is faster in **B**, because the amino group can act as a general-base catalyst to make water a better nucleophile.

42. **a.** The cis reactants each undergo an S_N2 reaction. Because the acetate displaces the tosylate group by back-side attack, each cis reactant forms a trans product.

b. The acetate group in a trans reactant is positioned to be able to displace the tosylate leaving group by an intramolecular S_N2 reaction. Acetate ion then attacks in a second S_N2 reaction from the back side of the group it displaces, so trans products are formed. Because both trans reactants form the same intermediate, they both form the same products. Because the acetate ion can attack either of the carbons in the intermediate equally easily, a racemic mixture will be formed.

c. The trans reactants are more reactive because the tosylate leaving group is displaced in an intramolecular reaction, and the resulting positively charged cis intermediate is considerably more reactive than the neutral cis reactants. Thus, a trans isomer undergoes two successive S_N2 reactions, each of which is faster than the one step S_N2 reaction that a cis isomer undergoes.

43. Reduction of the imine linkage with sodium borohydride causes fructose to become permanently attached to the enzyme because the hydrolyzable imine bond has been lost. Acid-catalyzed hydrolysis removes the two phosphate groups and hydrolyzes the two peptide bonds, so the radioactive fragment that is isolated after hydrolysis is the lysine residue (covalently attached to fructose) that originally formed the imine.

$$CH_2OPO_3^{2-}$$

(reaction scheme top left, starting material)

$$^{14}C=N-(CH_2)_4-CH$$

with groups: $CH_2OPO_3^{2-}$, $^{14}C=N-(CH_2)_4-CH$, NH (top right on CH), HO—H, H—OH, H—OH, $CH_2OPO_3^{2-}$, and C=O, NH

$$\xrightarrow[\text{2. } H_3O^+]{\text{1. } NaBH_4}$$

(product top right)

$$^{14}HC-NH-(CH_2)_4-CH$$

with groups: $CH_2OPO_3^{2-}$, NH, HO—H, H—OH, H—OH, $CH_2OPO_3^{2-}$, C=O, NH

$$\downarrow HCl \mid H_2O$$

(product lower)

$$^{14}HC-NH-(CH_2)_4-CH$$

with groups: CH_2OH, $^+NH_3$, HO—H, H—OH, H—OH, CH_2OH, C=O, OH

44. **a.** 3-Amino-2-oxindole catalyzes the decarboxylation of an α-keto acid by first forming an imine and then losing a proton from the α-carbon. Now the compound can be decarboxylated, because the electrons left behind can be delocalized onto an oxygen.

(reaction mechanism scheme)

3-amino-2-oxindole + α-keto acid $\xrightarrow{\text{imine formation}}$ (imine intermediate)

$\xrightarrow{\text{decarboxylation}}$ (intermediate) $+ CO_2$

$\downarrow$ enolization

(enol imine) $\xrightarrow{\text{imine hydrolysis}}$ 3-amino-2-oxindole + aldehyde

b. 3-Aminoindole would not be as effective a catalyst, because the electrons left behind when CO_2 is eliminated cannot be delocalized onto an electronegative atom.

3-aminoindole α-keto acid   imine formation

45. a. Intramolecular nucleophilic attack on an alkyl halide occurs more rapidly than intermolecular attack on an alkyl halide, because the reacting groups are tethered together in the former, which makes it more likely that they will find each other in order to react. The intramolecular reaction is followed by another relatively rapid S_N2 reaction, because the strain in the three-membered ring causes it to break easily.

:Nu^- = guanine

b.

46.

general-base
catalyst

dihydroxyacetone
phosphate

general-acid
catalyst

glyceraldehyde-
3-phosphate

Chapter 23 Practice Test

1. Indicate whether each of the following statements is true or false:

    a. A catalyst increases the equilibrium constant of a reaction.   T   F

    b. An acid catalyst donates a proton to the substrate and a base catalyst
        removes a proton from the substrate.   T   F

    c. The reactant of an enzyme-catalyzed reaction is called a substrate.   T   F

    d. Complexing with a metal ion increases the pK_a of water.   T   F

2. In lysozyme, glutamate 35 is a catalyst that is active in its acid form. Explain its catalytic function.

3. a. Draw a pH-rate profile for an enzyme that has one catalytic group at the active site with a $pK_a = 5.7$ that functions as a general-base catalyst.

    b. Draw a pH-rate profile for an enzyme that has one catalytic group at the active site with a $pK_a = 5.7$ that functions as a general-acid catalyst.

4. Using arrows, show the mechanism arrows for the first step in the mechanism for chymotrypsin.

    a. What kind of catalyst is histidine in this step?

    b. What kind of catalyst is serine in this step?

    c. How does aspartate catalyze the reaction?

CHAPTER 24
The Organic Chemistry of the Coenzymes, Compounds Derived from Vitamins

Important Terms

apoenzyme	an enzyme (that needs a cofactor to catalyze a reaction) without its cofactor.
biotin	the coenzyme required by enzymes that catalyze carboxylation of a carbon adjacent to an ester or a keto group.
coenzyme	a cofactor that is an organic molecule.
coenzyme A	a thiol used by biological organisms to form thioesters.
coenzyme B_{12}	the coenzyme required by enzymes that catalyze certain rearrangement reactions.
cofactor	an organic molecule or a metal ion that an enzyme needs in order to catalyze a reaction.
competitive inhibitor	a compound that inhibits an enzyme by competing with the substrate for binding at the active site.
dehydrogenase	an enzyme that carries out an oxidation reaction by removing hydrogen from the substrate.
electron sink	a site to which electrons can be delocalized.
flavin adenine dinucleotide (FAD)	a coenzyme required in certain oxidation reactions. It is reduced to $FADH_2$, which is a coenzyme required in certain reduction reactions.
heterocyclic compound	a cyclic compound in which one or more of the ring atoms is an atom other than carbon.
holoenzyme	an enzyme plus its cofactor.
lipoate	a coenzyme required in certain oxidation reactions.
mechanism-based inhibitor (suicide inhibitor)	an inhibitor that inactivates an enzyme by undergoing part of the normal catalytic mechanism.
molecular recognition	the recognition of one molecule by another as a result of specific interactions.
nicotinamide adenine dinucleotide (NAD^+)	a coenzyme required in certain oxidation reactions. It is reduced to NADH, which is a coenzyme required in certain reduction reactions.
nicotinamide adenine dinucleotide phosphate ($NADP^+$)	a coenzyme that is reduced to NADPH, which is a coenzyme required in certain reduction reactions.

731

nucleotide   a heterocycle attached in the β-position to the anomeric carbon of a phosphorylated ribose.

pyridoxal phosphate   the coenzyme required by enzymes that catalyze certain transformations of amino acids.

suicide inhibitor (mechanism-based inhibitor)   a compound that inactivates an enzyme by undergoing part of its normal catalytic mechanism.

tetrahydrofolate (THF)   the coenzyme required by enzymes that catalyze a reaction that donates a group containing a single carbon to its substrate.

thiamine pyrophosphate (TPP)   the coenzyme required by enzymes that catalyze a reaction that transfers an acyl group from one species to another.

transamination   a reaction in which an amino group is transferred from one compound to another.

transimination   the reaction of a primary amine with an imine to form a new imine and a primary amine derived from the original imine.

vitamin   a substance needed in small amounts for normal body function that the body cannot synthesize or cannot synthesize in adequate amounts.

vitamin KH$_2$ the coenzyme required by the enzyme that catalyzes the carboxylation of glutamate side chains.

Solutions to Problems

1. The metal ion (Zn^{2+}) makes the carbonyl carbon more susceptible to nucleophilic addition, increases the nucleophilicity of water by making it more like a hydroxide ion, and stabilizes the negative charge on the transition state.

2. The alcohol is oxidized to a ketone.

3. The ketone is reduced to an alcohol.

4. **a.** FAD has seven conjugated double bonds (indicated by *).

   b. FADH$_2$ has three conjugated double bonds (on the left of the molecule). It also has two conjugated double bonds (on the right of the molecule that are isolated from the other three.

5.   The nitrogen that is the stronger base is the one most apt to be protonated. N-1 is a weaker base than N-5, because the π electrons that belong to N-1 can be delocalized onto an oxygen, so protonation on nitrogen will not occur when a nucleophile adds to C-10a. The π electrons that belong to N-5 can be delocalized only into the benzene ring, so protonation on nitrogen will occur when a nucleophile adds to C-4a. Nucleophilic addition therefore occurs at the position (4a) that results in the stronger base (N-5) being protonated.

6.   Solved in the text.

7.   When a proton is removed from the methyl group at C-8, the electrons that are left behind can be delocalized onto the oxygen at the 2-position or onto the oxygen at the 4-position.

When a proton is removed from the methyl group at C-7, the electrons that are left behind can be delocalized only onto carbons that, being less electronegative than oxygen, are less able to accommodate the electrons.

8.

pyruvate-TPP
intermediate

a β-keto acid

9.　Notice that the only difference in the mechanisms for pyruvate decarboxylase and acetolactate synthase is the species the two-carbon fragment is transferred to: a proton in the case of pyruvate carboxylase and pyruvate in the case of acetolactate synthase.

acetolactate

10.   Notice that the only difference in this reaction and that in Problem 9 is that the species to which the two-carbon fragment is transferred has an ethyl group in place of the methyl group.

11.   Solved in the text.

12.   The α-keto group that accepts the amino group from pyridoxamine is converted into an amino group.

a.

pyruvate   alanine

b.

oxaloacetate   aspartate

13. For clarity, the substituents are not put on the three bonds attached to the pyridinium ring in Problems 13, 14, and 16.

14. In the first step, a proton is removed from the α-carbon; in the next step, the leaving group is eliminated from the β-carbon. Transimination produces an enamine. The enamine tautomerizes to an imine that is hydrolyzed to the final product.

15. The compound on the right is more easily decarboxylated because the electrons left behind when CO_2 is eliminated are delocalized onto the positively charged nitrogen of the pyridine ring. The electrons left behind if the other compound is decarboxylated cannot be delocalized.

16. The first step (after transimination) in all amino acid transformations catalyzed by PLP is removal of a substituent from the α-carbon of the amino acid. The electrons left behind when the substituent is removed are delocalized onto the positively charged nitrogen of the pyridine ring. If the ring nitrogen was not protonated, it would be less attractive to the electrons. In other words, it would be a less effective electron sink.

17. The hydrogen of the OH substituent forms a hydrogen bond with the nitrogen of the imine linkage (see Problem 16). This puts a partial positive charge on the nitrogen, which makes it easier for the amino acid to add to the imine carbon in the transimination reaction that attaches the amino acid to the coenzyme. It also makes it easier to remove a substituent from the α-carbon of the amino acid. If the OH substituent is replaced by an OCH_3 substituent, there is no longer a proton available to form the hydrogen bond.

18. The mechanism is the same as that shown in the text for dioldehydrase. The tetrahedral intermediate that is formed as a result of the coenzyme B_{12}–catalyzed isomerization is unstable and loses ammonia to give acetaldehyde, the final product of the reaction.

19.

20. They differ only in the circled part of the molecule.

folic acid   aminopterin

21. The methyl group in thymidine comes from the methylene group of N^5,N^{10}-methylene-THF, followed by the addition of a hydride ion from the coenzyme. The methylene group of N^5,N^{10}-methylene-THF comes from the CH_2OH group of serine by means of a PLP-catalyzed C_α—C_β cleavage (see page 1152 in the text).

22. Two thiol groups are oxidized in each of the two reactions that overall reduces vitamin K epoxide to vitamin KH_2. Because dihydrolipoate has two thiol groups, each thiol oxidation involves an intramolecular reaction. When thiols such as ethanethiol or propanethiol are used, each thiol oxidation involves an intermolecular reaction. These thiols react more slowly than dihydrolipoate, because the two thiol groups are not in the same molecule. Therefore, they have to find each other in order to react.

23.
   a. thiamine pyrophosphate
   b. FAD oxidizes dihydrolipoate back to lipoate.
   c. NAD^+ oxidizes $FADH_2$ back to FAD.
   d.

   e. thiamine pyrophosphate and pyridoxal phosphate
   f. Thiamine pyrophosphate is used for the decarboxylation of α-keto acids.
       Pyridoxal phosphate is used for the decarboxylation of amino acids.
   g. biotin and vitamin KH_2
   h. Biotin carboxylates a carbon adjacent to a carbonyl group (that is, an α-carbon).
       Vitamin KH_2 carboxylates the γ-carbon of a glutamate.

24.
   a. NAD^+, $NADP^+$, FAD, thiamine pyrophosphate, carboxybiotin, pyridoxal phosphate
   b. thiamine pyrophosphate, lipoate, biotin, pyridoxal phosphate, coenzyme B_{12}
   c. vitamin KH_2
   d. tetrahydrofolate

25.
   a. acetyl-CoA carboxylase; biotin
   b. dihydrolipoyl dehydrogenase; FAD
   c. methylmalonyl-CoA mutase; coenzyme B_{12}
   d. lactate dehydrogenase; NADH (This enzyme is named according to the reaction it catalyzes in the reverse direction.)
   e. aspartate transaminase; pyridoxal phosphate
   f. propionyl-CoA carboxylase; biotin

26.

27. **a.** thiamine pyrophosphate, lipoate, coenzyme A, FAD, NAD$^+$, indicated by numbers 1–5 in the mechanism below

b.

α-keto-glutarate

thiamine pyrophosphate
1

lipoate
2

+ CO$_2$

At the top of the page is a reaction scheme:

$$\text{(substrate) } \xrightarrow[\textbf{3}]{\text{CoASH}} \text{ succinyl-CoA} + \text{(dithiol)}$$

$$\text{(dithiol)} \xrightarrow[\textbf{4}]{\text{FAD}} \text{(disulfide)}$$

$$\text{H}^+ + \text{NADH} + \text{FAD} \xleftarrow[\textbf{5}]{\text{NAD}^+} \text{FADH}_2 + \text{(disulfide)}$$

28.

The structures shown (ketones):

CH$_3$—C(=O)— ; HOCH$_2$—C(=O)— ; $^-$OOCCH$_2$CH$_2$—C(=O)—

CH$_3$CH(CH$_3$)—C(=O)— ; CH$_3$CHCH$_2$(CH$_3$)—C(=O)— ; CH$_3$CH$_2$CH(CH$_3$)—C(=O)—

29.

A mechanism sequence is shown:

AdCH$_2$—Co(III), CH$_3$—C(H)(COO$^-$)—C(=O)—SCoA → AdCH$_2$, CH$_2$—C(H)(COO$^-$)—C(=O)—SCoA, Co(II) → AdCH$_3$, CH$_2$—C(H)(COO$^-$)—C(=O)—SCoA, Co(II)

→ CH$_2$—C$^\bullet$(H)(COO$^-$)—C(=O)—SCoA, AdCH$_2$—H, Co(II) → CH$_2$—CH$_2$—C(=O)(COO$^-$)—SCoA, AdCH$_2^\bullet$, Co(II) → AdCH$_2$—Co(III)

30. The product of transamination of an amino acid has a carbonyl group in place of the amino group.

a.

CH$_3$CH(CH$_3$)—C(=O)—C(=O)—O$^-$

derived from Val

CH$_3$CHCH$_2$(CH$_3$)—C(=O)—C(=O)—O$^-$

derived from Leu

CH$_3$CH$_2$CH(CH$_3$)—C(=O)—C(=O)—O$^-$

derived from Ile

b.

$$CH_3CH(CH_3)-C(=O)-SCoA \qquad CH_3CHCH_2(CH_3)-C(=O)-SCoA \qquad CH_3CH_2CH(CH_3)-C(=O)-SCoA$$

c. The reaction catalyzed by the enzyme is identical to the reaction catalyzed by the pyruvate dehydrogenase complex (Section 24.3). Therefore, they both require the same coenzymes: thiamine pyrophosphate, lipoate, coenzyme A, FAD, NAD$^+$.

d. The disease can be treated by a diet low in branched-chain amino acids.

31. The products of the reaction are shown below.

$$Ad-CHT(Co(III)) \ + \ CH_3CHT-C(=O)-T \ + \ H_2O$$

The following mechanism explains the products:

dehydration

$$H-C(CH_3)T-C(=O)-T \ + \ H_2O$$

If there is only a limited amount of coenzyme, the second H of the coenzyme will be replaced by T in a subsequent round of catalysis, which means that the final product will be:

$$Ad-CT_2 \quad + \quad CH_3CHT \quad C(=O) \quad T \quad + \quad H_2O$$

with Co(III) on the Ad—CT₂.

32.

ribose-5'-phosphate   ribose-5'-phosphate   ribose-5'-phosphate

33.   For clarity, the substituents are not put on the three bonds attached to the pyridinium ring in Problems 33–36. Each reaction in Problems 33–35 starts after the amino acid has undergone transimination with the lysine-bound coenzyme.

transimination

tautomerization

HCl | imine
H₂O | hydrolysis

34.

35.

36.

37. Nonenzyme-bound FAD has a greater oxidation potential than NAD^+. Consequently, FAD oxidizes NADH to NAD^+. When the enzyme binds FAD, it changes its oxidation potential, causing it to become less than that of NAD^+. Consequently, NAD^+ can oxidize enzyme-bound $FADH_2$ to FAD.

In addition, there will be much more NAD^+ relative to enzyme-bound $FADH_2$ in the mitochondria (where the oxidation of $FADH_2$ by NAD^+ takes place). Therefore, the equilibrium favors the oxidation of $FADH_2$ by NAD^+ (Le Châtelier's principle).

$$\underset{\text{excess}}{\text{E-FADH}_2 \;+\; NAD^+} \;\rightleftharpoons\; \text{E-FAD} \;+\; \text{NADH} \;+\; H^+$$

38.

Chapter 24 Practice Test

1. What two coenzymes put carboxyl groups on their substrates?

2. What are the three one-carbon groups that tetrahydrofolate coenzymes put on their substrates?

3. Show the mechanism for NADPH reducing its substrate.

4. Draw the structure of the compound obtained when the following amino acid undergoes transamination:

5. What is the first step in the reaction of the substrate with coenzyme B_{12} in an enzyme-catalyzed reaction that requires coenzyme B_{12}?

6. What coenzyme is required for each of the following enzyme-catalyzed reactions?

7. Show the enzyme-catalyzed reaction that requires vitamin KH_2 as a coenzyme.

8. Draw the product of the enzyme-catalyzed reaction that requires biotin and whose substrate is acetyl-CoA.

9. **a.** Other than the substrate, enzyme, and coenzyme, what three additional reagents are needed by a reaction that requires biotin as a coenzyme?

   b. What is the function of each of these reagents?

10. What is the function of FAD in the pyruvate dehydrogenase system?

11. Draw the structures of the two products obtained from the following transamination reaction:

12.   Indicate whether each of the following statements is true or false:

    a.   Vitamin B_1 is the only water-insoluble vitamin that has a coenzyme function.   T   F

    b. $FADH_2$ is a reducing agent.   T   F

    c.   Thiamine pyrophospate is vitamin B_6.   T   F

    d.   Cofactors that are organic molecules are called coenzymes.   T   F

    e.   Vitamin K is a water-soluble vitamin.   T   F

    f.   Lipoic acid is covalently bound to its enzyme by an amide linkage.   T   F

CHAPTER 25
The Organic Chemistry of the Metabolic Pathways · Terpene Biosynthesis

Important Terms

acyl adenylate	
acyl phosphate	
allosteric activator/ inhibitor	a compound that activates/inhibits an enzyme by binding to a site on the enzyme other than the active site
anabolism	the reactions living organisms carry out that result in the synthesis of complex biomolecules from simple precursor molecules.
catabolism	the reactions living organisms carry out to provide energy and simple precursor molecules for synthesis.
citric acid cycle	a series of reactions that convert the acetyl group of acetyl-CoA into two molecules of CO_2 and a molecule of CoASH.
cholesterol	a steroid that is the precursor of all other steroids.
dimethylallyl pyrophosphate	a compound needed for the biosynthesis of terpenes and biosynthesized from isopentenyl pyrophosphate.
diterpene	a terpene that contains 20 carbons.
feedback inhibitor	a compound that inhibits a step at the beginning of the pathway for its biosynthesis
glycolysis	the series of reactions that converts glucose into two molecules of pyruvate.
gluconeogenesis	the synthesis of glucose from pyruvate
high-energy bond	a bond that releases a great deal of energy when it is broken.
isopentenyl pyrophosphate	the starting material for the biosynthesis of terpenes.
metabolism	reactions living organisms carry out in order to obtain the energy they need and to synthesize the compounds they require.
monoterpene	a terpene that contains 10 carbons.
β-oxidation	a repeating series of four reactions that convert a fatty acyl-CoA molecule into molecules of acetyl-CoA.

751

oxidative phosphorylation   the fourth stage of catabolism in which NADH and $FADH_2$ are oxidized back to NAD^+ and FAD: for each NADH that is oxidized, 2.5 ATPs are formed; for each $FADH_2$ that is oxidized, 1.5 ATPs are formed.

phosphoanhydride bond   the bond that holds two phosphoric acid molecules together.

phosphoryl transfer reaction   the transfer of a phosphate group from one compound to another.

protein prenylation   the process of attaching farnesyl and geranylgeranyl groups to proteins to allow them to become anchored to membranes.

regulatory enzyme   an enzyme that catalyzes an irreversible reaction near the beginning of a pathway, thereby allowing independent control over degradation and synthesis.

sesquiterpene   a terpene that contains 15 carbons.

squalene   a triterpene that is a precursor of steroid molecules.

terpene   a lipid isolated from a plant that contains carbon atoms in multiples of five.

terpenoid   a terpene that contains oxygen.

tetraterpene   a terpene that contains 40 carbons.

triterpene   a terpene that contains 30 carbons.

Solutions to Problems

1.   **a.**   Solved in the text.

   b.   Because pH 7.4 is much more basic than the pK_a values of the first two ionizations of ADP, these two groups will be in their basic forms at that pH, giving ADP two negative charges. We can determine the fraction of the group with a pK_a of 6.8 that will be in its basic form at pH 7.4 using the method shown in part **a**.

$$\text{fraction of group present in basic form} = \frac{K_a}{K_a + [H^+]} = \frac{1.6 \times 10^{-7}}{1.6 \times 10^{-7} + 4.0 \times 10^{-8}}$$

$$= \frac{1.6 \times 10^{-7}}{1.6 \times 10^{-7} + 0.4 \times 10^{-7}} = 0.8$$

$$\text{total negative charge of ADP} = 2.0 + 0.8 = 2.8$$

   c.   At pH 7.4 the OH group with a pK_a of 1.9 will account for one negative charge and the OH group with a pK_a of 12.4 will have no negative charge. We need to calculate the fraction of the group with a pK_a value of 6.7 that will be negatively charged at pH 7.4.

$$\text{fraction of group present in basic form} = \frac{K_a}{K_a + [H^+]} = \frac{2.0 \times 10^{-7}}{2.0 \times 10^{-7} + 0.4 \times 10^{-7}} = 0.8$$

$$\text{total negative charge on phosphate} = 1.0 + 0.8 = 1.8$$

2.

3.

4.   The resonance contributor on the right shows that the β-carbon of the α,β-unsaturated carbonyl compound has a partial positive charge. The nucleophilic OH group, therefore, is attracted to the β-carbon.

5.   Because palmitic acid has 16 carbons and the acyl group of acetyl-CoA has 2 carbons, 8 molecules of acetyl-CoA are formed from 1 molecule of palmitic acid.

6.   seven: one mole of NADH will be obtained from each of the following rounds of β-oxidation.

7.

fructose-6-phosphate   ATP   fructose-1,6-bisphosphate   ADP

8.   **a.**   conversion of glucose to glucose-6-phosphate (the first step)
conversion of fructose-6-phosphate to fructose-1,6-bisphosphate (the third step)

b.   conversion of 1,3-bisphosphoglycerate to 3-phosphoglycerate (the seventh step)
conversion of phosphoenolpyruvate to pyruvate (the tenth step)

9.   The reaction that follows the oxidation of glyceraldehyde-3-phosphate to 1,3-bisphosphoglycerate (the conversion of 1,3-bisphosphoglycerate to 3-phosphoglycerate) is highly exergonic. Therefore as 1,3-bisphosphoglycerate is converted to 3-phosphoglycerate, glyceraldehyde-3-phosphate will be converted to 1,3-bisphosphoglycerate to replenish it.

10.   two; each molecule of D-glucose is converted to two molecules of glyceraldehyde-3-phosphate, and each molecule of glyceraldehyde-3-phosphate requires one molecule of NAD^+ for it to be converted to one molecule of pyruvate.

11.   acetaldehyde reductase

12.   a ketone

13.   thiamine pyrophosphate

14.

15. pyridoxal phosphate

16.

alanine   pyruvate

17. Protonated histidine (pK_a = 6.0) is not strong enough an acid to fully protonate the OH group to make it a good leaving group (H_2O) that would be able to leave in the first step of the elimination reaction, which is required for an E1 reaction. Therefore, it is an E2 reaction with protonated histidine acting as a general-acid catalyst to protonate the OH group as it departs.

18. a secondary alcohol

19. citrate and isocitrate (and the alkene intermediate generated during the conversion of citrate to isocitrate)

20.

21. From the name of the enzyme, we know that it adds a phosphate group to a nucleotide diphosphate.

22. succinyl-CoA synthetase: this enzyme catalyzes a reaction of succinyl-CoA; the reverse reaction would be the synthesis of succinyl-CoA.

23. **a.** The conversion of one molecule of glycerol to dihydroxyacetone phosphate consumes one molecule of ATP. The conversion of dihydroxyacetone phosphate to pyruvate produces two molecules of ATP. Therefore, one molecule of ATP is obtained from the conversion of one molecule of glycerol to pyruvate.

b. One NADH is formed from the conversion of one molecule of glycerol to dihydroxyacetone phosphate, and one NADH is formed from the conversion of dihydroxyacetone phosphate to pyruvate. Each NADH forms 2.5 ATP in the fourth stage of catabolism. So when the fourth stage of catabolism is included, six molecules (2.5 + 2.5 + 1 = 6) of ATP are obtained from the conversion of one molecule of glycerol to pyruvate.

24. **a.** glycerol kinase   **b.** phosphatidic acid phosphatase

25.   The first step is an S_N2 reaction; the second step is a nucleophilic addition elimination reaction.

ATP   glutamate

ADP

HB⁺

glutamine

26.   The fact that the tail-to-tail linkage occurs in the exact center of the molecule suggests that the two halves are synthesized (in a head-to-tail fashion) and then joined together in a tail-to-tail linkage.

tail-to-tail linkage

27.   Squalene, lycopene, and β-carotene are all synthesized in the same way. In each case, two halves are synthesized (in a head-to-tail fashion) and then joined together in a tail-to-tail linkage.

lycopene

β-carotene

28.

hydroxymethylglutaryl-CoA

29.   One equivalent of NADPH is required to reduce the thioester to an aldehyde. The second equivalent of NADPH is needed to reduce the aldehyde to an alcohol.

30.

mevalonic acid + ATP

mevalonyl pyrophosphate + ADP

31.   Solved in the text.

32.

geranyl
pyrophosphate

α-terpineol

33.

E isomer

rotate about
single bond

Z isomer

34. It tells you that the reaction is an S$_N$1 reaction, because the fluoro-substituted carbocation is less stable than the non-fluoro-substituted carbocation (due to the strongly electron-withdrawing fluoro substituent), so it would form more slowly.

less stable   F   more stable

If the reaction had been an S$_N$2 reaction, the fluoro-substituted compound would have reacted more rapidly than the non-fluoro-substituted carbocation, because the electron-withdrawing fluoro substituent would make the compound more susceptible to nucleophilic attack.

35.

farnesyl pyrophosphate

$+$ H$_3$O$^+$

36. Solved in the text.

37.   There are two 1,2-hydride shifts and two 1,2-methyl shifts. The last step is elimination of a proton.

protosterol cation

1,2-hydride shift

1,2-hydride shift

1,2-methyl shift

1,2-methyl shift

+ HB⁺

38.   **a.** catabolic   **b.** catabolic

39.

D-galactose   ATP

D-galactose-1-phosphate   ADP

40.   The hydrogen on the α-carbon.

pyruvate → lactate

41.   **a.** reactions 1 and 3   **b.** reactions 2, 5, and 8   **c.** reaction 6   **d.** reaction 9

42.   The conversion of citrate to isocitrate.
The conversion of fumarate to (S)-malate.

43.

44.

α-ketoglutarate

45.   The label will be on the phosphate group that is attached to the enzyme (phosphoglycerate mutase) that catalyzes the isomerization of 3-phosphoglycerate to 2-phosphoglycerate.

46.   If you examine the mechanism for the isomerism of glucose-6-phosphate to fructose-6-phosphate on page 1125 of the text, you can see that C-1 in D-glucose is also C-1 in D-fructose.

Now, if you examine the mechanism for the aldolase-catalyzed cleavage of fructose-1,6-bisphosphate to form glyceraldehyde-3-phosphate on page 1127 of the text, you can see which carbons in D-glucose correspond to the carbons in dihydroxyacetone phosphate and D-glyceraldehyde-3-phosphate.

Then, if you look at page 1180, you will see how the carbons in dihydroxyacetone phosphate correspond to the carbons in D-glyceraldehyde-3-phosphate.

Finally we see how the carbons in D-glyceraldehyde-3-phosphate correspond to the carbons in pyruvate. Therefore, C-3 and C-4 of glucose each becomes a carboxyl group in pyruvate.

47. Pyruvate loses its carboxyl group when it is converted to ethanol. Because the carboxyl group is C-3 or C-4 of glucose, half of the ethanol molecules contain C-1 and C-2 of glucose and the other half contain C-5 and C-6 of glucose.

48. At the beginning of a fast, blood glucose levels would be normal.
     After a 24-hour fast, blood glucose levels would be very low because both dietary glucose and glycogen have been depleted and glucose cannot be synthesized because of the deficiency of fructose-1,6-bisphosphatase.

49. The conversion of pyruvate to lactate is a reversible reaction. Lactate can be converted back to pyruvate by oxidation.

The conversion of pyruvate to acetaldehyde is not a reversible reaction because it is a decarboxylation. The CO_2 cannot be put back onto acetaldehyde.

50. The β-oxidation of a molecule of a 16-carbon fatty acyl-CoA will form 8 molecules of acetyl-CoA.

51. Each molecule of acetyl-CoA forms 2 molecules of CO_2. Therefore, the 8 molecules of acetyl-CoA obtained from a molecule of a 16-carbon fatty acyl-CoA will form 16 molecules of CO_2.

52.   No ATP is formed from β-oxidation.

53.   Each molecule of acetyl-CoA that is cleaved from the 16-carbon fatty acyl-CoA forms 1 molecule of $FADH_2$ and 1 molecule of NADH. Since a 16-carbon fatty acyl-CoA undergoes 7 cleavages, 7 molecules of $FADH_2$ and 7 molecules of NADH are formed from the 16-carbon fatty acyl-CoA.

54.   Because each NADH forms 2.5 molecules of ATP and each $FADH_2$ forms 1.5 molecules of ATP in oxidative phosphorylation, the 7 molecules of NADH form 17.5 molecules of ATP and the 7 molecules of $FADH_2$ form 10.5 molecules of ATP. Therefore, 28 molecules of ATP are formed.

55.   We have seen that each molecule of acetyl-CoA that enters the citric acid cycle forms 10 molecules of ATP (Section 25.11). A molecule of a 16-carbon fatty acid will form 8 molecules of acetyl-CoA. These will form 80 molecules of ATP. When these are added to the number of ATP molecules formed from the NADH and $FADH_2$ generated in β-oxidation, we see that 108 molecules of ATP are formed from complete metabolism of a 16-carbon saturated fatty acyl-CoA.

56.   Each molecule of glucose, while being converted to two molecules of pyruvate, forms two molecules of ATP and two molecules of NADH.

The two molecules of pyruvate form two molecules of NADH while being converted to two molecules of acetyl-CoA.

Each molecule of acetyl-CoA that enters the citric acid cycle forms three molecules of NADH, one molecule of $FADH_2$, and one molecule of ATP. Therefore the two molecules of acetyl-CoA obtained from glucose form six molecules of NADH, two molecules of $FADH_2$, and two molecules of ATP.

Therefore, each molecule of glucose forms four molecules of ATP, ten molecules of NADH (2 + 2 + 6), and two molecules of $FADH_2$.

Since each NADH forms 2.5 molecules of ATP and each $FADH_2$ forms 1.5 molecules of ATP, one molecule of glucose forms 4 + (10 × 2.5) + (2 × 1.5) molecules of ATP. That is, each molecule of glucose forms 32 molecules of ATP.

57.   Pyruvate can be converted to alanine (transamination), oxaloacetate (carboxylation), lactate (reduction), and acetyl-CoA (by the pyruvate dehydrogenase complex).

58. The conversion of propionyl-CoA to methylmalonyl-CoA requires biotin (vitamin H).
The conversion of methylmalonyl-CoA to succinyl-CoA requires coenzyme B_{12} (vitamin B_{12}).

59. In Problem 46 we saw how the carbons in D-glyceraldehyde-3-phosphate correspond to the carbons in pyruvate.

D-glyceraldehyde-3-phosphate   pyruvate

Now we can answer the questions. The label in pyruvate is indicated by *.

a.
```
  COO⁻
  |
  C=O
  |
* CH₃
```

c.
```
* COO⁻
  |
  C=O
  |
  CH₃
```

e.
```
  COO⁻
  |
* C=O
  |
  CH₃
```

b.
```
  COO⁻
  |
* C=O
  |
  CH₃
```

d.
```
* COO⁻
  |
  C=O
  |
  CH₃
```

f.
```
  COO⁻
  |
  C=O
  |
* CH₃
```

60.

pyruvate   pyruvate

pyruvate carboxylase   pyruvate dehydrogenase complex

citrate synthase

oxaloacetate   +   acetyl-CoA

H_2O

citrate

61. A Claisen condensation between two molecules of acetyl-CoA forms acetoacetyl-CoA that, when hydrolyzed, forms acetoacetate.

Acetoacetate can undergo decarboxylation to form acetone, or it can be reduced to 3-hydroxybutyrate.

62. The mechanism for the conversion of fructose-1,6-bisphosphate to glyceraldehyde-3-phosphate and dihydroxyacetone phosphate is shown on page 1127 of the text.

The mechanism for the conversion of dihydroxyacetone phosphate to glyceraldehyde-3-phosphate is shown on page 1180 of the text.

From these mechanisms, you can see that the label (*) was at C-1 in glyceraldehyde-3-phosphate.

63. **a.** UDP-galactose and UDP-glucose are C-4 epimers. NAD$^+$ oxidizes the C-4 OH group of UDP-galactose to a ketone. When NADH reduces the ketone back to an OH, it attacks the sp^2 carbon from above the plane, forming the C-4 epimer of the starting material.

OH
CH$_2$OH
H
HO   O—UDP
OH
UDP-galactose

NAD$^+$ NAD—H

O
H
HO   CH$_2$OH
B$^+$ O—UDP
OH

NAD$^+$

:B

CH$_2$OH
HO
HO   O—UDP
OH
UDP-glucose

b. The enzyme is called an epimerase because it converts a compound into an epimer (in this case, a C-4 epimer).

64. Because the compound that would react in the second step with the activated carboxylic acid group is excluded from the incubation mixture, the reaction between the carboxylate ion and ATP will come to equilibrium.

If radioactively labeled pyrophosphate is put into the incubation mixture, ATP will become radioactive if the mechanism involves attack on the α-phosphorus because pyrophosphate is a reactant in the reverse reaction that forms ATP.

ATP will not become radioactive if the mechanism involves attack on the β-phosphorus because pyrophosphate is not a reactant in the reverse reaction that forms ATP. (In other words, because pyrophosphate is not a product of the reaction, it cannot become incorporated into ATP in the reverse reaction.)

attack on the α-phosphorus

[reaction scheme: carboxylate + ATP ⇌ acyl–phosphate + pyrophosphate]

pyrophosphate

attack on the β-phosphorus

[reaction scheme: carboxylate + ATP ⇌ product + AMP]

AMP

65.   If radioactive AMP is added to the reaction mixture, the results will be opposite. If the mechanism involves attack on the α-phosphorus, ATP will not become radioactive, because AMP is not a reactant in the reverse reaction that forms ATP. If the mechanism involves attack on the β-phosphorus, ATP will become radioactive, because AMP is a reactant in the reverse reaction that forms ATP.

66.   A sesquiterpene is synthesized from farnesyl pyrophosphate. The carbons that would be labeled in farnesyl pyrophosphate, if it is synthesized from acetate with a ^{14}C-labeled carbonyl carbon, are indicated by an asterisk. (See Problem 36.)

farnesyl pyrophosphate

67.   First draw the starting material (geranyl pyrophosphate) so that its structure is as similar as possible as that of the target molecule.

geranyl pyrophosphate

α-pinene

α-pinene

68. The mechanism for the conversion of isopentenyl pyrophosphate to geranyl pyrophospate to farnesyl pyrophosphate is shown on page 1199 of the text. Now we need to show how farnesyl pyrophosphate is converted to eudesmol.

farnesyl pyrophosphate

eudesmol HB⁺

Chapter 25 Practice Test

1. Draw a structure for each of the following:

   a. the intermediate formed when a nucleophile (RO⁻) attacks the γ-phosphorus of ATP.

   b. an acyl pyrophosphate.

   c. pyrophosphate.

2. Fill in the six blanks in the following scheme:

3. Which of the following are not citric acid cycle intermediates: fumarate, acetate, citrate?

4. Which provide energy to the cell: anabolic reactions or catabolic reactions?

5. What compounds are formed when proteins undergo the first stage of catabolism?

6. What compound is formed when fatty acids undergo the second stage of catabolism?

7. Indicate whether each of the following statements is true or false:

   a. Each molecule of $FADH_2$ forms 2.5 molecules of ATP in the fourth stage of catabolism.   T   F

   b. $FADH_2$ is oxidized to FAD.   T   F

   c. NAD^+ is oxidized to NADH.   T   F

   d. Acetyl-CoA is a citric acid cycle intermediate.   T   F

   e. A diterpene contains 20 carbons.   T   F

8.   Mark off the isoprene units in squalene.

squalene

9.   Draw the mechanism for the following reaction:

($^-$OPP$_i$ = pyrophosphate)

10.   How many carbon atoms are there in a sesquiterpene?

CHAPTER 26
The Chemistry of the Nucleic Acids

Important Terms

anticodon	the three bases at the bottom of the middle loop in a tRNA.
base	a nitrogen-containing heterocyclic compound (a purine or a pyrimidine) found in DNA and RNA.
codon	a sequence of three bases in mRNA that specifies the amino acid to be incorporated into a protein.
deamination	a hydrolysis reaction that results in the removal of ammonia.
deoxyribonucleic acid (DNA)	a polymer of deoxyribonucleotides.
deoxyribonucleotide	a nucleotide where the sugar component is D-2-deoxyribose.
dinucleotide	two nucleotides linked by a phosphodiester bond.
double helix	the term used to describe the secondary structure of DNA.
exon	a stretch of bases in DNA that are a portion of a gene.
gene	a segment of DNA.
gene therapy	a technique that inserts a synthetic gene into the DNA of an organism defective in that gene.
genetic code	the amino acid specified by each three-base sequence of mRNA.
genetic engineering	recombinant DNA technology, where DNA molecules are attached to DNA in a host cell and allowed to replicate.
human genome	the total DNA of a human cell.
major groove	the wider and deeper of the two alternating grooves in DNA.
minor groove	the narrower and more shallow of the two alternating grooves in DNA.
nucleic acid	a chain of five-membered ring sugars linked by phosphodiester groups with each sugar bearing a heterocyclic amine at the anomeric carbon in the β-position. The two kinds of nucleic acids are DNA and RNA.
nucleoside	a heterocyclic base (purine or pyrimidine) bonded to the anomeric carbon of a sugar (D-ribose or D-2-deoxyribose) in the β-position.
nucleotide	a nucleoside with one of its OH groups bonded to a phosphate group via an ester linkage.
oligonucleotide	three to ten nucleotides linked by phosphodiester bonds.

phosphodiester	a species in which two of the OH groups of phosphoric acid have been converted to OR groups.
polymerase chain reaction (PCR)	a method that amplifies segments of DNA.
polynucleotide	many nucleotides linked by phosphodiester bonds.
primary structure	the sequence of bases in the nucleic acid.
pyrosequencing	an automated technique used to sequence DNA; it detects the identity of the base that adds to the DNA primer.
recombinant DNA	DNA that has been incorporated into a host cell.
replication	the synthesis of identical copies of DNA.
replication fork	the position on DNA where replication begins.
restriction endonuclease	an enzyme that cleaves DNA at a specific base sequence.
restriction fragment	a fragment that is formed when DNA is cleaved by a restriction endonuclease.
ribonucleic acid (RNA)	a polymer of ribonucleotides.
ribonucleotide	a nucleotide where the sugar component is D-ribose.
ribosomal RNA (rRNA)	the structural component of ribosomes, the particles on which protein synthesis takes place.
RNA splicing	the step in RNA processing that cuts out nonsense bases and splices informational pieces together.
semiconservative replication	the mode of replication that results in a daughter molecule of DNA having one of the original DNA strands plus a newly synthesized strand.
sense strand	the strand in DNA that is not read during transcription; it has the same sequence of bases as the synthesized mRNA strand, except that the mRNA has Us in place of the Ts in DNA.
stacking interactions	van der Waals interactions between the mutually induced dipoles of adjacent pairs of bases in DNA.
stop codon	a codon that says "stop protein synthesis here."
template strand	the strand in DNA that is read during transcription.
transcription	the synthesis of mRNA from a DNA blueprint.
transfer RNA (tRNA)	a single-stranded RNA molecule that carries an amino acid to be incorporated into a protein.
translation	the synthesis of a protein from an mRNA blueprint.

Solutions to Problems

1. The ring is protonated at its most basic position. In the case of a purine, this is the 7-position. In the next step, the bond between the heterocyclic base and the sugar breaks, with the anomeric carbocation being stabilized by the ring oxygen's nonbonding electrons.

The mechanism is exactly the same for pyrimidines except that the initial protonation takes place at the 3-position.

2. a.

dCDP

b.

dTTP

c.

dUMP

e.

guanosine 5′a-triphosphate
GTP

d.

UDP

f.

adenosine 3′-monophosphate
AMP

3. The hydrolysis of cyclic AMP forms adenosine 5′-phosphate and adenosine 3′-phosphate.

$$\xrightarrow[\text{H}_2\text{O}]{\text{HCl}}$$

adenosine 5′-phosphate   +   **adenosine 3′-phosphate**

4. A hydrogen bond acceptor is indicated by **A**; a hydrogen bond donor is indicated by **D**.
    The **A** and **D** designations show that the maximum number of hydrogen bonds that can form are two
    between thymine and adenine and three between cytosine and guanine. Notice that uracil and thymine
    have the same **A** and **D** designations.

uracil   thymine   adenine

cytosine   guanine

5. If the bases existed in the enol form, no hydrogen bonds could form between the bases unless one of the
    bases were shifted vertically.

uracil   thymine   adenine

cytosine   guanine

6.   **a.** 3′—C—C—T—G—T—T—A—G—A—C—G—5′
         b. guanine

7.   Notice that when a nucleophile attacks the phosphorus of a diester, the π bond breaks. However, when a nucleophile attacks the phosphorus of an anhydride, an σ bond breaks in preference to the π bond.

8.

9.   Thymine and uracil differ only in that thymine has a methyl substituent that uracil does not have (thymine is 5-methyluracil). Because thymine and uracil both have the same groups in the same positions that can participate in hydrogen bonding, they will both call for the incorporation of the same purine. Because thymine and uracil form one hydrogen bond with guanine and two with adenine, they will both incorporate adenine in order to maximize hydrogen bonding.

10.   Because methionine is known to be the first base incorporated into the heptapeptide, the mRNA sequence is read beginning at AUG, since that is the only codon that codes for methionine.

Met-Asp-Pro-Val-Ile-Lys-His

11.   Met-Asp-Pro-Leu-Leu-Asn

12.   It does not cause protein synthesis to stop because the sequence UAA does not occur within a triplet. The reading frame causes the triplets to be AUU and **AA**A. In other words, the U is at the end of a triplet and the next triplet starts with AA.

13.   A change in the third base of a codon would be least likely to cause a mutation because the third base is variable for many amino acids. For example, CUU, CUC, CUA, and CUG all code for leucine.

14.   The sequence of bases in the template strand of DNA specifies the sequence of bases in mRNA, so the bases in the template strand and the bases in mRNA are complementary. Therefore, the sequence of bases in the sense strand of DNA is identical to the sequence of bases in mRNA, except wherever there is a U in mRNA, there is a T in the sense strand of DNA.

5′—G—C—A—T—G—G—A—C—C—C—C—G—T—T—
A—T—T—A—A—A—C—A—C—3′

15.

	Met	Asp	Pro	Val	Ile	Lys	His
<u>codons</u>	AUG	GAU	CCU	GUU	AUU	AAA	CAU
		GAC	CCC	GUC	AUC	AAG	CAC
			CCA	GUA	AUA		
			CCG	GUG			

<u>anticodons</u>

Note that the anticodons are stated in the 5′ → 3′ direction. For example, the anticodon of AUG is stated as CAU (not UAC).

<u>codons</u>   5′ A U G 3′

<u>anticodons</u>   3′ U A C 5′

CAU	AUC	AGG	AAC	AAU	UUU	AUG
	GUC	GGG	GAC	GAU	CUU	GUG
		UGG	UAC	UAU		
		CGG	CAC			

16.

17. Deamination involves hydrolyzing an imine linkage to a carbonyl group and ammonia, so cytosine can be deaminated to uracil.

Thymine does not have an amino substituent on the ring, which means that it cannot form an imine and therefore it cannot be deaminated.

18. **a** is the only sequence that has a chance of being recognized by a restriction endonuclease because it is the only one that has the same sequence of bases in the $5' \rightarrow 3'$ direction that the complementary strand has in the $5' \rightarrow 3'$ direction.

$$^{5'}\text{A—C—G—C—G—T}^{3'}$$
$$^{3'}\text{T—G—C—G—C—A}^{5'}$$

19. **a.** guanosine 3'-monophosphate   **c.** 2'-deoxyadenosine 5'-monophosphate
      b. cytidine 5'-diphosphate   **d.** 2'-deoxythymidine

20. Lys-Val-Gly-Tyr-Pro-Gly-Met-Val-Val

21.   5′—GAC—CAC—CAT—TCC—GGG—GTA—GCC—AAC—TTT—3′

22.   5′—AAA—GTT—GGC—TAC—CCC—GGA—ATG—GTG—GTC—3′

23.   **a.** Ile   **b.** Asp   **c.** Val   **d.** Val

24.   The third base in each codon has some variability.

mRNA   5′-GG(UCA or G)UC(UCA or G)CG(UCA or G)GU(UCA or G)CA(U or C)GA(A or G)-3′

or AG(U or C)   AG(A or G)

DNA   3′-CC(AGT or C)AG(AGT or C)GC(AGT or C)CA(AGT or C)GT(A or G)CT(T or C)-5′
template

or TC(A or G)   TC(T or C)

sense   5′-GG(TCA or G)TC(TCA or G)CG(TCA or G)GT(TCA or G)CA(T or C)GA(A or G)-3′
or AG(T or C)   AG(A or G)

Notice that Ser and Arg are two of three amino acids that can be specified by six different codons.

25.

26.   A segment of DNA with 18 base pairs has 36 bases. If there are 7 cytosines, there are 7 guanines, accounting for 14 bases. Therefore, 22 (36 − 14 = 22) are adenines and uracils.

a. 11 uracils   **b.** 7 guanines

27.

28.

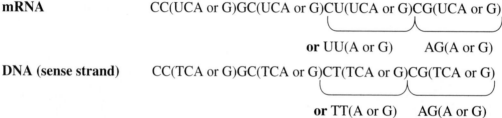

mRNA   CC(UCA or G)GC(UCA or G)CU(UCA or G)CG(UCA or G)

                 or UU(A or G)   AG(A or G)

DNA (sense strand)   CC(TCA or G)GC(TCA or G)CT(TCA or G)CG(TCA or G)

                 or TT(A or G)   AG(A or G)

Note that because mRNA is complementary to the template strand of DNA, which is complementary to the sense strand, mRNA and the sense strand of DNA have the same sequence of bases (expect DNA has a T where RNA has a U). Also note that Leu and Arg are each specified by six codons.

29.   **a.**   CC and GG   **c.**   CA and TG

CA and TG are formed in equal amounts, since A pairs with T and C pairs with G. (Remember that the dinucleotides are written in the $5' \rightarrow 3'$ direction. For example, part **f** is not a correct answer because it has A pairing with A and T pairing with T.)

$$5' \ C \ A \ 3'$$
$$3' \ G \ T \ 5'$$

30.   AZT is incorporated into DNA when the 3'-OH group of the last nucleotide incorporated into the growing chain of DNA attacks the α-phosphorus of AZT-triphosphate instead of a normal nucleotide. When AZT is thus incorporated into DNA, DNA synthesis stops because AZT does not have a 3'-OH group that can react with another nucleotide triphosphate.

31.   UUU = phenylalanine   UUG = leucine   GGU = glycine
         GGG = glycine   UGU = cysteine   GUG = valine
                                     GUU = valine   UGG = tryptophan

32. The number of different possible codons using four nucleotides is $(4)^n$, where n is the number of letters (nucleotides) in the code.

$$\text{for a two-letter code:} \quad (4)^2 = 16$$
$$\text{for a three-letter code:} \quad (4)^3 = 64$$
$$\text{for a four-letter code:} \quad (4)^4 = 256$$

Because there are 20 amino acids that must be specified, a two-letter code would not provide enough codons.
A three-letter code provides enough codes for all the amino acids and also provides the necessary stop codons.
A four-letter code provides many more codes than would be needed.

33. In the first step of the reaction, the imidazole ring of one histidine acts as a general-base catalyst, removing a proton from the 2′-OH group to make it a better nucleophile. The imidazole ring of the other histidine acts as a general-acid catalyst, protonating the leaving group to make it a weaker base and therefore a better leaving group. In the second step of the reaction, the roles of the two imidazole rings are reversed.

34. The normal and mutant peptides would have the following base sequence in their mRNA:

normal: CA(AG) UA(UC) GG(UCAG) AC(UCAG) CG(UCAG) UA(UC) GU(UCAG)
mutant: CA(AG) UC(UCAG) GA(AG) CC(UCGA) GG(UCGA) AC(UCAG)

a. The middle nucleotide (A) in the second triplet was deleted. This means that an A was deleted in the sense strand of DNA or a T was deleted in the template strand of DNA.

b. The mRNA for the mutant peptide has an unused 3′-terminal two-letter code, U(UCAG). The last amino acid in the octapeptide of the normal fragment is leucine, so its last triplet is UU(AG) or CU(UCAG).

This means that the triplet for the last amino acid in the mutant is U(UCAG)(UC) and that the last amino acid in the mutant is one of the following: Phe, Ser, Tyr, or Cys.

35.   If deamination does not occur, the mRNA sequence will be:

   AUG–UCG–CUA–AUC, which will code for the following tetrapeptide:
   Met-Ser-Leu-Ile

Deamination of a cytosine results in a uracil.
If the cytosines are deaminated, the mRNA sequence will be:

   AUG–UUG–UUA–AUU, which will code for the following tetrapeptide:
   Met-Leu-Leu-Ile

The only cytosine that will change the amino acid that is incorporated into the peptide is the first one. Therefore, this is the cytosine that could cause the most damage to an organism if it were deaminated.

36.   5-Bromouracil is incorporated into DNA in place of thymine because of their similar size. Thymine pairs with adenine via two hydrogen bonds. 5-Bromouracil exists primarily in the enol form. The enol cannot form any hydrogen bonds with adenine, but it can form two hydrogen bonds with guanine. Therefore, 5-bromouracil pairs with guanine. Because 5-bromouracil causes guanine to be incorporated instead of adenine into newly synthesized DNA strands, it causes mutations.

thymine   adenine   5-bromouracil   H   guanine

37.   In an acidic environment, nitrite ion is protonated to nitrous acid. We have seen that the nitrosonium ion is formed from nitrous acid (Section 19.23).

The nitrosonium ion reacts with a primary amino group to form a diazonium ion, which can be displaced by water (Section 19.21).

cytosine   **uracil**

38.   It requires energy to break the hydrogen bonds that hold the two chains together, so an enormous amount of energy would be required to unravel the chain completely. However, as the new nucleotides that are incorporated into the growing chain form hydrogen bonds with the parent chain, energy is released, and this energy can be used to unwind the next part of the double helix.

39. Ca^{2+} decreases the pK_a of water, forming calcium-bound hydroxide ion, which is a better nucleophile than water. The positively charged nitrogen (of the guanidinium group) of arginine stabilizes the negatively charged oxygen formed when the $P{=}O$ π bond of the diester breaks, and therefore makes it easier to form. Glutamic acid protonates the OR oxygen, thereby making it a weaker base and therefore a better leaving group when the $P{=}O$ π bond reforms.

40.   The ribosome, the particle on which protein synthesis occurs, has a binding site for the growing peptide chain and a binding site for the next amino acid to be incorporated into the chain.

peptide binding site   **amino acid binding site**

In protein synthesis, all peptide bonds are formed by the reaction of an amino acid with a peptide, except formation of the first peptide bond, which has to be formed by the reaction of two amino acids. Therefore, for the synthesis of the first peptide bond, the first (N-terminal) amino acid has to have a peptide bond that will fit into the peptide binding site.

The formyl group of *N*-formylmethionine will provide the peptide group that will be recognized by the peptide binding site, and the second amino acid will be bound in the amino acid binding site.

***N*-formylmethionine**

Chapter 26 Practice Test

1. Is the following compound dTMP, UMP, dUMP, or dUTP?

2. If one of the strands of DNA has the following sequence of bases running in the $5' \rightarrow 3'$ direction, what is the sequence of bases in the complementary strand?

$$5'-A-C-T-T-G-C-A-T-3'$$

3. What base is closest to the $5'$-end in the complementary strand?

4. Indicate whether each of the following statements is true or false:

a.	Guanine and cytosine are purines.	T	F
b.	The $3'$-OH group allows RNA to be easily cleaved.	T	F
c.	The number of As in DNA is equal to the number of Ts.	T	F
d.	rRNA carries the amino acid that will be incorporated into a protein.	T	F
e.	The template strand of DNA is the one transcribed to form RNA.	T	F
f.	The $5'$-end of DNA has a free OH group.	T	F
g.	The synthesis of proteins from an RNA blueprint is called transcription.	T	F
h.	A nucleotide consists of a base and a sugar.	T	F
i.	RNA contains Ts, and DNA contains Us.	T	F

5. Which of the following base sequences would most likely be recognized by a restriction endonuclease?

    1. ACGCGT   **3.** ACGGCA   **5.** ACATCGT

    2. ACGGGT   **4.** ACACGT   **6.** CCAACC

6.   What would be the sequence of bases in the mRNA obtained from the following segment of DNA?

sense strand

5′—G–C–A–T–G–G–A–C–C–C–C–G–T—3′

3′—C–G–T–A–C–C–T–G–G–G–G–C–A—5′

template strand

7.   Why is NADPH needed in order to convert uridines into thymidines?

8.   Which of the following pairs of dinucleotides will occur in equal amounts in DNA?

(Remember nucleotides are always written in the 5′ → 3′ direction.)

CA and GT   CG and AT

CG and GG   CA and TG

CHAPTER 27
Synthetic Polymers

Important Terms

addition polymer (**chain-growth polymer**)	a polymer that is made by adding monomers to the growing end of a chain.
alpha olefin	a monosubstituted ethylene.
alternating copolymer	a copolymer in which two monomers alternate.
anionic polymerization	a chain-growth polymerization where the initiator is a nucleophile; the propagation site, therefore, is an anion.
aramide	an aromatic polyamide.
atactic polymer	a polymer in which the substituents are randomly oriented on the extended carbon chain.
biodegradable polymer	a polymer that can be degraded by microorganisms.
biopolymer	a polymer that is synthesized in nature.
block copolymer	a copolymer in which there are blocks of each kind of monomer within the polymer chain.
cationic polymerization	a chain-growth polymerization where the initiator is an electrophile; the propagation site, therefore, is a cation.
chain-growth polymer	see addition polymer.
chain transfer	a growing polymer chain reacts with a molecule XY in a manner that allows X· to terminate the chain, leaving behind Y· to initiate a new chain.
condensation polymer (**step-growth polymer**)	a polymer that is made by combining two molecules while removing a small molecule (usually water or an alcohol).
conducting polymer	a polymer that can conduct electricity down its backbone.
copolymer	a polymer formed using two or more different monomers.
cross-linking	connecting polymer chains by bond formation.
crystallites	regions of a polymer in which the chains are highly ordered.
elastomer	a polymer that can stretch and then revert back to its original shape.

epoxy resin	a resin that is formed by mixing a low molecular weight prepolymer with a compound that forms a cross-linked polymer.
graft copolymer	a copolymer that contains branches of a polymer of one monomer grafted onto the backbone of a polymer made from another monomer.
head-to-tail addition	the head of one molecule is added to the tail of another molecule.
homopolymer	a polymer that contains only one kind of monomer.
isotactic polymer	a polymer in which all the substituents are on the same side of the fully extended carbon chain.
living polymer	a nonterminated chain-growth polymer that remains active. Therefore, the polymerization reaction can continue upon addition of more monomer.
materials science	the science of creating new materials that will have practical applications.
monomer	a repeating unit in a polymer.
oriented polymer	a polymer obtained by stretching out polymer chains and putting them back together in a parallel fashion.
plasticizer	an organic molecule that dissolves in a polymer and allows the polymer chains to slide by each other.
polyamide	a polymer in which monomers are connected by amide groups.
polycarbonate	a step-growth polymer in which the monomers are connected by carbonate groups.
polyester	a polymer in which monomers are connected by ester groups.
polymer	a large molecule made by linking monomers together.
polymer chemistry	the field of chemistry that deals with synthetic polymers; part of the larger discipline known as materials science.
polymerization	the process of linking up monomers to form a polymer.
polyurethane	a polymer in which monomers are connected by urethane groups.
propagating site	the reactive end of a chain-growth polymer.
radical polymerization	a chain-growth polymerization where the initiator is a radical; the propagation site, therefore, is a radical.
random copolymer	a copolymer with a random distribution of monomers.
ring-opening polymerization	a chain-growth polymerization that involves opening the ring of the monomer.

step-growth polymer
(condensation polymer)
a polymer that is made by combining two molecules while removing a small molecule (usually water or an alcohol).

syndiotactic polymer
a polymer in which the substituents regularly alternate on both sides of the fully extended carbon chain.

synthetic polymer
a polymer that is not synthesized in nature.

thermoplastic polymer
a polymer that has both ordered crystalline regions and amorphous non-crystalline regions.

thermosetting polymers
cross-linked polymers that, after they are hardened, cannot be re-melted by heating.

urethane
(carbamate)
a compound with a carbonyl group that is both an amide and an ester.

$$\underset{RO}{\overset{\displaystyle \overset{O}{\underset{\|}{C}}}{}}\diagdown NHR$$

vinyl polymer
a polymer in which the monomer is ethylene or a substituted ethylene.

vulcanization
increasing the flexibility of rubber by heating it with sulfur.

Ziegler-Natta catalyst
an aluminum-titanium initiator that controls the stereochemistry of a polymer.

Solutions to Problems

1.   **a.** $CH_2{=}CHCl$ **b.** $CH_2{=}CCH_3$ **c.** $CF_2{=}CF_2$

$$\underset{\displaystyle \underset{OCH_3}{\overset{\displaystyle C=O}{|}}}{|}$$

2.   Poly(vinyl chloride) would be more apt to contain head-to-head linkages, because a chloro substituent is less able (compared with a phenyl substituent) to stabilize the growing end of the polymer chain by electron delocalization.

3. $-CH_2CH-CHCH_2CH_2CH-CHCH_2CHCH_2CH_2CHCH_2CH-$

4.

5.   Because branching increases the flexibility of the polymer, beach balls are made from more highly branched polyethylene.

6.

7.   Decreasing ability to undergo cationic polymerization is in the same order as decreasing stability of the carbocation intermediate. (Electron donation increases the stability of the carbocation.)

CH$_2$=CH—OCH$_3$ > CH$_2$=CHCH$_3$ > CH$_2$=CH—CCH$_3$

donates electrons   **withdraws electrons**
by resonance   **by resonance**

A tertiary benzylic cation is more stable than a secondary benzylic cation.

8.   Decreasing ability to undergo anionic polymerization is in the same order as decreasing stability of the carbanion intermediate. (Electron withdrawal increases the stability of the carbanion.)

a.

withdraws electrons   **donates electrons**
by resonance   **by resonance**

b. CH$_2$=CHC≡N   >   CH$_2$=CHCl   >   CH$_2$=CHCH$_3$

withdraws electrons   **withdraws**
by resonance   **electrons inductively**

9.   Methyl methacrylate does not undergo cationic polymerization because the carbocation propagating site would be very unstable, since the ester group is strongly electron withdrawing.

a strongly electron-withdrawing group

10. In anionic polymerization, nucleophilic attack occurs at the less-substituted carbon because it is less sterically hindered; in cationic polymerization, nucleophilic attack occurs at the more-substituted carbon because the ring opens to give the more stable partial carbocation (Section 11.7).

position of nucleophilic attack in cationic polymerization

position of nucleophilic attack in anionic polymerization

11. **a.**

b.

12. **a.** $CH_2=CCH_3 + BF_3 + H_2O$
  $\ \ \ \ \ \ \ \ \ \ |$
  $\ \ \ \ \ \ \ \ \ CH_3$

b. $CH_2=CH + BF_3 + H_2O$

c.
$+ CH_3O^-$

d. $CH_2=CH + BuLi$

13.   3,3-Dimethyloxacyclobutane undergoes cationic polymerization by the following mechanism:

14.

15.

16.   **a.** $-NHCH_2CH_2CH_2\overset{O}{\overset{||}{C}}NHCH_2CH_2CH_2\overset{O}{\overset{||}{C}}-$

b. $-NH(CH_2)_4NH\overset{O}{\overset{||}{C}}CH_2CH_2\overset{O}{\overset{||}{C}}NH(CH_2)_4NH\overset{O}{\overset{||}{C}}CH_2CH_2\overset{O}{\overset{||}{C}}-$

17. $-\overset{O}{\overset{||}{C}}(CH_2)_4\overset{O}{\overset{||}{C}}\left[NH(CH_2)_6NH-\overset{O}{\overset{||}{C}}(CH_2)_4\overset{O}{\overset{||}{C}}\right]_n NH(CH_2)_6NH-$

$\Big\downarrow H_2SO_4$

$n\ HO\overset{O}{\overset{||}{C}}(CH_2)_4\overset{O}{\overset{||}{C}}OH \quad + \quad n\ H_3\overset{+}{N}(CH_2)_6\overset{+}{N}H_3$

18.

19. They hydrolyze to give salts of dicarboxylic acids and diols.

Kodel

Dacron

20.   a.

b.

H$_2$NCH$_2$CH$_2$NHCH$_2$CH$_2$N̈H$_2$ H$_2$N̈CH$_2$CH$_2$NHCH$_2$CH$_2$NH$_2$

-4 H$^+$

21. The polymer is stiffer because glycerol cross-links the polymer chains.

22. Formation of a protonated imine between formaldehyde and one amino group, followed by reaction with a second amino group, accounts for formation of the linkage that holds the monomers together.

a dimer of Melmac

23.

monomer used in the polymerization

Bakelite

24.   a.

b.

25.   **a.**   —CH₂CHCH₂CHCH₂CH—   chain-growth polymer

$$—CH_2CHCH_2CHCH_2CH—$$

with F, F, F substituents   chain-growth polymer

b.   —CH₂CHCH₂CHCH₂CH—   chain-growth polymer

with CO₂H, CO₂H, CO₂H substituents

c.   —O(CH₂)₅CO(CH₂)₅CO(CH₂)₅C—   step-growth polymer

(with three C=O groups)

d.   —NH(CH₂)₅NHC(CH₂)₅CNH(CH₂)₅NHC(CH₂)₅C—   step-growth polymer

(with four C=O groups)

e.   —OCNH⟨ring, CH₃⟩NHCOCH₂CH₂OCNH⟨ring, CH₃⟩NHCO—   step-growth polymer

26.   **a.**   —CH₂CH₂OCH₂CH₂N⟨piperazine ring⟩N—

b.   —⟨benzene ring⟩—OCH₂CH₂CH₂O—⟨benzene ring⟩—N=CHCH=N—

c.   —CH=⟨cyclohexane ring⟩=CH—⟨benzene ring⟩—

27.   **a.**   CH₂=CHCH₂CH₃

b.   ⟨epoxide⟩ with CH₃

c.   CH₂=CH—⟨pyridine ring, N⟩

d.   ClSO₂—⟨benzene ring⟩—SO₂Cl + H₂N(CH₂)₆NH₂

e.   CH₂=CCH=CH₂ (with CH₃ on the central carbon)

f.   HO(CH₂)₅COH (with C=O)

g.   CH₂=CCH₃ (with phenyl group)

h.   HOC—⟨benzene ring⟩—COH (each with C=O) + HOCH₂CH₂OH

a, **b**, **c**, **e**, and **g** are chain-growth polymers.

d, **f**, and **h** are step-growth polymers.

28. Whether a polymer is isotactic, syndiotactic, or atactic depends on whether the substituents are all on one side of the carbon chain, alternate on both sides of the chain, or are random with respect to the chain. Because a polymer of isobutylene has two identical substituents on each carbon in the chain, different configurations are not possible.

$$\begin{array}{ccccc} & CH_3 & CH_3 & CH_3 & CH_3 \\ & | & | & | & | \\ -CH_2C & CH_2C & CH_2C & CH_2C- \\ & | & | & | & | \\ & CH_3 & CH_3 & CH_3 & CH_3 \end{array}$$

polymer of isobutylene

29. **a.** $\underset{\displaystyle CH_3 \quad\quad CH_3 \quad\quad CH_3 \quad\quad CH_3}{CH_3OCH_2CHOCH_2CHOCH_2CHOCH_2CHO-}$

b.
$$\begin{array}{cccc} CH_3 & CH_3 & CH_3 & CH_3 \\ | & | & | & | \\ -CH_2C- & CH_2C- & CH_2C- & CH_2C- \\ | & | & | & | \\ CHCH_3 & CHCH_3 & CHCH_3 & CHCH_3 \end{array}$$

c.
$$\begin{array}{cccc} -CH_2CH- & CH_2CH- & CH_2CH- & CH_2CH- \\ | & | & | & | \\ COCH_3 & COCH_3 & COCH_3 & COCH_3 \\ \| & \| & \| & \| \\ O & O & O & O \end{array}$$

d.
$$\begin{array}{cccc} -CH_2CH- & CH_2CH- & CH_2CH- & CH_2CH- \\ | & | & | & | \\ OCH_3 & OCH_3 & OCH_3 & OCH_3 \end{array}$$

30.   **a.**   Because it is a polyamide (and not a polyester), it is a nylon.

b. H_2N—⬡—CH_2—⬡—NH_2 and $HOC(CH_2)_6COH$ (with two C=O groups)

31.   A copolymer is composed of more than one kind of monomer. Because the initially formed carbocation can rearrange, two different monomers are involved in formation of the polymer.

$$-CH_2\overset{+}{C}HCCH_3 \quad \xrightarrow{\text{1,2-methyl shift}} \quad -CH_2CH\overset{+}{C}CH_3$$

with CH_3 groups attached

unrearranged monomer   rearranged monomer

32.   The polymer in the flask that contained a high molecular weight polymer and little material of intermediate molecular weight was formed by a chain-growth mechanism, whereas the polymer in the flask that contained mainly material of intermediate molecular weight was formed by a step-growth mechanism.

In a chain-growth mechanism, monomers are added to the growing end of a chain. This means that at any one time there will be polymeric chains and monomers.

Step-growth polymerization is not a chain reaction; any two monomers can react. Therefore, high molecular weight material will not be formed until the end of the reaction when pieces of intermediate molecular weight can combine.

33.   **a.**   Vinyl alcohol is unstable; it tautomerizes to acetaldehyde.

$$CH_2=CH\overset{|}{\underset{}{OH}} \quad \xrightarrow{\text{tautomerization}} \quad CH_3CH\overset{O}{\underset{}{\parallel}}$$

vinyl alchol   acetaldehyde

b.   It is not a true polyester. It has ester groups as substituents **on** the backbone of the chain so it does have "polyester" groups, but it does not have ester groups **within** the backbone of the polymer chain. A true polyester has ester groups within the backbone of the polymer chain.

34. Each of the following five carbocations can add the growing end of the polymer chain.

35. Because 1,4-divinylbenzene has substituents on both ends of the benzene ring that can engage in polymerization, the polymer chains can become cross-linked, which increases the rigidity of the polymer.

1,4-divinylbenzene

cross-link

36. The presence of three OH groups in glycerol allows for extensive cross-linking during polymerization. Glyptal gets its strength from this cross-linking.

terephthalic acid   glycerol

37.   The five-membered ring and the four-membered ring react to form the species that will polymerize to form the alternating copolymer.

38.   Both compounds can form esters via intramolecular or intermolecular reactions. The product of the intramolecular reaction is a lactone; the intermolecular reaction leads to a polymer.

5-Hydroxypentanoic acid reacts intramolecularly to form a six-membered-ring lactone, whereas 6-hydroxyhexanoic acid reacts intramolecularly to form a seven-membered-ring lactone.

5-hydroxypentanoic acid

6-hydroxyhexanoic acid

The compound that forms the most polymer will be the one that forms the least lactone because the two reactions compete with one another.

The six-membered-ring lactone is more stable and therefore has a more stable transition state for its formation, compared to a seven-membered-ring lactone. Because it is easier for 5-hydroxypentanoic acid to form the six-membered-ring lactone than for 6-hydroxyhexanoic acid to form the seven-membered-ring lactone, **6-hydroxyhexanoic acid** will form more polymer.

39. Rubber contains cis double bonds. Ozone (O_3), which is present in the air, oxidizes double bonds to carbonyl groups, which destroys the polymer chain. Polyethylene does not contain double bonds, so it is not air-oxidized.

40.

41. The plasticizer that keeps vinyl soft and pliable can vaporize over time, causing the polymer to become brittle. For this reason, high boiling materials are preferred as plasticizers over low boiling materials.

42. **a.** Because the negative charge on the propagation site can be delocalized onto the carbonyl oxygen, the polymer is best prepared by anionic polymerization.

b. The carboxyl substituent is in position to remove a proton from water, making water a better nucleophile.

43.   Hydrolysis converts the ester substituents into alcohol substituents, which can react with ethylene oxide to graft a polymer of ethylene oxide onto the backbone of the alternating copolymer of styrene and vinyl acetate.

44.   The desired polymer is an alternating copolymer of ethylene and 1,2-dibromoethylene.

$$CH_2=CH_2 \qquad \overset{\mid}{\underset{Br}{CH}}=\overset{\mid}{\underset{Br}{CH}}$$

45.   **a.**

b.   Delrin is a chain-growth polymer because it is made by adding monomers to the end of a growing chain.

Important Terms

antarafacial bond formation	in a pericyclic reaction, where formation and/or cleavage of two σ bonds occurs on opposite sides of the π system of the reactant.
antarafacial rearrangement	a rearrangement where the migrating group moves to the opposite face of the π system.
antibonding π molecular orbital	a molecular orbital that results when two parallel atomic orbitals with opposite phases interact. Electrons in an antibonding orbital decrease bond strength.
antisymmetric molecular orbital	a molecular orbital in which the left half is not a mirror image of the right half but would be if one half of the MO were turned upside down.
bonding π molecular orbital	a molecular orbital that results when two parallel atomic orbitals with the same phase interact. Electrons in a bonding orbital increase bond strength.
Claisen rearrangement	a $[3,3]$ sigmatropic rearrangement of an allyl vinyl ether.
conrotatory ring closure	a ring closure that achieves head-to-head overlap of p orbitals by rotating the orbitals in the same direction.
conservation of orbital symmetry theory	a theory that explains the relationship between the structure and stereochemistry of the reactant, the conditions under which a pericyclic reaction takes place, and the stereochemistry of the product.
Cope rearrangement	a $[3,3]$ sigmatropic rearrangement of a 1,5-diene.
cycloaddition reaction	a reaction in which two π-bond-containing molecules react to form a cyclic compound.
disrotatory ring closure	a ring closure that achieves head-to-head overlap of p orbitals by rotating the orbitals in opposite directions.
electrocyclic reaction	an intramolecular reaction in which a new σ bond is formed between the ends of a conjugated system.
excited state	a description of which orbitals the electrons of an atom or molecule occupy when an electron in the ground state has been moved to a higher energy orbital.
frontier orbital analysis	an analysis that determines the outcome of a pericyclic reaction using frontier orbitals.
frontier orbitals	the HOMO and the LUMO of the two reacting species in a pericyclic reaction.
frontier orbital theory	a theory that, like the conservation of orbital symmetry, explains the relationship between reactant, product, and reaction conditions in a pericyclic reaction.

807

ground state	a description of which orbitals the electrons of an atom or molecule occupy when they are all in their lowest energy orbitals.
highest occupied molecular orbital (HOMO)	the molecular orbital of highest energy that contains an electron.
linear combination of atomic orbitals (LCAO)	the combination of atomic orbitals to produce a molecular orbital.
lowest unoccupied molecular orbital (LUMO)	the molecular orbital of lowest energy that does not contain an electron.
molecular orbital (MO) theory	a theory that describes a model in which the electrons occupy orbitals as they do in atoms but with the orbitals extending over the entire molecule.
pericyclic reaction	a concerted reaction that occurs as a result of a cyclic reorganization of electrons.
photochemical reaction	a reaction that takes place when a reactant absorbs light.
polar reaction	the reaction between a nucleophile and an electrophile.
radical reaction	a reaction in which a new bond is formed using one electron from one reactant and one electron from another reactant.
selection rules	the rules that determine the outcome of a pericyclic reaction.
sigmatropic rearrangement	a reaction in which a σ bond is broken in the reactant, a new σ bond is formed in the product, and the π bonds rearrange.
suprafacial bond formation	in a pericyclic reaction, where formation and/or cleavage of two σ bonds occurs on the same side of the π system of the reactant.
suprafacial rearrangement	a rearrangement where the migrating group remains on the same face of the π system.
symmetric molecular orbital	a molecular orbital in which the left half is a mirror image of the right half.
symmetry-allowed pathway	a pathway that leads to overlap of in-phase orbitals.
symmetry-forbidden pathway	a pathway that leads to overlap of out-of-phase orbitals.
thermal reaction	a reaction that takes place without the reactant having to absorb light.
Woodward-Hoffmann rules	a series of selection rules for pericyclic reactions.

Solutions to Problems

1. **a.** electrocyclic reaction   **c.** cycloaddition reaction

    b. sigmatropic rearrangement   **d.** cycloaddition reaction

2. **a.** bonding orbitals $= \psi_1, \psi_2, \psi_3$; antibonding orbitals $= \psi_4, \psi_5, \psi_6$

    b. ground-state HOMO $= \psi_3$; ground-state LUMO $= \psi_4$

    c. excited-state HOMO $= \psi_4$; excited-state LUMO $= \psi_5$

    d. symmetric orbitals $= \psi_1, \psi_3, \psi_5$; antisymmetric orbitals $= \psi_2, \psi_4, \psi_6$

    e. The HOMO and LUMO have opposite symmetries.

3. **a.** Eight π molecular orbitals   **b.** ψ_4 **c.** Seven nodes (There is also a node that passes through the nuclei, that is, through the centers of the p orbitals.)

4. **a.** 1,3-Pentadiene has two conjugated π bonds, so it has the same π molecular orbital description as 1,3-butadiene, a compound that also has two conjugated π bonds. (See Figure 28.2 on page 1270 of the text.)

    b. The π bonds in 1,4-pentadiene are isolated, so its π molecular orbital description is the same as ethene, a compound with an isolated π bond. (See Figure 28.1 on page 1270 of the text.)

    c. 1,3,5-Heptatriene has three conjugated π bonds, so it has the same π molecular orbital description as 1,3,5-hexatriene, a compound that also has three conjugated π bonds. (See Figure 28.3 on page 1271 of the text.)

    d. 1,3,5,8-Nonatetraene has three conjugated π bonds and an isolated π bond. The three conjugated π bonds are described by Figure 28.3 and the isolated π bond by Figure 28.1.

5. **a.** 2, 4, or 6 conjugated double bonds   3, 5, or 7 conjugated double bonds

    b. Under thermal conditions, electrocyclic ring closure involves the HOMO of the polyene (a compound with several double bonds). If the polyene has an even number of double bonds, conrotatory ring closure will result in in-phase overlap of the terminal p orbitals in the HOMO. If the polyene has an odd number of double bonds, disrotatory ring closure will result in in-phase overlap of the terminal p orbitals in the HOMO.

    Under photochemical conditions, electrocyclic ring closure involves the LUMO of the polyene, which will have the opposite symmetry of the HOMO and so will require the opposite mode of ring closure to that seen under thermal conditions.

6. **a.** (2E,4Z,6Z,8E)-2,4,6,8-Decatetraene has an even number of conjugated π bonds (4). Therefore, under thermal conditions, ring closure will be conrotatory.

    b. The substituents point in opposite directions, and conrotatory ring closure of such substituents will cause them to be trans in the ring-closed product.

   c. Under photochemical conditions, a compound with four conjugated π bonds will undergo disrotatory ring closure.

   d. Because ring closure is disrotatory and the substituents point in opposite directions, the product will have the cis configuration.

7.   **a.** correct   **b.** correct   **c.** correct

8.   **1. a.** conrotatory   **b.** trans   **2. a.** disrotatory   **b.** cis

9.   The reaction of maleic anhydride with 1,3-butadiene involves three π bonds in the reacting system, and such a reaction under thermal conditions involves suprafacial ring closure.

1,3-butadiene

The reaction of maleic anhydride with ethylene involves two π bonds in the reacting system, and such a reaction under thermal conditions involves antarafacial ring closure, which cannot occur with a four-membered ring.

ethene

10.   Solved in the text.

11.   Because we are dealing with formation of a small ring, ring closure will have to be suprafacial. Under photochemical conditions, suprafacial ring closure requires an even number of π bonds in the reacting system. Thus, a concerted reaction will occur but it will use only one of the π bonds of 1,3-butadiene.

12.   **a.**   **1.** [1,7] sigmatropic rearrangement

          2. [1,5] sigmatropic rearrangement

          3. [5,5] sigmatropic rearrangement

          4. [3,3] sigmatropic rearrangement

b. 1.

2.

3.

4.

13.   a. and **b.**

14. If a nondeuterated reactant had been used, the product would have been identical to the reactant. Therefore, the rearrangement would not have been detectable.

15. A suprafacial rearrangement can take place under photochemical conditions if there are an even number of electrons in the reacting system. Therefore, a 1,3-hydrogen shift occurs involving four electrons.

a [1,3] sigmatropic
migration of deuterium

A suprafacial rearrangement can take place under thermal conditions if there are an odd number of electrons in the reacting system. Therefore, a 1,5-hydrogen shift occurs involving six electrons.

a [1,5] sigmatropic
migration of deuterium

16. Solved in th

17. [1,3] Sign
membered t

[1,3] Sign
the required

18. **a.** Because
number
inversic

b. Because
odd nun
retentio

19. Because the
number (4)
cyclic trans

20. Because th
photochem
opposite di
directions i

21. Chorismate

22. Were you a

23. **a.**

b.

c.   H₃C

H₃C

H

25. The hydrogens that end up at the ring juncture in the first reaction point in opposite directions in the reactant. Because ring closure is disrotatory (odd number of π bonds, thermal conditions), the hydrogens in the ring-closed product are cis. (See Table 28.2 on page 1278 of the text.)

In the second example, the hydrogens that end up at the ring juncture point in the same direction in the reactant. Ring closure is still disrotatory, so the hydrogens in the ring-closed product are trans.

26.

27. **1.** Because the compound has an even number of π bonds, it will undergo conrotatory ring closure under thermal conditions and disrotatory ring closure under photochemical conditions. Because the two methyl substituents point in opposite directions, they will be trans in the ring-closed product when ring closure is conrotatory and cis in the ring-closed product when ring closure is disrotatory.

a.

b.

2. Because the compound has an even number of π bonds, it will undergo conrotatory ring closure under thermal conditions and disrotatory ring closure under photochemical conditions. Because the two methyl substituents point in the same direction, they will be cis in the ring-closed product when ring closure is conrotatory and trans in the ring-closed product when ring closure is disrotatory.

a.

b.

28. **a.**

b.

c.

tautomerization

d.

e.

tautomerization

29.   **a.**   This is a $[1,5]$ sigmatropic rearrangement, so it can take place by a concerted mechanism under thermal conditions.

b.   This is a $[1,3]$ sigmatropic rearrangement, so it can take place by a concerted mechanism under photochemical conditions.

30.   **B** is the product. Because the reaction is a $[1,3]$ sigmatropic rearrangement, antarafacial ring closure is required. Carbon, therefore, must migrate using both of its lobes. This means that the configuration of the migrating carbon will undergo inversion. The configuration of the migrating carbon has been inverted in **B** (the Hs are cis to one another but were trans to one another in the reactant) and retained in **A**.

the H's are trans
to one another

retention

A

the H's are trans
to one another

inversion

B

the H's are cis
to one another

31.   At first glance it is surprising that the isomerization of Dewar benzene (a highly strained and unstable molecule) to benzene (a stable aromatic compound) is so slow. However, the isomerization requires conrotatory ring opening, which is symmetry forbidden under thermal conditions. The reaction, therefore, cannot take place by a concerted pathway and has to take place by a much slower stepwise process.

32.   Hydrogen cannot undergo a $[1,3]$ sigmatropic rearrangement, because it cannot migrate by an antarafacial pathway that is required for a sigmatropic rearrangement involving an even number of pairs of electrons under thermal conditions. Carbon can undergo a $[1,3]$ sigmatropic rearrangement because it can migrate by a suprafacial pathway if it uses both of its lobes. Therefore, the first compound can undergo only a 1,3-methyl group migration.

The second compound can undergo the 1,3-methyl group migration that the first compound undergoes, and the *sec*-butyl group can also undergo a [1,3] sigmatropic rearrangement. The migrating *sec*-butyl group will have its configuration inverted due to the antarafacial nature of the rearrangement.

33.   An infrared absorption band is indicative of a carbonyl group. A [3,3] sigmatropic rearrangement of the reactant leads to a compound with two enolic groups. Tautomerization of the enols results in keto groups. The ketone carbonyl groups give the absorbance at 1715 cm^{-1}.

34.   The reaction is a [1,7] sigmatropic rearrangement. Since the reaction involves four pairs of electrons, antarafacial rearrangement occurs. Thus, when H migrates, because it is above the plane of the reactant molecule, it ends up below the plane of the product molecule. When D migrates, because it is below the plane of the reactant molecule, it ends up above the plane of the product molecule.

35.   **a.**

   b.

36. Disrotatory ring closure of (2E,4Z,6Z)-octatriene leads to the trans isomer, which can exist as a pair of enantiomers. One enantiomer is formed if the "top lobes" of the *p* orbitals rotate toward one other, and the other enantiomer is formed if the "bottom lobes" of the *p* orbitals rotate toward one other.

(2E,4Z,6Z)-octatriene

In contrast, disrotatory ring closure of (2E,4Z,6E)-octatriene leads to the cis isomer, which is a meso compound and, consequently, does not have a nonsuperimposable mirror image. Therefore, the same compound is formed from the "top lobes" of the *p* orbitals rotating toward one other and from the "bottom lobes" of the *p* orbitals rotating toward one other.

(2E,4Z,6E)-octatriene

37.

a.

b.

c.

d.

38.   Under thermal conditions a compound with two π bonds undergoes conrotatory ring closure. Conrotatory ring closure that results in a ring-closed compound with the substituents cis to one another requires that the substituents point in the same direction in the reactant. Therefore, the product with the methyl substituents pointing in the same direction is obtained in 99% yield.

1%	99%
Methyl groups point in opposite directions.	Methyl groups point in the same directions.

39.   Because the reactant has two π bonds, electrocyclic ring closure is conrotatory. Two different compounds, **X** and **Y**, can be formed because conrotatory ring closure can occur in either a clockwise or a counterclockwise direction.

Each of the compounds (**X** and **Y**) can undergo a conrotatory ring-opening reaction in either a clockwise or a counterclockwise direction to form either **A** or **B**. **A** and **B** are the only isomers that can be formed; formation of **C** and **D** would require disrotatory ring closure.

40.

41. Because the compounds that undergo ring closure to give **A** and **B** have two π bonds, ring opening of **A** and **B** under thermal conditions will be conrotatory. Because the hydrogens in **A** and **B** are cis, they must point in the same direction in the ring-opened product. To have the two hydrogens pointing in the same direction, one of the double bonds in the ring-opened compound must be cis and the other must be trans.

An eight-membered ring is too small to accommodate conjugated double bonds with one cis and the other trans, so **A** will not be able to undergo a ring-opening reaction under thermal conditions. A 10-membered ring can accommodate a trans double bond, so **B** will be able to undergo a ring-opening reaction under thermal conditions.

Both double bonds are cis.   **A**

cis double bond

trans double bond   **B**

42. The compound undergoes a 1,5-hydrogen shift of D or a 1,5-hydrogen shift of H. In each case, an unstable nonaromatic intermediate is formed that undergoes a subsequent 1,5-hydrogen shift to form an aromatic product.

nonaromatic   aromatic

nonaromatic   aromatic

43. Because the ring-opened compound formed in the first step has three conjugated π bonds involved in an electrocyclic reaction, conrotatory ring opening of the reactant will occur under photochemical conditions, and the trans hydrogens in the reactant require that the hydrogens point in the opposite direction in the ring-opened compound. Thermal electrocyclic ring closure of a three π bond system is disrotatory in step two, and disrotatory ring closure of a compound with hydrogens that point in opposite directions will cause those hydrogens to be cis in the ring-closed product.

44.   A Diels-Alder reaction is followed by a reverse Diels-Alder reaction that eliminates CO_2. Loss of a stable gas molecule (CO_2) and formation of a stable aromatic product provide the driving force for the second step.

45.

Answers to Chapter 1 Practice Test

1. **a.** a carbon-fluorine bond **c.** a carbon-hydrogen bond in ethane **e.** a sigma bond

 b. one **d.** the bond angle in ammonia

2. $^{+}CH_3$ $^{-}CH_3$ $^{\cdot}CH_3$
 sp^2 sp^3 sp^2

3. $H : \overset{..}{\underset{..}{O}} : \overset{\overset{..}{\overset{..}{O}}}{C} : \overset{..}{\underset{..}{O}} :^{-}$

4. CH_2Cl_2 $\boxed{CH_3CH_3}$ CH_3Cl $H_2C{=}O$ CCl_4

5. $^{+}NH_4$

6. **a.** $^{+}CH_3$ **b.** $H :^{-}$ **c.** $: \overset{..}{\underset{.}{Br}} \cdot$ **d.** CH_3CH_3

7. $CH_3CH_2CH_2CH{=}CH_2$ **or** $CH_3CH_2CH{=}CHCH_3$ **or** $CH_3\underset{\underset{CH_3}{|}}{CH}CH{=}CH_2$

8. $CH_3CH_2\overset{sp}{C}{\equiv}\overset{sp}{N}$

 $CH_3\underset{\underset{CH_3}{|}}{\overset{sp^2}{C}}{=}\overset{sp^2}{N}CH_3$

 $CH_3\overset{\overset{O}{\|}}{\underset{sp^2}{C}}CH_3 \quad sp^2$

 $\overset{sp}{O}{=}\overset{sp^2}{C}{=}\overset{sp^2}{O}$

9. **a.** $1s^2\,2s^2\,2p_x2p_y$ **b.** $1s^2\,2s\,2p_x\,2p_y\,2p_z$

10. **a.** $109.5°$ **b.** $180°$ **c.** $120°$ **d.** $104.5°$

11. $H\overset{\overset{O}{\|}}{C}OH \quad sp^2$ $HC{\equiv}N \quad sp$ $CH_3OCH_3 \quad sp^3$ $CH_3CH{=}CH_2 \quad sp^3$

12. **a.** A pi bond is stronger than a sigma bond. F
 b. A triple bond is shorter than a double bond. T
 c. The oxygen-hydrogen bonds in water are formed by the
   overlap of an sp^2 orbital of oxygen with an s orbital of hydrogen. F
 d. A double bond is stronger than a single bond. T
 e. A tetrahedral carbon has bond angles of $107.5°$. F

Answers to Chapter 2 Practice Test

1. **a.** $CH_3\underset{\underset{F}{|}}{CH}CH_2OH$ **b.** HI **c.** $CH_3\underset{\underset{Cl}{|}}{\overset{\overset{Cl}{|}}{C}}CH_2OH$ **d.** NH_3

2. **a.** $CH_3CH_2NH_2$ **b.** F^{-}

3. $CH_3\overset{\overset{O}{\|}}{C}O^{-}$ CH_3CO^{-} $\boxed{CH_3OH}$ $\left(CH_3\overset{-}{N}H\right)$ CH_3NH_2

823

4. CH_3COO^- CH_3CH_2OH CH_3OH $CH_3CH_2\overset{+}{N}H_3$

5.   **a.** $CH_3NH_2 + H_2O \rightleftharpoons CH_3\overset{+}{N}H_3 + HO^-$ **b.** reactants

6.   **a.** $^-NH_2$ **b.** $^+NH_4$

7.   6.2

8.   **a.** $CH_3OH + {^+NH_4} \rightleftharpoons CH_3\overset{+}{\underset{H}{O}}H + NH_3$ **b.** reactants

9.   between 8 and 9

10. CH_3CH_2OH $CH_3CH_2NH_2$ CH_3CH_2SH $CH_3CH_2CH_3$
             2   **3**   **1**   **4**

11.   formic acid/sodium formate

12.   **a.** HO^- is a stronger base than $^-NH_2$.   F
             b. A Lewis acid is a compound that accepts a share in a pair of electrons.   T
             c. CH_3CH_3 is more acidic than $H_2C{=}CH_2$.   F
             d. The weaker the acid, the more the conjugate base.   F
             e. The larger the pK_a, the weaker the acid.   T
             f. The weaker the base, the more stable it is.   T

Answers to Chapter 3 Practice Test

1.   **a.** 3-methyloctane   **b.** 2-methyl-1-heptanol   **c.** 3-octanol

2.   **a.**   **b.**   **c.**

3.   **a.** *sec*-butyl chloride, 2-chlorobutane   **c.** cyclopentyl bromide, bromocyclopentane

          b. isohexyl alcohol, 4-methyl-1-pentanol

4.   **a.** $CH_3CH_2CH_2CH_2CH_2Br$ $CH_3CH_2CH_2Br$ $CH_3CH_2CH_2CH_2Br$
                          1   **3**   **2**

          b. $CH_3CH_2CH_2CH_2CH_3$ $CH_3CH_2CH_2CH_2OH$ $CH_3CH_2CH_2CH_2Cl$
                          3   **1**   **2**

          c. $CH_3CH_2CH_2CH_2CH_2CH_2CH_2CH_3$
                          3   **1**   **2**

5. **a.** 6-methyl-3-heptanol

c. 1-bromo-3-methylcyclopentane

b. 3-ethoxyheptane

d. 1,4-dichloro-5-methylheptane

6.

7. *cis*-1-isopropyl-3-methylcyclohexane

8. **a.** butyl alcohol **c.** hexane **e.** ethyl alcohol

b. 1-butanol **d.** pentylamine

9. **a.** isopentyl bromide **b.** isopentyl alcohol **c.** isopentylamine
1-bromo-3-methylbutane 3-methyl-1-butanol 3-methyl-1-butanamine

10. **a.**

b.

c.

11. **a.** CH_3CHCH_3 **d.** CH_3CHCH_3 **e.** $CH_3CH_2CH_2OH$ CH_3CHOH $CH_3CH_2OCH_3$
      | | |
      Br CH_3 CH_3

b. $CH_3CH_2NHCH_3$

c.

12. **a.** 2,5-dimethylheptane

d. 4-bromo-2-chloro-1-methylcyclohexane

b. 7-bromo-2-heptanol

e. 1-butoxy-2-methylpentane

c. 2-chloro-4-heptanol

f. 3-methyl-*N*-propyl-1-pentanamine

Answers to Chapter 4 Practice Test

1. **a.** a pair of enantiomers   **b.** a pair of enantiomers   **2.** -3.0

3.

4. $CH_3CH_2CH_2CH_2Cl$ $CH_3CH_2CHCH_3$ CH_3CHCH_2Cl CH_3CCH_3

5. **a.**

b.

or

or

c. no stereoisomers

d.

or

or

e.

f.

6.

7. (−)-2-Methylbutanoic acid has the *S* configuration.

8. −16

9.

10. **a.** identical

b.   **c.**   **d. 1.**

2.

11.
a. Diastereomers have the same melting points.	F
b. 3-Chloro-2,3-dimethylpentane has two asymmetric centers.	F
c. Meso compounds do not rotate polarized light.	T
d. 2,3-Dichloropentane has a stereoisomer that is a meso compound.	F
e. All compounds with the *R* configuration are dextrorotatory.	F
f. A compound with three asymmetric centers can have a maximum of nine stereoisomers.	F

12. 4-methyl-2-hexene and 2-bromo-3-hexene

Answers to Chapter 5 Practice Test

1. **a.** 4-methyl-1-hexene   **c.** 7-methyl-3-nonene

b. 4-bromocyclopentene   **d.** 4-chloro-3-methylcyclohexene

2.

3. **a.** 2-pentene **c.** 3-methyl-2-pentene

b. 3-methyl-1-hexene **d.** 1-methylcyclohexene

4. **a.** Increasing the energy of activation increases the rate of the reaction. F

b. Decreasing the entropy of the products compared to the entropy of the reactants makes the equilibrium constant more favorable. F

c. An exergonic reaction is one with a $-\Delta G°$. T

d. An alkene is an electrophile. F

e. The higher the energy of activation, the more slowly the reaction will take place. T

f. Another name for *trans*-2-butene is Z-2-butene. F

g. A reaction with a negative $\Delta G°$ has an equilibrium constant greater than one. T

h. Increasing the energy of the reactants increases the rate of the reaction. T

i. Increasing the energy of the products increases the rate of the reaction. F

j. The magnitude of a rate constant is not dependent on the concentration of the reactants. T

5. **a.** Z **b.** Z

6. **a.** $CH_2{=}CHCH_2OH$ **c.**
$$CH_3CH_2 \qquad CH_2CH_2CH_3$$
$$C{=}C$$
$$H \qquad H$$

b.
CH₃

d. $CH_2{=}CHBr$

7. 5 **8.** $CH_3CH{=}CH_2 \ + \ H{-}\ddot{C}l{:} \ \rightleftharpoons \ CH_3\overset{+}{CH}{-}CH_3 \ + \ {:}\ddot{C}l{:}^- \ \longrightarrow \ CH_3CH{-}CH_3$
$\overset{\ \ \ \ \ \ \ \ \ \ \ \ \ \ \ \ \ \ }{:}\ddot{C}l{:}$

9. a negative $\Delta H°$, a positive $\Delta S°$, and a high temperature

10. **a.** 4 kcal/mol **b.** 35 °C **c.** one reactant forms two products

11. **a.**
$$BrCH_2 \quad CH_3 \qquad BrCH_2 \quad CH_2CH_3$$
$$C{=}C \qquad\qquad C{=}C$$
$$CH_3 \quad CH_2CH_3 \quad CH_3 \quad CH_3$$
$$E \qquad\qquad Z$$

b.
$$\overset{CH_3}{\underset{|}{CH_3CH}} \quad CH_3 \qquad \overset{CH_3}{\underset{|}{CH_3CH}} \quad CH_2CH_3$$
$$C{=}C \qquad\qquad C{=}C$$
$$CH_3 \quad CH_2CH_3 \quad CH_3 \quad CH_3$$
$$E \qquad\qquad Z$$

12.

Progress of the reaction

Answers to Chapter 6 Practice Test

1.

a. $CH_3\overset{+}{\underset{\underset{\displaystyle CH_3}{|}}{C}}CH_3$

b. $CH_3CH_2\overset{+}{C}HCH_3$

c. $CH_3CH_2\overset{+}{C}H_2$

d.

e.

2. $CH_3CH_2CH_2CH=CH_2$

3. B

4.

a. $CH_3\underset{\underset{\displaystyle Br}{|}}{\overset{\overset{\displaystyle CH_3}{|}}{C}}CH_2CH_3$

b. $CH_3\underset{\underset{\displaystyle Cl}{|}}{\overset{\overset{\displaystyle CH_3}{|}}{C}}CH_2CH_3$

c. $CH_3CH_2\underset{\underset{\displaystyle OH}{|}}{C}HCH_2Cl$

d. $CH_3\underset{\underset{\displaystyle Br}{|}}{\overset{\overset{\displaystyle CH_3}{|}}{C}}-\underset{\underset{\displaystyle CH_3}{|}}{C}HCH_3$

e. $CH_3\overset{\overset{\displaystyle O}{\diagup\diagdown}}{CH}-CH_2$

f.

5.

a. $CH_3\underset{\underset{\displaystyle OH}{|}}{C}H\overset{\overset{\displaystyle CH_3}{|}}{C}HCH_3$

b.

6.

a. $CH_3\underset{\underset{\displaystyle CH_3}{|}}{\overset{\overset{\displaystyle CH_3}{|}}{C}}CH=CH_2 \xrightarrow[Pd/C]{H_2} CH_3\underset{\underset{\displaystyle CH_3}{|}}{\overset{\overset{\displaystyle CH_3}{|}}{C}}CH_2CH_3$

b.

c.

d.

7. $CH_3CH_2\overset{+}{C}H\overset{\overset{\displaystyle CH_3}{|}}{C}HCH_3$

8. **a.** The addition of Br_2 to 1-butene to form 1,2-dibromobutane is a concerted reaction.   F
　　b. The reaction of 1-butene with HCl will form 1-chlorobutane as the major product.   F
　　c. 2,3-Dimethyl-2-pentene is more stable than 3,4-dimethyl-2-pentene.   T
　　d. The reaction of HBr with 3-methylcyclohexene is more highly regioselective
　　　　than the reaction of HBr with 1-methylcyclohexene.   F
　　e. The reaction of an alkene with a carboxylic acid forms an epoxide.   F
　　f. A catalyst increases the equilibrium constant of a reaction.   F
　　g. The addition of HBr to 3-methyl-2-pentene is stereospecific.   F
　　h. The addition of HBr to 3-methyl-2-pentene is stereoselective.   F
　　i. The addition of HBr to 3-methyl-2-pentene is regioselective.   T

9. **a.**
$$CH_3CH_2-\overset{Cl}{\underset{H}{|}}-CH_3 \;+\; CH_3-\overset{Cl}{\underset{H}{|}}-CH_2CH_3$$
　　c.
$$H-\overset{CH_2CH_3}{\underset{}{|}}-Br$$
$$H-\underset{CH_2CH_3}{|}-Br$$

　　b.
$$CH_3CH_2CH_2-\overset{Br}{\underset{H}{|}}-CH_3 \;+\; CH_3-\overset{Br}{\underset{H}{|}}-CH_2CH_2CH_3$$
$$+\; CH_3CH_2\underset{\underset{Br}{|}}{CH}CH_2CH_3$$
　　d.
$$H-\overset{CH_2CH_3}{|}-Br \qquad\qquad Br-\overset{CH_2CH_3}{|}-H$$
$$H-\underset{CH_2CH_2CH_3}{|}-Br \;+\; Br-\underset{CH_2CH_2CH_3}{|}-H$$

10. **a.** (cyclopentane structures)   **c.** (cyclopentane structure)

　　b. (cyclopentane structures)   **d.** (cyclopentane structures)

Answers to Chapter 7 Practice Test

1. **a.** **1.** R_2BH/THF **2.** HO^-, H_2O_2, H_2O
　　b. H_2/Lindlar catalyst

2. **a.** (cyclohexene—OH)   **b.** (two enol structures) and

3. **a.** $CH_3CH_2\underset{\underset{CH_3}{|}}{CH}C\equiv C\underset{\underset{CH_3}{|}}{CH}CH_2CHCH_3$ **b.** (methylcyclohexadiene with CH_3)

4. **a.** A terminal alkyne is more stable than an internal alkyne.   F
　　b. Propyne is more reactive than propene toward reaction with HBr.   F
　　c. 1-Butyne is more acidic than 1-butene.   T

d. An sp^2 hybridized carbon is more electronegative than an sp^3 hybridized carbon.   T

e. The reactions of internal alkynes are more regioselective than the reactions of terminal alkynes.   F

f. Alkenes are more reactive than alkynes.   T

5. **a.** 1-bromo-5-methyl-3-hexyne   **b.** 5-methyl-3-hexyn-1-ol

6. **a.** $CH_3CH_2CH_2C \equiv CH$ **b.** $CH_3CH_2C \equiv CCH_2CH_3$

7.

NH_3	$CH_3 \equiv CH$	CH_3CH_3	H_2O	$CH_3CH = CH_2$
3	**2**	**5**	**1**	**4**

8.

$$CH_3CH_2CH_2\overset{\overset{\displaystyle O}{\|}}{C}CH_2CH_3$$

(cyclohexanone with CH_3 substituent and $O=$)

9. **a.** $CH_3CH_2C \equiv CH$

$\downarrow$ $^-NH_2$

$CH_3CH_2C \equiv C^-$ $\xrightarrow{CH_3CH_2Br}$ $CH_3CH_2C \equiv CCH_2CH_3$ $\xrightarrow[\uparrow]{H_2/Pt}$ $CH_3CH_2CH_2CH_2CH_2CH_3$

b. $CH_3CH_2C \equiv CH$

$\downarrow$ 1. $^-NH_2$ 2. CH_3CH_2Br

$CH_3CH_2C \equiv CCH_2CH_3$ $\xrightarrow[\text{Lindlar catalyst}]{H_2}$ $CH_3CH_2CH = CHCH_2CH_3$ $\xrightarrow[\uparrow]{HBr}$ $CH_3CH_2CH_2\underset{\underset{\displaystyle Br}{|}}{C}HCH_2CH_3$

c. $CH_3CH_2C \equiv CH$

$\downarrow$ $^-NH_2$

$CH_3CH_2C \equiv C^-$ $\xrightarrow{CH_3CH_2Br}$ $CH_3CH_2C \equiv CCH_2CH_3$ $\xrightarrow[\uparrow]{H_2O \mid H_2SO_4}$ $CH_3CH_2\overset{\overset{\displaystyle O}{\|}}{C}CH_2CH_2CH_3$

Answers to Chapter 8 Practice Test

1. **a.** (benzyl cation) $\overset{+}{C}H_2$ **c.** $CH_3\overset{-}{C}H\overset{\overset{\displaystyle O}{\|}}{C}CH_3$ **e.** (cyclohexenyl) $\overset{+}{C}HCH_3$

b. $CH_3\overset{-}{C}HC \equiv CH$ **d.** $CH_2 = CH\overset{+}{C}H_2$

2. **a.** $CH_3CH = CH - \overset{..}{\underset{..}{O}}CH_3 \longleftrightarrow CH_3\overset{-}{C}H - CH = \overset{+}{O}CH_3$

b. $CH_3CH=CH-CH=CH-\overset{+}{C}H_2$ ⟷ $CH_3CH=CH-\overset{+}{C}H-CH=CH_2$

⥮

$CH_3\overset{+}{C}H_2-CH=CH-CH=CH_2$

c. $^-CH_2-CH=CH-\overset{O}{\overset{\|}{C}}H$ ⟷ $CH_2=CH-\overset{-}{C}H-\overset{O}{\overset{\|}{C}}H$ ⟷ $CH_2=CH-CH=\overset{O^-}{\overset{\|}{C}}H$

3. $CH_3\overset{CH_3}{\underset{+}{\overset{\|}{C}}}CH_2CH=CH_2$ $CH_2=CHCH_2CH=CH_2$ $CH_3CH_2NHCH_2CH=CHCH_3$

4. **a.** + Br_2 ⟶ +

b. H_3C— + $\overset{O}{\underset{CH_2}{\overset{\|}{C}}}\overset{}{\underset{}{}}CHCCH_3$ ⟶ +

5. $CH_3\overset{O}{\overset{\|}{C}}OH$ and $CH_3\overset{O^-}{\overset{\|}{C}}=\overset{+}{O}H$ **6.**

7. **a.** —NH_2 **b.** —CH_2O^-

8. **a.**

b.

c.

9. a.   b.

10. a. A compound with four conjugated double bonds will have four molecular orbitals.   F

    b. ψ_1 and ψ_2 are symmetric molecular orbitals.   F

    c. If ψ_3 is the HOMO in the ground state, ψ_4 will be the HOMO in the excited state.   T

    d. If ψ_3 is the LUMO, ψ_4 will be the HOMO.   F

    e. If the ground-state HOMO is symmetric, the ground-state LUMO will be antisymmetric.   T

    f. A conjugated diene is more stable than an isomeric isolated diene.   T

    g. A single bond formed by an sp^2—sp^2 overlap is longer than a single bond formed by an sp^2—sp^3 overlap.   F

    h. The thermodynamically controlled product is the major product obtained when the reaction is carried out under mild conditions.   F

    i. 1,3-Hexadiene is more stable than 1,4-hexadiene.   T

11.

13. $CH_3CH=CH\overset{+}{C}CH_3$ > $CH_3CH=CH\overset{+}{C}HCH_3$ > $CH_3CH=CH\overset{+}{C}H_2$ > $CH_3CH=CHCH_2\overset{+}{C}H_2$

15. $\xrightarrow[\text{H}_2\text{O}]{\text{Br}_2}$ HO—⟨⟩—Br

16.

17. **a.** CH₂=CH—C(CH₃)(Cl)—CH₃

product of kinetic control

CH₂—CH₂=C(CH₃)—CH₃
       |
       Cl

product of thermodynamic control

b.

product of kinetic control

product of thermodynamic control

18.   **19.**   **20.**

21. **a.**   **b.**   **c.**   **d.**

Answers to Chapter 9 Practice Test

1. **a.** —CHCH₂CH₃
              |
              Br

b. CH₃CH=CHCHCH₃
                     |
                     Br

2. **a.** CH₃CH₂CHBr
                |
                CH₃

b. —CH₂Br

3. **a.**

b.
CH₂CH₃
|
C····H
|
OCH₃

R

+

CH₂CH₃
|
H····C
|
CH₃O

c.
CH₃
|
HO——H
|
CH₂CH₃

d.
CH₃
|
H····C
|
CH₃NH   CH₂CH₃

e. CH₃CH=CHCH₂OCH₃   +   CH₃CHCH=CH₂
                                              |
                                              OCH₃

R and *S*

4. **a.** Increasing the concentration of the nucleophile favors an S_N1 reaction over an S_N2 reaction.   F
  b. Ethyl iodide is more reactive than ethyl chloride in an S_N2 reaction.   T
  c. In an S_N1 reaction, the product with the retained configuration is obtained in greater yield.   F
  d. The rate of a substitution reaction in which none of the reactants is charged will increase
    if the polarity of the solvent is increased.   T
  e. An S_N2 reaction is a two-step reaction.   F
  f. The pK_a of a carboxylic acid is greater in water than it is in a mixture of dioxane and water.   F
  g. 4-Bromo-1-butanol will form a cyclic ether faster than will 3-bromo-1-propanol.   T

5. **a.** CH_3O^- **b.** CH_3S^-

6. **a.** $CH_3CH_2CH_2Cl + HO^-$ **c.** $CH_3CH_2CH_2Br + HO^-$ **e.** $BrCH_2CH_2CH_2CH_2NHCH_3$

  b. $CH_3CH_2CH_2I + HO^-$ **d.** $CH_3\overset{\underset{|}{Br}}{C}HCH_3 \xrightarrow[CH_3OH]{CH_3O^-}$

7. All are aprotic solvents except ethanol.

8. **a.** The rate of the reaction would increase.   **d.** The pK_a would decrease.

  b. The rate of the reaction would increase.   **e.** The pK_a would decrease.

  c. The rate of the reaction would decrease.

Answers to Chapter 10 Practice Test

1. **a.** $-CH_2\overset{\underset{|}{Br}}{C}HCH_3$ **b.** $CH_2{=}CHCH_2\overset{\underset{|}{Br}}{C}HCH_3$ **2.** $CH_3\overset{\underset{|}{Br}}{\overset{CH_3}{C}}HBr$

3. $CH_3CH_2-\overset{\overset{OCH_3}{|}}{\underset{\underset{H}{|}}{C}}-CH_3$
                                                   major   minor

4. **a.** $CH_3CH_2\overset{\overset{CH_3}{|}}{\underset{\underset{CH_3}{|}}{C}}O^- + CH_3CH_2CH_2Br$ **b.** $-Br$ + $-O^-$ **c.** $-O^- + CH_3Br$

5. **a.** $CH_3\overset{\underset{|}{Cl}}{C}HCH_3 + HO^-$ **c.** $CH_3CH_2CH_2Br + HO^-$ **e.** $CH_3\overset{\overset{CH_3}{|}}{\underset{\underset{Br}{|}}{C}}CH_3 + HO^-$

  b. $CH_3CH_2CH_2I + HO^-$ **d.** $CH_3\overset{\underset{|}{Br}}{C}HCH_3 \xrightarrow[CH_3OH]{CH_3O^-}$

6.   **a.**   **b.**   **c.**   **d.**

7.   *cis*-1-bromo-2-methylcyclohexane

Answers to Chapter 11 Practice Test

1.   HBr   **2.** $SOCl_2$

3.   **a.** $HOCH_2\overset{\underset{\displaystyle CH_2CH_3}{\mid}}{\underset{\underset{\displaystyle OCH_3}{\mid}}{C}}CH_2CH_3$ **b.** $CH_3OCH_2\overset{\underset{\displaystyle CH_2CH_3}{\mid}}{\underset{\underset{\displaystyle OH}{\mid}}{C}}CH_2CH_3$

4.   **a.** $CH_3\overset{\underset{\displaystyle CH_3}{\mid}}{C}HCH_2\overset{\underset{\displaystyle CH_3}{\mid}}{N}$ + $CH_2{=}CHCH_2CH_3$ **b.**   **c.**

5.   **a.** $CH_3CH_2\overset{\underset{\displaystyle CH_3}{\mid}}{C}{=}\underset{\underset{\displaystyle CH_3}{\mid}}{C}CH_3$ **b.** $CH_3CH_2CH_2\overset{\underset{\displaystyle CH_3}{\mid}}{C}{=}\underset{\underset{\displaystyle CH_3}{\mid}}{C}CH_3$ **c.** $CH_3CH_2CH{=}CHCH_3$ **d.**

a., b., c. will form both trans and cis, but more trans.

6.   **a.** Tertiary alcohols are easier to dehydrate than secondary alcohols.   T
    b. Alcohols are more acidic than thiols.   F
    c. Alcohols have higher boiling points than thiols.   T
    d. The acid-catalyzed dehydration of a primary alcohol is an S_N1 reaction.   F
    e. The Hofmann elimination reaction is an E2 reaction.   T

7.   **a.** $CH_3CH_2\overset{\underset{\displaystyle CH_3}{\mid}}{\underset{\underset{\displaystyle CH_3}{\mid}}{C}}{-}I$ + CH_3OH **b.** $-OH$ + ICH_2-

8.   **a.** $CH_3CH_2CH_2NHCH_2CH_3$ **b.** $CH_3CH_2CH_2Cl$

    c. $CH_3CH_2CH_2CH{=}CHCH_3$ + $CH_3CH_2CH{=}CHCH_2CH_3$

    d. $CH_3CH_2CH_2\overset{\displaystyle O}{\overset{\displaystyle \|}{C}}H$ **e.** $CH_3CH_2CH_2\overset{\displaystyle O}{\overset{\displaystyle \|}{C}}OH$

Answers to Chapter 12 Practice Test

1. an organocadmium compound

2. and

3. **a.** **b.** **c.**

4. **a.** **b.** **c.**

5. **a.** $CH_2{=}CH_2$ +

b. $CH_2{=}CH_2$ +

c. $CH_2{=}CH_2$ +

6. **a.**

b.

7. **a.** $CH_3CH_2CH_2CH_2CH_2OH$ **b.**

8.

9. **a.** **1.** **2.**

b. **1.**

2.

10.   **a.**   **b.**

11.

Answers to Chapter 13 Practice Test

1.   **a.** 4   **b.** 4   **c.** 5   **d.** 4   **e.** 3

2. $CH_3CH_3 + \cdot Cl \longrightarrow CH_3\dot{C}H_2 + HCl$

3.

4. $CH_3CH_3 + Cl\cdot \longrightarrow CH_3\dot{C}H_2 + HCl$ $\Delta H° = 101 - 103 = -2$ kcal/mol

$CH_3\dot{C}H_2 + Cl_2 \longrightarrow CH_3CH_2Cl + Cl\cdot$ $\Delta H° = 58 - 85 = -27$ kcal/mol

5. $\dot{C}H_2CH_2CH_2CH\!=\!CH$ $CH_3CH_2\dot{C}HCH\!=\!CH_2$
　　　　　3　　　　　　　　　　**1**

　　$CH_2CH_2CH_2CH\!=\!\dot{C}H$ $CH_3\dot{C}HCH_2CH\!=\!CH_2$
　　　　　4　　　　　　　　　　**2**

6.   **a.**   **b.** $CH_3CHCH\!=\!CH_2 + CH_3CH\!=\!CHCH_2Br$
　　　　　　　　　　　　　　　　　　　　　　$\underset{Br}{|}$

7.   **a.**

b.   **c.**

8.   23%

9.   **a.** CH₃CH₂CCH₂CH₃ with CH₃ on top, Br below

$$CH_3CH_2\underset{Br}{\overset{CH_3}{C}}CH_2CH_3$$

c. $CH_3CH_2\underset{Cl}{\overset{CH_3}{C}}CH_2CH_3$

b. $CH_3\underset{Br}{CH}CHCH_2CH_3$ with CH₃ top

$$CH_3\underset{Br}{CH}\overset{CH_3}{CH}CH_2CH_3$$

d. $CH_3CH_2\underset{Cl}{\overset{CH_3}{C}}CH_2CH_3$

10.   **b**

Answers to Chapter 14 Practice Test

1.   **a.** butanal structure   ~2770 cm⁻¹

e. methyl ester structure   ~1050 or ~1250 cm⁻¹

b. amide structure   ~3300 cm⁻¹

f. $CH_3CH_2CH{=}CHCH_3$ ~1600 cm⁻¹ ~3100 cm⁻¹ $CH_3CH_2C{\equiv}CCH_3$ ~2100 cm⁻¹

c. $CH_3CH_2CH_2CH_2OH$ ~3600–3200 cm⁻¹

g. $CH_3CH_2C{\equiv}CH$ ~3300 cm⁻¹

d. cyclohexyl–CHCH₃ with OH   ~1380 cm⁻¹

2.   **a.** The O—H stretch of a concentrated solution of an alcohol occurs at a higher frequency than the O—H stretch of a dilute solution.   F
   b. Light of 2 μm is of higher energy than light of 3 μm   T
   c. It takes more energy for a bending vibration than for a stretching vibration.   F
   d. Propyne will not have an absorption band at 3100 cm⁻¹, because there is no change in the dipole moment.   F
   e. Light of 8 μm has the same energy as light of 1250 cm⁻¹.   T
   f. The M + 2 peak of an alkyl chloride is half the height of the M peak.   F

3. $CH_3CH_2\underset{OH}{\overset{CH_3}{C}}CH_2CH_2CH_3$

4.　　　Absorbance $=$ molar absorptivity $\times$ concentration $\times$ length of light path (in cm)

　　　　　　$0.75 =$ molar absorptivity $\times\ 3.8 \times 10^{-4} \times 1$

　　molar absorptivity $= 2000\ \mathrm{M^{-1}\ cm^{-1}}$

5.　**a.**

~3100 cm^{-1}
~1500 cm^{-1}

d.

~1700 cm^{-1}　　　~1050 cm^{-1}

b.

~2700 cm^{-1}　　　~2900 cm^{-1}
　　　　　　　　　~1380 cm^{-1}

e.

~3100 cm^{-1}　　　~2900 cm^{-1}
　　　　　　　　　~1380 cm^{-1}

c.

~3300–2500 cm^{-1}　　~2900 cm^{-1}
　　　　　　　　　　　~1380 cm^{-1}
　　　　　　　　　　　~1050 cm^{-1}

6.　**a.**　　　　　**b.**　　　　　**c.**

7.　　　Absorbance $=$ molar absorbtivity $\times$ concentration $\times$ length of light path (in cm)

　　　　　　$0.76 = 1200 \times$ concentration

　　concentration $= 6.3 \times 10^{-4}\ \mathrm{M}$

8.　$\dfrac{60}{13} = 4$ with 8 left over　　$C_4H_{12} = C_3H_8O$　　$CH_3CH_2CH_2OH$ and CH_3CHCH_3
　　|
　　　　　　　　　　　　　　　　　　　　　　　　　　　　　　　　　　　　　　　OH

9.　　A bond between a carbon and an atom of similar electronegativity breaks homolytically, whereas a bond between a carbon and a more electronegative atom breaks heterolytically.

Answers to Chapter 15 Practice Test

1.

$\underset{\textbf{4}}{CH_3CH_2CH_2} \overset{\overset{\displaystyle O}{\|}}{C} CH_3$ $\underset{\textbf{3}}{CH_3CH_2}\underset{\underset{\displaystyle Cl}{|}}{CH}CH_2CH_3$ $\underset{\textbf{4}}{CH_2}=CH \overset{\overset{\displaystyle O}{\|}}{C} H$

(benzene ring)—NO_2 **3**   (benzene ring with Cl, Cl) **3** $CH_3\underset{\underset{\displaystyle CH_3}{|}}{CH}CH_2\underset{\underset{\displaystyle CH_3}{|}}{CH}CH_3$ **3**

2.

$CH_3CH_2\overset{\overset{\displaystyle O}{\|}}{C}CH_3$ H—(benzene ring)—NO_2 $CH_3CH_2\overset{\overset{\displaystyle O}{\|}}{C}OCH_2CH_3$ ← triplet

↑ quartet   ↑ triplet

doublet of doublets → $\underset{H}{\overset{H}{\diagdown}}C=C\underset{Cl}{\overset{H}{\diagup}}$ $CH_3\underset{\underset{\displaystyle CH_3}{|}}{CH}CH_2Cl$ $ClCH_2CH_2CH_2OCH_3$

↑ doublet   ↑ multiplet

$CH_3OCH_2CH_2CH_2OCH_3$ $BrCH_2CH_2Br$

↑ quintet   ↑ singlet

3.

$\underset{CH_3}{}\overset{\overset{\displaystyle O}{\|}}{C}\underset{OCH_2CH_3}{}$ $\underset{CH_3CH_2}{}\overset{\overset{\displaystyle O}{\|}}{C}\underset{OCH_3}{}$ $\underset{H}{}\overset{\overset{\displaystyle O}{\|}}{C}\underset{OCH_2CH_2CH_3}{}$

3 signals   **3** signals   **4** signals

The signal at the highest frequency (farthest downfield) is a quartet.   The signal at the highest frequency (farthest downfield) is a singlet.

4.   **a.** The peaks on the right of an NMR spectrum are deshielded compared to the peaks on the left.   F

b. Dimethyl ketone has the same number of signals in its ^1H NMR spectrum as in its ^{13}C NMR spectrum.   F

c. In the ^1H NMR spectrum of the compound shown below, the lowest-frequency signal (the one farthest upfield) is a singlet and the highest-frequency signal (the one farthest downfield) is a doublet.   T

O_2N—(benzene ring)—CH_3

d. The greater the frequency of the signal, the greater its chemical shift in ppm.   T

5.

a. CH₃CH₂CH₂Cl

triplet

3 signals

b. CH₃CH₂COCH₃ (with C=O)

singlet

3 signals

c. CH₃CHCH₃ ── septet
    |
    Br

2 signals

6.

a. CH₃CH₂CH₂Cl

3 signals

triplet

b. CH₃CH₂COCH₃ (with C=O)

4 signals

singlet

c. CH₃CHCH₃
    |
    Br

2 signals

doublet

Answers to Chapter 16 Practice Test

1.

a. CH₃—C(=O)—OCH₃

b. CH₃—C(=O)—O—C₆H₅ (phenyl)

c. CH₃—C(=O)—O—C₆H₄—NO₂

d. CH₃—C(=O)—Cl

2.

a. *N*-ethylpentanamide

b. 3-methylpentanoic acid

c. methyl 4-phenylbutanoate

d. ethanoic propanoic anhydride

3.

a. CH₃CH₂—C(=O)—O—C(=O)—CH₂CH₃

b. CH₃—C(=O)—Cl + H₂O ⟶ CH₃—C(=O)—OH + HCl

Any reaction in which a reactant is cleaved as a result of reaction with water.

c. CH₃—C(=O)—OCH₃ + CH₃CH₂OH ⇌ (HCl) CH₃—C(=O)—OCH₂CH₃ + CH₃OH

d. CH₃—C(=O)—OCH₃ + CH₃NH₂ ⟶ (Δ) CH₃—C(=O)—NHCH₃ + CH₃OH

4.

a. CH₃—C(=O)—OH

b. CH₃—C(=O)—OCH₃

c. CH₃—C(=O)—NH₂

d. CH₃—C(=O)—OH

5.

a. CH$_3$CH$_2$—C(=O)—OH + $\overset{+}{\text{N}}$H$_4$

d. CH$_3$CH$_2$—C(=O)—NHCH$_2$CH$_3$ + Cl$^-$ + CH$_3$CH$_2\overset{+}{\text{N}}$H$_3$

b. CH$_3$CH$_2$—C(=O)—O$^-$ + CH$_3$CH$_2$OH

e. cyclohexyl—C(=O)—OCH$_2$CH$_3$ + CH$_3$OH

c. CH$_3$CH$_2$CH$_2$—C(=O)—OH + CH$_3$CH$_2$—C(=O)—OH

6.

a. CH$_3$CH$_2$CH$_2$Br $\xrightarrow{\ ^-\text{C}\equiv\text{N}\ }$ CH$_3$CH$_2$CH$_2$C$\equiv$N $\xrightarrow[\text{Raney Ni}]{\text{H}_2}$ CH$_3$CH$_2$CH$_2$CH$_2$NH$_2$

b. CH$_3$CH$_2$CH$_2$Br $\xrightarrow{\ ^-\text{C}\equiv\text{N}\ }$ CH$_3$CH$_2$CH$_2$C$\equiv$N $\xrightarrow[\Delta]{\text{HCl, H}_2\text{O}}$ CH$_3$CH$_2$CH$_2$—C(=O)—OH

$\downarrow$ SOCl$_2$

CH$_3$CH$_2$CH$_2$—C(=O)—O—C(=O)—CH$_3$ $\xleftarrow{\ \text{CH}_3\text{—C(=O)—O}^-\ }$ CH$_3$CH$_2$CH$_2$—C(=O)—Cl

7.

a. CH$_3$CH$_2$CH$_2$—C(=O)—OCH$_2$CH$_2$CH$_3$

d. CH$_3$CH$_2$CH$_2\overset{+}{\text{N}}$H$_3$

b. CH$_3$CH$_2$—C(=O)—OH + CH$_3\overset{+}{\text{N}}$H$_3$

e. CH$_3$—C(=O)—OCH$_2$CH$_3$ + CH$_3$—C(=O)—OH

c. CH$_3$CH$_2$—C(=O)—OH + HO—phenyl

Answers to Chapter 17 Practice Test

1.

a.

  + H₂O

b.

  +   H₂O

c. CH₃CH₂CH₂—C(=O)—OH

d.

  + H₂O

e. CH₃CH₂—C(OH)(CH₃)—CH₂CH₃

f.

g. CH₃CH₂—C(OH)(CH₂CH₃)—C≡N

h. CH₃CH—CH₂—C(=O)—CH₃ with SCH₃

i.

2. CH₃CH₂—C(OH)(CH₃)—CH₂CH₂CH₃

3.   **a.** ethyl 4-hydroxyhexanoate   **b.** 4-oxoheptanal   **c.** 4-formylhexanamide

4.   **a.**

c.

e.

b. CH₃CH₂CH(OCH₃)(OCH₃)

d. CH₃CH₂CH(OH)(OCH₃)

5.   **a.** butanal   **b.** 2-heptanone

6.   **a.** CH₃CH₂CH₂Br $\xrightarrow{Mg}$ CH₃CH₂CH₂MgBr $\xrightarrow{CO_2}$ CH₃CH₂CH₂—C(=O)—O⁻

$\downarrow$ SOCl₂

CH₃CH₂CH₂—C(=O)—OCH₂CH₃ $\xleftarrow{CH_3CH_2OH}$ CH₃CH₂CH₂—C(=O)—Cl

b. CH_3 — C(=O) — OCH_3 $\xrightarrow[CH_3MgBr]{excess}$ $CH_3\overset{\overset{\displaystyle O^-}{|}}{\underset{\underset{\displaystyle CH_3}{|}}{C}}CH_3$ $\xrightarrow{H_3O^+}$ $CH_3\overset{\overset{\displaystyle OH}{|}}{\underset{\underset{\displaystyle CH_3}{|}}{C}}CH_3$

c. CO_2 $\xrightarrow{CH_3MgBr}$ CH_3 — C(=O) — O^- $\xrightarrow[H_3O^+]{LiAlH_4}$ CH_3CH_2OH

d. (cyclohexanone) $\xrightarrow[H_3O^+]{LiAlH_4}$ (cyclohexanol, OH) $\xrightarrow[pyridine]{SOCl_2}$ (cyclohexyl chloride, Cl) $\xrightarrow{^-C\equiv N}$ (cyclohexanecarbonitrile, $C\equiv N$)

Answers to Chapter 18 Practice Test

1.

CH_3 — C(=O) — CH_2 — C(=O) — OCH_3 **2**

CH_3 — C(=O) — CH_2 — C(=O) — CH_3 **1**

CH_3 — C(=O) — CH_3 **3**

2.

a. $CH_3\overset{\overset{\displaystyle OH}{|}}{C}=CH$ — C(=O) — CH_3

b. CH_3 — C(=O) — CH_2 — C(=O) — OCH_3

3.

a. CH_3CH_2 — C(=O) — CH_3 + CO_2

d. CH_3 — $\overset{\overset{\displaystyle CH_2CH_2-C(=O)-CH_3}{|}}{\underset{\underset{\displaystyle}{}}{CH}}$ with two C(=O) groups and CH_3

g. $CH_3CH_2CH_2CH_2$ — C(=O) — OH

b. $CH_3CH_2\overset{\overset{\displaystyle Br}{|}}{\underset{\underset{\displaystyle Br}{|}}{C}}$ — C(=O) — H + $2\ Br^-$

e. (cyclopentanone with Br)

c. $CH_3CH_2CH_2$ — C(=O) — $\overset{}{\underset{\underset{\displaystyle CH_2CH_3}{|}}{CH}}$ — C(=O) — OCH_3

f. (cyclohexanone with Cl, Cl, Cl, Cl)

4.

a. 2 CH₃CH₂—C(=O)H $\xrightarrow{\text{HO}^-}$ CH₃CH₂CH(OH)CH(CH₃)—C(=O)H

b. 2 CH₃CH₂—C(=O)H $\xrightarrow{\text{HO}^-}$ CH₃CH₂CH(OH)CH(CH₃)—C(=O)H $\xrightarrow[\Delta]{\text{H}_2\text{SO}_4}$ CH₃CH₂CH=C(CH₃)—C(=O)H

c. 2 CH₃CH₂—C(=O)OCH₃ $\xrightarrow[\text{2. H}_3\text{O}^+]{\text{1. CH}_3\text{O}^-}$ CH₃CH₂—C(=O)—CH(CH₃)—C(=O)OCH₃ + CH₃OH

d. CH₃O—C(=O)—CH₂CH₂CH₂CH₂CH₂—C(=O)—OCH₃ $\xrightarrow[\text{2. H}_3\text{O}^+]{\text{1. CH}_3\text{O}^-}$ (2-oxocyclohexane)—C(=O)OCH₃

e. CH₃CH₂O—C(=O)—CH₂—C(=O)—OCH₂CH₃ $\xrightarrow[\substack{\text{2. CH}_3\text{CH}_2\text{Br} \\ \text{3. H}^+, \text{H}_2\text{O}, \Delta}]{\text{1. CH}_3\text{CH}_2\text{O}^-}$ CH₃CH₂CH₂—C(=O)OH

f. CH₃—C(=O)—CH₂—C(=O)—OCH₂CH₃ $\xrightarrow[\substack{\text{2. CH}_3\text{CH}_2\text{Br} \\ \text{3. H}^+, \text{H}_2\text{O}, \Delta}]{\text{1. CH}_3\text{CH}_2\text{O}^-}$ CH₃CH₂CH₂—C(=O)—CH₃

5.

CH₃CHCH₂CH₂CH(OH)CH(—C(=O)H)   CH₃CHCH₂CH₂CH(OH)CH(—C(=O)H)
  | | | |
  CH₃   CH₂CH₃   CH₃   CH₂CHCH₃
                                                              |
                                                            CH₃

CH₃CH₂CH₂CH(OH)CH(—C(=O)H)   CH₃CH₂CH₂CH(OH)CH(—C(=O)H)
                   | |
              CH₂CH₃   CH₂CHCH₃
                                                     |
                                                   CH₃

6. **a.** $CH_3CH_2CH_2$—C(=O)—OCH_3 **b.** CH_3O—C(=O)—$CH_2CH_2CH_2CH_2CH_2$—C(=O)—OCH_3

Answers to Chapter 19 Practice Test

1. **a.** *meta*-nitrotoluene
      3-nitrotoluene

   b. 1,2,4-tribromobenzene

   c. *ortho*-ethylbenzoic acid
      2-ethylbenzoic acid

   d. *para*-chlorophenol
      4-chlorophenol

2.

3.

4. **a.** *para*-bromonitrobenzene   **b.** *para*-bromoethylbenzene

5.

6. **a.** Benzoic acid is more reactive than benzene toward electrophilic aromatic substitution.   F
   b. *para*-Chlorobenzoic acid is more acidic than *para*-methoxybenzoic acid.   T
   c. A $CH{=}CH_2$ group is a meta director.   F
   d. *para*-Nitroaniline is more basic than *para*-chloroaniline.   F

7.

CH_3CH_2 O CH_2CH_3

8.

$HO-NO_2$ + $H-OSO_3H$ $\rightleftharpoons$ $HO-NO_2$ $\rightleftharpoons$ H_2O + $^+NO_2$

9.

Answers to Chapter 20 Practice Test

1. **a.** 2,4-dimethyl pyrrolidine
2,4-dimethylazacyclopentane

b. *N*-methylpiperidine
N-methylazacyclohexane

c. 3-ethyltetrahydrofuran
3-ethyloxacyclopentane

d. 3-bromopiperidine
3-bromoazacyclohexane

2. **a.**

$+$ $CH_3CH_2\overset{+}{N}H_3$ Cl^-

b.

$+$ H_2O

c.

d.

e.

3. **a.**

b.

c.

d.

4. **a.** 4-Chloropyridine is more reactive toward nucleophilic aromatic substitution than is 3-chloropyrrole.   T
b. Pyrrole is more reactive toward electrophilic aromatic substitution than is furan.   T
c. Pyrrole is more reactive toward electrophilic aromatic substitution than is benzene.   T
d. Pyridine is more reactive toward electrophilic aromatic substitution than is benzene.   F

Answers to Chapter 21 Practice Test

1.

a.

c.

b.

d.

2.
  a. Glycogen contains α-1,4′ and β-1,6′-glycosidic linkages.   F
  b. D-Mannose is a C-1 epimer of D-glucose.   F
  c. D-Glucose and L-glucose are anomers.   F
  d. D-Erythrose and D-threose are diastereomers.   T
  e. Wohl degradations of D-glucose and D-gulose form the same aldotetrose.   F

3.

4. D-mannose and D-glucose

5.

6. D-altrose

7. Amylose has α-1,4′-glycosidic linkages, whereas cellulose has β-1,4′-glycosidic linkages.

8. D-gulose and D-idose   **9.** D-allose

10.

Answers to Chapter 22 Practice Test

1.

a.

d.

b.

e.

c.

2.

a.

c.

b.

d.

3.   **a.** alanine, because it is farther away from its pI.   **c.** leucine and isoleucine

b. glycine   **d.** aspartic acid

4.   The electron-withdrawing protonated amino group causes the carboxyl group of alanine to have a lower pK_a.

5.   **a.** A cigar-shaped protein has a greater percentage of polar residues than a spherical protein.   T

  b. Naturally occurring amino acids have the L-configuration.   T

  c. There is free rotation about a peptide bond.   F

6.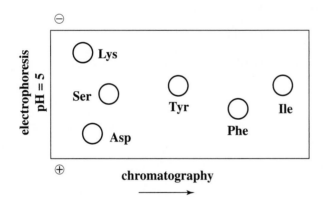

7.   **a.** The sequence of the amino acids and the location of the disulfide bonds in the protein.

  b. The three-dimensional arrangement of all the atoms in the protein.

  c. A description of the way the subunits of an oligomer are arranged in space.

8.

```
electrophoresis pH = 5

⊖

    ◯ Lys

Ser ◯         ◯              ◯
           Tyr      ◯      Ile
    ◯ Asp        Phe

⊕        chromatography
         ⟶
```

9.   **a.** $\dfrac{2.16 + 9.18}{2} = \dfrac{11.34}{2} = 5.67$ **b.** $\dfrac{9.04 + 12.48}{2} = \dfrac{21.52}{2} = 10.76$

10.   Ala   Ser   Arg   Gly   Arg   Met   His   Phe   Lys   Ile

11.

$$\underset{H}{\overset{CH_3}{>}}C{=}O \xrightarrow[NH_3]{trace\ acid} \underset{H}{\overset{CH_3}{>}}C{=}NH \xrightarrow[HCl]{^-C{\equiv}N} CH_3{-}\underset{^+NH_3}{\underset{|}{CH}}{-}C{\equiv}N \xrightarrow[\Delta]{HCl,\ H_2O} CH_3{-}\underset{^+NH_3}{\underset{|}{CH}}\overset{O}{\overset{\|}{C}}{-}OH$$

12. $R{-}SH \underset{H_2O}{\overset{HO^-}{\rightleftharpoons}} R{-}\ddot{S}{:}^- \xrightarrow{Br{-}Br} R{-}\underset{Br^-}{S{-}Br} \xrightarrow{R{-}\ddot{S}{:}^-} R{-}S{-}S{-}R \ + \ Br^-$

Answers to Chapter 23 Practice Test

1.   **a.** A catalyst increases the equilibrium constant of a reaction.   F
     b. An acid catalyst donates a proton to the substrate, and a base catalyst removes a proton
         from the substrate.   T
     c. The reactant of an enzyme-catalyzed reaction is called a substrate.   T
     d. Complexing with a metal ion increases the pK_a of water.   F

2.   It protonates the leaving group to make it a better leaving group.

3.   **a.**   **b.**

4.

     a. base catalyst

     b. nucleophilic catalyst

     c. it stabilizes the positive charge on histidine

Answers to Chapter 24 Practice Test

1. biotin and vitamin KH_2

2. methyl (CH_3), methylene (CH_2), formyl ($HC=O$)

3.

4.

5. The first step is removing the hydrogen atom that is going to change places with a group on an adjacent carbon.

6. first coenzyme = biotin
   second coenzyme = coenzyme B_{12}

7.

8.

9. a. ATP, Mg^{2+}, HCO_3^-

   b. ATP activates bicarbonate (HCO_3^-) by putting a good leaving group (phosphate) on it. Mg^{2+} complexes with ATP in order to reduce the negative charge on ATP so it can react with a nucleophile.

   HCO_3^- is the source of the COO^- group that is put on the reactant.

10.   FAD oxidizes dihydrolipoate to lipoate.

11.

12.   **a.** Vitamin B_1 is the only water-insoluble vitamin that has a coenzyme function.   F
b. $FADH_2$ is a reducing agent.   T
c. Thiamine pyrophosphate is vitamin B_6.   F
d. Cofactors that are organic molecules are called coenzymes.   T
e. Vitamin K is a water-soluble vitamin.   F
f. Lipoic acid is covalently bound to its enzyme by an amide linkage.   T

Answers to Chapter 25 Practice Test

1.   **a.**   **b.**   **c.**

2.

3.   acetate

4.   catabolic reactions

5.   amino acids

6. acetyl-CoA

7.
a. Each molecule of FADH$_2$ forms three molecules of ATP in the fourth stage of catabolism.   F
b. FADH$_2$ is oxidized to FAD.   T
c. NAD$^+$ is oxidized to NADH.   F
d. Acetyl-CoA is a citric acid cycle intermediate.   F
e. A diterpene contains 20 carbons.   T

8.

squalene

9.

10. 15

Answers to Chapter 26 Practice Test

1. dUMP

2. 5′—A—T—G—C—A—A—G—T—3′

3. A

4.
a. Guanine and cytosine are purines.   F
b. The 3′-OH group allows RNA to be easily cleaved.   F
c. The number of A's in DNA is equal to the number of T's.   T
d. rRNA carries the amino acid that will be incorporated into a protein.   F
e. The template strand of DNA is the one transcribed to form RNA.   T

f. The 5′-end of DNA has a free OH group.   F

g. The synthesis of proteins from an RNA blueprint is called transcription.   F

h. A nucleotide consists of a base and a sugar.   F

i. RNA contains T's, and DNA contains U's.   F

5.   only #1

6.   5′—G—C—A—U—G—G—A—C—C—C—C—G—U—3′

7.   In the process of converting uridines to thymidines, tetrahydrofolate is oxidized to dihydrofolate. NADPH is used to reduce dihydrofolate back to tetrahydrofolate.

8.   CA and TG